G. A. S. PALMER.

THE MIR SPACE STATION
A Precursor to Space Colonization

WILEY-PRAXIS SERIES IN SPACE SCIENCE AND TECHNOLOGY
Series Editor: John Mason, B.Sc., Ph.D.

This series reflects the significant advances being made in space science and technology, including developments in astronautics and space life sciences. It provides a forum for the publication of new ideas and results of current research in areas such as spacecraft materials, propulsion systems, space automation and robotics, spacecraft communications, mission planning and management, and satellite data processing and archiving.

Aspects of space policy and space industrialization, including the commercial, legal and political ramifications of such activities, and the physiological, sociological and psychological problems of living and working in space, and spaceflight risk management are also addressed.

These books are written for professional space scientists, technologists, physicists and materials scientists, aeronautical and astronautical engineers, and life scientists, together with managers, policy makers and those involved in the space business. They are also of value to postgraduate and undergraduate students of space science and technology, and those on space-related courses (including psychology, physiology, medicine and sociology) and areas of the social and behavioural sciences.

For further details of the books listed below and ordering information, why not visit the Praxis Web Site at http//www.praxis-publishing.co.uk

METALLURGICAL ASSESSMENT OF SPACECRAFT PARTS, MATERIALS AND PROCESSES
Barrie D. Dunn, Head of Metallic Materials and Processes Section, ESA-ESTEC, Noordwijk, The Netherlands

SATELLITE CONTROL: A Comprehensive Approach
John T. Garner, Aerospace Consultant, formerly Principal Ground Support Engineer, Communications Satellite Programmes, ESA-ESTEC, Noordwijk, The Netherlands

THE MIR SPACE STATION: A Precursor to Space Colonization
David M. Harland, formerly Visiting Professor, University of Strathclyde, UK

LIVING AND WORKING IN SPACE: Human Behavior, Culture and Organization, Second edition
Philip Robert Harris, Executive Editor, *Space Governance* Journal; Vice President, United Societies in Space, Inc., USA

THE NEW RUSSIAN SPACE PROGRAMME: From Competition to Collaboration
Brian Harvey, M.A., H.D.E., F.B.I.S.

Forthcoming titles in the series are listed at the back of the book.

THE MIR SPACE STATION
A Precursor to Space Colonization

David M. Harland
formerly Visiting Professor at the University of Strathclyde

JOHN WILEY & SONS
Chichester • New York • Weinheim • Brisbane • Singapore • Toronto

Published in association with
PRAXIS PUBLISHING
Chichester

Copyright © 1997 Praxis Publishing Ltd
The White House,
Eastergate, Chichester,
West Sussex, PO20 6UR, England

Published in 1997 by
John Wiley & Sons Ltd
in association with Praxis Publishing Ltd

Wiley Editorial Offices

John Wiley & Sons Ltd, Baffins Lane,
Chichester, West Sussex, PO19 1UD, England

John Wiley & Sons, Inc., 605 Third Avenue,
New York, NY 10158-0012, USA

Wiley-VCH Verlag GmbH, Pappelallee 3,
D-69469 Weinheim, Germany

Jacaranda Wiley Ltd, G.P.O. 33 Park Road, Milton,
Queensland 4001, Australia

John Wiley & Sons (Asia) Pte Ltd, 2 Clementi Loop #02-01,
Jin Xing Distripark, Singapore 12981

John Wiley & Sons (Canada) Ltd, 22 Worcester Road,
Rexdale, Ontario, M9W 1L1, Canada

Library of Congress Cataloguing-in-Publication Data
 Harland, D. M. (David Michael), 1955–
 The Mir space station: a precursor to space colonization / David M. Harland.
 p. cm. – (Wiley-Praxis series in space science and technology)
 Published in association with Praxis Publishing, Chichester.
 Includes bibliographical references and index.
 ISBN 0-471-97587-7 (alk. paper)
 1. Mir (space station) 2. Space colonies. I. Title. II Series.
T1 789.8.S652M564 1997
629.44'2'0947—dc21   97-19025
                                                                    CIP

A catalogue record for this book is available from the British Library

ISBN 0-471-97587-7

Printed and Bound in Great Britain by Hartnolls Ltd, Bodmin

In memory of

Dick Scobee
Mike Smith
Judy Resnik
Greg Jarvis
Ron McNair
Ellison Onizuka
and
Christa McAuliffe

*who perished attempting to reach space,
and of*

Vladimir Komarov
Georgi Dobrovolsky
Viktor Patsayev
and
Vladislav Volkov

who perished attempting to return to Earth.

STS-76 view of Mir. Backdropped against the waters of Cook Strait near New Zealand's South Island, the Mir space station is seen from the aft flight deck of the space shuttle Atlantis. The two spacecraft were in the process of making their third docking in Earth-orbit.

altitude   400 km ?

The dream of yesterday is the hope of today and the reality of tomorrow
Robert Goddard—American rocket pioneer

The Earth is the cradle... but you cannot live in the cradle forever
Konstantin Tsiolkovsky—Russian rocket pioneer

Mir [is] an awesome sight
Bill Readdy—STS-79 commander

Never in my wildest dreams did I ever think that I would [be] on the Russian space station
Shannon Lucid—Mir research astronaut

Space station Mir is an incredible laboratory and workshop... I really enjoy it here!
John Blaha—Mir research astronaut

The Mir station has been on orbit for 10 years now; that's an amazing feat in terms of duration in a harsh environment
Rick Searfoss—STS-76 pilot

You have to understand that initially we planned to fly [Mir] for only 3 years...
Yelena Kondakova—Mir flight engineer

Table of contents

PART 2: MIR

PART 4: CONCLUSIONS

Author's preface

Throughout history, the human race has expanded its physical horizon from its African Rift Valley origin southward into Africa, north through the Middle East onto the European and Scandinavia peninsulas, across the Asian vastness and then on through the Americas, across the Pacific and finally to the poles. As a result, almost every corner of the Earth has been settled. By pushing into space, this relentless expansion has effectively crossed the final frontier.

Colonising space will not be like the Founding Fathers settling America, as although they ventured into *terra incognita*, it was a familiar and inherently benign environment. Making the first move into space is more akin to the first fish venturing out of the sea onto dry land. In that case, the species adapted to the intrinsically unfamiliar environment by biogenetic evolution. This, however, is a process that takes generations to effect a significant change. Although evolution eventually enabled the fish to leave its watery habitat, it is unlikely that it could enable us to adapt to live in a weightless, radiation-soaked vacuum. In our case, we shall have to take with us as much of our cosy environment as strictly necessary. This will be possible only by virtue of human ingenuity; by the creation of technology, a process that facilitates far more rapid change than is feasible by biological evolution. We fly through the air, after all, only as a result of the discovery of aerodynamic principles and the appliance of technology. Our ingenuity has been such that one of our creations, the computer, may one day become self-aware, in which case, in a very real sense, we will have introduced a form of life to the planet entirely without precedent in its biogenetic root; we live in a remarkable time indeed.

Space colonisation will involve the entire spectrum of disciplines, from psychology to philosophy and from legal to financial, but technology will be the key; without technology, colonising space is but a dream. A consequence of technological progress is that what was once impossible becomes first plausible, then achievable with great difficulty, then routine, and finally is taken entirely for granted as an insignificant step in some greater process. Just as in the days before computer modelling, when test pilots literally risked death performing flight trials to assess whether newly-designed aircraft were viable, pioneering in space has been an *empirical* exercise in engineering during which crews have risked their lives to enhance the technological state-of-the-art. This slow and deliberate rate of progress was strikingly demonstrated by Apollo 10 in 1969; a vehicle that had been specifically designed to land on the Moon was ordered on a mission to descend to within a dozen miles of the surface, then simulate an abort and return home so that this option would have been shown viable in case it should prove necessary on the following flight, which would be assigned the final step in the process – the landing itself.

In the immediate aftermath of the last Apollo Moon landing in 1972, Gerard O'Neill, a physics professor at Princeton University, began to investigate factors which might limit the colonisation of space in the immediate Earth environment. To his surprise he concluded that really large habitats made the most sense. Furthermore, he realised that solar energy would facilitate processes which on Earth were deemed to be too inefficient to be cost effective, so it would be possible not only to make such habitats self-sufficient but also net producers of energy, which could be beamed down to Earth. With the world just plunged into an energy crisis, it seemed to O'Neill that a power-station-in-the-sky that produced clean energy held out every prospect of being the 'killer application' that would attract the necessary funding; commerciality was the key. O'Neill published his initial conclusions in 1974, in a technical paper boldly entitled *The Colonisation of Space*.

O'Neill's proposal was so provocative that a conference was held in 1975 to investigate its viability. This conference was sponsored jointly by the National Aeronautics and Space Administration (NASA), the American Institute for Aeronautics and Astronautics (AIAA), and the National Science Foundation (NSF). It studied launch vehicles, the assembly of vast space structures, the possibility of exploiting lunar materials, orbits for the habitat, techniques for beaming solar energy to Earth and shielding to protect the inhabitants from cosmic rays; and it also explored socio-cultural issues and space law in addition to the basic commerciality of the proposal.

O'Neill's habitat was to be a massive hollow structure, several kilometres across; it was to be pressurised and spun to create Earth-equivalent gravity at its inner surface. As many as 10,000 people would be housed in towns on the inner surface and work on the farms or the recycling systems, or in the power station. By providing a home for entire families who would spend their lives aboard, the habitat would be a true colony in space.

Amazingly, it was concluded that work on constructing the first habitat would be able to begin as soon as the space shuttle entered service. At that time, it was widely expected that the shuttle would fly before the end of the decade, that when it became operational it would fly once a week, and that it would so reduce the cost of placing payload into low Earth orbit that access to space would be both routine and cheap. Nevertheless, it was realised that this would not be cheap enough to launch the materials needed to construct the first habitat, so a heavy-lift booster derived from the space shuttle propulsion system was proposed, one able to place over 100 tonnes in orbit at a fraction of the cost of a shuttle flight. By starting in 1982, it was calculated, it should be possible to have the first habitat in place near the Moon by the end of the century. The investment (about $100 billion in 1975 terms) was to be recouped in a decade from beaming power down to American consumers! But here we are in the latter half of the final decade of this century, and O'Neill's colony is not even on the agenda for the first quarter of the next century. So what happened?

By the year 2001 the shuttle will have been in service for 20 years and will have carried out 100 missions. Despite initial promises, it will never fly on a weekly basis, and it will never provide cheap access to space; but if assessed in terms of what was practicable, it has clearly been successful because it routinely carries payloads which would otherwise not be feasible. But the economics of operating the shuttle is not the reason why O'Neill's colony is not now under construction. Even if the required orbital transportation infrastructure had been available, it is unlikely that assembly would have begun.

Operating an enormous self-sustaining habitat will simply not be feasible until the basic technology has been developed; until the engineering knowledge needed to sustain this has been learnt; until a full regenerative atmospheric processor has been developed; until food can be grown in space; and until a comprehensive waste recycler has been developed. The first step is to learn to sustain the technology.

The Mir complex is a tentative first step towards an orbital habitat. It is a tiny precursor, a *technology demonstrator*; the issue of scaling it up will come later, once the requisite technology is fully understood. That, along with the evaluation of the human organism in a weightless state, is Mir's principal *raison d'être*. Of course, a range of activities, including life sciences (both plants and animals), materials processing (semiconductor, metallic and biological) and observation (Earth, astronomical and space physics) were performed aboard, but these, taking the long-term view, were simply opportunistic exploitation of the fact that the orbital complex existed. Mir is often referred to as an 'ageing' space station, but to say this not only reflects misunderstanding of how it has been progressively built up over the years, it also serves to *confirm* that Mir has *succeeded* in its main mission. Some commentators scoff that up to 40 per cent of a crew's work is devoted to maintenance, implying that this is time wasted; yet learning to sustain its systems may well turn out to be the most valuable result. Although it is becoming increasingly common to assess Mir in terms of its commerciality, the financial value of its tangible output is debatable. But when Mir is assessed in terms of its strategic objective of permanent occupancy, it is an indisputable success: it has supported a succession of crews for more than a decade, and one cosmonaut spent 15 months aboard. In fact, Mir has been augmented by so many new modules and so much scientific apparatus that it is now acknowledged to be a priceless international resource, and accommodation for guest researchers is fully booked until the end of the century.

This book will focus on the technology developed during the Salyut and Mir programmes, on the supporting infrastructure, and on the long struggle (which has cost four cosmonauts their lives) to construct and operate a continuously inhabited research facility in low orbit. Although many of the scientific activities undertaken will be discussed, together with issues of living and working in space, this is primarily an assessment of Mir from an engineering viewpoint, and an attempt to put it into the context of what is yet to come.

Kelvinbridge, Glasgow
July 1997
David M. Harland

Acknowledgements

I am delighted to acknowledge the sterling service provided in print over many years by Neville Kidger, Gordon Hooper, Craig Covault and Peter Gualtieri.

Undoubtedly, the most comprehensive record of space activities has been provided by the publications of the British Interplanetary Society.

I would also like to thank the Broadcast and Imaging Branch at NASA Headquarters, the Image Library at NASA's Lyndon B. Johnson Space Center, and the Russian Space Agency for the many photographs which they supplied. With the exception of the schematic diagrams on pages 277 and 297, all line artwork is © David M. Harland.

Finally, I would like to thank John Mason, Brian Harvey and Philip Harris for helpful suggestions in the formative stage, Flo McGuire and Alex Williams for reading preliminary versions of the manuscript, and Clive Horwood for his enthusiastic support throughout.

Notes

The term 'weightlessness' is used because it is popular. It is not intended to indicate the absence of the force of gravity, rather its cancellation by the free-fall of orbital flight, and is used to describe the experience felt by human beings (the term microgravity is reserved for scientific experiments).

The launch site in Kazakhstan is referred to as the Baikonur Cosmodrome, even though the actual site at Tyuratam is 350 km south-west of the town of Baikonur; the reason for this misidentification is, of course, a long-standing sore point with the responsible authority.

The various classificatory systems for rockets have been eschewed. There are only two rockets that matter in the context of this book: the Semyorka, derived from the world's first intercontinental ballistic missile, which is used to launch cosmonauts, and the Proton, which is used to launch 20-tonne space station modules.

The metric system is used unless it is obvious that figures quoted in the imperial system had been chosen to be rounded to convenient sizes.

Unless the time of an event is particularly pertinent, the times of launches, dockings, undockings and recoveries have been omitted from the text; they are all listed in the relevant tables.

Unless otherwise stated, times are Moscow Time. Note, however, that Moscow Summer time (which is one hour ahead) has been used without specific mention during the times that it applied (That is, from the final Sunday in March to the final Sunday in September). The Kazakh cosmodrome and the recovery zone are two hours ahead of Moscow.

Space shuttles use Central Standard (or Daylight) Time to synchronise with Houston's Johnson Space Center. The Kennedy Space Center in Florida operates Eastern Standard (or Daylight) Time. Note that Greenwich Mean Time is five hours ahead of Eastern Standard Time, which is one hour ahead of Central Time.

List of illustrations and tables

Tables

Part 1: Salyut

1

Into the unknown

On 12 April 1961, after a 108-minute flight which took him all the way around the Earth, Yuri Gagarin landed in the Soviet Union to a hero's welcome. This historic test flight proved the Vostok spacecraft's basic systems, but how long could the human body spend in weightlessness? Several months later, Gherman Titov spent a day in space. He felt sick when he moved about in the cabin, but he suffered no lasting effects, and was certainly not incapacitated, as some had predicted. A year later, Andrian Nikolayev extended the record to four days, and a year after that Valeri Bykovsky flew for five days. Was there no limit?

In 1965, as America prepared to send men to the Moon, Frank Borman and James Lovell spent 14 days orbiting the Earth in Gemini 7. This verified that astronauts would be able to survive a trip to the Moon and back. Two weeks sitting in the cramped capsule had been no fun, however, and it really had been an *endurance* mission.

The Gemini record still stood in 1969, when the Soviets switched the emphasis of their space programme from the Moon to the establishment of a space station in Earth orbit. Being able to survive a fortnight in space was sufficient for sojourns to the Moon, but it represented only a small step towards living in space. In 1970, with the launch of their first station still a year away, the Soviets decided to fly a record-breaking mission in the Soyuz spacecraft to demonstrate that cosmonauts would be able to serve the anticipated tour of duty. Andrian Nikolayev, now head of the cosmonaut corps, assigned himself as commander and selected Vitali Sevastyanov as his flight engineer. Sevastyanov was well known as the presenter of a television programme that explained scientific discoveries for a general audience.

1 June had been a blisteringly hot day at the launch site on the open desert plain, but a welcome chill had followed the setting of the Sun. It was already dark when, two hours before launch, the two men rode the elevator up the side of the service structure and settled into the capsule. The last time Nikolayev had flown, he had worn a bulky pressure suit. The Soyuz supported a 'shirt-sleeved' environment, so this time he and Sevastyanov wore lightweight track suits. The only sign that they were about to make a flight was that they wore communications headsets.

The preparation of the rocket was controlled from the nearby blockhouse, so there was little for the cosmonauts to do. In fact, the R-7, or Semyorka, designed by Sergei Korolev in the 1950s, was the world's first intercontinental ballistic missile. It had been developed to send a nuclear warhead on a ballistic arc. It was a 'parallel-stage' rocket, due to the fact that the technology to air-start an in-tandem stage had not been available when it had been designed. Its four conical strap-on boosters were to fire together with the cylindrical core for lift off. It had been used to launch Sputnik and later, with an upper stage, Vostok. The powerful upper stage with which it was now fitted was sufficient to lift the 7.5-tonne Soyuz spacecraft. It was the same pad too: a concrete pier on the edge of a massive pear-shaped flame pit.

(a)

(b)

(c)

Preparations for launch. (a) The venerable Semyorka rocket carrying a Soyuz spacecraft makes its way from the assembly building to the launch pad across the steppe at the Baikonur Cosmodrome. (b) The rocket arrives at the pad. (c) The rocket is supported by the four short arms.

With 30 minutes to go, the twin sections of the service structure split the eight levels of wrap-around walkway and then swung down to leave the slender rocket exposed on the pad. Ten minutes later, the topping-off of the liquid oxygen tanks ceased, the kerosene tanks were pressurised, and nitrogen gas was pumped through the propellant feed lines to clear them and to blow off the covers of the 20 nozzles. With only 40 seconds to go, the rocket was switched to internal power. 20 seconds later, the umbilical arm was disconnected and swung down. A turbopump in each of the five segments began to feed fuel and oxidiser into the four reaction chambers. After a final check, pyrotechnic charges fired simultaneously. As the engines slowly built up thrust, the flame belched into the pit beneath. The rocket was not actually supported at its base; the core was held up by four arms just above the top of the strap-ons. As soon as the thrust overcame its 450-tonne mass, the rocket started to climb. This released the supporting arms, which immediately swung out like the petals of a flower to clear the way for the protruding strap-ons. One way or another, Nikolayev and Sevastyanov were now committed.

At the moment of launch the spacecraft activated its flight sequencer. As the rocket pitched to a 51-degree (51.6-degree in fact) inclination, the globe of the tiny moving map display on the spacecraft's control panel showed its progress along its north-easterly ground track, and as it climbed up through the atmosphere it flew almost directly over the town of Baikonur. (The launch site in the Soviet Central Asian Republic of Kazakhstan was actually 350 km south-west of Baikonur. When Sputnik was launched it was the nearest large population centre on the track, so the launch site was given its name. Afterwards, a town sprung up to house the growing rocket community. This 'closed' town, built at the Tyuratam rail junction, was called Leninsk.)

At 115 seconds (at 46 km altitude) the ring of solid-rocket motors at the top of the 6-metre tall escape tower fired to pull it free of the shroud; three seconds later, the 20-metre long strap-ons shut down and were jettisoned. After so many Semyorka launches, this part of the Kyzyl-Kum desert 400 km downrange was littered with spent rocket debris.

The core kept thrusting. Now that half of its propellant had been consumed, and the dead weight of the empty strap-ons was gone, the acceleration built up rapidly. At 165 seconds (at an altitude of 80 km, above most of the atmosphere, and at the official edge of space) the shroud which had protected the spacecraft from aerodynamic loads was jettisoned; it fell to Earth 525 km downrange. At 288 seconds (at an altitude of 175 km) the 30-metre long core shut down and was jettisoned; it re-entered the atmosphere 1,500 km downrange, and burnt up. The four-chambered engine of the 8-metre long upper stage started immediately, and by the time it had built up to its full 35-tonne thrust the framework interstage had been jettisoned too. At an altitude of about 200 km, as the upper stage continued to accelerate to a peak of 3.5 g, it pitched over and accelerated to a horizontal speed of 7.8 km per second, sufficient for low orbit. At 530 seconds, by now far north of China, the upper stage shut down and was jettisoned. In this 175 × 225 km orbit, the air drag was sufficient to cause such a large lightweight object to re-enter the atmosphere within a few hours, and so the spacecraft – now officially designated Soyuz 9 – manoeuvred to lift its perigee to avoid this fate. As soon as it was clear that the spacecraft was in good condition, it was announced that its objective was to investigate human adaptation to weightlessness on a long-duration mission. An earlier report had claimed that Soyuz's life-support system could sustain a 30-day flight; this was presumed to be with a single

(a)

(b)

(c)

Launch of Soyuz spacecraft. (a) The wrap-around eight-level gantry splits apart and swings down. (b) Lift-off! (c) The empty pad.

cosmonaut, so it seemed reasonable to expect that Soyuz 9 would stay up for two weeks or so.

Despite the late launch, and the time spent over the next few orbits checking out the systems, the cosmonauts were to operate a daily schedule synchronised with the control centre at Yevpatoria, in the Crimea. The tracking network of ground stations and ships enabled them to maintain communications for substantial periods of favourable passes, and they slept for eight hours a day during the extended period when the orbit did not offer a favourable ground track. Though spacious in comparison to earlier spacecraft, Soyuz was rather cramped. Its descent module barely accommodated the couches, but the orbital module provided room to stretch, so the crew spent most of their time there, and sleeping bags were set across the storage lockers. During the first night, they had some difficulty adapting to the absence of gravity, but slept well during the rest of the flight.

Nikolayev and Sevastyanov were to follow a strict exercise regime, perform a range of biomedical tests, carry out some scientific experiments and make Earth observations. They were to test a variety of exercise apparatus which, it was hoped, would tone up their cardiovascular system. It was also hoped to counter the muscle atrophy and avoid the significant loss of bone calcium that had been observed in dogs following 22 days in a Voskhod capsule, flown as Cosmos 110, a few years earlier.

The spacecraft's orbital module had been fitted with a gymnasium. A special 'load-suit' was fastened to the floor by a pair of expanders which created a load of 20–40 kg, to simulate the force of gravity; this exercised specific muscles by transmitting skeletal loads while walking, running and jumping in place against its action, and a 10-kg chest expander was also used to stress the arm and chest muscles. If these proved effective, they were to

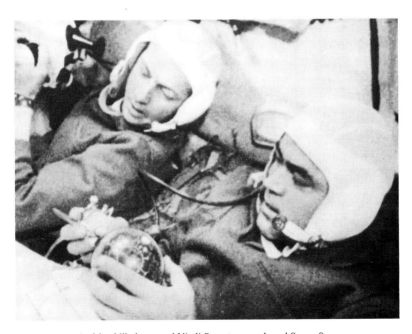

Andrian Nikolayev and Vitali Sevastyanov aboard Soyuz 9.

be used aboard the orbital station. The state of health of the cosmonauts was methodically assessed by measuring heart and respiration rates, measuring arterial pressure, testing the ability of the vestibular system in the inner ear to maintain a sense of balance in weightlessness, and testing the sensitivity of the eye to different levels of illumination and contrast. All of this followed up on cosmonauts' reports after earlier flights. Typical symptoms were a sensation of falling backwards, blood rushing to the head, a stuffy nose, a puffy face, decreased appetite and motion sickness, but although most of these symptoms disappeared after a few days, the sensation of excess blood to the head lasted rather longer, and returned during periods of strenuous exercise. Visual acuity tended to degrade too, making it difficult to resolve extremely fine detail at long range, but this was believed to be reversible.

The orbital module was also equipped with a small electric oven, to prepare hot food and drink, and a foldaway toilet. The cosmonauts had two sets of towels (one damp and the other dry) with which to sponge themselves down, but they were restricted to ablutions once a day, and changed their underwear weekly. In addition to a traditional wet shaving kit, they evaluated an electric razor, and a hand-held vacuum cleaner to clear the air of particulates that could result in choking or blindness. Compared with Borman and Lovell's Gemini ordeal, however, this was sheer luxury.

One of Sevastyanov's immediate recommendations was that crews assigned to long flights be provided with a stock of thick socks – because the altered distribution of blood induced by weightlessness caused the extremities to grow cold – together with a darning needle and a supply of wool so that they could repair the holes resulting from knocking their toes against the walls as they moved about within the spacecraft.

The primary scientific research involved taking multispectral pictures. Their results were to be used to develop a technique to interpret imagery from future missions which would permit variations in the moisture content of glaciers to be measured, the stocks of plankton in the sea to be assessed, and different rock and soil types to be determined in order to help in assessing and exploiting the vast, but as yet essentially unappraised, natural resources of the more remote regions of the Soviet Union. (Despite the fact that the Siberian wilderness supplied 10 per cent of the gas, 25 per cent of the oil and 30 per cent of the coal for the Soviet economy, it had barely been tapped). Another device was an RSS hand-held spectrometer which, by measuring reflectance spectra, could distinguish soils with different moisture content. Its results were used to map salt deposits in the Kara–Bogaz–Gol Bay area.

As the cosmonauts prepared to return to Earth, some doctors feared that they would black out from orthostatic intolerance from the deceleration loading imposed by re-entry, even though they had adhered to the regime of an hour's strenuous physical exercise twice each day to maintain a high level of fitness; some even warned that they might die from the shock of loading after so long in weightlessness. The *real* test of the mission, therefore, would come in its final phase.

On their final orbit, the cosmonauts ran through the deorbit checklist and aligned the spacecraft with its main engine facing in the direction of motion, to serve as a brake. The deorbit manoeuvre was performed while passing on a north-easterly track across the South Atlantic, towards Africa. This slowed it down by about 150 metres per second, which was just sufficient to lower its trajectory into the atmosphere while low over Soviet territory.

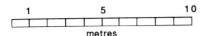

metres

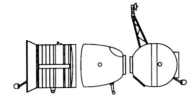

Soyuz spacecraft – service, descent and orbital modules

(The recoveries of Soviet spacecraft were governed by several parameters: the deorbit manoeuvre had to be undertaken while in the southern hemisphere, and it had to be in sunlight so that the cosmonauts could visually check that the spacecraft was properly aligned, and the landing itself had to be in daylight, so that the recovery team could visually find the capsule on the Kazakh steppe. If it was early morning over the South Atlantic, it was late afternoon in Kazakhstan. Another requirement was that the spacecraft's ground track pass over the recovery zone, and only one or two orbits per day were suitable. Finally, the spacecraft had to be at the proper place at the proper time to make the deorbit manoeuvre. All of these factors led to long flights being recovered a few hours before sunset, local time.)

About half an hour later, explosive bolts detonated to release the orbital and service modules. Four minutes after that, passing 100 km above the Horn of Africa, the descent module began to feel significant atmospheric drag. Although it was still moving at 7 km per second, and was entering the atmosphere at an angle of about 30 degrees, the deceleration developed slowly. As soon as the cosmonauts noticed themselves being eased back into their couches (at about 0.1 g), they settled as best they could and tightened their straps.

If the capsule had been spherical, as had the Vostok, it would have followed a steep ballistic trajectory that would have imposed a peak deceleration load of 10 g. In fact, the Soyuz descent module was bell-shaped, and dug into the atmosphere with its blunt base facing forward. The automated control system exploited the fact that the capsule had an offset centre of mass with a 0.28 lift-to-drag ratio, and selectively fired small hydrogen peroxide thrusters to control its roll, pitch and yaw, to vary the lift vector. It had a pair of nozzles set 60 degrees apart near the upper rim of the capsule to control the roll rate, two set between these to control the pitch rate, and two near the broad base to control the yaw rate. By flying almost horizontally in a narrow 100-km long corridor through the upper atmosphere, the tiny capsule was able to extend its trajectory by about 450 km, and to refine its lateral motion so as to aim at a specific landing site. In addition to refining the trajectory, this dynamic re-entry also reduced the g-load on the occupants. Flying heads-up with 0-degree bank imposed a peak of only 3 g, and even when the capsule rolled into a maximum 60-degree bank to one side or the other to refine its track this load rose to only 4 g. Nevertheless, after being weightless for so long, it came as quite a shock to the two cosmonauts.

The heat shield around the base of the aluminium capsule was a 2-cm thick titanium honeycomb treated with a fibrous asbestos binder. This was an ablative material, so the outer surface blistered and peeled off as it was heated to a peak of 3,000°C. Each flake transferred frictional heat to the atmospheric sheath, however, which prevented it from penetrating the cabin; the internal temperature did not exceed 28°C. As the capsule dug deeper into the atmosphere, the air was unable to move out of its way rapidly enough, so it created a shock wave immediately in front of the blunt base. This extremely hot dense plasma wrapped itself around the capsule, but did not actually touch it (if it had, it would have seared through the wall of the capsule, and incinerated the inside, as it had several times in early tests). Radio communication with Yevpatoria was blocked for several minutes by the free electrons in the ionised air surrounding the capsule, so the cosmonauts were completely alone during this most dangerous part of the descent.

An external pressure sensor detected the capsule's descent through 10 km altitude and explosively ejected a tiny drogue parachute to stabilise the capsule; its 850 km per hour rate of descent was not diminished, but it was violently wrenched upright. A few seconds later, a larger drogue was ejected and this yanked out the main parachute, which had a catchment area of 1,000 m^2. (It was at this point during the 1967 flight that a tangled chute condemned Vladimir Komarov to death.) The cosmonauts were slammed down into their couches as the sink rate was instantly cut from 35 metres per second to just eight metres per second, and a radio beacon automatically switched on to help the recovery forces track its descent. Nitrogen was pumped into the shock absorbers in the couch supports, raising them by about 10 cm, and at an altitude of 5 km a tiny valve opened to equalise the internal and external pressures. At an altitude of 3 km, the base of the heatshield was jettisoned, and the remaining hydrogen peroxide in the thruster system was vented. Some 52 minutes after initiating the deorbit burn, an indicator on the control panel illuminated to warn the crew to brace themselves for landing. When 2 metres off the ground two pairs of solid-propellant retrorockets at the edges of the newly-exposed underside of the capsule fired for a fraction of a second to slow it. The 3-tonne capsule hit the ground at only 2 metres per second, and the shock absorbers in the couches absorbed this impact. Since the hatch was in the narrow neck of the capsule, if it remained upright the crew usually waited to be hauled out by the recovery team, but if it rolled over onto its side they scrambled out by themselves; or at least they had after flights lasting only a few days.

The re-entry trajectory had been so accurate that the heliborne recovery team watched the capsule descend, and they were on the scene within minutes. In this case, it settled upright. Although Nikolayev and Sevastyanov were exuberant, their muscles were so overwhelmed by 1 g that they had to be lifted out of the hatch. Evidently, fears that space-farers would not survive the descent were false, but the two men were clearly in poor shape. As their bodies recovered, the doctors confirmed that neither had suffered any irreversible effects. Dr Boris Yegorov, who had flown aboard Voskhod 1 in 1964, wrote in Izvestia a week later that, with proper exercise, cosmonauts should be able to stay in space for at least a month. In fact, this 18-day flight had shown that the problem of extended flight was less to do with survival in weightlessness than later readaptation to Earth's gravity. As longer flights would subsequently reveal, although the body did change while adjusting to weightlessness, this eased off after three to four weeks, then more or less stabilised. Belatedly, it was realised that Nikolayev and Sevastyanov had flown *just* long

(a)

(b)

Recovery of Soyuz. (a) The descent module slowly floats to Earth under its single massive parachute. (b) The capsule fires a solid propellant charge to soften the impact of touchdown.

enough for their bodies to adapt. Ironically, therefore, longer-duration crews would later return in considerably *better* condition than had this intrepid pair. In recognition of their contribution to the effects of the space environment on the human organism, Nikolayev and Sevastyanov were issued the 1970 Award of the International Academy of Astronautics.

This flight had also tripled the endurance record for the Soyuz spacecraft's systems, thereby demonstrating that it could function for an extended period in space, as it would have to once docked with an orbital station.

Writing in *Pravda* afterwards, Professor Vasili Pavlov said that the condition of the cosmonauts after the extended flight indicated that the programme would now concentrate on "flights of increasing duration". Professor Mstislav Keldysh, the president of the Academy of Sciences, said that within a few years it was intended to establish an orbital station operated "in the interests of the national economy", and that it would be manned by a succession of crews.

2

Initial engineering development

While Korolev had been charged with developing a lunar programme to compete with NASA, which had eagerly accepted John Kennedy's challenge to be the first to land a man on the Moon, Vladimir Chelomei, Korolev's long-time rival in the Soviet rocketry establishment, had secured support from the military to build a reconnaissance platform in low-Earth orbit. He already had a bigger rocket (designated the UR-500, informally known as the Proton) than the Semyorka, so he designed a 20-tonne platform and a large spacecraft to service it.

Chelomei's platform, which he called Almaz, was a single pressurised unit having a stepped-cylinder configuration. Its main section was a cylinder 4.15 metres in diameter, and the 3-metre wide forward part was mated with a capsule mounted at the front. It was called Merkur for some perverse reason, as there had already been an American spacecraft with this name. It was a blunt cone rather like its NASA namesake, but it retained the lower part of its launch escape tower in orbit, because this contained the re-entry control system and the parachute. The crew of three were to enter the capsule by a hatch in the side, but in orbit they were to open a hatch in the base (in the heat shield) and pass through a tunnel to another hatch which opened into the module behind.

To serve as a reconnaissance platform, the main vehicle was to be equipped with an ocean-surveillance radar and a high resolution optical imaging system which developed its own film and fed it through a scanner which periodically transmitted the imagery to Earth

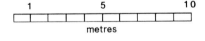

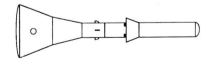

Merkur capsule with escape tower.

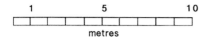

metres

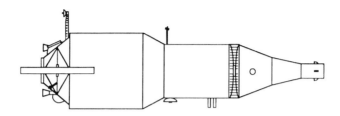

Almaz with Merkur capsule.

electronically. Almaz had a pair of large solar panels to generate sufficient power to run this equipment in addition to the life-support system. The main compartment was to be split into two sections divided by a control panel behind which was to be the small ward-room, and the main section was dominated by the conical bulk of the long focal length folded-optics camera system with its massive film canisters. It would offer a utilitarian existence, but compared with being cooped up in Korolev's Soyuz it was spacious.

Obviously, such an expensive facility could not simply be used once and discarded. It had to be sufficiently well stocked to support a crew for several months, and, in addition, provision was made to resupply it. It had a docking port at its rear, and Chelomei designed a cargo ferry which was almost as large as the Almaz platform itself. Because the technology for automated rendezvous and docking had not been proven, this was to be flown by a crew, presenting the possibility of exchanging crews so that the work of the platform could be carried on without interruption. The ferry, which served the combined function of crew transport and resupply, was known simply as the TKS (an acronym for the Russian words describing its rôle).

The TKS would be launched on a Proton. In orbit, the crew would transfer back to the main compartment, and because the docking system was on the other end, they would thereafter fly the vehicle backwards. During the final approach to Almaz, they would observe through a porthole adjacent to the docking unit, and once the vehicles were joined together the two docking assemblies would be disengaged and swung aside to expose a hermetically sealed tunnel. Once the handover was complete, the retiring crew would board the Merkur capsule on the ferry, separate it, and then return to Earth. The TKS would remain attached to the Almaz until its consumables had been consumed and then it would be jettisoned. This exchange process would be continued until the final crew powered down the equipment in the main platform, retreated to its capsule and returned to Earth.

The Almaz/TKS/Merkur concept offered great potential, but unfortunately the Chelomei Bureau had no previous experience with spacecraft, and its development soon fell far behind schedule. The date set for the test flight – 1968 – came and went.

Meanwhile, Korolev's Soyuz spacecraft suffered a series of failures culminating in the loss of its first pilot, Vladimir Komarov, in April 1967. An ambitious test involving the

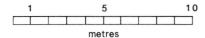

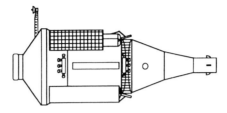

TKS with Merkur capsule.

docking of two spacecraft (so that cosmonauts could transfer from one vehicle to another in space) was cancelled when Komarov encountered serious problems with a faulty attitude control system and a jammed solar panel which limited his spacecraft's power. After a successful re-entry, he died when the parachute failed to deploy properly and the descent module smashed into the ground. Korolev had not lived to witness this disaster, as a year earlier he had died of complications following an operation to remove a colonic tumour.

The dream of landing a man on the Moon in 1967, to mark the fiftieth anniversary of the Revolution, was evidently impossible, so the target date was pushed back first to 1968, then to 1969. Although in January 1969 Soyuz 4 and Soyuz 5 docked and a pair of cosmonauts spacewalked from one vehicle to the other to test the procedure to be employed in transferring to and from the vehicle which was to make the lunar landing, the lunar programme suffered a setback the following month when, during the first test, the huge new rocket (the N-1) exploded a minute into its flight. The second flight, in early July, a mere two weeks before Apollo 11 was due to attempt the first manned landing, fared even worse: the rocket exploded and toppled back onto the pad, obliterating the entire facility.

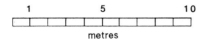

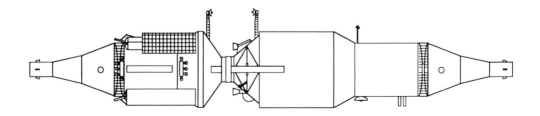

Chelomei's full Almaz configuration.

With the lunar programme spectacularly stalled, it was decided to switch the emphasis and pursue a space station in Earth orbit. With Chelomei's spacecraft still to fly its initial re-entry trials, and the Proton rocket yet to demonstrate the reliability needed to trust it to carry a crew, the rival bureaux were told to work together to use the Soyuz launched on the trusty Semyorka to service the Almaz.

The Korolev Bureau, now run by Vasili Mishin, Korolev's former deputy, therefore adapted the Almaz platform. The first part to go was the Merkur capsule. Chelomei had intended to employ a pair of engines, mounted either side of the aft-mounted axial docking port. This engine block was replaced by a Soyuz propulsion module, but this prevented rear access, and so a short transfer compartment was added at the front, and the docking port relocated to that end, to provide access to the inner hatch which had previously connected to the Merkur capsule. Rather than risk unproven solar panel deployment frames, a pair of Soyuz panels were mounted at each end. To ensure that there was the minimum of delay in effecting the redesign, the fabrication work was done on the same assembly line at the Khrunichev factory as had been established for Almaz, and the stepped-cylinder configuration of the main compartment was retained so as to be able to mate with the existing attachment ring on the Proton rocket. This new configuration was, therefore, a 14.6-metre long structure with a 2.2-metre propulsion module at the rear of the main compartment of the Almaz (the 4.15-metre and 3-metre diameter stepped-cylinder) with a 2.2-metre diameter docking adapter on the front. The energy in solar insolation at the Earth's distance from the Sun is about 1 kW/m^2, and each pair of Soyuz panels had an area of 12 m^2. The transducers were only 10 per cent efficient, and so generated a peak of 1.2 kW (falling off sinusoidally with divergence from the ideal perpendicular Sun angle). Much of this was lost in the low-voltage (28 volt) direct-current distribution lines, and even with the contribution from a docked ferry, the 3.6 kW peak would be barely sufficient to run the environmental systems required to support a crew.

In its new configuration, the space station would be launched unmanned, and its crew would follow a few days later in a Soyuz ferry. It was hoped that the first station would be ready within two years, and this hybrid would certainly be ready much earlier than the Almaz would have been.

Chelomei continued work on the reconnaissance platform and, even though it was no longer strictly necessary, went on to test the TKS with its Merkur capsule. The TKS was destined to play a key role in the future development of the space station programme.

THE WORLD'S FIRST SPACE STATION

In early 1971, in anticipation of launching its first orbital station, the Soviet Union began to hype the expected benefits. Speaking on Radio Moscow on 9 April, Academician Boris Petrov, the chairman of the Intercosmos Council of the Academy of Sciences, said that small orbital platforms which would be manned by several specialists would appear in the near future. He said that these would operate for periods of between one month and one year, and would later be superseded by large laboratories which would be assembled in orbit and remain in use for several years. On Cosmonaut Day, 12 April, he emphasised the importance of establishing an orbital station for sustained operations, and claimed that "all of our manned flights ... are aimed at the ultimate achievement of this goal". The next

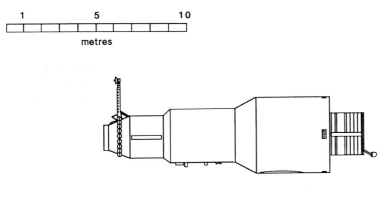

1 5 10
metres

Salyut 1.

day, a Prague news agency reported Anatoli Filipchenko, who had flown Soyuz 7, as saying that the launch of an orbital station was imminent and that it would be visited by successive crews. Andrian Nikolayev said that an orderly programme had been devised which would lead to more sophisticated orbital stations, and Georgi Beregovoi, who had flown Soyuz 3, predicted that by the end of the decade the operation of an orbital station would be a matter of routine. On 19 April a Proton rocket put Salyut 1 into orbit. (Many years later, pictures were released showing Salyut being prepared for launch. One of these revealed the name Zarya, the original choice of name, but this was the radio call sign of Yevpatoria, so the world's first space station was hastily renamed Salyut to celebrate the anniversary of Yuri Gagarin's pioneering flight a decade earlier.)

When the station was not specifically orientated to facilitate observations of the sky or of the Earth, it was placed into a slow rotation for stability, with its solar panels facing the Sun so as to keep its batteries charged. To receive its crew, it had to be orientated with its front facing the approaching ferry.

FRUSTRATION

Soyuz 10, launched on 23 April 1971, carried Vladimir Shatalov, Alexei Yeliseyev and Nikolai Rukavishnikov, the latter, on his first flight, being a member of the team that had designed the station.

After being thoroughly checked out during its first orbit, Soyuz 10 began a series of manoeuvres designed to exploit energy-efficient (Hohmann) transfers to reach Salyut 1. At each stage, radars in a network spread across the Soviet Union determined its orbit, and the computer at Yevpatoria computed the parameters for the next manoeuvre. On its 22nd orbit, the ferry's final transfer brought it within a few kilometres of its target, and then the cosmonauts took over. Shatalov was the most experienced cosmonaut available for making this vital final step, as he had docked Soyuz 4 with Soyuz 5 in 1969. (On that occasion, Yeliseyev had been one of two cosmonauts who had spacewalked between vehicles, and later that same year they had flown together in Soyuz 8 to try out a docking system fitted with an internal transfer tunnel such as the one Soyuz 10 later carried, but they had failed to mate with their target.) Compared with approaching another Soyuz, Shatalov described approaching the 18-tonne station as being "like a train entering a terminus", when its huge bulk appeared in his viewfinder. In the final manoeuvres, which were reportedly far more demanding of the pilot than the approach to another Soyuz, the television camera transmitted pictures of the docking unit. The slow rendezvous had resulted in the final approach being made whilst making an optimum pass over the radar tracking network.

Although the atmosphere was tense, the docking was achieved without a hitch. The probe on the nose of the ferry slipped straight into the conical drogue, and the capture latches engaged to hold it in a 'soft-docked' state. The probe was then retracted to draw the 1.5-metre wide annular collars together and achieve the 'hard docking'. It had been feared that unless it was perfectly aligned when its probe retracted, the smaller vehicle might be shaken so badly that the shock would damage the connectors holding the three modular components of the ferry together, but in the event there was no such thrashing. Unfortunately, the electrical connections within the mated collars failed to link up, and the cosmonauts were unable to swing back the ferry's docking assembly to access the 0.8-metre wide hermetically sealed tunnel to transfer to the station. After only 5.5 hours docked, during which, as *Tass* put it, it had "completed its planned experiments", Soyuz 10 undocked. For an hour, the cosmonauts visually inspected the station and sent back more television pictures of its docking unit, after which they withdrew and returned to Earth.

Over the next few days, robust support was given to the programme. *Pravda* reported Academician Anatoli Blagonrovov's view that orbital stations would permit space research to be elevated to a qualitatively new level in Earth sciences, biology and astronomy. He suggested that observations made from orbit would help solve complex hydrological problems such as the assessment of rainfall intensity, the location of subsurface water and the analysis of the moisture content of soil, and would greatly assist in the study of surface and marine topography. He also predicted that the continuous monitoring of the Earth and its atmosphere on a global scale, when combined with observations of the Sun, the solar wind and the Earth's magnetic field, would increase our understanding of the solar–terrestrial relationship. *Tass* reported Academician Feodor Chukhrov as saying that orbital stations would be able to give continuous measurements of the total thermal, radiational and gravitational properties of the Earth. *Tass* also quoted Konstantin Feoktistov, who had flown on Voskhod 1 and had played a leading rôle in designing the station, as saying that the primary activities of cosmonauts aboard an orbital station would be determined by their economic benefit to the national economy. He said that activities would concentrate on the assessment of crop yields, the manufacture of extremely pure crystals and metal structures in microgravity and the collection of data for oceanographical and geological studies. Feoktistov emphasised that the development of an orbital station to be used in the interest of the national economy was "top of the agenda" in the space programme.

AN EXCELLENT START

Soyuz 11 was launched on 6 June. In comparison to Shatalov and Yeliseyev, the new crew were relatively inexperienced: Commander Georgi Dobrovolsky was on his first flight, as was Viktor Patsayev, and only Vladislav Volkov had flight experience (on Soyuz 7). It was subsequently revealed that they were the backup crew, as Alexei Leonov's crew had

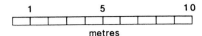

metres

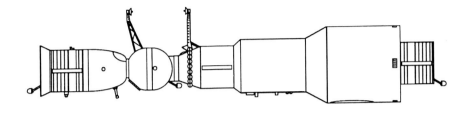

Salyut 1 with Soyuz 11.

been dropped following the pre-flight medical because Valeri Kubasov was found to have symptoms of a lung infection.

As with Soyuz 10, Soyuz 11 followed a slow rendezvous and docked with Salyut 1 on its second day. This time there was no problem, the hatch was easily opened, and the cosmonauts entered the station. The television downlink from the automatic camera showed the three men performing somersaults in the enormous (by comparison with the tiny ferry) main compartment. The first task was to start up the environmental systems, and as soon as the station was verified fully operational, Soyuz 11 was powered down.

The objective of the mission was to stay in space for a month, during which a range of biomedical experiments to study the effects of prolonged exposure to weightlessness were to be performed. These included measuring bone density, studying the reaction of the vestibular system to different stimuli, measuring energy expenditure and measuring the capacity of the respiratory system to assess stamina, taking electrocardiographs to monitor the electrical activity of the heart, measuring the flow of blood within the major blood vessels by taking seismocardiograms, measuring arterial pressure, measuring the force and rhythm of the heart, and collecting blood samples for post-flight analysis. In effect, the cosmonauts were acting as guinea pigs to prove that the human organism could survive a tour of duty aboard an orbital station and return safely to Earth. Looking back from a quarter of a century later, it is now difficult to appreciate just how truly pioneering this mission was.

It had been decided to operate an eight-hour shift cycle, whereby while one cosmonaut slept the other two would be on duty. Their staggered daily cycles involved eight hours on duty, eight hours off duty, and eight hours of sleep. In earlier, smaller, spacecraft this had not been practicable because the cosmonauts would have been in each other's way, but the compartmentalisation of the station's working spaces meant that the sleeping crewman could be isolated (the ferry's orbital module was used as the dormitory). The off-duty time included all toiletries, meals and exercises. Following the results of Soyuz 9, the exercise regime included a 40-minute session after breakfast and a further 80 minutes accumulated during the day. The gymnasium included a variety of elasticated expanders and a moving-belt treadmill (called the KTF) built into the floor across the centre of the main compartment (an elasticated harness was worn to draw the cosmonaut against it to provide the necessary traction). It was hoped that a balanced exercise programme would stave off the muscle atrophy which, despite their efforts, had afflicted Nikolayev and Sevastyanov, but it incurred a significant overhead, reducing the time available for work, although if that was the price of surviving in space, then so be it. Psychologists monitored the video down-link to study the crew, and quickly realised that weightlessness rendered assessment of interrelationships based on body language impractical because these were based on postures and gestures appropriate to physical orientations imposed by gravity. Despite the fact that the station had been given a strict sense of up and down, in space the crewmen floated in random orientations, which made it difficult to assess their facial expressions; this was made worse by the pooling of blood in the head, which had the disturbing effects of distorting facial features. In addition, the pitch of the voice was changed, and for a while the psychologists mistakenly believed that the cosmonauts were under a great deal of stress.

As the days passed, the crew made television broadcasts of their activities, showing off the station and its apparatus, and they were featured nightly on the domestic news, en-abling the audience to see that although Patsayev shaved, Dobrovolsky and Volkov had opted to grow beards. A number of scientific experiments were performed, including the growing of plants, making protein from chlorella algae, and a genetic study involving drosophilae fruit flies. As part of a study of the adaptation of the vestibular system to weightlessness, frogs eggs were developed into tadpoles, which developed receptors for sensing gravity even though they were weightless. After all of the tadpoles had devel-oped, they were frozen for return to Earth. It was hoped that the experiment would provide data which could be applied to human beings, because frogs use similar organs for balance.

Viktor Patsayev, Georgi Dobrovolsky and Vladislav Volkov, the crew of Soyuz 11, in the descent module simulator.

Earth observations were conducted, and a portable spectrograph was used to study aspects of the atmosphere, including airglow. The cosmonauts reported seeing noctilucent clouds, a fairly rare luminous silvery high-altitude cloud. This was the first time these had been seen from space, and the view from orbit was so good that meteorologists requested future crews to sketch and photograph structural detail whenever possible. Whereas from the ground noctilucent clouds are almost exclusively observed between latitudes of about 60 and 80 degrees during the summer months, cosmonauts frequently detected them nearer the equator (even though they were not observable from the ground), often reporting them to extend over large areas and, on rare occasions, over almost an entire hemisphere. Formed at altitudes in excess of about 80 km, where there is very little water vapour present in the atmosphere, the exact mechanism of their formation is poorly understood. They are probably formed by the condensation of water vapour around small nuclei. The precise nature of the condensation nuclei is unknown, but suggestions have included concentrations of silicon and iron particles from violent volcanic eruptions and residual particles from meteoroid showers. Measurements by later Salyut crews revealed that these clouds formed at three distinct altitudes, each of which displayed a different characteristic temperature ranging from −130°C to −150°C. In another experiment, coverage of water vapour cloud was assessed, and this data was correlated with that from the Meteor weather satellites in high orbit, to compare studies conducted at different altitudes under identical lighting conditions. Whenever the station passed beneath a suitable satellite during a given period, a visual assessment was made.

Another major experiment involved using a multispectral camera to photograph the reflectance spectrum of the Earth's surface. The Caspian Sea region was selected because similar data were being collected by a pattern of aircraft flying at 300 metres and 8,000 metres in order to calibrate the station's sensors and assist in the interpretation of the results. It was hoped that the characteristic reflectance spectra of different soil conditions and different crops would be discernible, so that such imagery would provide a way to survey agriculture from orbit. Once the orbital platform's system had been calibrated, it could be used to scan broad swathes of the nation, measuring the extent and the rate of growth of different crops, identifying diseased crops, and enabling the likely yield to be forecast. (The ERTS-Landsat which NASA planned to launch the following year was intended to perform a similar function.) Clearly, the space station really was intended to undertake work which would make a significant contribution to the national economy.

Advantage was taken of being above the atmosphere to also undertake astronomical research. The Anna-3 telescopic spectrometer measured the gamma-ray energy spectrum of various known sources. A forward-planning experiment was conducted to determine how the space environment affected optical materials intended to be used on a later telescope. Gamma-rays, and heavy ion, neutron and charged particle fluxes were measured in the immediate vicinity of the station, to assess cumulative exposure to the various forms of radiation. The high frequency electron-resonance effects of the low-temperature plasma at orbital altitude on the transmission antennas were measured to establish the conditions under which resonance developed, to measure the deterioration of the signal, and to measure the distortion of the antennas' propagation characteristics. All this produced important engineering data to assist in designing later orbital stations, as resonance seriously degrades the efficiency of communications antennas. In addition, engineering tests were conducted with the station's orientation system to determine how well it would maintain its attitude when rotating at specific rates in all three axes. It was a busy mission for the three men.

Life seemed to have settled down to a routine, when a small electrical fire in a cable was discovered on 18 June. The cosmonauts became so alarmed that they urged ground controllers to let them return to Earth immediately. In weightlessness, however, a fire in an oxygen–nitrogen mix is not so great a danger, as there is no convection to draw away the resultant carbon dioxide, so it suffocates. Nevertheless, they powered up their ferry just in case, a fact which did not escape the Western observers monitoring the telemetry signals. As soon as it became evident that there was indeed no danger, the cosmonauts re-entered the station and resumed work. Unfortunately, their pioneering spirit appeared to have been broken by this scare, and a week later it was decided to let them return to Earth early. On 29 June, therefore, they packed the film canisters and the results of their experiments into the cramped descent module of their ferry, and returned the station to its automatic flight regime. The transfer tunnel was overpressurised to check for a leaky hatch, and then vented. They undocked and, laughing and joking because they were happy to be on their way home, aligned the spacecraft for the deorbit manoeuvre. There was a premature loss of signal, but the recovery team that gathered around the capsule on the steppe did not know this, so when they opened the hatch they were appalled to find that Dobrovolsky, Patsayev and Volkov were dead in their couches.

An enquiry eventually determined that a tiny valve in the descent module, designed to equalise the pressure during the final stage of the parachute descent, had been shaken open by the shock from the explosive bolts which had jettisoned the orbital and service modules immediately after the deorbit burn. Analysis of the telemetry revealed that even though the leaking air had rotated the descent module, the automatic system had acted to correct this. Although the cosmonauts would have been alerted to a malfunction by the violent separation, the status of this valve was not monitored, so they would have been distracted whilst the air had vented. Cosmonauts did not wear pressure suits for launch and re-entry, so they first succumbed to unconsciousness resulting from asphyxiation as the pressure fell, then died from embolism of the blood in vacuum. They were exposed to vacuum for 12 minutes before, upon entering the lower atmosphere, the capsule was repressurised. Fortunately, the re-entry manoeuvres and the parachute deployment were automatic, so the capsule landed safely precisely on target.

It took almost two weeks for the official announcement to be issued. In the interim, there was much speculation that the cosmonauts had been killed by the 3 g deceleration forces imposed by re-entering the atmosphere. The pessimists pointed out that the deaths of all three men proved that there was indeed a limit to the time that the human organism could spend in weightlessness; however, there was so little difference between this 23-day flight and an 18-day flight that it was generally accepted that some other fate must have befallen the unfortunate crew. When it was realised that they had died *prior* to re-entry, it became clear that deceleration forces had played absolutely no part in their demise.

Dobrovolsky, Patsayev and Volkov, familiar faces to the nation as a result of their nightly television presentations, were ceremonially laid out in Moscow the following day, and their ashes were subsequently entombed in the Kremlin Wall, together with their fallen colleagues Yuri Gagarin and Vladimir Komarov, and former chief spacecraft designer Sergei Korolev.

ABANDONED STATION

On 4 July, *Tass* reported Boris Petrov's remark that, in the 1970s, orbital stations with changing crews would be used for a broadly based programme that would combine both a regular programme of research with opportunistic experiments. It also predicted that although stations like Salyut would continue to be used for some time, more complex multipurpose and specialised stations would eventually be developed. At the end of that month, Petrov told a Japanese reporter in Moscow that despite the loss of the Soyuz 11 crew, spacesuits would be carried only for working outside a spacecraft, not for launch and re-entry. By October, however, it had become evident that the modifications to the Soyuz ferry required to make it safe were so extensive that it would take a year or more to prepare it. This ruled out the possibility of making another visit to Salyut 1, so it was deorbited over the Pacific. During six months in space, its systems had performed well. On 22 October, *Izvestia* reported Konstantin Feoktistov as predicting that future orbital stations would be capable of automatic operation, so that the crew could devote their time to scientific and economic work rather than to housekeeping tasks. Evidently, the modification of the Soyuz was a secondary issue; the main point was that the station had performed well, and there was no evidence that there was a limit to the time a crew could spend in space.

Table 2.1. Salyut 1 docking operations

Spacecraft	Docking		Port	Undocking		Days
	Date	MT		Date	MT	
Soyuz 10	24 Apr 1971	0447	front	24 Apr 1971	(1030)	0.24
Soyuz 11	7 Jun 1971	0755	front	29 Jun 1971	2128	22.56

Table 2.2. Salyut 1 crewing

Cosmonaut	Role	Spacecraft		Duration days
		Arrive	Depart	
Georgi Dobrovolsky	CDR	Soyuz 11	Soyuz 11	23.76
Vladislav Volkov	FE	Soyuz 11	Soyuz 11	23.76
Viktor Patsayev	FE	Soyuz 11	Soyuz 11	23.76

FROM BAD TO WORSE

Things went quiet for a while, but on 4 January 1972 Petrov said that the emphasis in space for the foreseeable future would include geodesy, agriculture and prospecting for minerals in the interests of the national economy. On 11 April, *Tass* quoted Valeri Bykovsky as saying that the main task of the manned space programme was still to develop orbital stations. That same day, Vladimir Shatalov hinted that further manned missions would probably be launched later in the year. The next day he was quoted by *Izvestia* as predicting that cosmonauts in an orbital station would be able to track meteorological phenomena and provide timely warnings of danger, prospect for minerals, assess the state of crops and warn of disease, spot forest fires, locate plankton in the open sea and steer fishing fleets to the richest fish stocks. He also predicted that scientists would later join the cosmonaut crews serving aboard future, much larger, orbital stations. In March, French sources had suggested that a new orbital station was to be launched in May, but this did not take place. Significantly, these reports predicted that it would be visited by pairs of cosmonauts, which suggested that the modifications to the Soyuz were rather more substantial than had been admitted. Since it was Soviet practice to evaluate a new configuration in orbit before using it operationally, the fact that this had not been done suggested the reports that another station was about to be launched were premature. In late June, however, the modifications were tested in the guise of Cosmos 496. With the modifications verified, the new space station was launched, but the Proton's second stage malfunctioned and the payload plunged back into the atmosphere. A photograph released several years later showed a vehicle labelled 'Salyut 2' being prepared; it had the same configuration as Salyut 1. An analysis of the orbital data by Phillip Clark suggested that if this station had achieved orbit, the same mission sequence would have been followed as had been planned for its predecessor: two crews each spending a month onboard. In this case, orbital dynamics suggested the first crew would have boarded in the second week of August, the second in mid-October.

The Khrunichev factory, meanwhile, was turning out stations on a production line, and it already had another one ready. This was set up on the pad in September, and the preparations for its launch started; but a fault was discovered, and it was returned to the assembly building to be stripped down.

On 4 October *Pravda* marked the anniversary of the launch of Sputnik by quoting Vitali Sevastyanov as saying that cosmonauts in orbital stations would be able to carry out vital hydrological studies by assessing the extent and depth of snow and ice cover, and by monitoring the level of water in rivers. It was yet another argument evinced in support of a programme which would directly benefit the national economy. This was in marked contrast to Skylab, which was being sold by NASA primarily as a laboratory in the sky, and whose primary instrumentation was to be a sophisticated solar observatory.

The programme finally seemed to get off the ground on 3 April 1973, when a Proton successfully placed a large object into orbit. As soon as it had been checked out, it was designated Salyut 2. By this point, Khrunichev had delivered one each of the two types of station, and it had been decided, at the highest level, to test the Almaz first. The very fact that there was a military reconnaissance variant was a secret, and the simple ploy of calling it Salyut, and talking up the benefits to the economy that an orbital station would bring, served to disguise its true mission. Suspicions in the West were piqued, however, by the fact that the radio frequencies used by its predecessor were silent, although it was soon realised that the new station produced telemetry resembling that emitted by spy satellites.

Salyut 2 had essentially the same configuration as Chelomei had envisaged, but the Merkur capsule on the front had been deleted from the design and the transfer hatch sealed. It therefore comprised only the stepped-cylinder main compartment, and it had the original rear-mounted docking unit, between the peripheral engines, with the two large steerable solar panels mounted close alongside. The major change, of course, was that the crew would fly up to it in a Soyuz rather than a TKS ferry.

Yevgeni Khrunov said in early April that his fellow cosmonauts were preparing for a flight in the near future. A week later, on 10 April, Boris Petrov, speaking at Helsinki University, said that a greater degree of automation had been built into Salyut 2, so it would require only a crew of two to operate it. This confirmed that the modifications to the Soyuz had reduced its crew capacity. He added that one was at that moment being prepared to pay the new station a visit. Unfortunately, at this point fate intervened, and on 14 April, just after performing a manoeuvre to refine its orbit, Salyut 2 suffered a catastrophic malfunction. Several days later, with the embarrassment of yet another failure, it was officially denied that Salyut 2 had ever been intended to host a crew. In fact, even as the station tumbled out of control, Pavel Popovich and Yuri Artyukhin had been preparing to fly to it. The loss was eventually attributed to an electrical fault in the new propulsive unit, which apparently started a fire which rapidly engulfed the internal compartment. The sudden rise in pressure ruptured the hull, and the resultant venting sent the spacecraft spinning out of control. As the spin rate increased, the solar panels sheared off and the structure broke up. Considering the suddenness of the station's demise, it was fortunate that the cosmonauts had not already boarded it.

Given the clear evidence of a major systems failure, the West was astonished on 11 May when a Proton launched another space station. In fact, because this was a Korolev configuration, which did not employ the Almaz propulsion block, the failure of that on Salyut 2 was no reason to delay launching this second station delivered by Khrunichev. Unfortunately, after boosting itself clear of its carrier rocket, as it manoeuvred to orientate itself, a fault resulted in the new station draining its attitude-control propellant, and this left it tumbling. This time the problem developed so soon after reaching orbit that it was immediately evident that the vehicle would have to be written off; the useless hulk was disguised as Cosmos 557. Although it was not acknowledged, it is likely that instead of having the small solar panels on the transfer compartment and the propulsion unit, this new station had three larger panels on the narrower part of the main compartment.

It is believed that Alexei Leonov and Valeri Kubasov, who had planned to make the inaugural visit to Salyut 1, had been due to fly to this new station, but following this failure they were reassigned as the prime crew on the joint Apollo–Soyuz mission arranged for 1975.

There was speculation in the West that the immediate 'replacement' of Salyut 2 by Cosmos 557 indicated that it had been intended to operate both stations simultaneously, to upstage the forthcoming American Skylab. This is unlikely, because sustained parallel operations would have been beyond the capability of the control centre at Yevpatoria, and the facility at Kaliningrad, near Moscow, was still under construction at that time. It is conceivable, however, that it had been intended to alternate missions between the two stations, with only one occupied at any given time, but there is no evidence of this. The most outrageous speculation was that it had been intended to link up the two stations. Given the state of the technology, this would have been impracticable, and, again, there is no evidence that such a union had ever been considered; rather the opposite, in fact, as the two programmes were distinctly divergent at that point.

FILLING THE GAP

The Khrunichev factory's production line drew to a halt while the faulty elements of each configuration were redesigned. Since it would clearly be some time before another station could be completed, it was decided to make the best of what was available. This meant independent missions using the Soyuz, the revised version of which had not yet been flown by a crew.

Accordingly, on 27 September Soyuz 12, with Vasili Lazarev and Oleg Makarov on board, was placed into orbit. Its basic mission was to test the new systems. Lightweight pressure suits were now mandatory for launch, orbital manoeuvring and re-entry, and the one-piece suits had been designed to be easy to don in weightlessness, so that they could be put on in an emergency. In essence there was a zip from the chin to the crotch, sealed by an internal rubber bladder. The feet were inserted first, then one arm, then the head was eased through the integral metal collar, and finally the other arm was inserted. An integral hood at the rear incorporated the visor, which locked onto the collar at the front. The life support system was so bulky that it ousted the third seat.

Other changes had been made to the spacecraft to dedicate it to the role of serving as a station ferry. The solar panels had been deleted, and electrical power was provided by chemical storage batteries. This restricted it to just 2.3 days of independent flight while making its rendezvous with the station. Once linked to the station's panels, the batteries could be recharged, providing another two days after undocking for the return to Earth. At the time, this seemed to be an adequate rendezvous margin, because the link-up was meant to take place on the seventeenth orbit, at the end of the first day. Given this limit, it was immediately announced that the mission would last only 48 hours (undoubtedly to pre-empt criticism in the West that some failure had forced an 'early' return). It made a series of manoeuvres to rendezvous with a hypothetical station in an orbit at 340 km, which was considerably (almost 100 km) higher than the operating altitude of Salyut 1. This was taken as an indication that future stations would operate in higher orbits, so as not to decay so rapidly, thereby reducing the propellant consumed periodically boosting the orbit, and, consequently, increasing the operational life of the station. When it was out of range of the tracking stations and communications ships, Soyuz 12 relayed voice and telemetry via the network of Molniya communications satellites which spent most of each orbit thousands of kilometres above the Soviet Union, and so were visible from low orbits virtually across the entire eastern hemisphere.

During this flight the opportunity was taken to test a more advanced multispectral camera to measure the reflectance spectrum of the Earth's surface at different wavelengths, in order to reveal information of economic value. Set up in the orbital module, this had nine lenses, each of which had a different filter. Any given three were chosen for each observation: two for film sensitive to visible light and the other sensitive to infrared wavelengths. The imagery had a resolution of about 100 metres. As before, data was collected by aircraft and ground teams for comparison. Measurements of reflectance spectra had been made on earlier flights, and a means of interpreting the photographs devised, so it was vital to assess whether viewing the surface through the atmosphere introduced any distortions which would have to be allowed for in the calibration process.

As advertised, 48 hours after launch Soyuz 12's descent module returned to Earth. The flight had been a total success.

An indication of future expectations was revealed following the launch of Cosmos 613 in late November. This Soyuz was powered down for 60 days to demonstrate that even though it was limited to a 48-hour period of sustaining a crew, once it had docked with a station, and had been powered down, it could survive a two-month tour. In essence, its purpose was to test whether its systems degraded, because a 'service life' would set a limit to the length of a tour aboard an orbital station. Evidently, it returned to Earth in satisfactory condition.

Another fill-in mission was flown in December by Soyuz 13, with Pyotr Klimuk and Valentin Lebedev. It now became evident that there were in fact two types of Soyuz in operation, incorporating different modifications resulting from the Soyuz 11 tragedy. In this case, although the descent module carried only two pressure-suited cosmonauts, the service module retained the solar panels. It was the first flight to be controlled from Kaliningrad.

Soyuz 13 was to test apparatus intended for future stations. Its main instrument was the Orion-2 telescope, an improved form of the apparatus tested aboard Salyut 1. The cosmonauts had received extensive training in using it at the Byurakan Observatory in Armenia. With it mounted on the nose instead of the docking unit, the spacecraft was orientated for each observation by Klimuk, at his seat in the descent module, and then the instrument was operated by Lebedev in the orbital module. Earth-resources studies were continued using a multispectral camera similar to that carried on Soyuz 12, and the KSS-2 spectrograph, which was used to study the atmosphere by observing the day and twilight horizons to assess the vertical distribution of water vapour in the atmosphere.

Throughout, the Oasis-2 closed-cycle biological cultivator manufactured protein. In a related experiment, chlorella algae absorbed carbon dioxide and liberated oxygen at a much higher rate than expected (it was found that a litre of algae produced 50 litres of oxygen per day). Although it would be feasible to use this process to top up the cabin's atmosphere, on a future station it would be better applied to food production. A human cannot digest chlorella, but a fish can, so the rapid rate of growth of chlorella in space suggested that a closed-cycle fish farm might be feasible. The fish would consume the algae and the oxygen, the algae would consume the carbon dioxide exhaled by the fish, and the human crew would eat the fish once they had matured.

After eight days, having caught up with crucial experiments lost on the failed stations, Klimuk and Lebedev returned to Earth.

The flight of Cosmos 573 (which had made a puzzling two-day flight in June) now made sense. As Cosmos 496 had tested the type with solar panels, now being flown as Soyuz 13, Cosmos 573 had demonstrated the battery-powered form used by Soyuz 12. Something else also now became apparent. Since Cosmos 496 (in June 1972) pre-dated all three of the failed stations, and Cosmos 573 had post-dated them, all of the stations would have been served by the variant with solar panels rather than the stripped-down ferry. This is deduced from the simple truth that it would have been folly to launch the station before proving the ferry type that was to deliver its crew, because that test might have revealed a design fault that would have put the station beyond reach. In the case of the Salyut 1 copy which failed to reach orbit, it would have been necessary to use the variant with the solar panels, because the ferry was required to augment the output from the tiny panels on the hastily developed Korolev configuration.

Interestingly, a painting had been released earlier in the year depicting a station with three large rotating solar panels flying in formation with a ferry equipped with solar panels. With hindsight it was clear that this was intended to be the Korolev Bureau's revised configuration (written off as Cosmos 557) with a ferry of the Cosmos 496 type, because we know that the station lost in the launch failure was of the original type and that Salyut 2 was an Almaz configuration. Not until Salyut 4 was this triple-panel configuration finally revealed, and by that time the stripped-down ferry was being used exclusively. In fact, from this point, until the more advanced Soyuz-T was introduced in 1979, the only Soyuz equipped with solar panels were those assigned to independent missions.

RECONNAISSANCE FROM ORBIT

On 25 June 1974, Salyut 3 was launched. Considering the fate of its predecessors, little was announced until it was certain that the new station was functioning properly.

Salyut 3's 11.6-metre body comprised only the two-element stepped-cylinder of the main compartment. The docking port was at the rear, set between the two engines, and a pair of large solar panels were stowed beside the transfer tunnel. These first unfolded straight out and then released fore-and-after flaps, and they could rotate through 180 degrees in order to track the Sun. Between them, the panels produced a peak of 5 kW.

The new station was sufficiently similar to Salyut 1 to suggest to Western observers that it was an evolutionary development, whose primary improvement was an increased power generating capacity. It was, in fact, as Salyut 2 had been, the Almaz stripped of the superfluous Merkur capsule.

Salyut 3 was permanently orientated with the instruments on its underside facing the Earth. This was feasible because, so long as the station's main axis intersected the Sun, its solar panels were able to produce power. (Salyut 1, in contrast, had been orientated so that its fixed panels faced the Sun, which restricted its ability to hold other orientations.) Its main instrument was a 6-metre focal length folded optics telescopic camera for Earth photography. This produced high-resolution images on large-format film which had to be returned to Earth for developing. Lower-resolution images could be processed so that the

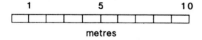

1   5   10

metres

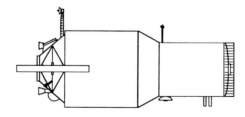

Salyut 3.

cosmonauts could study the result, and there was a scanner so that interesting items could be transmitted to Earth for analysis. It was announced that this camera was to be used to make scientific studies of the Earth, and undoubtedly this was indeed intended, but when it was noticed that the telemetry was the same as that of the lost Salyut 2, the US Department of Defense (DoD) concluded that its real purpose was to monitor US naval deployments, much as its own automated Big Bird satellites monitored the Soviet fleet.

Tass claimed that the objective was to perfect a specialised automated orbital station that could be periodically visited by crews who would adjust experimental equipment, perform repairs and retrieve and replace film and other data storage media. A hatch in the wall of the docking port's transfer tunnel provided access to a small capsule for the return of film to Earth, but this fact was not announced. Compared with the contemporary Soviet photographic reconnaissance satellites, which employed the Vostok spacecraft and operated only for short periods (typically less than two weeks), a man-tended vehicle on an extended flight was considered to be a significant advance. In contrast, DoD's Big Bird satellites, the first of which had been launched two years earlier, combined an extended mission (at that time, typically four months) with a Sun-synchronous polar orbit. In addition to scanning low-resolution images for transmission to Earth, they carried a stock of six small capsules for returning important high-resolution film to Earth without interrupting their coverage. In fact, the promise of the Big Bird had been so outstanding that the decision to develop it had pre-empted the Manned Orbiting Laboratory (MOL), the US Air Force's equivalent of Almaz. Salyut 3's systems were capable of sustaining a maximum of three months of occupancy over a period of six months, and it was hoped to send up two or three pairs of cosmonauts.

The first crew, Pavel Popovich and Yuri Artyukhin, both military cosmonauts, was launched in Soyuz 14 on 3 July. The following day, when its transfer orbit brought the spacecraft within a few kilometres of the station, the automatic rendezvous system was activated. When the ferry had closed to 150 metres, Popovich took over, completed the approach, and successfully docked at the first attempt. He remained at the controls while Artyukhin opened the hatches and went to check out the station's life support system, and then Soyuz 14 was powered down and its batteries were switched over for recharging from the station's power supply. Considering that their previous assignment had been to fly to Salyut 2, simply docking and boarding this new station represented a significant improvement in the fortunes of the programme.

The layout of Salyut 3 was different from that of Salyut 1. An instrumentation and storage unit formed a wall between the main compartment and the living compartment at the front, and a small corridor on the right provided access. The main compartment was dominated by the massive conical camera housing which sat in the centre and extended almost to the roof. Their living compartment, although small, was unobstructed. It was arranged with lockers and two seats, and a table that incorporated recesses and storage bins for food preparation. Psychologists had insisted that a crew would adapt more easily to weightlessness if a distinct notion of up and down was provided, so the floor and the ceiling were different colours and there was a carpet on the floor made of velcro so that the cosmonauts could stand in place and walk. In addition, based on the experience of previous flights, all protruding apparatus had been wrapped in padding to prevent the cosmonauts injuring themselves by banging into objects. In the living room they had four

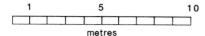

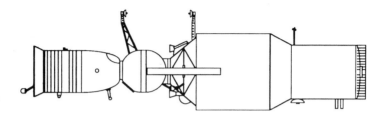

Salyut 3 with Soyuz 14.

portholes, which, in addition to providing a less claustrophobic atmosphere, were used for making navigational observations, photography and filming. Instead of sleeping in the ferry (as their predecessors on Salyut 1 had done) there were *bona fide* berths in the living compartment. Although the crews were not expected to remain aboard for more than a few weeks at a time, a primitive deployable shower had been fitted. A thin plastic curtain device, it was mounted on the roof and was drawn down for use. It used a jet of air to force the water from the spray head to the suction cup at the far end, but it was to prove awkward to use (it took several hours to set up, use, clean and pack away again). In weightlessness, water is dominated by surface tension, and the drops combine to create free-floating globules that simultaneously oscillate on random axes, and these tended to cling to the occupant. The centrepiece of the gymnasium was a treadmill, and the daily cycle called for two hours of exercise. There was a stove to heat canned food, and a spigot to supply hot water to reconstitute dehydrated food. As before, the cosmonauts operated staggered eight-hour shifts but, with only two of them present, round-the-clock operations were not possible.

Salyut 3's reconnaissance mission required its camera to be kept aimed at the Earth. One way of making a spacecraft track the Earth is to orientate it vertically, so that the force of gravity would tug slightly more on its bottom end than on its top and thereby provide 'gravity-gradient' stability. For this to be useful, however, the instrumentation would have had to have been mounted axially; Salyut 3's camera was in the floor of the main compartment, so the station would have to be kept level, and to keep it facing the Earth a force had to be applied to make it undertake one axial rotation per orbit. Rather than use thrusters, Salyut 3 had an electrically-powered gyroscopic attitude control system which could adjust the orientation of the station without expending propellant. Because the gyrodynes used electricity, it was also necessary to keep the solar panels facing the Sun. Because they rotated, the panels could track the Sun, but only if the Sun lay more or less perpendicular to their rotation axis. The station had therefore to be orientated with both of these considerations in mind: with its panels aimed at the Sun, and with its camera aimed at the Earth. This sometimes required it to rotate at different rates in all three axes simultaneously, and as the geometry altered, the rates required were different at different times of the year. The gyrodynes proved to be very responsive to control inputs, and they were able to lock the station with respect to the ground track. Power management was critical. High-capacity batteries were charged during a daylight pass, and then provided power when the station was in the Earth's shadow. Operations had to be scheduled with this limitation in mind.

Other new systems were introduced and tested, including one for improved thermal regulation and a system for condensing water vapour from the air. This reclaimed 1 litre per day per cosmonaut, for hygienic use and food preparation. It was a significant step towards making a station self-sufficient. In contrast to Salyut 1, which had been able to communicate with Yevpatoria only when directly over the communications network, Salyut 3 communicated via the Molniya relay satellites, which spent most of their time high over the Soviet Union, so that the imagery from its camera could be downlinked. To test the resolution of the camera, special photographic targets were displayed at the cosmodrome.

It was not entirely a reconnaissance mission, however. Geological structures and atmospheric phenomena were studied to gain information in the interest of the national economy. Geological structures were photographed to determine areas likely to contain mineral deposits, to identify land subject to salination, and to assess the condition of the soil in pre-specified areas. Earth-resources observations concentrated on Soviet Central Asia, the Pamirs, the eastern shore of the Caspian Sea, the Caucasus and the Ustyurt Plateau. Leonid Sedov later announced that this imagery had identified 67 sites in the Caspian Sea region which were likely to yield oil and natural gas, and similar observations revealed 84 other sites in Uzbekistan. In addition, oceanic observations of the Atlantic were made for the Tropex-74 Project, in concert with science ships and Meteor weather satellites.

The major non-imaging experiment involved the RSS-2 spectrograph, a hand-held device designed to study the processes affecting the Earth's thermal environment, and to gain data to assess trends in climatic changes caused by the saturation of the atmosphere with aerosols resulting from industrial smoke, dust and chemical pollutants. This was important work because it was believed that the disruption of the optical and radiation properties of the upper atmosphere would create the conditions necessary for climatic changes. The RSS-2 could be used either to observe the atmosphere at times of orbital sunrise or sunset by measuring the selective absorption of sunlight passing through the layers of the atmosphere, or to make observations of the Earth's surface and the oceans by measuring the reflectance spectrum under solar insolation. By measuring the humidity of soil, subsurface water reservoirs were identified in the parched lands by the Caspian Sea. If tapped, these could support intensive oasis farming.

Several purely engineering tests were performed. One (called Resonance) measured the vibration modes of the combined structure under different conditions. To carry out these measurements, the cosmonauts set up a variety of spring-loaded masses, then monitored their oscillations. These measurements were vital for two reasons: firstly to identify sources of vibrations which might blur the camera's imagery, and secondly to discover potentially dangerous vibrations that might resonate and threaten the integrity of the seals around the portholes and the docking collar. Another test evaluated the ferry's ability to control the complex.

On 19 July Soyuz 14 undocked and returned to Earth. Despite having been in space for a fortnight, Popovich and Artyukhin scrambled out of their capsule without the assistance of the recovery forces, in marked contrast to the Soyuz 9 crew. When they arrived, the doctors pronounced both men to be in good shape. This was encouraging, because it implied that the strict exercise regime had ameliorated the atrophying effects of adaptation to weightlessness.

MORE FRUSTRATIONS

A month later, Soyuz 15 set off with the second crew. As before, both cosmonauts, Gennadi Sarafanov and Lev Demin, were military officers, and neither had flown in space before. All went well during the rendezvous, but then, although the spacecraft's automatic system attempted several approaches, the cosmonauts had to abort it with just 40 metres to go each time because its closing rate was too high. When Sarafanov finally attempted to make the docking manually he ran low on propellant and was forced to withdraw. As they completed their 22nd orbit, having given up their efforts to link up to the station, and with their ground track now far west of the recovery zone, the cosmonauts settled down to contemplate their lost opportunity. The 48-hour battery had been provided for precisely this eventuality. Just as it took 24 hours for a favourable pass to provide the extended period of communications required for Kaliningrad to supervise a docking, it took another 24 hours for the track to pass over the recovery zone in Kazakhstan. The announcement of Soyuz 15's successful recovery did not include the standard phrase, "following the completion of its planned experiment programme".

A few weeks later, Vladimir Shatalov, visiting America as part of the preparations for the Apollo–Soyuz flight, reported that a cargo/tanker ferry was under development, that Soyuz 15 had been testing a fully-automated rendezvous and docking system, and that only after this had repeatedly failed to manoeuvre properly did the crew demand to dock manually, as had always been the case; but by then there was insufficient propellant for the extensive manoeuvres required and it had been decided to abort the mission. Phillip Clark's analysis of the orbital data suggests that a three-week mission had been intended.

On 23 September, a month after the aborted attempt to place a second crew aboard, Salyut 3 released a capsule which had been loaded by the first crew. Although this was successfully retrieved, its payload was not reported, and for many years the configuration of this capsule was kept secret. In the absence of specific information, some speculated that there might have been a large Vostok capsule on the nose of the station, others that it was more likely to have been a small capsule similar to those employed by the lunar sample-return probes; and later still, once its existence was known, it was argued that a Merkur spacecraft might have been installed. In fact, it was a 400-kg, 0.85-metre wide drum-shaped spin-stabilised capsule. It could return 120 kg of compact payload, which was far more than could be ferried back in a Soyuz descent module. Several years later, a reconnaissance variant of the basic Soyuz vehicle was introduced which was able to carry a number of small film-return capsules, and it is likely that these were similar to those released by the military Salyut. The Almaz platform proved the technology later exploited by the automated satellites.

No further attempts were made to launch crews to Salyut 3. Despite its potential not having been fully exploited, it had to be seen as a success. It continued in its automated flight regime until the end of the year so that the deterioration of its systems could be evaluated, and it was then powered down.

A GLIMPSE OF THE FUTURE?

Shatalov's reference to the development of an automated cargo/tanker ferry caused considerable speculation, because for it to be practicable an orbital station would have to incorporate two docking ports. It would be almost two years, however, before Andrian Nikolayev told Peter Smolders that future Salyuts would be equipped with two ports, (although he did not say where the additional port would be located) and another year before the first such station was launched. Clearly, the testing of this automatic docking system so early reflected considerable forward planning. Since the actual configuration of future stations was not announced, there was speculation that a multiple docking module was to be launched to link together two Salyuts and two Soyuz ferry craft. This view was reinforced by the frequent official statements that the future orbital complexes would be built up from modular components. The launching of two stations within one month during the previous year had demonstrated that a production line was already operating.

By this point, however, it had also become clear that the Salyut program was more complex than the straightforward series of incremental steps which had typified earlier Soviet engineering developments in space. Given the dramatic differences between the two successful stations, it was evident that these represented two types: one a scientific research station, the other a reconnaissance platform staffed by military officers. Apart from the publicity given to the scientific station, and the almost total absence of it in the case of the military platform, it proved remarkably easy to tell the two apart immediately they reached orbit: the civilian station's telemetry used continuous-wave pulse-duration modulation, and transmitted on the traditional Soyuz frequencies; the military used spy-satellite frequencies with frequency-shift-keyed pulse-duration modulation. Of course, the Soviets did not announce this. Western 'space buffs' discovered it through diligent observation.

Table 2.3. Salyut 3 docking operation

Spacecraft	Docking		Port	Undocking		Days
	Date	MT		Date	MT	
Soyuz 14	4 Jul 1974	2351	rear	19 Jul 1974	1203	14.51

Table 2.4. Salyut 3 crewing

Cosmonaut	Role	Spacecraft		Duration days
		Arrive	Depart	
Pavel Popovich	CDR	Soyuz 14	Soyuz 14	15.73
Yuri Artyukhin	FE	Soyuz 14	Soyuz 14	15.73

AN ORBITAL LABORATORY

Placed into orbit on 26 December 1974, Salyut 4 was one of the Korolev Bureau's vehicles. It had the same configuration as the failed Cosmos 557 (the service module at the rear stripped of solar panels and three large steerable ones mounted on the narrower section of its main compartment). Its panels had a total area of 60 m^2, provided a peak of 4 kW, and could be rotated through 340 degrees to face the Sun whilst permitting the station flexibility in orientation. Regardless of the improved operational flexibility, the overall power supply was barely greater than that of Salyut 1, because the attached ferry could not make a contribution.

Salyut 4 could support up to three months' occupancy, over a period of about six months, by teams of two cosmonauts. Although the Korolev and Chelomei bureaux were long-standing rivals, there was a degree of cooperation. Salyut 3's water reclamation system had been fitted to maximise the use of this valuable resource; even so, water was to be taken up in small flasks by successive crews. The food was preloaded. In anticipation of equipment failure, many of the internal systems had been configured to permit access and replacement. Equipment such as filters for the air conditioning system and canisters for carbon dioxide extraction (which had to be replaced every few days) would likewise have to be brought up by each new crew, along with any spare parts for repairs. If the life support system suffered a major failure while they were aboard, the station would probably have to be evacuated and the task of effecting a repair, if this was feasible, left to the following crew. Nevertheless, this new station clearly embodied the optimism of its designers. The gyrodyne attitude control system had not been fitted, so consumption of the unreplenishable propellant supply would be a major factor in planning work that required manoeuvring. Although a resupply ferry was under development, it would not be possible to use it to resupply Salyut 4 because it did not have the plumbing needed to pump fluid aboard. Clearly, therefore, although Salyut 4 was a significant improvement in capability, it was just one step in a long development process which was advancing on several fronts.

Salyut 4 tested the Delta semi-automatic navigational system. This was a significant advance over previous stations, which had required data on its path to be measured on every revolution by a network of tracking stations. The new system used a set of Sun sensors to keep track of the station's position in its orbit by monitoring the rising and setting of the Sun on the Earth's horizon in order to determine the period of each orbit. It incorporated the Argon 16 computer, which combined this with the readings from the radio altimeter to compute the station's orbital parameters, monitor its progress around the globe and compute the times during which it would be within communications range of the various ground stations. This could pinpoint the station's location over the surface of the Earth to within 3 km, and its altitude to within a few hundred metres. The position was displayed by the Globus navigational display on the main control panel. The new station also tested the Kaskad attitude control system. This employed a pair of infrared sensors to determine its attitude relative to the Earth, together with an ion sensor which determined its orientation with respect to its direction of flight through the ambient environment. It could align and maintain the station within 5 degrees of a specified orientation. Between them, the Delta and Kaskad systems both simplified the ground-support system and reduced the crew's workload. (Salyut 1's crew had spent up to 30 per cent of their time on

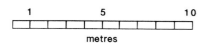

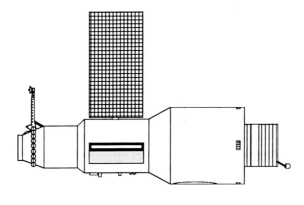

Salyut 4.

tasks related to orientating the station). Importantly, Kaskad resulted in a significant reduction in fuel consumption. A teleprinter (Stroka) enabled Kaliningrad to send up
lengthy communications without requiring a cosmonaut to write it all down, so as to further free the crew for more productive work.

It also tested a sophisticated thermal regulation system (STR). The outer surface was
covered with 'screen vacuum heat insulation' composed of layers of synthetic film sprayed
with aluminium that minimised heat loss. The station incorporated an intricate set of radiators which collected solar heat on the sunward side and radiated excess heat on the shaded
side. These enabled the thermal regulation system to control heat transfer within a wide
range of temperatures. Individual elements of the multiple-loop system, which comprised
heating and cooling units, had three or more backups to provide a high degree of redundancy, and so safety. The computer commanded the entire system, correlating the heaters
and coolers with the station's orientation with respect to the Sun and the Earth's shadow.
Water could be discharged into space, so that the evaporation would rapidly reduce the
thermal energy of the station in an emergency. While a ferry was powered down docked
to the station, its thermal regulation systems were put under the station's control.

With missions planned to last several months, Salyut 4 provided extensive facilities for
its crews. There was a table in the main compartment, between the control panel and the
telescope, and this supplied hot and cold water for food preparation. The telescope took
up most of the space at the rear of the compartment, so the medical and exercise equipment was on the roof above the telescope and the toilet was in the floor at the rear. The
cosmonauts put their sleeping bags on the roof, above the control panel. Despite the sophistication of the new station's facilities, life aboard would still be rather spartan, especially as the duration of missions increased.

Salyut 4 was a significant orbital laboratory for Earth and astrophysical observations,
with the OST-1 solar telescope as its primary instrument. This used the same port as Almaz's reconnaissance camera, and was contained in a similar conical mount in the floor of

the main compartment. It had a 25-cm diameter mirror with a 2.5-metre focal length. It had been developed by the Crimean Astrophysical Observatory specifically for Salyut 4, and the cosmonauts had been extensively trained by the staff of the observatory in how to use it to study dynamic processes such as the flocculi and prominences on the solar disk. It was similar to but not as capable as the ATM solar telescope flown on the American Skylab a year earlier. The OST could be used in conjunction with the CDS-1 diffraction spectrometer. This could produce an ultraviolet spectrum in the range 800–1,300 Å with an average resolution of 2 Å. Observing the Sun at sunrise and sunset, the spectrometer was to make the first comprehensive survey of aerosol distribution in the upper atmosphere. It could be used to measure the strength of the ozone's absorption lines (ozone forms a thin layer in the upper atmosphere, at a height of 20–50 km), and these data proved to be timely. Following the British Antarctic Survey's discovery in 1985 that the ozone layer above its base was severely depleted, NASA examined data accumulated by its Nimbus 7 weather satellite since it had been launched in 1978, and then announced that there was a seasonal 'ozone hole' over the southern pole. There did not seem to be any comparable variability at other latitudes, but by including Salyut 4 data this retrospective analysis was able to be extended to 1975. (These data could not confirm the depletion of polar ozone, because the station's orbital inclination was too low, but it set limits to the variability of ozone in sub-polar latitudes, which assisted in characterising the peculiar phenomenon over Antarctica).

The ITS-K telescopic infrared spectrometer could be used to observe the sky, but its primary objective was to study the Earth's atmosphere. The infrared sensors carried by previous stations had used conventional compressor-based cooling units, which had used large amounts of power, and had been unreliable. In this case, however, a far more advanced cryogenic system was employed which used an icy coating of solid nitrogen. This had been designed by the Kharkov Physical–Technical Institute of Low Temperatures. Not only did it draw less power, it could be run for extended periods at −223°C. Observing at sunrise and sunset with its diffraction slit aligned along the horizon, it determined the density and distribution of water vapour and measured the temperature of the layers of the upper atmosphere.

The photometer and spectrometers of the Emissiya experiment were clustered at the rear of the station to scan the Earth's horizon and measure the luminescence intensity of the red atomic oxygen spectral line, and to study processes in the atmosphere at altitudes between 250 km and 270 km (the most Sun-sensitive part of the ionosphere, where the electrons trapped in the Earth's magnetic field interact with the rarefied gas of the upper atmosphere). Combined with data from geophysical ground stations, the results were to facilitate forecasting of short-term variability in the upper atmosphere and, eventually, contribute to an understanding of global climate change.

The Spektr experiment involved a variety of apparatus installed around the station to measure the properties of the atmosphere at orbital altitudes. This expanded on studies made during previous flights. The instruments measured the density, composition and temperature of neutral gas and plasma encountered by the station in its orbit. (The hull of a spacecraft became charged as it passed through this, and it interfered with some of the communications systems. The results of this study were to improve the design of future stations).

In addition to this comprehensive atmospheric study, Salyut 4 was equipped for astronomical work. The Filin telescopic X-ray spectrometer was sensitive in the range 1–60 Å. The main telescope, mounted outside the station, was boresighted to a smaller sighting telescope. The spectrometer was essentially autonomous, so when it detected a signal the operator would use the sight to note the target's position against the star field. The RT-4 telescope was sensitive to the 'soft' end of the energy range sampled by the Filin. Its 20-cm wide parabolic mirror illuminated a sophisticated photon counter, but its field of view was not wide enough for it to be used to scan the sky for new sources; it was used to make further studies of sources discovered by the Filin.

Finally, the Silya experiment used a light nuclear-isotope spectrometer to record the isotopic and chemical composition of cosmic rays. This apparatus established a major new field of investigation.

Early in the new year, Salyut 4 raised its orbit to about 350 km, which was almost 100 km higher than the operating altitude of its predecessors. Although this consumed considerable propellant, such a high orbit meant that it would not suffer so much drag from the isolated molecules that occur at that height, which in turn meant that it would not need to fire its engines so often to overcome orbital decay. Overall, therefore, this was a saving which would enable it to remain serviceable far longer than had been possible before.

FIRST CREW

Soyuz 17 was launched on 11 January 1975. Alexei Gubarev and Georgi Grechko were both making their first flight, and their objectives were announced to be "observation of geological and morphological objects and of atmospheric formations" to gather data "in the interest of the national economy", together with studies of "physical processes and phenomena in outer space", and continued testing of the station's systems and the adaptation of its crew to weightlessness. The following day, when 100 metres from the station, Gubarev took control. The docking was achieved so easily that the crew's heart rates were lower than they had been during simulations. They found the station a little chilly, so Grechko turned up the heaters to raise it to a comfortable 23°C (it would later be noticed that crews tended to overheat a station during their first few days aboard).

Then followed a lengthy process of activating the systems. An air hose was fed from the transfer compartment into the descent module of the powered-down ferry to ensure that it was properly ventilated (on previous long missions it had been found that the air in the ferry had become rather stale), baseline biomedical data was sampled (Polynom-2 was used to assess the condition of their cardiovascular systems), and then the two men settled into a daily routine, this time adopting the same duty cycle. However, their sleep cycle was scheduled to exploit the periods during which they had sustained radio contact with the ground, and the fact that the orbit precessed led to a corresponding migration in their sleep pattern. Over an extended period, starting work slightly earlier every day – in effect operating a 'day' slightly shorter than 24 hours – proved to be very tiring. In general, they operated a six-day working week, and then took one day off.

To ease the process of adapting to weightlessness they had the Tchibis rubberised leggings from which air could be pumped to make a negative pressure that drew blood into the lower body. This proved to be useful in the initial adaptation to weightlessness,

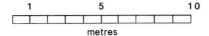

metres

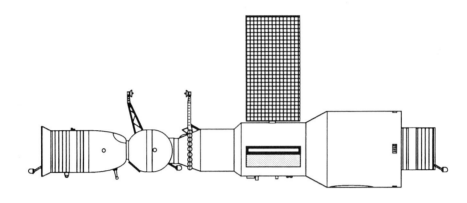

Salyut 4 with Soyuz 17.

because it prevented blood pooling in the brain, and immediately prior to returning to Earth, when it could increase the capacity of the circulatory system. It was a cumbersome garment which prevented movement, so there was little to do whilst using it but perform upper body exercises. Each cosmonaut used the Tchibis an hour a day in the first week or so, and thereafter once a week, until the final ten days of the mission, at which time the cycle returned to an hour a day. They were to follow a strict exercise cycle to reduce decalcification of their bones, muscle atrophy and the loss of capacity in their cardiovascular system (the most debilitating effects suffered by their predecessors due to prolonged exposure to weightlessness). This cycle was repeated every four days: the first built up speed and strength, the second concentrated on exertion, the third emphasised endurance and the fourth was 'anything goes', when they could do whatever they liked. In addition to the standard treadmill, they were provided, for the first time, with a stationary bicycle (a veloergometer, the VTL). Most cosmonauts initially find exercise to be psychologically advantageous because it serves as a tonic, but they suffer in the process because, in the absence of a brisk wind to enhance evaporation, surface-tension ensures that sweat sticks to the body and pools in hollows (such as the chest) forming a puddle of moisture a centimetre or so in depth. Any clothes worn during vigorous exercise immediately become thoroughly soaked (this is a real problem aboard a Salyut because the wardrobe is rather limited), and every exercise period has to be followed by an all-over towel wash to clean up.

In weightlessness, the fully relaxed human body assumes a crouching position, with the knees bent, the shoulders hunched over, the elbows bent, and the forearms floating in front of the chest. The doctors had devised two types of load-inducing suit to exploit this posture. Gubarev tried Atlet, in which elasticated straps linked his shoes to a waist corset and braces ran over his shoulders, while Grechko's Penguin suit had elasticated straps sewn into its fabric to impose compressional loads on the muscle groups likely to suffer from lack of use in weightlessness. Both were designed to make the wearer's muscles work to straighten out. Because they were to be worn as everyday working clothes, the suits were made as comfortable as possible.

In the two weeks before the crew's arrival, the OST-1 had been operated by remote control, and at some point a sensor in the mechanism that aligned the secondary mirror had malfunctioned, disabling its aiming system and preventing its further use. Because this was the main instrument, its repair was crucial. Although they could not see the mechanism, the cosmonauts realised that by listening to the servo, and timing its travel, they could estimate its position sufficiently accurately to align the mirror, so they set it up and locked it in place. Although this rendered the telescope useable, the fact that the secondary mirror was fixed meant that the station itself would have to rotate to scan the field of view across its target. Without their ingenious intervention, the telescope would have remained unusable, and a significant portion of the scientific programme would have been lost.

Another advantage of having a crew tending to the telescope was demonstrated later, when it became necessary to respray the mirror; after several weeks of exposure to atomic oxygen in the rarefied upper atmosphere, its reflective coating had tarnished. Taking advantage of the near vacuum, a remote-control apparatus was used to respray the surface. An electric current was passed through a tungsten wire to melt a sample of aluminium in such a way that a vaporised spray of this metal would recoat the mirror. This itself was an experiment, and since it proved to be so simple to restore the mirror's reflectivity, it opened the way for the installation of more sophisticated mirror-based apparatus on future stations.

Two weeks into their mission, *Izvestia* reported that Gubarev and Grechko were in excellent spirits and were working well together as a crew. In fact, they were working so hard that they had to be ordered to rest, and they had developed eager appetites (they were eating four meals a day, instead of the assigned three). This took its toll, though, and they had to ease off for a few days to catch up on much-needed sleep. In their third week, they reported that they had fully adjusted, and felt quite at home. On 2 February, they broke the 23-day record set by the ill-fated Soyuz 11 crew; this fact was allowed to pass without comment, however, no doubt in order not to evoke unpleasant memories.

A week later, Soyuz 17 undocked and returned to Earth. With Salyut 4 operating at a much higher altitude than its predecessors, the departing ferry made its descent in two stages. Firstly, immediately after undocking high over the Soviet Union, it performed a breaking burn which resulted in a 200-km perigee on the far side of the Earth, over the South Atlantic. Once in place, it followed through with the standard deorbit procedure, leading to a late afternoon recovery.

Both men were found to be fit and in excellent spirits. They had set a new record of 30 days in space. For the first few days, they spent time in a lower-body *positive*-pressure apparatus similar to Tchibis, but with the pressure set to prevent blood pooling in the legs. Increasing the flow of blood in the upper torso helped the cardiovascular system adapt to gravity. Both men were able to walk without difficulty, but Grechko suffered intermittent chest pains (it turned out that his heart had shifted slightly in the absence of gravity). It was concluded that the exercise program had been inadequate to prepare their bodies for readaptation to gravity, so future crews would increase their exercises in the final week, and evaluate a variety of stamina-increasing drugs.

ABORT!

Soyuz 18 was launched on 5 April 1975 to deliver Salyut 4's second crew, Vasili Lazarev and Oleg Makarov. All went well until the signal was issued to jettison the core stage, almost five minutes into the flight, at which time the short trestle interstage ring, which linked it to the upper stage, failed to separate properly.

The sequence of events during the staging operation was for the exhausted stage to separate from the interstage, so that the upper stage's engines could ignite, then for the interstage to be jettisoned. In this case, however, when the core was within seconds of shutting down, vibrations triggered a relay which detonated three of the six bolts on the upper rim of the interstage, partially separating it from the upper stage, and severing cables which were, moments later, to command the firing of the bolts on the lower rim. When the upper stage's engines ignited, therefore, the core was still attached by the latches on one side of the rim of the interstage and it was deflected by the efflux. The asymmetric loading tilted the upper stage over. The computer noticed the misalignment and shut off the engines. The vehicle was already at an altitude of about 180 km, so it was far above most of the atmosphere. It was travelling at 5 km per second, which was considerably short of orbital velocity. The escape tower, intended for aborts in the lower atmosphere, had already been jettisoned. The procedure for escaping a malfunctioning rocket on the verge of space called firstly for the second stage to be jettisoned, then for the descent module to separate from the rest of the spacecraft stack. This had been done several times by capsules carried by errant Moon-bound rockets a decade earlier, but never with a crew aboard. The procedure worked, and the descent module's momentum carried it on a high arc. After following a suborbital trajectory, it plunged into the atmosphere some 1,600 km downrange.

The ballistic re-entry imposed extremely high deceleration forces on the occupants of the capsule; the peak of around 15 g imposed almost enough force on the chest to stifle breathing. As they were recovering from this, they hardly noticed the jolt as the capsule landed in snow, but they were alarmed to feel it slither down a slope for some distance before the parachute snagged on a tree. When they clambered out, they were shocked to see a sheer cliff further down the hill. The did not know it, but they were in the Altai in western Siberia. The Sun was very low on the horizon, so they lit a fire to keep warm until the recovery forces could find them. Their capsule contained survival rations for a few days, but it had no heating. In fact, the radar network had tracked their descent to a spot just south-west of Gorno-Altaisk, and the rescue forces soon arrived. Spaceflight is an inherently dangerous business.

The designation 'Soyuz 18' was reassigned to the next flight, and the launch abort became simply 'the 5th April anomaly'. Yet again, therefore, it was proving difficult to revisit a station. Phillip Clark's analysis of its orbital data suggests that a 60-day tour of duty had been intended. After the fault with the rocket had been rectified, it was decided to send up another crew, and on 24 May another rocket, designated Soyuz 18, lifted off, this time carrying Pyotr Klimuk and Vitali Sevastyanov, who had served as Lazarev's and Makarov's backups.

A SECOND TOUR

When the automatic system had closed to just 150 metres from Salyut 4, Klimuk took command, made a straight-in approach, and docked at the first attempt. This marked a significant milestone, as it was the first time that a station had been revisited. Because they had already had a long day, they just verified that the station was functioning properly and then retired for a good night's sleep.

The experiment programme was, necessarily, a continuation of that performed by their predecessors. This time, however, they were to attempt to schedule their work so that several days would be assigned to a specific type of experiment, before moving on to another experiment. In part, this was an effort to economise on the use of propellant in reorientating the station to suit different types of work, but it would also prove to be a significant improvement in working methodology, because using the same apparatus on consecutive days eliminated the time that had previously been spent first setting up and then stripping down apparatus from one day to the next.

Their first task, however, reflecting the novelty of settling down in a station which was already well into its service life, was to perform maintenance chores. This included changing the air filters, installing a gas analyser, replacing one of the six condensers in the water reclamation system with a hand-operated pump, and repairing all experimental equipment (such as the Silya spectrometer) which had previously broken down and for which spare parts had been brought up. Design work to make apparatus accessible now paid off.

Although biological experiments were set up with plants and insects (these had been evolved on the ground from those which had been returned by the previous crew), and solar and astronomical observations were made, the main target of their studies was the Earth. They gathered extensive spectrometric data that detected pollutants in the upper atmosphere, their work being greatly assisted by the Kaskad attitude control system that provided automatic and continuous maintenance of the orientation so that the instrument scanned the horizon. They also took Earth-resources imagery of European Russia, the TransCaucasus area, the Primoye, the Ukraine, along the length of the Volga, the Oren-burg area, the Kuril Islands, northern Kazakhstan and across the Central Asian Republics. One specific objective was to take photographs to assist in the planning of the route for the 3,500-km long railway which was then being built from Lake Baikal, across Eastern Siberia, to the river Amur on the Manchurian border (their data revealed previously unsuspected environmental and tectonic features that would have to be taken into account in laying the precise route, and in designing its many bridges and tunnels).

In mid-June, when the Crimean Astrophysical Observatory reported an increase in solar activity, Klimuk and Sevastyanov were asked to turn to solar observations. They took a series of spectrograms and photographed a large prominence on the limb of the solar disk. This eruption was significant because it took place towards the minimum of the solar cycle. Unfortunately, although the station had a powerful solar telescope, this did not continuously monitor the state of the Sun, so it could not automatically alert the crew to an interesting development. The inclination of the station's orbit was sufficient to facilitate the first comprehensive photographic and spectrographic study of aurorae in the polar regions (their view was magnificent, because they looked obliquely *down* on the streamers as they zig-zagged for thousands of kilometres across the surface of the Earth). Since this was caused by greater than usual energy deposition from the magnetosphere into the ionosphere during a magnetic storm (such as followed the eruption on the Sun), a special effort was made over the next few weeks to record aurorae, both by snapping pictures and by sketching.

The 30-day endurance record of their immediate predecessors was exceeded a week later. By the end of June, even though Salyut 4 was six months old, it was in excellent condition. At a press conference for the build-up to the Apollo–Soyuz docking mission, Alexei Leonov reported that his colleagues had done so well that they had been ordered to remain in space for the full two-month tour, which meant that they would be in space at the same time as his Soyuz 19 flight. Konstantin Bushuyev, the technical director for the Apollo–Soyuz project, said that simultaneous missions would be possible because while Kaliningrad was occupied by the ASTP flight Salyut 4 would be controlled from Yevpatoria.

Sevastyanov's 40th birthday was on 8 July, and Klimuk's 33rd occurred two days later. They celebrated with a feast of spring onions which had grown in the Oasis. In fact, the onions represented a rare success, as in general, plants had developed well for the first few weeks and then withered and died after a month or so.

Soyuz 19 was launched on 15 July with Alexei Leonov and Valeri Kubasov. A few hours later, Apollo followed with Tom Stafford, Vance Brand and Deke Slayton on the last Saturn 1B rocket. The next day, in a brief communications link with Salyut 4, Leonov offered to fly across and fix any broken apparatus, a remark which reflected his frustration at having been prevented by one reason or another from visiting three of the Salyut stations. Sevastyanov observed that it was the second time that seven men had been in orbit at the same time on three spacecraft (the first time had been in 1969, when Soyuz 6, Soyuz 7 and Soyuz 8 had flown; Kubasov had flown on Soyuz 6). The link-up with Apollo was made on 17 July, symbolically whilst over West Germany. On the following orbit, the hatches were opened and Stafford and Leonov shook hands. It was to be an all-too-brief lull in the Cold War. The two spacecraft undocked two days later and, after conducting independent missions, returned to Earth.

For almost a week, the plane of Salyut 4's orbit kept it under continuous insolation, because it no longer intersected the Earth's shadow. Whilst this guaranteed power, the absence of shadow risked the powered-down ferry overheating, so additional air hoses were strung through the hatch from the station to help ventilate its descent module.

Radio Moscow announced that the cosmonauts were 'rounding off' their work and would soon prepare the station for autonomous operation. After having examined the Soyuz 17 crew, the doctors had suggested that during their final ten days Klimuk and Sevastyanov should increase their exercise programme and increase their intake of water and salt to help rehydrate their bodies and increase the capacity of their cardiovascular systems, in the hope that this would better prepare them for return to Earth. They started to pack up the results of their many experiments. In all, they loaded 50 kg of materials, films and logbooks stuffed into storage bags. This was the maximum which the descent module could carry as return cargo, and it had to be stored carefully so as not to disturb the capsule's centre of mass. On 23 July, in the midst of the preparations, Klimuk and Sevastyanov exceeded the 56 days spent in space by the second Skylab crew. Only the still-extant 84-day record of the last Skylab crew remained to be beaten. That, however, would have to be left to another crew, on another station, because by this time the state of Salyut 4's environmental system had deteriorated to the extent that the portholes had fogged over with condensation and rich green mould had covered the walls. Klimuk and Sevastyanov were not sorry to be leaving. In fact, they had suggested curtailing the flight just before the ASTP mission, but they had been urged to continue with the plan.

Soyuz 18 used its engine on 24 July to boost the station's orbit to 350×370 km; this was the first time that a docked ferry had been used to raise a station's orbit, as previously the station's own engines had been used. Finally, on 26 July, after 63 days in space, Soyuz 18 returned to Earth. Refusing help, the two men walked to the medical tent. For Sevastyanov, who had been debilitated by 18 days on Soyuz 9, this was a satisfying outcome. For the doctors, this was dramatic proof that the exercise regime, and in particular the readaptation process during the final week of the flight, had worked. This was excellent news for cosmonauts hoping to fly even longer missions. It was announced that the success of the flight had shown that orbital stations manned by rotating crews was feasible. It was a crucial milestone because, as Boris Petrov later put it, "the construction of an orbital station served by replacement crews was man's main road to outer space". A few days later, Valentin Glushko was quoted by *Izvestia* as saying that the psychological issues associated with weightlessness were the main problem facing further development of orbital stations. A tour of several months in a tin can would be an ordeal by any standards, no matter how interesting the assigned work.

In fact, Klimuk and Sevastyanov had been extremely productive: in addition to 9 days starting up, maintaining, and powering down the station and 10 days off, they had devoted 15 days to geophysical and atmospheric work, 13 to astrophysical, and 10 to biomedical. Recorded highlights from the video downlink had been shown by domestic television every evening. Analysis of the photographs returned by the Soyuz 17 crew had demonstrated that the makeshift repair to the OST was satisfactory, so Klimuk and Sevastyanov had taken about 600 pictures of the Sun with it. They had also expanded the Earth-resources programme and taken 2,000 photographs. It was hoped that comparing pictures taken during winter with those taken in summer would reveal new information. Many of their pictures had been taken to reveal details of mountain ranges, currents and shelves in shallow seas and alluvial deposits at river mouths, and later analysis of their photography revealed a pair of closely spaced tectonic faults in the River Ob region of Siberia, between which a rich oil field was discovered. Surveys at the intersections of other newly discovered faults revealed coal and metallic ore deposits. In all, Salyut 4's data identified two dozen previously unrecognised ring structures (often the result of an impact event) and faults for further study as likely sources of valuable mineral deposits.

The ultraviolet and infrared observations of the distribution of water vapour, ozone, and nitrogen oxides in the upper atmosphere, established baseline data for subsequent studies. These substances condition atmospheric processes involved in the Earth's heat flow to space, and the absorption of the infrared and ultraviolet components of incident solar energy. The ITS-K data showed that the upper atmosphere ranged from 400°C to 1,700°C, the heating being due to absorption of most of the Sun's ultraviolet radiation. Such raw data was the key to achieving an understanding of the global weather system and thereafter of long-term climatic changes. This was the first time that the atmosphere had been studied on a global basis with a spatial and temporal resolution sufficient to permit trends to be recorded in detail. Many specialised automated satellites would be deployed in the years to come to extend this pioneering work.

Salyut 4 had been a tremendous success, but although it could not support any more crews, its usefulness was not over.

A TESTING TIME

In early November 1975, Salyut 4 made an adjustment to its orbit which presented favourable launch opportunities every two days. After having had its orbit boosted by Soyuz 18, it was now back down to a 350-km circular orbit. An unmanned Soyuz was launched on 17 November, and significantly was named Soyuz 20, rather than being incorporated into the catch-all Cosmos series, as most other crewless Soyuz craft had been. Soyuz 20 pursued a slow rendezvous and then docked. Whereas spacecraft delivering crews usually made the docking on their 17th orbit, the slow automated rendezvous had delayed it to the 34th orbit, and this would eventually become the norm. Soyuz 20 was immediately powered down. It had verified the approach sequence for the long-awaited cargo/tanker variant of the Soyuz, and an opportunity was taken to verify that the propulsion system would survive an extended period in space.

Soyuz 20 returned to Earth on 20 February 1976. Detailed examination revealed that some of the apparatus had substantially deteriorated from the baking and freezing which accompanied flying in and out of sunlight, and this prompted a 90-day limit being imposed on the time that a Soyuz could spend in space. Clearly, if a crew was to stay aboard an orbital station for longer than this, a way would have to be found to replace its ferry. The simplest means of doing this would be to have a station with two docking ports, as in any case a second port would be necessary to accommodate the cargo/tanker resupply craft (there would be no point sending up supplies unless there was a crew aboard the station to unload them). Commenting on Soyuz 20's successful docking, Konstantin Feoktistov noted that that future stations would have "several docking units".

At the end of 1974 the orbital space station programme had been in dire straits. The first station, in 1971, had made it to orbit, the first crew sent to it had docked but been unable to gain entry, the next crew had spent 23 days aboard the station and been killed on their way back to Earth, the second station had been lost before it reached orbit, the next two had been crippled by malfunctions soon after reaching orbit, and only Salyut 3 had successfully hosted a 16-day crew, although even in this case its second crew had been unable to reach it. In just 14 months, Salyut 4 had advanced the programme enormously: it had supported two crews during 30-day and 63-day tours, and it had played a part in testing the resupply ferry.

The next step was now clear: to launch a station with 'several' docking ports, at least one of which would incorporate the pipes needed for a tanker to replenish the station's propellants. Crews on very long-duration missions would have their ferry regularly replaced; this could be done by having it fly up automatically, or as a result of colleagues making brief visits. It would even be feasible for successive crews to hand the station over, so that it would remain continuously occupied for prolonged periods, possibly even for years. The technology required for this ultimate goal was obviously available.

Table 2.5. Salyut 4 docking operations

Spacecraft	Docking		Port	Undocking		Days
	Date	MT		Date	MT	
Soyuz 17	12 Jan 1975	0400	front	9 Feb 1975	0908	28.21
Soyuz 18	25 May 1975	(2300)	front	26 Jul 1975	1356	61.62
Soyuz 20	19 Nov 1975	(1800)	front	16 Feb 1976	(0200)	88.33

Table 2.6. Salyut 4 crewing

Cosmonaut	Role	Spacecraft		Duration days
		Arrive	Depart	
Alexei Gubarev	CDR	Soyuz 17	Soyuz 17	29.55
Georgi Grechko	FE	Soyuz 17	Soyuz 17	29.55
Pyotr Klimuk	CDR	Soyuz 18	Soyuz 18	62.97
Vitali Sevastyanov	FE	Soyuz 18	Soyuz 18	62.97

THE MILITARY'S TURN

Salyut 5 was launched on 22 June 1976, and its telemetry immediately revealed it to be a reconnaissance platform. Over the next two weeks, it manoeuvred into a circular orbit at 275 km altitude. Soyuz 21 followed on 6 July, with Boris Volynov and Vitali Zholobov, both of whom were military officers. Volynov had flown Soyuz 5 in 1969, and had made the first docking between two manned spacecraft. Although it was Zholobov's first trip into space, he had served with Volynov as backup for Soyuz 15. They docked the next day.

The military station had no requirement for replenishment, ferry exchange or crew handover, so it incorporated none of the features predicted for Salyut 4's successor. In fact, it was structurally identical to Salyut 3, and the main instrument was an improved folded-optics telescopic camera. In addition to military reconnaissance, it would also be used to make observations of geographical and morphological structures on the Earth's surface, and the cosmonauts had attended courses on geology. The handheld RSS-2M spectrometer was to be used both to make further measurements of the distribution of aerosols of the upper atmosphere and to monitor ecological conditions on the surface, and was also used to measure the distribution of water in the Volga basin. The Ministry of Land Improvement and Water Conservation had plans to divert some of the outflow of rivers in the northern Soviet Union into areas further south, so these observations were useful in assessing the extent to which the southern rivers could accept increased capacity. The Volga study subsequently influenced planning for several hydroengineering projects in that region, and timber on 2.5 million acres on the Augara river in the Ust-Ilim was studied to assess the number, type, age and quality of trees, to assist in planning a major woodworking complex.

In addition to environmental studies, a mission objective was to comprehensively photograph Soviet territory south of latitude 51° (the northerly limit of the orbital track) to facilitate the production of extremely detailed maps on the scales of 1:1,000,000 and 1:500,000 which would be used to identify tectonic structures that might have deposits of oil, gas and ores. As they flew overhead, the cosmonauts were able to photograph oil fields being developed from data returned by the crews of Salyut 3 and Salyut 4, besides taking pictures to help plan routes for the oil and gas pipelines needed to operate them. All of this was very practical work of immediate benefit to the national economy.

Several purely scientific materials-processing experiments were also carried out. The Sfera experiment was designed to study the process of smelting metals by passing ingots of bismuth, lead, tin and cadmium through a furnace in an effort to create perfect spheres upon solidification; Kristall immersed a seed-crystal in a solution of potash and alum to observe how crystals grew in microgravity; Diffusia produced a homogeneous alloy of dibenzyl and toluene; Potok evaluated the possibility of building microgravity capillary pumps to circulate liquids (such pumps would not require electricity); and Reaktsia used an exothermic chemical reaction to smelt nickel–manganese solder, and then joined stainless steel pipes to evaluate techniques for assembly work in space. The soldering proved extremely effective, but, surprisingly, Sfera revealed that metal spheres cast in microgravity did not in fact yield the smooth surface which had been expected.

Aviation Week & Space Technology argued that the scientific work on Salyut 5 was "window dressing" intended to divert attention from its real mission of reconnaissance. Certainly, neither the television nor printed media carried the detailed daily coverage of the cosmonauts' activities that they had during the Salyut 4 flights; even *Tass*'s reports were sparse and superficial. Drawing a parallel with Salyut 3, the American magazine predicted that Volynov and Zholobov would return to Earth after a few weeks; in this it would be proved wrong.

Although meteorological observations were not a key part of their programme, in early August the cosmonauts reported that the clear boundary that had separated the summer anticyclone in western Europe from the cyclones in the east, which had been stable ever since they had first arrived in space, had finally started to break up. Such opportunistic observations frequently provided timely input to the weather forecasting service. The prolonged period of extreme weather caused by these uncommonly stable atmospheric fronts had produced droughts throughout western Europe whilst producing torrential rains in eastern Europe. As it broke up, the cosmonauts could see the vortices produce intense thunderstorms. They predicted a period of extremely changeable weather across the whole of Europe, and were soon proved right.

Izvestia reported on 18 August that the cosmonauts appeared to be suffering from "sensory deprivation" and that psychologists monitoring their health had suggested that music be played to them over the voice uplink. Two days later Radio Moscow reported, in a very different tone, that solar radiation levels were "favourable" for "prolonged flight", prompting speculation that the cosmonauts might, after all, attempt to set a new endurance record. Then an early morning announcement by *Tass* on 24 August said that the cosmonauts were in the process of returning to Earth. Later that evening, Soyuz 21 undocked and touched down in darkness 250 km from the normal recovery zone in Kaza-khstan. As they clambered out of their capsule unaided, the cosmonauts were given a hearty greeting by astonished workers from the local Karl Marx collective farm. The doc-tors pronounced them to be "generally in satisfactory condition" after 49 days in space; this contrasted with the description of previous long-endurance crews as being in "good" health. In fact, their physical condition was worse than that shown by the Soyuz 18 crew, who had been in space longer. They had expected to spend two months in space, but they had been recalled about 10 days early, just before they were due to start their intensive readaptation exercises. Even so, they had recovered completely in about two weeks. Un-characteristically, Radio Moscow's only announcement was a brief sentence at the end of its next broadcast; usually a returning crew featured as the main item.

It was not officially acknowledged that the flight had been cut short, but the landing in darkness so far from the usual zone suggested that the descent had been prompted by an emergency. The two men appeared to be in reasonable physical shape, so a medical emer-gency did not seem to be the cause for such alarm, and it was unlikely that sensory depri-vation could have prompted an early return. Western speculation concluded that only a serious systems failure could have prompted such a sudden evacuation. When the next spacecraft was launched six weeks later, it was placed in a very different orbit from that of Salyut 5, seemingly confirming that the station had been written off.

A NEW CAMERA

Soyuz 22 was launched on 15 September 1976, with Valeri Bykovsky and Vladimir Aksy-onov. It was announced to be a joint effort with the German Democratic Republic. The spacecraft was the backup built for the ASTP mission; the androgynous docking collar had been replaced by the six-channel MKF-6 multispectral camera, a rather bulky 200-kg package which had been built by Carl Zeiss. The spacecraft had solar panels, so it could remain in orbit for a week. The primary objective of the flight was to evaluate the camera, and some of the imagery was to be correlated with data collected by aircraft and survey teams already working on an Earth-resources programme. From its low orbit (more or less circular at 260 km), the camera was able to image an area of 110 × 150 km with a resolu-tion of 15 metres, and there was a 60 per cent overlap between adjacent pictures to provide stereoscopic views. An advanced form of the MKF was under development for an orbital station, and it said a lot for its importance that a Soyuz flight was laid on specifically to try it out.

SALYUT 5 REPRISE – ALMOST

Less than a month later, on 14 October, Soyuz 23 set off with Vyacheslav Zudov and Valeri Rozhdestvensky to "continue scientific and technical work and to undertake experi-ments in conjunction with the orbital station Salyut 5", so the old station was not uninhab-itable after all. Unfortunately, during the 18th orbit it became apparent that there was a malfunction in the automatic rendezvous system. The station's transponder was confirmed to be working, but the ferry's system was unable to compute an approach. Vladimir Shatalov later emphasised that the crew's analysis of the system's difficulties had helped ground controllers to determine what had gone wrong, but he did not go on to say what this had been. Whatever the problem, it had prevented the spacecraft from completing its rendezvous, so the flight was cancelled. It was too late to return, so all of the non-essential systems were powered down and the cosmonauts settled down to wait for the following day's landing window. The capsule descended into a blizzard, was blown far off course, and splashed down in darkness into Lake Tengiz. Although all crews were trained for water recoveries, none had been required to test the procedure, and it would have been difficult to pick a worse time to do so. The capsule was 8 km offshore, in shallow freezing water that was so clogged by ice that the hastily requisitioned small boats could not reach it, and thick fog impeded the recovery helicopters. The capsule's batteries had been meant to supply power only for the descent through the atmosphere, a process which took under an hour after the service module had been jettisoned, so the cosmonauts were forced to spend the rest of the night with the heaters turned off so that they could sustain the lamp. It was dawn before a helicopter attached a line and dragged the capsule to the shore. It was fortunate that such an ordeal had not befallen Volynov and Zholobov in their weakened condition.

ANOTHER TRY

The failure of Soyuz 23 to reach Salyut 5 prompted the mounting of another flight; clearly, reboarding the station was a high priority.

On 7 February 1977, Soyuz 24 set off with Viktor Gorbatko and Yuri Glazkov. It was Glazkov's first flight, but Gorbatko had flown on Soyuz 7 in 1969. The docking was achieved without incident, even though it was done in darkness using a spotlight to illuminate the port. Although it was normal practice to open up the hatch and transfer to the station within an hour or so, on this occasion the cosmonauts stayed in their ferry and slept until the next day. It had been claimed in the West that Volynov and Zholobov had evacuated the station when an "acrid odour" had polluted its air, and that, during this unusual period of being docked but not entering the station, the air in the station was vented and replenished from new supplies delivered by the ferry. Alexei Yeliseyev later acknowledged that when the previous crew had left the station they had not deactivated the life support system, and that the air purifier had been turned on several days before the new crew had arrived. Salyut 5 was clearly habitable the following morning, when Gorbatko and Glazkov opened the hatch and floated through. Indeed, they reported it to be "cosy and warm", that it was "comfortable", and that it was "easy to breathe". After testing the systems, they transferred replacement apparatus and performed maintenance. A few days later, in a press conference with journalists at Kaliningrad, they noted how "pleasant" it was to work aboard the station.

One of the first technological experiments involved an investigation into diffusion in microgravity. This used a hermetically sealed capsule containing two substances in an originally solid state which were heated to 70°C to create a melt, permitted to diffuse for three days, then cooled to resolidify so that the state of the mix would be preserved. They followed this with another new experiment to study the diffusion of two organic compounds. On 16 February, *Tass* made the surprising announcement that the mission was now half over. The next day, both men developed slight head colds, an ailment rather more serious in space than on Earth because, in the absence of gravity, mucus collects in the nasal passages. Instead of taking the medication recommended, however, they spent fifteen minutes "sunbathing" in front of a porthole, which proved to be a remarkably successful treatment.

Intriguingly, on 21 February they tested a method of replacing the air in the station by venting air from one end of the main compartment while releasing air from tanks in the orbital module of the ferry, an operation which was televised. As the air was exchanged the crew reported feeling "a light breeze" blowing through the compartments, and in total 100 kg of air was replenished. This test served only to revive speculation that the first crew had been forced to leave in a hurry after the cabin had become contaminated by an "acrid odour". Three days later, Gorbatko and Glazkov completed their programme and prepared to return to Earth. They transferred the results of their researches to the Soyuz 24 descent module, tested the spacecraft's systems, and began to return the station to its autonomous operating mode. Vladimir Shatalov noted that they had undertaken much "preventative maintenance" to ensure that the station would be able to be used following their departure. Soyuz 24 returned to Earth on 25 February. At just 18 days, it had been a relatively short visit – the kind of flight that had been expected for the first crew on this reconnaissance platform.

The following day, Salyut 5 released a descent capsule containing "the materials of various researches". This had been loaded by the cosmonauts with the items for which there was no room in their Soyuz, immediately before they departed. It is likely that the motivation to revisit the station had been to retrieve the film that the first crew, in their hurry to evacuate the station, had not had time to load into the capsule. (As an interesting aside, this 'secret' capsule was publicly auctioned in New York in 1993.)

In March, Salyut 5 recircularised its orbit at about 255 km altitude. The most recent backup crew of Anatoli Berezovoi and Mikhail Lisun began preparations for a short visit, but their impromptu flight was cancelled.

By any measure, Salyut 5 had been a success. With the 'civilian' and the 'military' stations flying alternately, it was universally expected that the next would have several docking ports, would be replenished, and would involve resident and visiting crews. It was time for the orbital station programme to switch from basic engineering development to routine operations.

Table 2.7. Salyut 5 docking operations

Spacecraft	Docking		Port	Undocking		Days
	Date	MT		Date	MT	
Soyuz 21	7 Jul 1976	2133	rear	24 Aug 1976	1812	47.86
Soyuz 24	8 Feb 1977	2130	rear	25 Feb 1977	0912	16.49

Table 2.8. Salyut 5 crewing

Cosmonaut	Role	Spacecraft		Duration
		Arrive	Depart	Days
Boris Volynov	CDR	Soyuz 21	Soyuz 21	49.27
Vitali Zholobov	FE	Soyuz 21	Soyuz 21	49.27
Viktor Gorbatko	CDR	Soyuz 24	Soyuz 24	17.72
Yuri Glazkov	FE	Soyuz 24	Soyuz 24	17.72

3

Routine operations

Salyut 6, launched on 29 September 1977, incorporated the best elements of all the previous stations. Ignoring the differences between the two variants of the Almaz station configuration produced by the Chelomei and Korolev Bureaux, and the improvements made with each successive vehicle, the long-awaited introduction of the two-ended version offered such a significant advance in operational capability that it can justifiably be said to be a new *generation* of orbital station, marking the transition from engineering development to routine operations. It incorporated an improved form of the water reclamation system introduced by Salyut 3, and the same power system (three panels generating a peak of 4 kW), thermal regulation system (STR), and navigational systems (Delta to compute the station's orbit and display its track on a globe on the main control panel, and Kaskad to compute and control its orientation) as Salyut 4, plus gyrodynes to minimise propellant consumption. The increased automation would enable the cosmonauts to devote more time to experiments.

The addition of the rear-mounted docking port required that the propulsion system be redesigned. Chelomei's design for Almaz had required the docking port to be in the rear because the Merkur spacecraft was mounted on the front, so a twin-chamber engine had been designed to fit around the transfer tunnel. When it had been handed the design, and ordered to get it into service as soon as possible, the Korolev Bureau had rejected this unproven engine and had used its Soyuz service module instead. Chelomei had continued with his own configuration and, following the loss of Salyut 2, the twin-chamber engine had proven itself on Salyuts 3 and 5. The Korolev Bureau had adopted this peripherally-mounted engine in order to leave the axis free for the rear docking port and its transfer tunnel. Rather than have the engine and its propellant tanks exposed, as on Almaz, the entire assembly, together with other apparatus, was encased in an unpressurised bay forming an extension of the shell of the main compartment.

On earlier Korolev stations, the main engines had burned nitric acid and hydrazine fed by a hydrogen peroxide powered turbine, and the small thrusters had simply vented cold hydrogen peroxide. In this new configuration, this turbine had been discarded in favour of a system employing tanks pressurised by a bladder inflated by high-pressure nitrogen gas, and the engines burned unsymmetrical dimethyl hydrazine (UDMH) fuel in nitrogen tetroxide. As with previous reactants, these were hypergolic; they ignited on contact, eliminating the need for an igniter. The orbital manoeuvring engines and the attitude control thrusters burned the same propellants, drawn from a common supply. Unifying the propellant system meant that the new station would not get itself into the ridiculous situation in which it had to be abandoned because it had run out of attitude control propellant, even though there was plenty for the manoeuvring engines. It also tremendously simplified the task of replenishing the station's propellants. It was not announced, but only the rear port was equipped to accommodate a tanker, so operations would have to be carefully phased to ensure that the rear port was free when a resupply ferry was due. Salyut 6 introduced a new 'radio-technical' transponder system (Igla) designed to facilitate automatic dockings; this was to be employed not just by the new resupply craft, as crew ferries were also to dock automatically. The cosmonauts would take over only if this system failed. Like its predecessor, Igla would orientate the station to aim the appropriate port at the ferry, to simplify the manoeuvres required of the approaching spacecraft.

The main compartment was laid out like previous Korolev stations, except that there was now an axial tunnel behind the massive conical housing of the primary instrument leading to the rear docking port. Like Salyut 4, the forward transfer compartment could serve as an airlock, and it had an inward-opening side-hatch to facilitate extravehicular activity. The two semi-rigid spacesuits were for use by successive crews, so they were adjustable, and they were designed to facilitate greater mobility than earlier suits. The entire backpack was hinged, and a cosmonaut entered the suit by the large oval hinged hatch in its rear, which formed a hermetic seal when swung shut. In an emergency, the suit could be donned in just five minutes. The integrated backpack contained the life support system, and its facilities were directly accessible for maintenance once the hatch was open. It could support several hours work in space, but needed an umbilical for communications and power, so a spacewalking cosmonaut would never be completely free of the station and could not possibly drift away. One of the portholes in the forward transfer compart-

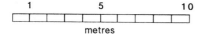

metres

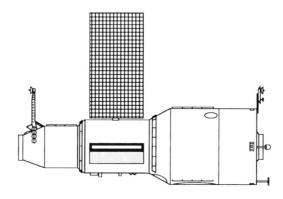

Salyut 6.

ment was set to pass ultraviolet light, so that this would act as a germicide to sterilise the air. It would also permit the cosmonauts to recover from any vitamin-D deficiency (but even better, synthesis of excess vitamin-D would assist calcium absorption in the intestines, and reduce the build up of calcium leached from bone tissue in response to prolonged weightlessness).

The primary instrument was the BST-1M submillimetre telescope. With a 1.5-metre wide mirror, this was actually larger than most ground-based instruments. It was set up for infrared, ultraviolet and submillimetre astronomical work. Although it operated in the vacuum of space, its sensors had to be cooled to −269°C. This was achieved using an improved form of the closed-cycle cryogenic unit tested on Salyut 4 with the ITS-K. Helium was produced using a compressor, a pair of refrigerating units and intermediate heat exchangers, and this was chilled by being fed through an expanding-throttle valve. Although this apparatus consumed 1.5 kW, once the sensor was chilled to its operating temperature it could be maintained at that level for extended periods at little extra cost. However, the telescope could be used only when in the Earth's shadow, and for the rest of the time it had to be protected by a cover.

The second major instrument was the 6-channel MKF-6M multispectral camera, used for Earth-resources observations. This was an improved form of the camera that had been tested on Soyuz 22. At Salyut 6's higher orbit, which was more or less circular at 355 km altitude, however, each photograph imaged an area of 165 × 220 km with a resolution of 20 metres. Simultaneous imagery was taken in six spectral bands (four visual and two in the infrared spectrum) in cassettes holding 1,200 frames. Because film deteriorated rapidly in the increased radiation environment in orbit, film stock would need to be replenished and exposed film would have to be returned before it could fog over, so long-term use of the camera would rely upon spacecraft visiting on a regular basis. Salyut 6 also had a KATE-140 stereoscopic topographical mapping camera (so designated because it had a 140-mm focal length) which operated in the visual and infrared spectrum. From the station's orbital altitude, its 85-degree field of view could photograph an area of 450 × 450 km with a resolution of 50 metres, and could produce individual frames or generate an extended strip image. Each cassette held sufficient film for 600 photographs. Once it had been properly orientated, it could be operated either by the cosmonauts or under remote control by the flight control centre. The Earth-imaging capability of the station, therefore, was formidable. The Ministry of Agriculture had established extensive test sites at Salsky, near Rostov in the Ukraine, and at Voronezh, near Lake Baikal, to help in assessing the capabilities of the multispectral apparatus. These had been planted with specific cereals, vegetables and grasses to determine the ability of the cameras to distinguish between various plants under different conditions.

Since Salyut 6 could be resupplied, a variety of biological and materials-processing apparatus was to be delivered to augment that installed for launch, so that, over time, its science capability could be greatly expanded. Regarded as a long-term facility, this new station was to become a space laboratory which would be inhabited by resident crews and support short visits by international researchers. It was hoped to use Salyut 6 for a long time (at least, compared to its predecessors) in order to support long-term projects. The Biosfera experiment was designed to study the state of the Earth's environment, and began by making visual observations of the oceans. What was seen was to be measured against a specially created comprehensive chromatic chart and used to determine optical properties of the atmosphere under different conditions, and observations were to be made by successive crews. It was hoped that the results would determine conditions under which different phenomena could best be viewed. Later, hand-held cameras were used, and various films were to be used to test the analysis. Later still, spectroscopic studies were added. All of this was designed to improve the observational effectiveness of an orbital station. As Biosfera data accumulated, it would map the boundaries between fertile and arid terrain, assess the capacity of rivers and run-off from glaciers and snowfields, and assess oceanic and atmospheric pollution. Its purpose was to provide a global overview for assessing various Earth-based ecological projects.

The station had been made as comfortable as possible, because cosmonauts were to spend periods of several months living in its cramped confines. The suggestions by the inhabitants of earlier stations had been heeded, and particular attention had been paid to soundproofing the many motors and pumps to reduce the general level of background noise. Previous crews had derived comfort from hearing the various noises, because they were indicative of the state of the different systems, but they had nevertheless argued for a much lower level of ambient noise. Similarly, although it was perfectly practicable simply to attach sleeping bags to the ceiling, the Almaz crews had felt better with specific sleep facilities, and so recognisable 'cots' had been installed (one on the roof and the other on a wall) to offer a sense of functionality. Also exploiting experience on Almaz, a shower had been installed. Located at the rear of the compartment, this had an erectable polythene structure and relied upon forced airflow and a suction device to direct and collect the spray of hot water, which was recycled, purified, mineralised, and then used to rehydrate prepackaged food. An improved menu enabled the cosmonauts to have a different meal each day of the week, and an improved diet provided a daily intake of 3,200 calories. Many items of food had been broken into bite-sized chunks to make them easier to handle in the absence of gravity, and flakey items such as cakes had been coated with an edible film to lock in crumbs. Some 65 types of food had been prepared prior to launch. To prevent dehydration (caused by increased urination as the body tried to remove what it perceived to be an excess of fluid in its upper torso), each cosmonaut was required to ingest 2 litres of water per day; this included that employed to rehydrate food. All consumables could be replenished by a cargo ferry as required. In the centre of the main compartment, between the control panel recessed into the floor and the bulk of the telescope, was the large table for food preparation, and to either side of the table were lockers. For relaxation, there was a stock of novels, a library of video films, and a chess set with magnetic pieces on a metal board.

The gymnasium included a veloergometer, a treadmill and a mass-meter (a spring-loaded apparatus mounted in the ceiling, used by the cosmonauts to weigh themselves). In the roof there were two small airlocks which could be used for scientific apparatus or to eject rubbish. The toilet was between the telescope and the transfer tunnel for the rear docking unit. Most of the 20 portholes were dedicated to specific equipment. There were several television cameras – one colour and two monochrome were for internal use – and there was a monochrome camera incorporated in each docking unit. The colour camera (the first on a Salyut) was fixed in the usual position above the forward hatch of the main compartment, to enable the flight controllers (and the psychologists) to see the crew during the brief periods when the station was within communications range. Most of the scientific apparatus, the shower and the gymnasium were set in the aft section of the compartment, and the front of the compartment contained the control panel and a variety of storage lockers.

Salyut 6's systems had been designed to operate for 18 to 24 months, during which time it was hoped that it would be able to host a 90-day, a 120-day and a 175-day crew, each of which would be visited by short Intercosmos missions as appropriate to ensure that the docked Soyuz ferry never expired its service life. Although a continuous orbital presence was the ultimate goal, given the state of the technology this was too much to expect at this stage, so the immediate objective was to occupy Salyut 6 for at least 50 per cent of its design life. Anything beyond this would be a bonus.

FRUSTRATION

Soyuz 25 set off on 9 October with Vladimir Kovalyonok and Valeri Ryumin, both of whom were making their first flight. The spacecraft followed the usual approach and, the next day, was close enough to Salyut 6 to activate the automated rendezvous system. When this paused at 120 metres, Kovalyonok took command and soft docked. Unfortunately, although the probe retracted, the latches in the annular collar did not engage, preventing the hard docking required to establish a hermetic seal in the transfer tunnel.

While the ferry remained loosely docked, the flight controllers considered possible failures and recovery procedures, and finally ordered Kovalyonok to withdraw and try the docking sequence again, which he did with no result. He undocked again and made another attempt, this time with greater force, but this too failed.

The extra manoeuvring had expended much more propellant than planned, so it was decided not to consume any more by making another approach to attempt to dock at the rear port. This would have been pointless in any case because the engineers thought the fault was in the ferry's docking collar rather than the station's (unfortunately, because the docking system on the orbital module could not be returned to Earth for study, there was no way to be sure why its latches had failed to engage). The dejected cosmonauts were ordered to return to Earth the following day. This failure was a rather poor start to operations with the new station.

SECOND TIME LUCKY

Yuri Romanenko and Georgi Grechko set off in Soyuz 26 on 10 December, and docked at the rear port on their first attempt the following day. This was Romanenko's first flight, but Grechko had spent a month on Salyut 4 and had been substituted for the inexperienced Alexander Ivanchenkov who, having served with Romanenko as backup for Soyuz 25, had hoped to fly next. Grechko's knowledge of the docking assembly on the station was key to ascertaining whether it had been damaged when Soyuz 25 struck it with greater than usual force in an attempt to force the latches to engage.

On 20 December, Romanenko and Grechko sealed themselves into the forward transfer compartment, donned the spacesuits and depressurised the compartment to begin the first extravehicular activity since Yeliseyev and Khrunov had transferred from Soyuz 5 to

Soyuz 4 in 1969, which was itself the first excursion since Alexei Leonov's historic space-walk in 1965. Although a hatch had been provided specifically for spacewalks, in this case Grechko cracked the seal of the front docking unit and swung back its drogue assembly. He examined the conical receptor and reported that it seemed to be in perfect condition. It was, he said, "without a single scratch" just as if it was newly delivered from the factory. Then he floated out through the tunnel to inspect the external facilities. He examined the Igla radar transponder and the collar of the docking unit to see if any damage had been caused by Soyuz 25's repeated attempts to link up, methodically examined all the various joints, sensors, guide pins, fasteners and seals, used special tools handed out by Roma-nenko and exercised the vital mechanisms, and then used a portable television camera to show it all to fellow engineers at Kaliningrad. Romanenko remained in the compartment throughout; he held onto Grechko's feet, to anchor him so that he would have both hands free. Having verified that there was no damage to the docking unit (in other words, that the problem had indeed been in the ferry's docking assembly), Grechko affixed the Medusa experimental cassette (containing prebiotic compounds) just outside the hatch. When Grechko returned, Romanenko, contrary to instructions, stuck his head and shoulders through the hatch to admire the view. Seeing that his commander's tether was loose, Grechko instinctively grabbed for it, but there was no need because their suits were perma-nently attached to the station by umbilicals. They neglected to tell the flight controllers of this breach of the rules. After their return to Earth, however, Grechko mischievously cre-ated the impression that he had only just managed to save his commander from drifting away to his death!

There was considerable concern amongst the flight controllers while the airlock was being repressurised, because the telemetry indicated that the valve used to vent the final vestiges of air so that the inward-opening hatch could be opened had not shut, thereby suggesting that the compressed air which was flooding into the airlock was venting into space. The fact that this ate into the limited reserve of pressurised air was not really the issue; unless they could pressurise the airlock, Romanenko and Grechko would not be able to re-enter the main compartment. With their ferry at the rear port, they would have nowhere to go. Depressurising the main compartment would be tricky. As it turned out, however, the telemetry was spurious, the valve was shut and the airlock repressurised; it was later determined that a cable in the airlock had generated the spurious signal. This rather anxious period lasted about half an hour. Later, Grechko said that his impromptu spacewalk was "very difficult" but a "tremendous pleasure". Vladimir Shatalov pointed out that external activities would form an integral part of orbital operations, and that all cosmonauts would be trained in a neutral buoyancy tank at Kaliningrad.

Salyut 6 cosmonauts followed a different daily cycle from that employed on Salyut 4. Adapting the cycle in order to make full use of the periods when the station was in radio contact with Kaliningrad had disrupted the circadian rhythm, which over an extended pe-riod had proven tiring. Instead, Salyut 6 crews employed a fixed cycle synchronised with Moscow Time. At 0800 they would wake up, shave and exercise. They would have break-fast at 0845, and at 0945 would perform a medical check up and inspect the station's systems. At 1045 they would have their early break before starting the work programme at 1100, and at 1230 would have lunch and exercise. The experiments would be resumed at 1500 and run right through to 1800, at which time they would break off for their evening meal. They would then do whatever was most urgent until retiring at 2300. Such a schedule was made possible by using sophisticated data recorders to store experimental data while out of radio contact, and then 'dumping' it to the communications ships, which in turn relayed it via the Molniya system to Kaliningrad. A system of relay satellites in geostationary orbit was planned to maintain continuous contact with future stations.

Two weeks into the mission, *Tass* announced that whereas Grechko was behaving in an "even tempered" manner, Romanenko's personality was far more "volatile", and whereas Grechko tended to relax in the forward transfer compartment making sketches and taking pictures of the Earth, Romanenko spent much of his free time monitoring the control panel. They were clearly different personalities but, as an improvised team, they were turning out to be rather well matched. They welcomed in the New Year with fruit juices, and since their orbit took them across the International Date Line sixteen times a day it was fortunate that alcohol was prohibited. (Romanenko and Grechko were the first cosmonauts to celebrate New Year in space; only the final Skylab crew had had this dubious pleasure before).

Since they were working far more effectively than expected, and were far ahead on the overall mission schedule, on 3 January 1978, having already finished the assigned technology experiments, they asked Kaliningrad to suggest more things for them to do. There was no such limit to the Earth observational programme, however, and they worked on this enthusiastically. They studied a strong ocean current skirting South America (its boundaries were easy to detect because of the sharp contrast in the colour of the water), and they suggested that their atlas of ocean currents was incorrect. They also observed glacial melting in the southern hemisphere, and noted that some glaciers appeared to be sky blue whereas others seemed to be striped. Most of their glacial studies were conducted over Soviet territory however, because their data was to be used to help forecast glacial flow into rivers so as to predict water levels, and to help evaluate the water flowing into the desert regions of the Republics of Soviet Central Asia. At one point, they reported spotting the snowy peak of Mount Fujiyama near Tokyo for the first time (until then it had been clouded over). Continuing earlier studies, they reported on noctilucent clouds observed at southern sub-polar latitudes.

VISITORS!

History was made on 11 January when a spacecraft approached an already occupied station. Soyuz 27 docked at the front port without incident. Some engineers had feared that because a ferry hits the docking port at about 0.3 metres per second, with a force of about 20 tonnes, the propagation of the brief but very sharp 40 g shock through the station's structure might crack the hermetic seal of the collar of the ferry docked at the far end. Romanenko and Grechko had sealed themselves into Soyuz 26, therefore, just in case. As soon as it was confirmed that the seal had held, however, they re-entered the station and prepared to greet their visitors. The usual post-docking procedures had to be followed though, so it was over two hours before the valve was opened to equalise the pressures so that the front drogue assembly could once more be swung back.

After a month with only each other for company, the residents were naturally very pleased to welcome their guests. What Vladimir Dzhanibekov and Oleg Makarov made of the smell that wafted through when they opened the hatch is not recorded, but one consequence of weightlessness is that stomach gases do not rise to the gullet; instead they pass through the intestines and give rise to intense flatulence of a highly aromatic nature. The olfactory senses of the residents would have become desensitised, and the air quality would have been very striking to the newcomers, the first people to visit an already inhabited station. A television camera recorded them being enthusiastically dragged in through the hatch, hugged and offered the traditional bread and salt, which was washed down with a toast of cherry juice.

Soyuz 27's arrival made the structure 30 metres long and increased its mass to just over 32 tonnes. It was the first time that an orbital complex had been assembled using three vehicles. With only two docking ports, and a crew of two cosmonauts per ferry, the four people aboard represented the station's maximum capacity. It was the first time that four people had been together in space.

The visitors performed medical experiments designed to measure blood flow as part of the general effort to study the process of adaptation to the absence of gravity during the first week in space. To provide a statistically valid database, each visiting crew was to undertake a similar battery of tests.

One engineering objective of this pioneering link-up was to determine the dynamic characteristics of the triple structure. Adding a ferry craft changed the station's centre of gravity, and it was important to verify that the Kaskad attitude control system was able to accommodate this change of mass distribution. It was also necessary to determine the stresses placed on the structure, particularly the docking collars, under different loads. The Resonance experiment, in its simplest case, required a cosmonaut to jump on the treadmill at intervals determined by a timing signal beamed up from Kaliningrad, so that instruments set up at various points within the complex could measure the propagation and damping of the resulting vibrations. This data would identify any resonance which, over an extended period, might lead to metal fatigue. Although the designers had made every effort to eliminate dangerous vibrations in ground trials, it was essential to verify that the structure was indeed sound. Dzhanibekov also did preventive maintenance on the electrical systems, and replaced a voice-channel transmitter that had failed early on.

Soyuz 27's orbital module had been stocked with fresh food, newspapers, books, mail and a variety of other cargo for the station, including the joint Soviet–French Cytos experiment. A variety of micro-organisms had been delivered in a thermostat which kept its contents at 8°C to inhibit their development in transit. The French experiment used paramecium protozoa and the Soviet experiment used proteidae. These were placed into the apparatus, which was maintained at 25°C to facilitate their development, and after 12 hours were chilled for return to Earth for analysis. It was a simple test designed to determine the effects of microgravity and space radiation on the kinetics of cell division, this being the very basis of the life process. It was part of an investigation of the long-term ability of organisms to live in space.

Although a ferry can be launched at any time when the station's orbital plane passes through the cosmodrome, ideal recovery windows occur only at two-monthly intervals. This interval is due to the fact that the Earth is not perfectly spherical but oblate, with the gravitational attraction of its equatorial bulge acting on a satellite's angular momentum in such a way as to make its orbital plane precess. The rate of this rotation is dependent on the inclination of the orbit; at the inclination used by the Salyuts this is 58 days. This arcane orbital dynamic dictated the timing of the visiting flights to the second-generation stations. Each recovery window lasted for about ten days. Visiting ferries were launched early in a window, and recovered towards its end, setting the limit on the time that such visits could last. Clearly, therefore, visits exploiting ideal recovery conditions could be mounted only at two-monthly intervals, and last only about one week. Romanenko and Grechko were to return to Earth in Soyuz 26, so they transferred their contoured couch liners and pressure suits from Soyuz 27, stowed the various packages that they were to take away, and then reset the adjustable weights on the couches to correct the capsule's centre of mass. On 15 January, Dzhanibekov powered up Soyuz 26 to verify that it had not been damaged during its month in space. He ran through all the procedures required to return to Earth, tested its attitude control thrusters and briefly fired the main engine to ensure that it produced the proper thrust. The following day, he and Makarov undocked and returned to Earth, leaving Romanenko and Grechko with a brand new ferry with which to continue their planned 90-day mission.

The failure of Soyuz 25 to dock at the front port had seriously disrupted the planned sequence of operations. Whilst Salyut 6 was occupied, it needed replenishment at roughly two-monthly intervals. It had expended a great deal of propellant climbing to its operating altitude. Also, much of the apparatus for science experiments was to be installed in orbit, so the first cargo delivery had been scheduled early. The automated ferry could replenish the propellant tanks only from the rear port, so the resupply could not be attempted until this was vacated; the exchange of Soyuz ferries had achieved this. In effect, therefore, Soyuz 27 had been an extra flight inserted into the schedule simply to restore the docking sequence.

Alexei Leonov later emphasised that Dzhanibekov and Makarov had performed an enormous amount of work. Valeri Kubasov added that the ability to send visitors to an occupied station not only meant that a scientist could be sent up to perform a specific experiment, but also that an engineer could be sent up to install new (or to repair faulty) apparatus, in parallel with the ongoing mission.

Romanenko and Grechko took the next day off, even though it was not a scheduled rest day, to recover from the intensive joint programme. To help them to relax, Kaliningrad relayed radio coverage of a Soviet–Canadian ice hockey match. They spent the next two days imaging vast tracts of the Central Asian Republics with the MKF-6M camera in a search for mineral deposits. Meanwhile, at the cosmodrome, the first supply ferry had been set up on the pad.

RESUPPLY

Launched on 20 January, the new vehicle was named Progress. As expected, it was a modified Soyuz. Since it was not required to return to Earth, the descent module with its heavy heat shield and landing system had been replaced by a cylindrical pressurised compartment containing tanks for 'wet' cargo and the navigation and control packages that were normally carried within the descent module. The 'dry' cargo was stowed in containers in the orbital module. This new configuration was about half a metre longer than the Soyuz ferry. Unlike the crew transfer ferry, it used a unified propellant system based on the ODU introduced by Salyut 6. The only difference to the carrier rocket was the omission of the escape tower; in an accident the spacecraft was expendable. Of its 7,020-kg total mass, fully 2,300 kg was cargo (1,000 kg was propellant for the station; the 1,300 kg of 'dry' cargo included compressed air, food, water and other consumables for the crew). Progress did not have solar panels but, because it did not need to support a crew, it could sustain up to eight days of independent operations. It pursued a two-day rendezvous which resulted in a much slower closing rate, and simplified the task of the automatic approach system. This procedure had been tested by Soyuz 20 on its approach to Salyut 4 three years earlier.

In order to enable the flight controllers to supervise the automated docking, and, if necessary, take control, there was a television camera on the front of the vehicle which presented a view similar to that through the descent module's Vzor periscope. It carried a spotlight to assist in docking in darkness.

It had originally been intended that the cosmonauts would retreat to their ferry, just in case the impact of the automatic cargo craft damaged the station, but after the success of Soyuz 27's docking it had been decided that they could remain. One control room at Kalin-ingrad monitored Salyut 6, and another the manoeuvring spacecraft. The cosmonauts manned the station's control panel and prepared to manoeuvre the station to compensate for deviations in the ferry's approach but, to general satisfaction, the newcomer made a perfect approach and docked on its first attempt. Although Romanenko and Grechko had been scheduled to take the rest of the day off, they were granted permission to open the hatch to retrieve parcels and letters.

Alexei Yeliseyev, former cosmonaut, now flight director, noted that although there were still "many problems" to be solved in operating an orbital station, this docking had proven that they had finally overcome the restrictions imposed by the limited amount of apparatus and consumables that could be installed in a station prior to launch. This was therefore "a really big step forward" in the exploration of space.

The 2.2-metre spheroidal compartment of the ferry's orbital module, which had a vol-ume of just 6 m^3, was criss-crossed by a metal framework in which large containers had been fastened. These could be released simply by turning the locking bolts through half a turn. Each contained a variety of individual items. Bulky items of apparatus were bolted directly onto the framework, and the compartment was so full that the hatch could barely be swung back. An inventory and a locator had been stowed near the hatch.

The cosmonauts squeezed in and closed the hatch behind them while they worked, so that loose metal, wood and rubber chips from the containers could not drift into the station. A small vacuum cleaner was used to suck up as much of the debris as possible prior to reopening the hatch. They wore goggles to protect their eyes, and face masks to prevent inhaling fragments. With two men working so hard in such a small volume, the cargo compartment soon became very humid and stuffy.

The inventory suggested the order for unloading. After transfer to the station, every item was either stored in a locker or set up on the appropriate experiment site. The main experiment on this first ferry was the Splav-1 furnace. A regular item on the manifest was to be replacement components, most notably 1-metre long cylindrical canisters of potas-sium superoxide for air regulation. Spare parts (in this case an orientation sensor) were included as required. Much of the food was fresh, and specialities such as apples, onions, caviar and garlic were much appreciated. Water was supplied in small spherical flasks,

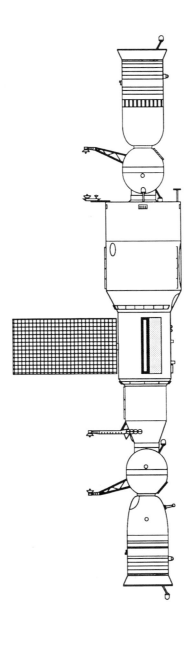

Salyut 6 with Soyuz 27 (front) and Progress 1 (rear).

metres

each of which contained 5 kg of water, but it was soon decided that this was not very satisfactory.

A set of pipes had been installed to feed propellant through the orbital module to the docking collar. Earlier stations had not been capable of replenishment, so this transfer system had never been tested. The first step was to verify the integrity of the pipes by pressurising them with nitrogen, a process which involved proving the hermetic seal of each manifold and valve in the path. This could be controlled either by Kaliningrad or by the crew, but on this occasion, to certify the system, the cosmonauts supervised the task.

The next step of the process was to reduce the pressure in the station's propellant storage system. Propellant was fed to the engines by metallic bladders within the tanks, driven by nitrogen gas at 20 atmospheres, but unless this pressure was reduced, the pump in the ferry would be unable to force-feed propellant into the tanks. A 1 kW compressor had been incorporated into the station to drive the nitrogen back into its bottle, to relax the bladders in the tanks so that more propellant could be pumped in. Unfortunately the nitrogen was stored at 220 atmospheres, so this little pump was actually a heavy power drain. This priming operation had to be timed so as not to place too great a load on the power supply, which meant that it could only be done while the station was in sunlight, and was usually staggered over several days. Although this was a slow process, it was simple and reliable, and the use of the bladder guaranteed absence of gas bubbles in the propellant supply (any gas in the flow would disturb combustion, and make the engine run rough). Once the pressure in the tanks was down to 3 atmospheres, the pump in the ferry, which was pressurised by a bladder at just 8 atmospheres, fed propellant into the station. After the oxidiser had been transferred the pipes had to be vented to clean them, and then reverified, before the process could be repeated for the fuel. If the station had used a lot of propellant, it might be necessary to top up each of its two oxidiser and fuel tanks. Clearly, therefore, given these constraints, it might easily take a week to pump aboard the tonne or so of propellant the ferry carried. Once the process was finished, a blast of high-pressure nitrogen was sent through the transfer pipes into the ferry's tank to clean the pipe, to ensure that when the ferry undocked it would not vent corrosives onto the rear of the station.

Even after it had been completely unloaded, the ferry was retained to serve as a 'tug' which would periodically boost the station's orbit, thereby further reducing propellant consumption. Before such manoeuvres, all loose apparatus had to be strapped down to prevent it crashing about. Before the ferry was jettisoned, dirty clothing, empty water flasks, food containers, clogged air filters and saturated carbon dioxide scrubbers were stowed in the racks in its orbital module. (Care had to be taken not to offset the vehicle's centre of mass, or its attitude control system would not be able to orientate itself accurately once it resumed independent flight). It undocked on 7 February, dropped some 14 km behind the station, and then activated its back-up rendezvous system and began to make a second approach; as soon as it was evident that this was working, it was once again commanded to withdraw. It was deorbited over the Pacific Ocean the following day and was destroyed in the upper atmosphere.

By early February the cosmonauts admitted that they were suffering from acute home sickness, and were looking forward to completing their mission. On 10 February they celebrated the 150th anniversary of the birth of Jules Verne by pointing out during their daily television broadcast that whereas one of the famous writer's characters had circled the world in 80 days, they completed an orbit in almost as many minutes. They also observed that they had circled the Earth more than 1,000 times. The next day they broke the 63-day endurance record set in 1975 by Klimuk and Sevastyanov. Dr Anatoli Yegorov, the head of the medical team at the flight control centre, said that their latest medical check-up had confirmed that both men had reached "a stable adaptation" to weightlessness. They received a radio message from Thor Heyerdahl, who was sailing across the Indian Ocean in *Tigris*, a reed boat that demonstrated the technological gulf between past and present modes of transport. After chatting with the previous holders of the record, Romanenko and Grechko resumed MKF-6M photography of Siberia and the Central Asian Republics. Amongst other things, the imagery helped in assessing the extent to which silkworm had spread through the vast Siberian forests, killing the trees. Diseased trees could be spotted more easily from space using multispectral techniques than they could be by visual examination on the ground. This information enabled the pesticides to be delivered far more effectively than would otherwise have been possible. In contrast to the Earth observations performed by the crews of earlier stations, 90 per cent of the pictures taken by Romanenko and Grechko were specifically to satisfy the requests of Earth-bound specialists.

In mid-February the Splav furnace was installed into one of the scientific airlocks. This apparatus took the form of a disk attached to a thin cylindrical stem, and was put in the airlock so that it could radiate excess heat to vacuum. It operated at 1,100°C, and its electronic control system could maintain this temperature to within 5°C. It incorporated a set of molybdenum reflectors to focus the heat onto a shaft containing three ampoules. The fused ampoules were automatically transferred to a 'cooling chamber' with a linear thermal gradient which ensured optimum conditions for the formation of a monocrystal. When the sample had cooled to 650°C, it was transferred into a third chamber which was maintained at this temperature to sustain the three-dimensional crystallisation process over several days, with the crystal growth being captured by time-lapse filming. During such experiments, the Kaskad attitude control system would be switched off so that its manoeuvres would not affect the experiment. Even so, the cosmonauts reported that they could see tiny imperfections in the crystal due to the vibrations caused by their moving about within the station. One experiment which the cosmonauts welcomed was Svezhest. Although this made the station slightly more comfortable by ionising the air, they also liked it because it released a refreshing scent of pine. Even though they were growing tired, their productivity was up about 10 per cent above what the flight plan required. A similar trend had been observed in the case of Salyut 4, when Klimuk and Sevastyanov had developed a 'second wind' towards the end of their flight.

A CZECHOSLOVAKIAN VISITOR

Despite the frustration of Soyuz 25, the ferry exchange and the replenishment flight had demonstrated that the technology to sustain an orbital station for an extended period existed. Although the automated approach and docking system made it possible to send up an unmanned replacement ferry with extra supplies, and then have the expired ferry return to Earth with the results of experiments, it had been decided to have a pilot fly up the new vehicle and take away the old one, and to offer the second couch on each flight to a guest nominated by the individual member states of Intercosmos, the international space research organisation run by the Soviet Union with fellow socialist states. With Romanenko's and Grechko's record-breaking flight going so well, it had been decided to fly the first of these Intercosmos visits just before the residents were due to pack up and return to Earth. On 2 March, therefore, Soyuz 28 took off with Alexei Gubarev and Vladimir Remek. To mark the fact that a Czechoslovak was to be the first person to fly in space who was neither a Soviet nor an American citizen, the event was watched by Jaroslav Kozesnik, president of the Czechoslovak Academy of Sciences, and by Boris Petrov, chairman of the Intercosmos organisation. Georgi Beregovoi, now heading the Gagarin Cosmonaut Training Centre, said that researchers from Poland and the German Democratic Republic would fly to Salyut 6 later in the year, and that candidates from Bulgaria, Hungary, Cuba, Mongolia and Romania were about to start training for later flights. The space station programme had clearly taken a giant step forward during the time that the Almaz platform had been in orbit; the difference in capability between Salyut 4 and Salyut 6 was marked.

The docking of Soyuz 28 at the rear port was a welcome reunion for Grechko and Gubarev, who had spent a month together on Salyut 4. The very next day, the residents exceeded the 84-day endurance record for a single mission that had been established by the final Skylab crew, so they now held the principal record. The flight controllers read them various congratulatory messages, including one from the former record holders.

The visitors performed a standard set of cardiovascular tests using the Polynom-2 multifunction apparatus to add another chapter to the file on the process of adaptation to weightlessness. Comparing data from different crews had shown that if the cosmonauts spent time in the Tchibis lower-body negative-pressure suit it inhibited blood pooling in

Vladimir Remek and Alexei Gubarev visit the Salyut 6 residents, Georgi Grechko and Yuri Romanenko.

the head, and speeded adaptation to weightlessness. They also performed Opros, which involved filling out a comprehensive questionnaire concerning their eating and sleeping habits, level of physical fitness and posture, their sense of smell, vision and hearing and other factors which would enable the psychologists to correlate their physical and mental health. The visitors filled this out every day, and the residents did it once each week.

Although in general the visiting crews would not be permitted to interfere with the resident's ongoing programme, Romanenko and Grechko had already completed the majority of their own work, so they helped their guests with some of their experiments. Remek used the Czech-built Oxymeter to measure the concentration of oxygen in body tissue in weightlessness. One materials-processing experiment in the Splav produced extremely pure semiconductor crystal. Another made glass-melt far more homogeneous than anything which could be produced on Earth but, although it yielded a large crystal, there were physical deformities (such as a helical surface) caused by microaccelerations due to ongoing operations on the station. One advantage of having so many visits was that once film which had been returned to Earth by the first visitors had been developed, prints could be sent up with the next visitors to indicate sites of special interest for further study by the residents before they themselves came home.

Gubarek and Remek returned to Earth on 10 March. They did not exchange ferries because the main mission was drawing to a close; if Soyuz 27 had not been added to the schedule, this international visit would probably have taken place earlier, and have swapped ferries. Alexei Yeliseyev said that the object of this first Intercosmos visit was to evaluate the potential for an international crew. Vladimir Shatalov noted that the two crews had worked well together. The operational constraint that the visiting mission fit within a single recovery window required that such visits be brief, which meant that the visitor could not be assigned complex tasks. Nevertheless, the public relations value of having an operational facility accommodating a succession of international researchers was significant, especially at a time when the great rival, America, was unable to mount any sort of space mission.

END OF A MARATHON

By mid-March, although the Earth-resources imagery work continued, Romanenko and Grechko started to deactivate Salyut 6's systems, to return it to its automated flight regime. They loaded their results into Soyuz 27's descent module, powered it up, and tested its engine. They returned to Earth on 16 March. Following a medical examination at the recovery site, the doctors announced that, considering what they had done, both men were in excellent health. A few days later, Dr Anatoli Yegorov said that the flight had demonstrated that cosmonauts would be able to "live and work in space for a year or longer". Dr Oleg Gazenko congratulated them on having completed all their tasks "with unflagging interest and even with inspiration", right through to the last day of the flight. They had extended the endurance record to 96 days, and Grechko's accumulated total had now risen to 125 days. He reported readaptation easier this time than he had following his first flight. During their first few days back on Earth the cosmonauts had difficulty coordinating their actions, such as when picking up a cup of tea. Dr Robert Dyakonov, their personal physician, said that this was because they were both "still up there in space, not only physically but also mentally", so when they awoke they tried to swim out of their beds. They had become so accustomed to the lack of gravity that most actions involved deliberate effort; even turning a radio dial required concentration. To inhibit blood draining from the torso (the opposite to what had happened when they had entered space), they used positive-pressure leggings for the first few days. This warded off orthostatic intolerance and enabled the heart to gradually regain its normal capacity.

Boris Petrov announced that no further missions would be sent up to Salyut 6 until the results of this latest flight had been thoroughly assimilated. So far, the new station had been an outstanding success. From this point on though, in an engineering sense, it was *terra incognita*.

A SECOND LONG-DURATION MISSION

In early June 1978, after three months virtually powered down, Salyut 6 reactivated its environmental systems and recircularised its orbit at about 350 km altitude to prepare for its next crew.

Soyuz 29 was launched on 15 June. It was commanded by Vladimir Kovalyonok, who had suffered the frustration of Soyuz 25's failure to link up. His flight engineer, Alexander Ivanchenkov, had been dropped from Soyuz 26 to make way for Grechko, and this was his first flight. They had been thrown together by fate to be Salyut 6's second resident crew. This time, Kovalyonok had the satisfaction of hearing the latches engage with the station's front port.

Kovalyonok and Ivanchenkov wore Penguin suits from the start, so as to exercise their muscles during the initial phase of adaptation to weightlessness, and Dr Dyakonov observed that they adapted much faster than their predecessors. Unlike Romanenko and Grechko, who had opted to miss sleep in order to get ahead of schedule in activating the station, and suffered as a result, the new crew had been ordered not to overwork during the initial phase of adaptation to weightlessness. Their research programme was a continuation of that before: specifically Earth-resources surveys, studies of atmospheric phenomena, astrophysical observations and materials-processing experiments. As part of their training they had flown at high altitude in an aircraft equipped with an MKF-6 camera, and specialists had pointed out the types of feature they were to seek when they made similar observations from orbit.

At the suggestion of Romanenko and Grechko it had been decided to alter the work schedule. Whereas their predecessors had evaluated a fixed schedule that matched their natural sleep patterns but fragmented their day, Kovalyonok and Ivanchenkov were to adopt a more conventional '9–5' working day and switch to a 5-day week. They could do whatever they wished in their free time. This crew-orientated approach contrasted with the way that the Skylab programme had evolved, where a study of how long it had taken the first crew to perform each task had enabled mission planners to produce a schedule which assigned a coordinated series of tasks to each astronaut for virtually each minute of each day of the follow-on missions. This mission-orientated approach had prompted the third crew to rebel in order to assert a degree of control over their schedule, to provide themselves with time off (time off was time wasted to the mission planners). However, on

Romanenko and Grechko's flight, which had broken the Skylab record, there had been no such problem because the flight controllers had left the cosmonauts alone. It was hoped that the second expedition would be able to further extend the record, and to ensure that the flight controllers understood the crew's mood, Romanenko was put in charge of the ground support team. In fact, cosmonauts were much more involved in mission operations than were astronauts. Many of the new batch of cosmonauts were engineers in the design group that had developed the Salyut, and veteran cosmonauts served as flight directors. The majority of NASA's astronauts were pilots first and foremost; only a few were scientists, and they were employed in the control centre only as 'capcom', whose sole role was to ensure that communication with a flight crew used familiar vernacular. There was less demarcation at Kaliningrad.

Once the station had been fully reactivated, the cosmonauts carried out maintenance. This included replacing the pressure indicator in the airlock that had caused Romanenko and Grechko such consternation after their historic spacewalk, and replacing a defective ventilator in the Splav furnace. They then orientated the station to exploit gravity-gradient stability, turned off the Kaskad attitude control system and began an extensive programme of materials-processing experiments in the repaired furnace. The station turned out to be remarkably stable in this position; its dorsal solar panel acted like a rudder, and the drag of the rarefied upper atmosphere eventually locked it in trail, so that the station travelled belly forward.

During the latter half of the month, the plane of the station's orbit was perpendicular to the Sun, so it flew in continuous sunlight. This condition was due to a combination of the station's orbit precessing around the Earth, the Earth's movement around the Sun, the offset of the Earth's axis to the plane of its orbit, and the inclination of the station's orbit to the equator. For a week or so, at six-monthly intervals when the angles reinforce, a spacecraft in a highly inclined orbit spends most of its time in sunlight. Although this maximised power generation, it put greater demand on the thermal regulation system, but it also offered a valuable opportunity for making prolonged atmospheric observations and for studying the surface at very low angles of illumination to emphasise terrain relief. In fact, the cosmonauts reported that under perfect conditions they could see undulations in the floor of shallow seas, which they were able to correlate with surface currents. No sooner had the residents settled down than their first Intercosmos visitors were making preparations to join them.

A POLISH VISITOR

Soyuz 30 was launched on 27 June. Pyotr Klimuk was accompanied by Miroslaw Hermaszewski, a Polish cosmonaut researcher. The next day, the new ferry docked at the rear port.

Most of Hermaszewski's luggage was Polish-built biomedical apparatus to perform a range of studies into space ergonomics and psychology. Smak studied changes in the sense of taste in space (some favourite foods had turned out to be unpleasant in space, whilst others had been found to be far tastier). It measured the electrical stimulation of the taste buds, so that the subjective sense of taste could be correlated with specific data. Hermaszewski had also been assigned several materials-processing experiments, one of which (Sirena), for scientists at the Warsaw Institute of Physics, was to make a monocrystal of cadmium–mercury–telluride (CMT) semiconductor, a material extremely sensitive to infrared radiation.

A standard assignment on Intercosmos visits was for a guest to take MKF-6M pictures of his own country for Earth-resources studies. On this occasion, however, a solid bank of cloud intervened, but the weather over the Ukraine and Kazakhstan was ideal, so these were imaged instead. There was a basic conflict between materials processing and Earth-resources; the furnace benefitted if the station was gravity-gradient stabilised, whilst the MKF-6M required the porthole in the floor through which it viewed to be aimed at the Earth. Dynamic scheduling to exploit weather windows had to acknowledge the fact that materials processing could not be interrupted without the risk of spoiling the crystal. Consequently, once a materials experiment requiring several days had begun, this more or less ruled out opportunistic imaging. Weather forecasting was therefore an important element of planning work aboard the station, although, of course, a specific picture could be taken only when the ground track and lighting conditions were both appropriate. Neither was it simply a matter of snapping pictures with a hand-held camera, as the MKF-6M had to be primed and the station had to be orientated so that it aimed the camera directly at the ground at the very moment that the picture sequence had to be taken. It could easily take several days before all these factors came together to allow a given image sequence to be taken, and it was a lucky visitor who managed to complete his programme. In gravity-gradient mode, however, all the portholes in the forward transfer compartment faced the horizon, providing an astounding panoramic vista for observations involving hand-held instrumentation. Persistent poor weather over a guest's home country was not a disaster, because the residents could usually fill in gaps in required coverage later.

On 5 July Klimuk and Hermaszewski undocked in their own ferry and returned to Earth. Hermaszewski's haul included two monocrystals of CMT, one with a mass of almost 50 grammes.

A NEW FURNACE

Progress 2 was launched on 7 July, and on 9 July docked at the newly vacated rear port. It had been loaded with only 600 kg of propellant in order to carry more dry cargo. In addition to the usual replacement components for the environmental systems, and sufficient consumables to last the two cosmonauts 50 days, it delivered several new experiments, including a new furnace (Kristall) and electrocardio monitors that could be worn continuously and transmit data over the telemetry link.

While the cosmonauts were busy with the unloading operation, the flight controllers remotely supervised the preliminary procedures for replenishing the station's propellant tanks. Although the first crew had controlled the fluid transfer, in this case, and in the future, this was to be done by Kaliningrad.

Unlike Splav, which had to be installed in a scientific airlock to radiate excess heat into vacuum, the Kristall furnace could be mounted inside the station. It was installed in the rear transfer tunnel. Splav maintained a temperature gradient during crystallisation, but Kristall employed uniform heating, and exposed its sample to a steady-state thermal zone at a temperature in the range 400–1,200°C. It could be used in four different ways: one produced monocrystals directly from the gaseous phase by sublimation, and evaporated the sample and then transported the gas to the cooling zone, where it settled on and enhanced a seed; a second method made homogeneous multi-layered films using chemical gas-transportation; a third made a monocrystal by using a 'moving-solvent' to make a high-temperature solution, and in the fourth method the temperature was slowly reduced in the crystallisation zone – one side of the sample was cooled while the other was kept hot, forming a seed on the cold side which, as it was slowly stretched across the zone, produced incremental crystallisation. Kristall controlled its temperature more accurately than did Splav, so its results were expected to be more homogeneous. During its first test, a monocrystal of gallium arsenide – a semiconductor used in the construction of highly efficient solar collectors – was grown by the high-temperature solution technique.

SPACEWALK: SAMPLE CASSETTES

Although Grechko had spacewalked, he had merely removed the front drogue to inspect the docking assembly. Kovalyonok and Ivanchenkov had been assigned a more demanding task. On 29 July, Ivanchenkov used a colour television camera to show his commander using special tools to dismantle experimental packages mounted outside the station. In addition to the Medusa biopolymer cassette which Grechko had deployed, he retrieved a passive micrometeoroid detector and cassettes that had been mounted before launch (these contained materials such as titanium, steel, glass, ceramic, paint, rubber and sealant compounds used in spacecraft construction). This done, he affixed several new cassettes and a radiation detector.

When the station passed into the Earth's shadow, Kovalyonok used lamps on either side of his helmet to continue working. His final act was to evaluate the effectiveness of a set of handholds positioned on the outside of the station to help a spacewalker move about. To their surprise they completed all their assigned tasks in less time than training in the hydrotank had suggested would be required. Although the flight director politely suggested that they move back into the station, because there was no more work to do, Kovalyonok retorted that he was in no hurry, as he wanted to admire the view because he had been cooped up inside for almost six weeks (the view through his curved visor was incomparably better than that through a flat porthole). He watched a meteor trail in the atmosphere, far below.

Konstantin Feoktistov later reported that the 6 cm^2 micrometeoroid detector that the cosmonauts had retrieved had 200 tiny craters in it – more than predicted. Later analysis revealed that many were the results of impacts with chips of paint and particles of aluminium flakes from solid fuel rocket motors. The Medusa biopolymer experiment revealed that exposure to sunlight had caused components of nucleic acids to develop into substances similar to nucleotides; this was a step in the process of creating nucleic acids, which were vital for the development of organic life.

Progress 2 departed on 2 August. Once it had withdrawn, Kaliningrad methodically tested its individual onboard systems during a period of autonomous flight, and it was then deorbited. Progress 3 took its place on 10 August. Since it followed so soon after its predecessor, no propellant had been loaded into its storage tanks, so its orbital module was packed to capacity. Viktor Blagov, deputy flight director, reported that whereas the first two resupply craft had served to replenish the station's consumables, in this case it was to build up a stockpile; it delivered 280 kg of food, 200 kg of water and 450 kg of compressed air. Ivanchenkov had asked Grechko to send a guitar, and this had been done. Fur boots had also been sent up, because they had complained that their feet were cold (a consequence of the reduced blood circulation in the lower body). While the cosmonauts unloaded the cargo, the station was left in gravity-gradient mode so that the furnaces could process the newly delivered samples. After being unloaded in record time, Progress 3 was discarded on 21 August.

AN EAST GERMAN VISITOR

A week later, Soyuz 31 docked at the busy rear port. It delivered Valeri Bykovsky and Sigmund Jahn, the first cosmonaut from the German Democratic Republic. The gift of fresh onions, milk, lemons, apples, honey, pork, peppers and gingerbread, ensured a warm welcome for the newcomers.

The visiting crew had a heavy biomedical, biotechnology and materials-processing programme. One biological experiment was designed to reveal how the space environment affected the operation of basic bioprocesses in cells in a tissue culture. Another, by the Institute of Aviation Medicine, studied how microorganisms could be 'sewn' together with organic polymers to determine whether different geometries could be formed in microgravity (on Earth, floccule structures formed) with a view to opening up new possibilities for the preparation of medicines. Of two experiments with the Splav furnace, one boiled beryllium–thorium to produce homogeneous glass of higher quality than was possible under the influence of gravity, and the other used a special quartz matrix to produce a monocrystal of a bismuth–antimony semiconductor. Experiments in the Kristall furnace grew a lead–telluride crystal, using the sublimation technique for the first time. It was also used to create a crystal of bismuth–antimony by stretching a seed; it was processed with the material sandwiched between two plates, within an ampoule, and produced a tree-structured crystal which was about five times larger than that which could be produced on Earth (this was done so that the result could be compared with that produced in the Splav). This kind of research was easily accommodated by the limited facilities aboard the station, but, as yet, there was no capacity for visitors to devise wholly novel experiments and send up heavy apparatus to undertake an advanced programme. However, Progress 3 had delivered a pair of modified cameras for an experiment to test different types of film for external and internal photography, and Jahn methodically worked through the Reporter test sequence. Another aspect of the Intercosmos programme began to develop when Jahn reran Hermaszewski's Smak experiment. Although each visitor could introduce new biomedical tests, existing ones would often be repeated to improve the database on their results.

Since he was to take away the old ferry, Bykovsky transferred the couch liners and pressure suits from Soyuz 31 to Soyuz 29, then stowed 100 accumulated samples from the furnaces and a cassette of film from the MKF-6M. A deceptively trivial test was to take a sample of the station's air. Even though the environmental system purified the air, there was concern that, after such a long period of inhabitation, toxic compounds might be building up. If this proved to be the case, then sufficient compressed air would have to be delivered by the next cargo ferry to purge the cabin (possibly using the procedure tested by Salyut 5). In the event, this test revealed no chemical, biological or bacterial pollutants.

SWAPPING ENDS

On 3 September Soyuz 29 returned to Earth, leaving Soyuz 31 at the rear port. The next cargo ferry would not be able to dock until this had been moved. A ferry had never been transferred from one port to the other before, and on 6 September, to prepare for this, Kovalyonok and Ivanchenkov verified that Soyuz 31 was fully functional. The next day they powered down most of the station's systems, just in case they were unable to redock and were forced to return to Earth; then Soyuz 31 undocked and pulled straight back 150 metres. As the two vehicles flew in formation, the station was commanded to pitch over through 180 degrees, to offer its front port to the ferry, so that it could move back in and redock. This simple rotation expended far less propellant than it would have taken to fly the ferry around it and line up for an approach from the opposite side, and placing the onus on the ferry would have consumed a significant fraction of its limited reserve of propell-ant. Kaliningrad wanted to monitor the operation, so it had to be done within a single 35-minute communications session running from the South Atlantic, north-east across Africa, then over the Soviet Union. The flight controllers observed through the television camera in the ferry's docking assembly, and about 30 minutes after undocking, Soyuz 31 redocked at the front port without incident, marking a significant milestone in the pro-gramme. Although this transfer could have been undertaken at any time, it had been de-cided that it would be done while the recovery window remained open. Transferring a new ferry to the front port within days of the old one leaving was to become standard practice.

Swapping ends was a significant operational overhead. The possibility that a crew might not be able to reboard the station meant that it had to be returned to its automated flight regime, and it took several days to put the station into hibernation and fully restart it. During this time, no power-hungry apparatus could be left running, so furnace work had to be carefully scheduled; with smelts often running as long as a week, an immi-nent ferry transfer clearly might affect operations over a ten-day period. All of this was necessary because the plumbing to replenish the station's fluids was available only at the rear port.

RECORD BREAKERS

On 20 September, Kovalyonok and Ivanchenkov exceeded the 96-day record set by their predecessors. Professor Nikolai Gurovsky, head of the Ministry of Health's space medicine department, announced he was fully satisfied with the crew's state of health, Dr Robert Dyakonov said that they were sustaining a high rate of work, and Professor Oleg Gazenko, a member of the Kaliningrad medical team, announced that they were to be permitted to complete 120 days, then return as soon as convenient after that time.

It had been observed that their choice of activity selected for time off had changed. Initially they had opted to perform extra medical tests, but they then turned to making visual observations, and later began to work with the BST-1M telescope. The purpose of the experiment with a schedule involving a 5-day week, with weekends off, had been to determine whether they were able to use the free time to recover both their physical and mental state after a full week of intensive work. Unfortunately, their circumstances were not particularly conducive to the pursuance of hobbies, and it was difficult for the cosmonauts to disengage from their daily activities and do something completely different. Until a large space station was constructed with a large staff, there was little hope of realistically mimicking free time.

FINAL RESUPPLY

Progress 4 set off on 4 October, the 21st anniversary of the launch of Sputnik 1. It docked two days later and immediately raised the station's orbit to 370 km. It delivered a full load of propellants and consumables, spare parts, samples for the furnaces, film, new clothes, suits, mittens to keep warm, a new tape player and a selection of cassettes to relieve the monotony of the physical exercise programme, and, on specific request, a set of screens that could be erected to create two tiny 'rooms' for improved privacy. When Ivanchenkov opened the box of chocolates sent by his wife, the sweets sprayed out and drifted throughout the station, and rediscovering them in the days to come became quite a treat.

Unloading Progress 4 was to be the cosmonauts' last major task, as they were to return to Earth at the end of the month. After replenishing the station's propellants, Progress 4 departed on 24 October. The heavy work of shifting cargo had helped to develop their stamina. One of the items delivered was eleutheroccus, a drug that acts as a tonic and stimulates the body to work harder, and they began to take it daily to help prepare to return to gravity. For most of the final month, they used the Tchibis suit twice a week to help increase the capacity of their cardiovascular systems, and in the final week they used it every two days. Throughout, they had worn their Penguin load-suits to keep their muscles in trim, and had followed their exercise schedule strictly, even occasionally resisting the temptation to postpone sessions in order to finish some other task. With their flight drawing to an end, they increased the daily regime to three hours. Furthermore, to help their bodies restore the fluids lost by the adaptation to space, they started to drink large quantities of salinated water. It was beginning to look as if there was no limit to the time the human organism could spend in weightlessness, so long as proper precautions were taken for the shock of returning to the Earth's gravity.

At the end of October, Kovalyonok and Ivanchenkov began the lengthy process of accounting for and transferring the results of their researches to Soyuz 31. Although the visiting crews had taken some of it away, they had accumulated a great deal of material. As flights became longer, and crews became more productive, simply returning results would be a major issue. One solution would be to send up an automated Soyuz, so that the resident crew could load it and dispatch it (an unoccupied ferry could return several hundred kilograms of cargo to Earth), but, strangely, this was rarely done.

It had been a highly productive mission. Some 18,000 photographs had been taken with the MKF-6M camera. It would have taken two years of surveying from aircraft, or a lifetime of fieldwork, to survey the same terrain conventionally. Continuing previous programmes, pictures had been taken of specific targets in the Crimea, the Caucasus, the southern Urals, the Caspian Basin, Kazakhstan, the Central Asian Republics, Siberia, the Soviet Far East, the Ukraine and Belorussia. Amongst their specific results were the discovery of an underground water reserve on a peninsula on the eastern shore of the Caspian (in an area that surface surveys had pronounced to be utterly devoid of water), a comprehensive glacial and snow-cover survey of the Pamirs, a survey of pasture lands in Turkmenia, an assessment of irrigation in Uzbekistan and a geobotanical map of the Lake Balkash area. In addition, the Salsky test site had been photographed regularly to continue to refine the ability of the multispectral camera to distinguish between a wide variety of different cereal, vegetable and grass under different conditions. Hundreds of drawings of atmospheric and surface phenomena had been sketched. The BST-1M had been used to make infrared observations of the Earth's atmosphere.

Kovalyonok and Ivanchenkov had tried to grow higher plants in Phyton, a cultivator of nutrient medium with a lamp to simulate sunlight. It was set up so that the root grew 'down' in the dark under a black sheet, while the shoots grew 'up' towards the lamp. The goal was to make a plant flower and yield a seed, to test the complete growth cycle from seed, to mature plant and back to seed, in weightlessness. This supplemented the results of experiments to investigate the development of individual cells. Arabidopsis (wallcress, a common weed) had been selected because it had a very short life cycle. It grew rapidly, but then died before it could flower. Onion quickly grew seedstalks (this was surprising because on Earth onion does not try to seed so soon) but it did not seed. A hardy strain of wheat characterised by a short stalk and a very heavy yield of grain was tried. Although on Earth its development cycle could be observed over a period of a few weeks, it grew much slower than expected. (When it was returned to Earth, it had only just reached the stage of producing grain). The pleasure of witnessing a plant complete its growth cycle in space would therefore fall to a future crew. Apart from these plant experiments, therefore, the mission had been a great success.

Soyuz 31 returned to Earth on 2 November 1978. The touchdown and recovery were carried live by domestic television. Kovalyonok and Ivanchenkov scrambled out of the capsule unaided, despite having set a new record of 140 days in space. Intoxicated by the brisk wind over the sundrenched steppe, Kovalyonok knelt to pick up a lump of dirt to celebrate his return. The doctors were delighted to find that they were in better health than earlier crews who had made extended flights. Significantly, because they had been in space for longer than 120 days (the life cycle of the erythrocytes which carry oxygen to body tissue) their blood stream contained only red cells produced in space, and analysis revealed that although these were rather smaller than normal, they functioned properly. On their second day back they were able to go for a walk in the park around the Cosmonaut Hotel at the cosmodrome, and after 10 days they had recovered sufficiently to play tennis. It was now evident that the readaptation process had been successfully decoupled from the time spent weightless, suggesting that a flight of a year or more would be possible. This would not be possible on Salyut 6, however, because its expected life of 24 months was already more than half over. The original station plan had called for only one more expedition, to extend the endurance record to six months. A problem had arisen, however, and the station's future was in doubt. A few days later, *Tass* pointed out that Salyut 6 had been occupied for 60 per cent of its time, and had therefore successfully completed its primary mission.

WRITE-OFF?

In mid-November, Konstantin Feoktistov said that in the coming months Salyut 6 would be thoroughly studied "to determine its possible further use". What he did not reveal was that while Progress 4 had been replenishing propellant, a problem had developed in one of the tanks in the unpressurised compartment at the rear of the station. An intensive effort was underway to understand the failure, and to devise a remedy, and telemetry analysis suggested that the pressurised bladder within the tank had developed a leak. Bubbles of nitrogen gas mixed in with the propellant would make the engine misfire, so the engineers had recommended not using it until the fault could be fixed. This was despite the ODU having two *independent* sets of tanks. If the engine was fired, there was a risk of the highly-corrosive UDMH seeping through the pressurisation system. If the damaged tank could be drained and isolated, the second set of tanks would be able to be used safely. However, the repair operation could not be started until a crew could be put aboard, and until an automated ferry could be docked to drain the propellant from the faulty tank. An operation such as this would not have been attempted on an earlier station; it would simply have been written off. The decision to rescue Salyut 6 therefore indicated how far the programme had come in such a short period.

Regardless of the leak in the propellant system, the station was regularly reported to be operating normally. It had not escaped the notice of Western observers that it had not manoeuvred since Progress 4 had undocked; its orbit had dropped to 307 × 330 km, well below the 350-km circular orbit that offered a launch window every two days. It was expected that the next crew would not be sent up until the station adjusted its orbit. The absence of manoeuvres was interpreted to mean that Salyut 6 was low on propellant, so it was supposed that a tanker would be sent up first, to raise its orbit.

In mid-January 1979, Vladimir Shatalov said that specialists were still considering the possibility of sending another crew to the station. A month later the fleet of tracking and communications ships set off for their relay points, and this suggested that a launch was imminent.

REPAIR CREW

Soyuz 32 finally set off with Vladimir Lyakhov and Valeri Ryumin on 25 February, by which time Salyut 6's orbit had decayed to 296 × 309 km. The ferry docked at the front port the next day, and upon opening the hatch the newcomers were greeted by a faint scent of pine, the residue of the Svezhest experiment. Before the ferry was powered down, its engine was used to raise the orbit to 308 × 338 km. This was possible only because the ferry had not expended as much fuel making its rendezvous as allowed for in the flight plan; the reserve had been put to good use. Raising the orbit further would not be possible until a tanker arrived, and *its* engine used.

Ryumin was particularly qualified for the task of trying to isolate the propellant tank, as he had been a member of the team that had designed the station, although his only experience in space was the frustration of Soyuz 25's failure to dock. For Lyakhov, it was a first flight.

Even though Salyut 6 had been in orbit for 18 months, its systems had proven to be extremely reliable, but many pieces of apparatus were approaching the end of their recommended service life, and these would have be replaced before the station would be able to support another long-duration mission. A variety of spare parts and specialised tools had been ferried up by Soyuz 32, but most of whatever would be required would have to be shipped up on the next cargo ferry, which could not be done until the cosmonauts submitted a list of their requirements. Accordingly, once they had reactivated the station, Lyakhov and Ryumin inspected every piece of apparatus to look for potential failures. They then replaced filters in the environmental unit, ventilators in the thermal regulation system, elements of the veloergometer and the treadmill, assorted cables and all the lights, and installed a sextant (for use by later crews, in determining their position relative to the stars rather than the Earth). The ability to repair faulty apparatus, to replace expired items and to replenish consumables demonstrated the maturity of the programme. The survey confirmed that (apart from the fault in its propulsion system) the station was in good shape.

In early March, when the maintenance that could be attempted had been completed, the station was orientated nose-down in the gravity gradient. Unlike their predecessors, Lyakhov and Ryumin had been assigned primarily visual observation tasks. With the portholes in the forward transfer compartment covering the entire horizon, they would be able to make comprehensive observations without expending propellant.

By the time Progress 5 arrived on 14 March, the station's orbit had decayed to 293 × 325 km. In addition to propellant, it delivered food, water, film, ampoules for the furnaces, clothing and "various other items", including improved smoke alarms, a linen dryer, a tape recorder, the Koltso walk-around communications system, a replacement Stroka teletype (the first had broken down), a monochrome video monitor to establish a television uplink, a new NiCd storage battery (to augment the existing batteries, which had degraded, as indeed had the solar panels) and parts to repair the water reclamation system. In addition, reflecting the confidence that the efforts to repair the engine would be successful, it delivered new equipment, including the Yelena-F gamma-ray detector, an improved Kristall furnace (the other had broken down after successfully performing 40 smelting operations), the Biogravistat, the Phyton-2 plant cultivator, the Spektr-15K spectrometer and the Duga electrophotometer. They opened 27 containers and unpacked over 300 individual items, each of which had to be stored in its proper place within the station. In the process, as the assorted stores and its packing floated around in the main work compartment, the station took on the appearance of a junk shop. Nevertheless, they finished the task in just four working days.

The repair of the station's propulsion system began on 16 March. The first step was to drain the faulty fuel tank. The Soyuz 32–Salyut 6–Progress 5 complex was placed in an end-over-end rotation, so that the centrifugal force would separate the liquid-gas mix within the damaged tank. The fuel was pumped into the second onboard tank, until this was full, and then the rest was pumped to an empty tank in Progress 5. Finally, the faulty tank's valves were opened to vent the residue behind the bladder. The ferry's thrusters were then used to stabilise the complex. To ensure that all the highly corrosive fuel was removed, the pipes were left exposed to vacuum for a week, then nitrogen was blasted out as a further guarantee. The faulty tank was filled with nitrogen, then sealed off from the other tank. The entire repair operation had been completed without incident. It was quite an achievement. Progress 5 recircularised the station's orbit at 350 km altitude, then left on 3 April.

The station's radio and television systems were overhauled and upgraded so that it could receive a video uplink from Kaliningrad. This was not as easy as might at first be expected, due to interference from radiation at orbital altitude and from the station's own systems. This marked a significant step forward in the operation of the station, because long-stay crews would now be able to watch their colleagues test procedures for unplanned external activities, scientists would be able to demonstrate their requirements, the crew would be able to view the results of their Earth photography and, of course, they would be able to see their families as well as hear them. One of the first things beamed up to them, however, was a rerun of their own launch. The Koltso walk-around communication system would enable the crew to talk to the ground without leaving their post to use the transceiver on the control panel.

With the station once again fully operational, Lyakhov and Ryumin settled down to a comprehensive programme involving Earth-resources, upper atmospheric and astrophysical observations, materials-processing (some of the samples had been supplied by France) and biological experiments (including starting a new line of research, by placing a quail egg into an incubator for an embryological study). The Yelena-F was set up to measure gamma-rays and charged particles in near-Earth space. Because this drew just 10 watts, it could undertake a long-term survey of the radiation in the station's environment. Not unexpectedly, it recorded its highest readings near the South Atlantic Anomaly.

SO NEAR AND YET SO FAR

On 10 April Nikolai Rukavishnikov and Georgi Ivanov, a Bulgarian cosmonaut, set off in Soyuz 33 to visit Salyut. (Remarkably, Ivanov's pulse did not rise above 74 beats per minute during the launch.) The next day, Soyuz 33's final transfer orbit brought it within 6 km of the station, and the Igla system was activated to make an automated approach. With 1,000 metres to go, as the ferry aligned itself with the station, the engine shut off part way through a six-second burn and the vehicle shook rather violently. Rukavishnikov reached out to steady the vibrating control panel. Ryumin, observing from the station, reported that he thought he had seen an unusual lateral plume from the engine, but this was behind the ferry so it was hard for him to be sure. Although Rukavishnikov later admitted to being "scared as hell" by the engine failure, at the time he asked permission to continue the approach using only the small thrusters. Kaliningrad, however, ordered him to abort. This must have been extremely frustrating for Rukavishnikov, as it was the second time he had been prevented from boarding a station (he had suffered the frustration of not being able to open Soyuz 10's hatch to enter Salyut 1). To abort, however, had been the only option, as although Rukavishnikov may well have been able to complete the docking using the secondary engines it would have been to no avail, because there would have been no point in leaving a damaged ferry at the station for the residents.

As usual, the first favourable landing opportunity was the following day, so the two men settled down to wait. Displaying his customary cool, Ivanov slept, but Rukavishnikov stayed awake. There was concern that the shaking that accompanied the main engine's misfire might have damaged the adjacent backup engine, which would be required for the deorbit burn. The following day – ironically Cosmonaut Day – was very tense. Unlike the main engine, which was restartable, the backup unit was intended to be used for just one long firing at full thrust to terminate orbital flight. Although it was set to shut off automatically, it did not, and Rukavishnikov had shut it down manually. The capsule landed in darkness, 15 km from the planned recovery site. By the time the recovery team reached it, Rukavishnikov had already scrambled out. He announced that he felt as if he had been in space for a month!

Unfortunately, because the service module did not return to Earth, the engineers had only the telemetry to go on in trying to determine what had gone wrong, and their task was made worse by the fact that the propulsive unit was not heavily instrumented. Even so, within days they were able to report that they had identified the fault, which was a part of the engine itself, not the control system as had initially been suspected. Fortunately, a pressure sensor in the combustion chamber had shut the engine down when it sensed that the engine had not achieved the correct operating pressure, thereby eliminating the danger of the chamber exploding from uncontrolled mixing of propellants. This engine configuration had been fired 1,000 times since its introduction in the mid-1960s, so this malfunction was a surprising discovery. It took a month to fully study the problem, but as soon as the exact cause of the failure had been identified, steps were taken to modify the spacecraft that had already left the production line.

Only Soyuz 32 could not be modified, and it was docked with Salyut 6, but whether its engine would fail was not the primary concern. It had been in space for six weeks, and was to have been returned to Earth by the Intercosmos crew. Their inability to reach the station meant that this old ferry would end its 90-day service life long before the next visiting crew could be sent up. Strictly speaking, at the end of May it would be deemed unserviceable, but in a sense this did not matter, as only in an emergency would the status of the ferry become a real issue. Its unserviceability did, however, rule out sending up the next Intercosmos crew, because they would not be able to return in the expired spacecraft. A Soyuz with a new engine would have to be launched empty as soon as possible. The Hungarian mission, scheduled for June, was postponed, because its launch window would now be used to send up the replacement ferry. The Soyuz 33 drama had completely disrupted the tightly interleaved schedule of visiting missions.

WORKING ON

Lyakhov and Ryumin took these events stoically. There was nothing they could do, and they were safe so long as they remained where they were, so they carried on with their work. They observed ocean currents, made semiconductor in the Kristall furnace and installed the Biogravistat cultivator. Even the planned Bulgarian programme had not really been lost. The equipment had been delivered by Progress 5, so they would be able to perform much of Ivanov's work, including the spectroscopic survey of the Earth by the Spektr-15K for the Balkan experiment. Using this spectrometer and an infrared camera, they discovered that they could detect a thin film of oil that formed over a large shoal of fish swimming in a warm ocean current, and they reported their observations to fishermen. On 1 May they took a holiday and watched the military parade through Red Square on their video uplink.

Progress 6 arrived on 15 May. The automated ferry was not affected by the failure of Soyuz 33's engine because it used the integrated ODU engine. Once again, it carried primarily dry cargo, including a variety of experimental apparatus (samples for the furnaces, samples for the Biogravistat, the Phyton-3 cultivator, and a soon-to-bloom tulip) and replacement hardware (higher-power television lights, and a control panel for the navigational system), in addition to the usual components for the environmental system. The replenishment of the propellant tanks was carefully monitored by Kaliningrad, to make sure that there was no further problem.

On 5 June, while Lyakhov and Ryumin passed their 100th day in space by loading assorted rubbish into Progress 6's orbital module, final preparations were being made to send up Soyuz 34. When launched the following day, it began a two-day rendezvous. The rear port was not vacated until it was established that the modified engine was fully functional. Progress 6 had been retained to ensure that the station would have a proven propulsion system (despite its being fixed, there was some reluctance to rely upon the big engine). Soyuz 34 docked at the rear port without difficulty. Its arrival assured the crew of a serviceable ferry for their return to Earth, and an opportunity had been taken to stock it with a variety of additional supplies.

A week later, Soyuz 32's descent module was loaded with 180 kg of cargo and sent back to Earth using the window that the Hungarian mission would have used. Although it was deemed to be unsafe for returning cosmonauts, the fact that Soyuz 32 was being trusted to return the results of their experiments demonstrated that it was fully expected to return safely; and it did. In fact, because the residents did not receive either of their scheduled Intercosmos visits, each of which was to have relieved them of about 50 kg of results, this empty Soyuz was the only way for them to return their haul. In addition, of course, the descent module was itself to form a valuable result because analysis of how well its systems had survived the extended mission (108 days) would be able to be compared with those of Soyuz 20, the degradation of which had resulted in defining the 90-day operating limit. Not all of Soyuz 32's return cargo had been experimental output, as in addition to the 50 kg of film and samples from the furnaces, it carried many items of expired equipment (such as burnt-out lamps and saturated filters) which could be examined by the engineers; these would otherwise have been jettisoned with Progress 6.

Soyuz 34 was transferred to the front port on 14 June to clear the way for the next cargo ferry. This was not scheduled for six weeks, but the transfer had to be done while the recovery window remained open. The station's propellant reserve was so high that Progress 7, which arrived on 30 June, carried mostly dry cargo. It delivered a number of new experiments including the Isparitel apparatus, the Resistance experiment (to measure their upper atmospheric drag) and the KRT-10 radio telescope. In the mail bag, the crew discovered a guidebook for the Moscow area, sent up by one of their colleagues to help alleviate symptoms of homesickness.

The Vaporiser experiment used the Isparitel apparatus supplied by the Institute of Electrical Welding in Kiev. This was installed in the scientific airlock. After some initial problems, the low-voltage electron beam melted tiny granules of aluminium and silver, then sprayed the vaporised melt onto carbon and titanium disks to condense and create a film. Depending on the desired film thickness, the test ran from one second to a maximum of 10 minutes; at its finest setting, a layer only a few microns thick could be deposited. These would subsequently be analysed to determine whether such coatings would be practicable as a way of protecting future stations from the deteriorating effects of the space environment. This experiment followed directly from the successful respraying of the mirrors in the OST-1 by Gubarev and Grechko on Salyut 4.

A novel engineering experiment (Deformatsiya) used optical instruments to measure the distortion of the complex's structure resulting from solar heating. The three-vehicle train was set in an attitude with respect to the Sun which provided a 300°C temperature differential between the illuminated and the shaded sides to bake one side and freeze the other. This could not be held for long without risk of damage, but the test indicated that even several hours of exposure deformed the axis by just a few tenths of a degree.

On 15 July, Lyakhov and Ryumin passed the 140-day record set by their immediate predecessors. Their objective was to achieve the psychologically important 'half a year' in orbit, and this would require them to remain aboard for at least another month. By this point, they required only about seven hours of sleep, rather than the 10 hours with which they had begun, but even at this late stage their working efficiency was extremely high. In addition, whereas their predecessors had undertaken an intensified exercise programme in the final weeks of their mission to help prepare for their return to gravity, Lyakhov and Ryumin had followed this regime throughout, so they were in much better shape. Only the stamina-inducing drugs, vitamins, saline rehydration and the Tchibis cardiovascular conditioning would be required to prepare for the return to Earth.

A RADIO TELESCOPE

The KRT-10 incorporated a directional antenna and five radiometers (four horns in the 12-cm band and a spiral antenna operating in the 72-cm band). As delivered, the 350-kg unit comprised a control unit and a parabolic antenna with an integral three-legged mount to hold the radiometers at its focus. A video had also been sent to show the cosmonauts how to operate it. The controller was to be retained inside, but the rest of the structure had to be set up outside, and this was to be done without requiring the crew to make a spacewalk. In mid-July, they 'docked' the tightly-furled assemblage to the rear docking drogue and left it projecting axially into Progress 7's orbital module (its docking system had been dismounted). When the ferry unlatched, the escaping air pushed it smoothly away from the collar, and as it withdrew it exposed the antenna of the KRT-10. The ferry was commanded to halt a few hundred metres away so that Kaliningrad could observe the opening of the 10-metre diameter dish, which unfolded exactly as intended, totally obscuring the station behind.

Because of the need to recharge the station's batteries, the radio telescope could be used only on alternate days. Over the next few weeks, it was used in conjunction with the 70-metre deep-space communications dish in the Crimea as an interferometer. These observations demonstrated mixed results, but it was only an early test of an orbital telescope; the very act of trying to use it had generated valuable design feedback. For the deep sky observations, it had been used in two modes: in one case the station had been stabilised so that the telescope gave continuous coverage of a given source, and in the other mode the station was set in a rotation which made the telescope scan along the Milky Way to map Galactic sources. When aimed at the Earth, it demonstrated that such an instrument would be able to reveal the structure of meteorological features, map sea state, measure water salinity and soil humidity, and monitor volcanic activity (it was able to observe an eruption of Mount Etna, in Sicily), even though it had a surface resolution of only 7,000 metres.

The KRT-10 was jettisoned to free the rear port, but it wobbled as it slowly drifted off, and became entangled in the apparatus which projected from the rear of the station. The station's engine was fired briefly to attempt to blast the structure away, but it was stuck fast. The station was rocked to try to shake it free, but this did not work either, and it was eventually decided that the cosmonauts would have to cut it free of whatever it had fouled. Since they were likely to be the final occupants, Lyakhov and Ryumin had been assigned a brief spacewalk at the end of their mission to retrieve the various packages that had been affixed near the hatch by an earlier crew. Although the doctors expressed concern that the two men might be too tired to perform such a complex unscheduled operation, this contradicted the earlier pronouncement that they were working extremely efficiently. When asked, Lyakhov and Ryumin said they would extend the excursion to try to free the telescope dish. With the decision taken, fellow cosmonauts set out to test procedures in the hydrotank. Meanwhile, the residents loaded experimental results into Soyuz 34 so that these would not be left aboard the station if it proved impossible to re-enter the main compartment after the spacewalk.

The big day was 15 August. As Ryumin pulled on the airlock hatch, he found that it was jammed, and it took considerable force to release it. Just after they finally emerged, the station moved into the Earth's shadow, so Kaliningrad told them to wait by the airlock. Back in sunlight, half an hour later, Ryumin left Lyakhov at the airlock, and made his way slowly to the rear of the main compartment. No cosmonaut had ever strayed so far from the hatch, but the handholds and the 20-metre umbilical he used had been provided for precisely this contingency. Once in place, Lyakhov paused to rest, and to assess the situation.

After studying the antenna, Ryumin decide how he would cut the frame free, but he had to wait to re-establish communication with Kaliningrad before proceeding. Although all that was required was a few quick snips with his metal cutters, this was a potentially dangerous operation because it required him to crawl over the rim, down in between the dish and the rear of the station. As a general rule, mission planners tend to regard items not specifically designed for use by spacewalking cosmonauts as inherently dangerous. Ryumin was reminded of this when he cut loose the first strand of the fouled mesh, and discovered that he had set the huge sharp-edged structure rotating towards himself. He persisted, however, and finally managed to shove it clear. On his way back, Ryumin examined some of the portholes. The glass had gradually become tarnished by exposure to space. He rubbed one with his glove in an attempt to collect some of this dust for analysis, but it appeared to be embedded in the glass rather than just accumulated on its surface. Lyakhov, meanwhile, had retrieved the experiment packages from around the airlock.

As with the isolation of the damaged propellant tank, this impromptu spacewalk had demonstrated just how rapidly the programme was maturing. It was not clear whether Salyut 6 would be able to sustain a further visit, but if it had proven impossible to clear the rear port this would have precluded docking cargo ferries, which would have meant that the station would have been able to support a crew only until its consumables were exhausted; if indeed it was deemed useable at all with the massive but lightweight dish hanging off its back end. Real-time planning in Kaliningrad, and determined action in space, had paid off.

HALF A YEAR

During their long mission Lyakhov and Ryumin had undertaken a wide variety of experiments. As before, the biological experiments had been disappointing; although the root vegetables had been successful, the flowering plants had rapidly withered, and the arabidopsis had grown but not yielded seeds. Various theories were put forward for the repeated failure of plants to produce seeds, including the possibility that the plant's mechanism for disposing of its waste products failed in the absence of gravity, causing it to poison itself. In addition to the ongoing multispectral Earth-resources and mapping photography, 54 samples had been processed in the furnaces, and ocean currents and plankton blooms had been monitored (Lyakhov was awarded a certificate proclaiming him to be an "outstanding worker" by the trawler skippers who he had regularly steered onto rich shoals of fish). The long-duration missions were becoming so productive that the researchers on Earth were beginning to be swamped by the flood of data from the science programme, and it was evident that special support teams would have to be formed for future stations.

Despite the limited return cargo capacity of the Soyuz, the broken Kristall furnace was dismantled and stowed so that its designers could examine it, as was a regeneration canister from the environmental system which had started to corrode. On 19 August, after 175 days in space, Lyakhov and Ryumin finally put Salyut 6 back in its automatic flight regime and returned to Earth. Because Soyuz 33 had pre-empted the Intercosmos series, they had been forced to complete the entire flight without the relief of company, so the crowd at the recovery site must have seemed quite overwhelming.

A month later, Konstantin Feoktistov reported that Salyut 6's systems were being studied to assess its future rôle. In July, Progress 7 had boosted it to the unprecedented altitude

Valeri Ryumin and Vladimir Lyakhov relax in recliners while the inflatable tent is set up for the medical check. As usually happened, the capsule came to rest on its side.

of 400 km, and by the end of November it had decayed to its nominal operating altitude at around 360 km. It had been decided that even though it had already lasted far beyond its design life, it should be able to support another expedition in the new year.

A NEW FERRY

Meanwhile, other developments were bearing fruit. A new variant of the Soyuz had been tested under the anonymity of the Cosmos series. Cosmos 869 and Cosmos 1001, launched in November 1976 and in April 1978 respectively, had performed a simulated rendezvous and then returned within a few days, but, earlier in the year, Cosmos 1074 had been powered down for 60 days to assess its suitability as a ferry.

On 16 December, this new configuration, launched crewless, was openly declared as Soyuz-T 1. After a two-day rendezvous it approached Salyut 6 by descending from above and ahead of the station, rather than from below and behind it as was usual, but it did not dock. Instead, it withdrew, and repeated the approach the next day. This time, it docked at the front port. It raised the station's orbit to 370 × 382 km to counter decay, and was powered down. In March 1980, having spent 100 days in space, it returned to Earth.

As with the previous two tests, the 1,260-kg orbital module was jettisoned prior to the deorbit burn in order to reduce the overall propellant requirement by 10 per cent. Reducing the propellant which had to be carried enabled the cargo capacity to be correspondingly increased. The orbital module re-entered the atmosphere a few months later.

Konstantin Feoktistov reported that although this new variant had a similar external appearance to the current Soyuz ferry, it incorporated significant internal modifications. The integrated ODU, which had been introduced by Salyut 6 and incorporated into the service module of the Progress cargo ferry, had been used for this new crew ferry. The most significant change was the installation of a digital computer that could control the spacecraft's attitude and perform all the manoeuvres necessary to reach and return from an orbital station automatically, and do so in the most propellant-efficient way. The use of advanced electronics also meant that more sophisticated telemetry could be produced. The Argon 16 computer (as used on Salyut 4) monitored this to present real-time data to the crew. Significantly, the inertial platform of the computerised navigational system permitted the new spacecraft to operate completely independently of Kaliningrad. Solar panels had been reintroduced; now if any spacecraft was unable to dock it would not be forced to make an emergency return to Earth within 48 hours (which had happened four times), and there should be time to analyse and hopefully overcome the problem. What was not revealed, however, was that this new configuration could carry *three* pressure-suited cosmonauts, thereby restoring the capacity lost by the revision following the loss of the Soyuz 11 crew in 1971.

YET ANOTHER EXPEDITION

The remarkable longevity of Salyut 6 enabled a fourth expedition to be mounted. In this case, the tanker was launched first. Progress 8 docked at the rear port of the vacant station on 29 March 1980, and Soyuz 35 docked at the front on 10 April.

Astonishingly, Leonid Popov was accompanied by none other than Valeri Ryumin. Valentin Lebedev had been assigned to fly the engineer's seat, but in early March he had badly injured a knee while trampolining. Their backups were not given to the mission due to their inexperience. Vyacheslav Zudov had flown only the aborted Soyuz 23, and Boris Andreyev had yet to fly. Ryumin, who had been giving the new crew the benefit of his experience of six months aboard, realised that he was best placed to restore Salyut 6 to life for another long-term mission, so he offered to take Lebedev's place. He therefore had the *déjà vu* experience of reading the note that he had left seven months earlier welcoming anyone visiting his old haunt. As a joke, he then pulled out a plastic cucumber, showed it to the ever-watching flight controllers, and told them that it had grown in his absence.

They were to continue the weekend-off work schedule, and use Moscow Time, but whereas the intense exercise regime had previously been enforced daily, this time every fourth day was to be a lazy day, to break the monotony. Ryumin's rapid recovery from his last flight had convinced the doctors that they understood the requirements for spending six months in space. Now it appeared that his body had learnt how to react to weightlessness, because he readapted almost immediately, and was able to offer Popov tips, so they were both able to settle into a productive cycle straight away.

In addition to the standard consumables and replacements for the thermal regulation and atmospheric systems, Progress 8 had delivered another NiCd battery, a new Splav (Splav-2) furnace, a Kristall (Kristall-3) furnace and the Malakhit plant cultivator. After about two weeks dedicated to maintenance, experimental work started. They had been assigned more smelting work than earlier crews, testing a wider variety of substances and greater range of crystallisation times, extending from the typical 10 hours to as long as 120 hours in the case of CMT samples, throughout which the station would be left to drift in gravity-gradient mode. Materials processing was one of the successful parts of the Salyut research programme. They had also been assigned Earth-observational research directed at the ocean and its biological productivity. The Yelena-F gamma-ray detector, which continuously sampled data on the near-Earth environment, was restarted (but it failed soon afterwards, so they took it apart and repaired it). The Malakhit cultivator was used to study the growth of orchids. but unfortunately, although the plants had been delivered after they had already begun to flower, they immediately wilted. However, there was much better luck with orchids planted after reaching orbit, as they grew roots long enough to protrude from the apparatus. Orchids had been chosen because they thrived in a dry atmosphere, and so might survive in the artificial environment within the station, but although they flowered, they did not seed. Samples were to be sent to Earth with visiting crews at 60 days, 110 days, and at the end of the mission, which was scheduled to last six months.

Progress 8 left on 25 April, and Progress 9 took its place four days later. This delivered a replacement motor for the centrifuge in the Biogravistat, and the Lotos experiment (an attempt to make polyurethane foam shapes); it vacated the rear port on 20 May.

BETTER LATE THAN NEVER: A HUNGARIAN VISITOR

It must have been with yet another sense of *déjà vu* that Ryumin awaited the launch of the first Intercosmos crew which would bring the Hungarian cosmonaut who was to have visited him the previous year. When Soyuz 36 docked at the rear port, on 27 May, Valeri Kubasov and Bertalan Farkas finally boarded the station.

The main experiment was Interferon, which was designed to evaluate the possibility of manufacturing interferon (a chemical, produced by human cells, that inhibits viruses) in commercial quantities. One aspect of this test involved injecting a vial of white blood corpuscles into an interferon-yielding substance to test whether the rate of production was different in microgravity. A related blood analysis experiment determined whether the generation of this chemical by the body was influenced by weightlessness. Another experiment evaluated whether existing interferon pharmaceutical preparations, delivered both in the liquid state and in a lyophillised gel, were rendered more or less effective by exposure to weightlessness. Obviously, although such experiments were undertaken as part of the collaborative Intercosmos programme, and may often have been inspired by the participating member, the results could be exploited only by the Soviet Union, because only it had regular access to space. In this case, the results proved so encouraging that a number of follow-on experiments were devised.

Of the Earth-observation experiments, Refraction used the VPA-1M to measure the optical polarisation of the horizon as part of a study of the upper atmosphere, Zarya used the Spektr-15K spectrometer to observe absorption lines at sunrise and sunset to measure the density and temperature of the stratosphere and the troposphere, and Utrof used the MKF-6M camera to compile a geomorphological map of the Carpathian Basin and the Tisza River Basin, to assess the effect of the Kishkere reservoir on soil salination in the inland waterways linked to the Danube, and to study the ecology of Lake Balaton.

HOTEL SALYUT!

On 3 June the visitors retreated to Soyuz 35, undocked and returned to Earth. They had been so active that they had caught no more than three hours of sleep a night, and they were exhausted; luckily a brief visiting flight did not tax the body that much. Kubasov immediately scrambled out of the capsule and supervised the unloading of his cargo, and the next day the residents transferred Soyuz 36 to the front port. The day after that, Soyuz-T 2 was launched with Yuri Malyschev and Vladimir Aksyonov.

This was the first of the new spacecraft to be trusted to a crew, and the fact that it could carry three people remained secret. The cosmonauts wore a new Sokol pressure suit, which in addition to being relatively lightweight (8 kg) was more comfortable than the suit that had been used since the loss of the Soyuz 11 crew. Aksyonov had been flight engineer on the solo Soyuz 22, which had tested the MKF-6 camera. This was Malyschev's first flight.

Unlike its predecessor (which had made a slow rendezvous) Soyuz-T 2 started its final approach after the standard 24-hour chase. The Argon computer was programmed to select the optimum sequence of manoeuvres for each phase of the flight but, when it was 180 metres from Salyut 6, Malyschev saw it start a manoeuvre with which he was not familiar, so he took command and manually docked at the rear port. Later analysis established that the route selected by the computer should have resulted in a successful docking, so the commander's intervention had been premature.

Malyschev and Aksyonov spent three days performing the routine biomedical tests, making spectrometric observations of the Sun on the horizon to continue the study of the upper atmosphere, and taking Earth-resources imagery with the MKF-6M, and then returned to Earth having achieved their primary objective of evaluating the performance of the new ferry. This was the first time that two visiting missions had been mounted in a single recovery window.

Everything had been going well, so it was a major inconvenience when, not long after Soyuz-T 2 departed, the Kaskad attitude control system malfunctioned. Popov and Ryumin managed to repair it, but doing so took considerable time, and the subsequent manoeuvring trials consumed a great deal of propellant.

Progress 10 docked on 1 July, and delivered a colour monitor for the video uplink (superseding the monochrome one), a Polaroid camera (one of the few objects onboard of Western origin), a new image converter for the Duga electrophotometer (to replace the one that projected an inverted image) and replacement components for the BST-1M. It replenished the propellants and then, after it adjusted the orbit, left on 18 July. Meanwhile, the crew took the day off to watch the opening ceremony of the Moscow Olympics on their new uplink, then sent a pre-scripted message that was beamed onto a large monitor in the Lenin Stadium. A few days later, the portholes began to fog over, suggesting that the environmental control system was deteriorating. They took another set of pictures to record the erosion on the outer surface of the glass; the mark left on one porthole by Ryumin's glove during his spacewalk a year earlier was still clearly visible. On 24 July, Soyuz 37 settled into the busy rear port, and Viktor Gorbatko and Pham Tuan, a Vietnamese citizen, floated in to become the latest visitors to Hotel Salyut.

Tuan's main task was the Kyulong experiment, using the MKF-6M and KATE-140 cameras and the Spektr-15K spectrometer to perform the first comprehensive survey of hydrological features on the Central Vietnamese Plateau and tidal flooding and silting in the deltas of the Mekong and Red rivers. The Halong experiment (named after a bay on the Vietnamese coast) used the Kristall furnace to produce bismuth–tellurium–selenium and bismuth–antimony–telluride alloys, and a gallium phosphide semiconductor crystal. The Azolla biological experiment tested a fast-growing nitrogen-rich water fern (*azolla pinnata*, commonly used as fertiliser in rice growing) to determine whether it would be suitable for use in a closed-cycle hydroponics farm on a future station, but it fared no better than any other higher plant to date. One great advantage of the frequency of flights was that a hypothesis could be tested quickly. The inability of plants to survive in space had perplexed the botanists, and they were delighted several weeks later, when an arabidopsis flowered. After it had started to grow, the cosmonauts had set it up beside a porthole to illuminate it with sunlight, in the hope that this might prove more satisfactory than the cultivator's lamp. Furthermore, it demonstrated that the plant had not been confused by the orbital illumination cycle. The next objective was to make one to complete the cycle and produce a seed.

Gorbatko and Tuan returned to Earth in Soyuz 36 on 31 July. The next day, Popov and Ryumin once more transferred their new ferry to the front port. They then resumed Earth observations of the Krasnoyarsk region, the Baikal area, the Caspian Sea, Soviet Central Asia and the Soviet Far East. After having tried so hard to hold the station still to minimise disturbance to smelting operations, a new experiment involved setting the station rotating to subject a sample in the Splav furnace to directional solidification. The comprehensive smelting tests had shown that the semiconductor monocrystals grown in microgravity were of "immeasurably better" quality than those made on Earth. Another novel experiment involved using a lens to project an image of the Sun onto a screen, so that, as the Sun dropped below the horizon, one camera filmed the Sun while another filmed its image on the screen to accurately measure the distortion of the solar disk by atmospheric refraction. Throughout most of August and September, the station was left in the gravity gradient to combine smelting with observations of the Sun on the horizon using the BST-1M. As a result, little propellant was used and there was no need to send up a new cargo ferry. In mid-September, its orbit having decayed considerably, Salyut 6 fired its big engine to boost itself back to 335 × 352 km; this was the first time that this engine had been used since the damaged fuel tank had been isolated.

Soyuz 38, which arrived on 19 September, brought Yuri Romanenko and Arnaldo Tamayo Mendez, a Cuban. As usual, the visitor's main task was the study of his home. The Tropics experiment involved using the MKF-6M to map Cuba's natural resources (in fact, this was Tropics-3, and it continued observations already made earlier of other locations). As always, the orbital photography was coordinated with data gathered by airborne and ground teams. Unfortunately, although the weather in the Caribbean was excellent throughout the flight, the timing of the flight meant that there were few suitable daylight passes, so the residents had to take most of the pictures after Tamayo Mendez had left. The results revealed that at both its eastern and western extremities, the island of Cuba is criss-crossed by intersecting networks of faults.

One novel Cuban contribution to the study of adaptation to weightlessness was the Support experiment. This involved Tamayo Mendez wearing a special adjustable shoe for six hours each day. It was designed to place a load on the arch of his foot to simulate the force the foot feels when standing on Earth, to test whether it affected the ability of his vestibular system to adapt to the absence of gravity. The experiment was based on the idea that on Earth the state of the muscles in the foot contributes directly to the sense of balance, and that there was a vestibular reaction to this sensory input being denied in space. Such a device was also to be used to investigate the recovery and readaptation of locomotive stability upon returning to gravity, because cosmonauts had been noticed to develop a distinctive gait and posture whilst recovering from long flights.

It must have been an interesting experience for Romanenko to revisit Salyut 6 after so many years of use, but he was probably glad to be making only a brief stopover. He and Tamayo Mendez returned to Earth on 26 September. The fact that they left in their own ferry indicated that the residents expected to follow them quite soon.

COMPLETING A SECOND SIX-MONTH TOUR

As Salyut 6 began its fourth year in orbit, Progress 11 slipped into the rear port. It delivered sufficient propellant to sustain the station during a period of automated flight, just in case it proved possible to mount another expedition. On 1 October, Popov and Ryumin exceeded the 175-day endurance record that Ryumin had established the year before.

During their residency, the crew had devoted nearly 35 per cent of their work time to Earth observations. In addition to a large number of synoptic pictures with hand-held cameras, almost 3,500 MKF-6M and 1,000 KATE-140 pictures had been taken. The MKF-6M had regularly photographed the test site at Salsky. They had reported on crop yields and forestry in preselected areas, and on plankton and large fish shoals, and their observations had been exploited to great effect. Of three faults which they had identified in Hungary, one had already been found to have oil and gas reserves. They had collected extensive data with the Spektr-15K, RSS-2M and VPA-1M for atmospheric studies, and infrared data from the BST-1M had revealed that there were small-scale variations in the submillimetre flux of the atmosphere in the zones in which cyclonic activity originated. The Yelena-F gamma-ray detector had mapped streams of high-energy electrons flowing in the South Atlantic Anomaly, and had revealed that the background flux was highly dependent on latitude (its strength varied by a factor of 10 between the equator, where it was weakest, and the latitude extremes of the station's orbital track). They had conducted variations of the Vaporiser experiment, one of which involved spraying small disks made of glass, carbon, titanium and various other metals with a variety of thin coatings of gold, silver, copper and aluminium; in all they produced 196 samples. Future crews would experiment with polymer coatings. In addition, they put almost 100 samples in the furnaces to produce semiconductors and a variety of alloys.

Overall, they spent about 25 per cent of their working time on routine maintenance and repairs, although most of the latter had been carried out at the beginning of the mission, to prepare the station for their own period of residency. In addition, they devoted a few days to unloading each cargo ferry. Their working efficiency had been extremely high, but had dropped markedly when they had entertained visitors. For Ryumin, who had spent 12 of the last 20 months aboard, the contrast between his tours must have been marked: on his first he had received no visitors, but he had played host four times on the second.

Popov and Ryumin undocked and returned to Earth on 11 October. Almost as soon as the recovery team arrived, they were astonished to see the crew scramble out of the hatch without assistance and walk to their reclining couches. Ryumin recovered much faster than he had after his previous flight, and he had actually *gained* a little weight (2 kg). Although the rigorous exercise regime had maintained the performance of his cardiovascular system and limited the degradation of his calf muscles, and the idle back and chest muscles had degraded, by moving so much cargo he had developed his arm and hand muscles. The results of the 185-day mission convinced the doctors that it would be feasible to mount an expedition of a year or more.

REPRISE?

Salyut 6 had been left with Progress 11 still docked to ensure an independent means of orbital manoeuvring. Before departing, Popov and Ryumin had undertaken routine maintenance to prevent vital equipment from expiring its service life and had made such repairs as they could. The station was in fairly good condition and, with further routine maintenance, it might possibly support another expedition. A factor limiting further use of Salyut 6 was that the hydraulics within its thermal regulation system were beginning to degrade. The station had been designed to operate for up to 24 months, and so far it had been in use for three years. The hydraulics had not been designed to be serviced, so if the thermal regulation system failed the station would immediately become uninhabitable. Another factor in deciding whether to send up another crew was that the output from the solar panels had degraded to the point that the station was now constantly short of power, so the research work would have to be planned with this in mind. The decision depended on the cosmonauts' debriefing and on analysis of the station's telemetry.

Having assessed the situation, and compiled a list of all the systems which required attention, in early November it was decided to send a maintenance crew for a short visit to see what could be done. It would be costly, but it was far cheaper than launching a new station. Interviewed by *L'Humanité* in Paris, Georgi Beregovoi reported that a three-man crew would be sent; this was the first indication that the new Soyuz had such a capability. In mid-November, Progress 11 replenished Salyut 6's propellant tanks, and then boosted its rapidly decaying orbit back up to 300 × 315 km; the orbit had been allowed to drop to this comparatively low level because the Soyuz-T ferry could not reach its standard 350-km operating altitude when carrying three cosmonauts.

Soyuz-T 3 was sent up on 27 November 1980, with Leonid Kizim, Oleg Makarov and Gennadi Strekalov. Konstantin Feoktistov had been heavily involved in the detailed planning for this mission and had intended to make the flight, but he had been banned by his doctor, and Strekalov had taken his place. Kizim and Makarov had been the backup crew for Soyuz-T 2, so they were well acquainted. This time, the Argon computer was left to choose the approach, and the following day it docked at the front port of the Salyut 6–Progress 11 complex without incident.

The station was immediately put into the gravity gradient to save propellant, and the crew set to work. It was standard practice to avoid heavy work during the initial phase of adaptation to weightlessness, but they ignored this. Because they were so busy, they let the strict exercise regime lapse, but as they did not intend to be aboard all that long this was not deemed to be a significant risk. Despite the problem of gaining access, the objective was to perform maintenance on the thermal regulation by replacing the pumps which circulated fluid through coolant loops running to radiators on the surface of the station. If they could complete this, they were to move on to other tasks. Having gained access to the enclosure, they had to saw through a metal supporting bracket to reach the pumps. This was tricky, because they had to be careful to catch the tiny metal shards to prevent them from floating into their eyes. They had brought with them a complete assembly, which had been preloaded with glycol (in effect, an antifreeze), and were very conscious of the fact that if they spilt any of this they would be ordered to abort and evacuate the station. The pipes running through the station's walls had to be depressurised before the new assembly could be installed, and it had to be set up so that the temperature in the station could be maintained at the required level. This was essential not just for the crew's comfort, but to preserve the service life of the many rubber gaskets that sealed the holes in the pressure shell through which the many pipes and cables ran to externally-mounted apparatus. With this pump assembly installed, the system was as good as new.

Having crossed off the most urgent item, they moved down the list in short order. They dismantled the telemetry system to replace some of its components, and although this was another system that had not been designed to be accessible, they repaired it without difficulty. Several timers in the station's control system were replaced. A commutator in the power distribution system was replaced, as was a transducer in a compressor in the ODU replenishment unit (it had failed while Progress 11 had been pumping into the tanks during the time that the station had been unmanned, the only time it had been done without a crew aboard to monitor the situation). Air circulation within the station was tested to ensure that there were no stagnant regions in which carbon dioxide could concentrate, and the Resonance and Amplitude experiments were performed to verify the structural integrity of the docking collars. Everything checked out.

The most significant new experiment involved using the KGA-1 camera, which used a helium–neon laser to make a series of holographic images. Romanenko was to have tested it, but it had not been ready in time for his flight, so it had been reassigned to this impromptu visit. This was tested by recording how the density of a dissolving salt crystal was distributed without convective flows. This imaging process is extremely susceptible to interference from vibration. On Earth, a massive supporting structure had to be used to eliminate interference, but in space this was not an issue. It had been built by the Physical–Technical Institute in Leningrad, and would later be used to record the deterioration of the station's portholes. In the future, it was hoped to employ such holographic cameras to study engine performance and other processes that would otherwise be difficult to record.

After raising the apogee of the station's orbit back to 370 km, Progress 11 departed on 9 December. The next day, Salyut 6 was returned to autonomous flight, and then the cosmonauts returned to Earth. Their recommendation was that Salyut 6 would be able to support another expedition.

The decision made, Progress 12 was sent up on 24 January 1981. It was hoped that its cargo, together with the stockpile of consumables already aboard the station, would be sufficient to sustain a crew for three months, so that successive launch windows could be used to fly the final two Intercosmos missions.

GREMLINS

In late February the motor of one of Salyut 6's solar panels failed, further reducing the already low power supply unless the station was orientated so that the jammed panel faced the Sun, but this would bake one side and freeze the other and thereby strain the thermal regulation system. The power deficiency meant that the heaters had to be turned down, so the internal temperature dropped to 10°C (cold enough to cause water vapour to condense, which in turn ran the risk of a short circuit). In early March, the station was continually reorientated to trade power for thermal control in an effort to increase the internal temperature.

ONE FINAL TOUR

Soyuz-T 4 slipped into the front port on 13 March, to deliver Vladimir Kovalyonok and Viktor Savinykh. Because of the unexpected longevity of the station, this team had been formed only when the repair crew had reported that a further occupancy should be possible. In 1978, when Savinykh had joined the cosmonaut corps, Kovalyonok had been aboard Salyut 6. The third seat had been removed so that extra equipment could be ferried up. The first task was to attend to the stuck solar panel, and to general delight, once the controller was replaced (an operation which took only a few hours), the panel rotated to face the Sun and, with the heaters restored, the cabin was soon a comfortable 22°C. Upon spotting a crater in a porthole caused by the impact of a micrometeoroid, they set up the KGA-1 and made a hologram of it.

Savinykh was able to benefit from Kovalyonok's experience, and they adapted to weightlessness extremely quickly. Progress 12 was soon unloaded. It delivered a new plant cultivator which was basically a Biogravistat with a magnetic field applied across its centrifuge compartment. In one of their television broadcasts, Popov and Ryumin had explained their difficulties growing higher plants and had appealed for ideas; the use of a magnetic field had been proposed, so it had been rigged. Appropriately, this apparatus was designated the Magnetobiostat. It was immediately planted with shoots and roots of crepis, and arabidopsis was put in the Phyton and orchids in the Malakhit. In their effort to unload Progress 12 rapidly, they simply strapped most of the cargo to the cabin walls using a series of bungee cords.

After boosting the station's orbit back to 345 × 358 km, Progress 12 undocked on 19 March, and Soyuz 39 took its place on 23 March. Vladimir Dzhanibekov was accompanied by Jugderdemidiyin Gurragcha, a Mongol. Coincidentally (because orbital dynamics was the ultimate deciding factor) it was the 60th anniversary of the Mongolian People's Republic. The visitors performed the usual biomedical tests and tested the Mongolian-supplied cervical shock absorber. Worn continuously during the first three days (except when sleeping), this applied pressure to the cervical part of the spinal column, restricted head movement and simulated local loads experienced on Earth. It was hoped that this would inhibit the motion sickness commonly suffered during the first stage of adapting to weightlessness.

The primary task was Earth-resources work. Mongolia is a vast, inaccessible and largely unexplored country, so orbital observation was clearly the most cost-effective method of surveying. The KATE-140 and MKF-6M were used to locate geological fault lines, identify subterranean water deposits, compile a map of soil conditions, and assess crops and forests. Spectroscopic studies were made with the Spektr-15K for the ongoing Biosfera project. The Emission experiment involved using dielectric detectors to record the flux of cosmic rays, and the tracks bored into the detector substance would be examined upon return to Earth. A similar detector, also Mongolian-built, had previously been flown on the Intercosmos 6 satellite, and this had revealed that energetic nuclei with charges up to 28 units constituted 90 per cent of the incident radiation at 10 MeV. One of the detectors was exposed in the scientific airlock, and another was kept inside the station to measure how well the hull blocked this radiation.

The visitors returned to Earth on 30 March. They did not swap ferries because none of the Intercosmos crews were familiar with the new Soyuz-T spacecraft.

As soon as things returned to normal, Kovalyonok and Savinykh set about routine preventive maintenance on the water reclamation system and on the fans in the thermal regulation system, serviced the Yelena-F gamma-ray detector and the MKF-6M camera, and replaced a compressor switching unit in the BST-1M's cryogenic system. Most of April was devoted to Earth-resources studies of the Soviet Union and making the observations required to complete the unfinished Intercosmos projects. At one point, with the station in the gravity gradient to observe the Earth panoramically through the various portholes, they spotted noctilucent clouds extending over almost the whole of one hemisphere, so they used the BST-1M to take data. One significant result of opportunistically using the BST-1M was the unexpected discovery of anomalously strong submillimetre emissions in the atmosphere high above thunderstorms. In 1994, a NASA research aircraft fitted with a low-light camera observed thin jets of high-velocity gas (which were immediately nick-named 'sprites') projecting above violent storms to an an altitude of 50 km which give rise to red glows. It has been speculated that this is a result of establishing a temporary dis-charge path from a thunderstorm to the ionosphere; a form of lightning to space rather than to the ground.

Throughout this period, while awaiting the final Intercosmos mission, smelting was performed while the crew slept (that is, while they were not moving around to disturb the microgravity environment). On two occasions, however, six-day experiments were run in the Splav to create CMT. Soyuz 40 arrived on 15 May with Leonid Popov and the Roma-nian Dumitru Prunariu. Although Prunariu's flight had been pending for some time, he had been paired with Popov only at the start of the year; the scheduling chaos which had foll-owed the problem with Soyuz 33's engine had led to his training with a variety of comman-ders, including Valeri Bykovsky and Yevgeni Khrunov.

Soyuz 40 delivered an experiment to determine the capacity of human lymphocytes to synthesise interferon. The Minidoza experiment measured the energy spectrum of the heavy nuclei making up cosmic rays. Two novel smelting experiments were performed: one formed a crystal of germanium in the Kristall furnace by exploiting capillary action within a molybdenum matrix, and the other made a disk-shaped silicon monocrystal to test the making of solar cells in space. The Pion apparatus facilitated crystallisation both in and out of solution, exploiting the fact that there were no convection currents in fluids in space. It was set up for the Struktura experiment, which investigated heat exchange and mass transfer during crystallisation in an aqueous solution, and the film of the process was returned to Earth by the visitors on 22 May. As Popov stood beside the charred capsule, the last of the old series, he praised its manufacturer on behalf of all the cosmonauts who had used it during the previous decade.

Alone again, Kovalyonok and Savinykh undertook a comprehensive medical check, then began to prepare to depart, which they did on 26 May 1981 after 75 days in space. A few days later, it was announced that no more cosmonauts would be sent to Salyut 6.

SALYUT 6: A CLEAR SUCCESS

Magnificently successful, Salyut 6 had been occupied for a total of 684 days. It had supp-orted five long-duration crews, four of which had established successively longer en-durance records, and it had supported 11 shorter visits. It had been sustained by propel-lant, consumables and experimental cargo delivered by a dozen Progress ferries, and of nine international flights, only Bulgaria's Soyuz 33 had failed to dock. Remarkably, the entire first round of Intercosmos visits had been hosted by this single station.

Although its rôle as a habitable station was over, Salyut 6 could still serve a useful purpose, and Kaliningrad had a brand new project planned.

Table 3.1. Salyut 6 docking operations

Spacecraft	Docking Date	MT	Port	Undocking Date	MT	Days
Soyuz 25	10 Oct 1977	0709	front	11 Oct 1977	(0800)	1.03
Soyuz 26	11 Dec 1977	0602	rear	16 Jan 1978	1422	36.35
Soyuz 27	11 Jan 1978	1706	front	16 Mar 1978	1100	63.75
Progress 1	22 Jan 1978	1312	rear	7 Feb 1978	0855	15.82
Soyuz 28	3 Mar 1978	2010	rear	10 Mar 1978	1325	6.72
Soyuz 29	17 Jun 1978	0058	front	3 Sep 1978	1123	78.43
Soyuz 30	28 Jun 1978	2008	rear	5 Jul 1978	1315	6.71
Progress 2	9 Jul 1978	1559	rear	2 Aug 1978	0757	23.66
Progress 3	10 Aug 1978	0300	rear	21 Aug 1978	-	11
Soyuz 31	27 Aug 1978	1937	rear	7 Sep 1978	1353	10.76
Soyuz 31	7 Sep 1978	1421	front	2 Nov 1978	1046	55.85
Progress 4	6 Oct 1978	0400	rear	24 Oct 1978	1607	18.50
Soyuz 32	26 Feb 1979	0830	front	13 Jun 1979	1251	107.18
Progress 5	14 Mar 1979	1020	rear	3 Apr 1979	1910	20.37
Progress 6	15 May 1979	0919	rear	8 Jun 1979	1100	24.07
Soyuz 34	8 Jun 1979	2302	rear	14 Jun 1979	1918	5.84
Soyuz 34	14 Jun 1979	(1950)	front	19 Aug 1979	1208	65.68
Progress 7	30 Jun 1979	1418	rear	18 Jul 1979	0650	17.69
Soyuz-T 1	19 Dec 1979	1705	front	24 Mar 1980	0004	94.29
Progress 8	29 Mar 1980	2301	rear	25 Apr 1980	1104	26.50
soyuz 35	10 Apr 1980	1816	front	3 Jun 1980	1447	53.85
Progress 9	29 Apr 1980	1109	rear	20 May 1980	2151	21.45
Soyuz 36	27 May 1980	2256	rear	4 Jun 1980	1938	7.86
Soyuz 36	4 Jun 1980	1808	front	31 Jul 1980	1455	56.86
Soyuz-T2	6 Jun 1980	1858	rear	9 Jun 1980	1224	2.73
Progress 10	1 Jul 1980	0853	rear	18 Jul 1980	0121	16.69
Soyuz 37	24 Jul 1980	2302	rear	1 Aug 1980	1943	7.86
Soyuz 37	1 Aug 1980	(2010)	front	11 Oct 1980	0930	70.56
Soyuz 38	19 Sep 1980	2349	rear	26 Sep 1980	1534	6.66
Progress 11	30 Sep 1980	2003	rear	9 Dec 1980	1323	69.72
Soyuz-T3	28 Nov 1980	1854	front	10 Dec 1980	0910	11.59
Progress 12	26 Jan 1981	1856	rear	19 Mar 1981	2114	52.09
Soyuz-T4	13 Mar 1981	2333	front	26 May 1981	-	74
Soyuz 39	23 Mar 1981	1928	rear	30 Mar 1981	1122	6.66
Soyuz 40	15 May 1981	2150	rear	22 May 1981	1337	6.66
Cosmos 1267	19 Jun 1981	1052	front	<permanently docked>		

Table 3.2. Salyut 6 crewing

| Cosmonaut | Role | Spacecraft | | Duration |
		Arrive	Depart	Days
Yuri Romanenko	CDR	Soyuz 26	Soyuz 27	96.42
Georgi Grechko	FE	Soyuz 26	Soyuz 27	96.42
Vladimir Dzhanibekov	CDR	Soyuz 27	Soyuz 26	5.96
Oleg Makarov	FE	Soyuz 27	Soyuz 26	5.96
Alexei Gubarev	CDR	Soyuz 28	Soyuz 28	7.93
Vladimir Remek	CR	Soyuz 28	Soyuz 28	7.93
Vladimir Kovalyonok	CDR	Soyuz 29	Soyuz 31	139.60
Alexander Ivanchenkov	FE	Soyuz 29	Soyuz 31	139.60
Pyotr Klimuk	CDR	Soyuz 30	Soyuz 30	7.92
Miroslaw Hermaszewski	CR	Soyuz 30	Soyuz 30	7.92
Valeri Bykovsky	CDR	Soyuz 31	Soyuz 29	7.87
Sigmund Jahn	CR	Soyuz 31	Soyuz 29	7.87
Vladimir Lyakhov	CDR	Soyuz 32	Soyuz 34	175.06
Valeri Ryumin	FE	Soyuz 32	Soyuz 34	175.06
Leonid Popov	CDR	Soyuz 35	Soyuz 37	184.84
Valeri Ryumin	FE	Soyuz 35	Soyuz 37	184.84
Valeri Kubasov	CDR	Soyuz 36	Soyuz 35	7.86
Bertalan Farkas	CR	Soyuz 36	Soyuz 35	7.86
Yuri Malyschev	CDR	Soyuz-T 2	Soyuz-T 2	3.93
Vladimir Aksyonov	FE	Soyuz-T 2	Soyuz-T 2	3.93
Viktor Gorbatko	CDR	Soyuz 37	Soyuz 36	7.86
Pham Tuan	CR	Soyuz 37	Soyuz 36	7.86
Yuri Romanenko	CDR	Soyuz 38	Soyuz 38	7.86
Arnaldo Tamayo Mendez	CR	Soyuz 38	Soyuz 38	7.86
Leonid Kizim	CDR	Soyuz-T 3	Soyuz-T 3	12.79
Oleg Makarov	FE	Soyuz-T 3	Soyuz-T 3	12.79
Gennadi Strekalov	FE	Soyuz-T 3	Soyuz-T 3	12.79
Vladimir Kovalyonok	CDR	Soyuz-T 4	Soyuz-T 4	74.77
Viktor Savinykh	FE	Soyuz-T 4	Soyuz-T 4	74.77
Vladimir Dzhanibekov	CDR	Soyuz 39	Soyuz 39	7.86
Jugderdemidiyin Gurragcha	CR	Soyuz 39	Soyuz 39	7.86
Leonid Popov	CDR	Soyuz 40	Soyuz 40	7.86
Dumitru Prunariu	CR	Soyuz 40	Soyuz 40	7.86

Table 3.3. Salyut 6 spacewalks

Date	Hours	Activity
20 Dec 1977	1.5	Romanenko and Grechko opened the front docking port, inspected the drogue and the external rendezvous apparatus to verify that there was no damage, and then deployed the Medusa cassette.
29 Jul 1978	2.1	Kovalyonok and Ivanchenkov exited the airlock hatch, retrieved the Medusa and other cassettes, installed new ones, retrieved a passive meteoroid detector, and deployed a radiation monitor.
15 Aug 1979	1.4	Lyakhov and Ryumin exited the airlock, and Lyakhov remained at the airlock and collected experiment cassettes while Ryumin made his way to the rear to cut loose the fouled KRT-10 radio telescope dish.

ANOTHER ALMAZ?

With the 'military' and the 'civilian' stations being alternated, it seemed reasonable to assume that while the lessons of Salyut 6 were assimilated, the next Almaz would be flown as Salyut 7. Even though the Chelomei Bureau had successfully flown its Almaz as Salyut 3 and Salyut 5, it had been forced to rely on Korolev's Soyuz for crew transport. Without the large TKS ferry that it had designed to complement the reconnaissance platform, there was no way to deliver bulk cargo. Throughout, therefore, Chelomei had continued with the development of its own spacecraft, so that it could finally fly the missions originally proposed for the military.

The first step was to test the re-entry characteristics of the Merkur capsule. A Proton rocket placed Cosmos 881 and Cosmos 882 into low orbit on 15 December 1976, and both returned after a single orbit, although they did not use the usual recovery site in Kazakhstan. Since a secondary objective was to advance the process of crew-rating the rocket, a pair of capsules had been mounted in tandem to exploit the Proton's lifting capability. They verified the Merkur's attitude control system, retrofire engine, heat shield and parachute system. The test was repeated with Cosmos 997 and Cosmos 998, in March 1978. But on 5 January 1979, the Proton malfunctioned; although the escape tower lifted the top capsule clear, thereby demonstrating that it functioned, the capsule tucked in underneath had to be abandoned. A rerun in May 1979, with Cosmos 1100 and Cosmos 1101, certified the Merkur capsule, but in keeping with the secrecy of the military space programme, the nature of these vehicles was not announced.

Encouraged by this success, another Proton launched the full TKS configuration as Cosmos 929 on 17 July 1977. During the next month, it performed a number of minor manoeuvres while holding a low perigee to enable tracking stations to assess its drag characteristics. On 17 August, one component of its sophisticated telemetry stream was observed by Western radio monitors to have ceased; the Merkur capsule had returned to

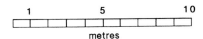

metres

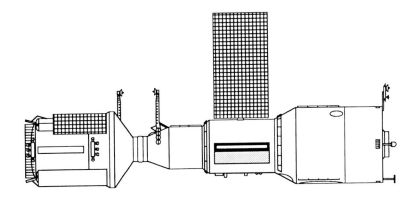

Salyut 6 with Cosmos 1267 (after it had released its Merkur capsule).

Earth. The next day, the main module of the TKS made a major manoeuvre, and boosted its orbit to 310 × 330 km. On 19 December, it raised this to 440 × 448 km, which was far higher than that used by Salyut stations. On 2 February 1978, after six months in space, it was deorbited over the Pacific Ocean. In addition to showing that it had a restartable engine, by imparting a total change of velocity of 300 metres per second it revealed that it had a large propellant tank. Clearly, it was not testing a mere 'escape' stage for geostation-ary or planetary probes, and the Royal Aircraft Establishment's Satellite Tracking Group suggested that it had tested a 'tug' that would deliver large laboratory modules to a future space station. This proved to be a prescient prediction.

With the TKS and its Merkur capsule successfully tested, the Chelomei Bureau had intended to launch another Almaz, but Salyut 6's unexpected longevity had resulted in its being repeatedly postponed, and then cancelled. If it had gone ahead, the crew of three would have ridden a Merkur on a TKS and been launched on a Proton, and a succession of such vehicles would have delivered replacement crews and supplies to sustain extended operations on the station. To permit crew exchanges, the new Almaz would have had a docking port at each end. When the programme was cancelled, all the flightworthy stations were placed in storage (they were later modified, and flown as automated radar-imaging satellites).

Instead, the TKS was launched as Cosmos 1267 on 25 April 1981, and, just as had Cosmos 929, it returned its Merkur capsule after a month. A month later, on 19 June, it docked at Salyut 6's front port. At 15 tonnes, the addition almost doubled the station's mass to 32 tonnes. Over the following 12 months, during which the complex remained unoccupied, Cosmos 1267's engine was fired periodically to sustain a nominal 350-km orbit while its systems were monitored to determine how they degraded. Meanwhile, as the next station was being prepared, Kaliningrad's facilities were upgraded and another control room added.

In late 1981, *Aviation Week & Space Technology* reported that the Department of Defense knew that Cosmos 1267 was surrounded by a ring of "infrared-homing guided interceptors" which transformed Salyut 6 into an "orbital battle station". The cylindrical objects that had been said to be interceptors were actually externally-mounted propellant tanks, but what was interesting was that this report proved that the DoD possessed a camera with sufficient resolution to reveal such detail on a satellite. As speculated, the new vehicle was announced to be the prototype of a module designed for the assembly of a large orbital station from modular components.

4

A step towards permanent occupancy

On Salyut 6 each two-man crew had, on average, consumed 20–30 kg of expendables per day. If a station was to be permanently occupied, therefore, the replenishment process would have to be capable of delivering 1,000 kg of consumables per month, requiring one resupply ferry every two months. The extended life of the Soyuz-T reduced the need for crews to visit to exchange ferries, and this would considerably relieve the hectic schedule that had been maintained during the periods while Salyut 6 had been occupied.

One objective of the next station would be to have one crew hand the station over to its successor, as a step towards permanent occupancy. This *could* have been done on Salyut 6, if a retiring crew had departed in its own ferry and left the new residents with their own. A hand-over in orbit would eliminate the time wasted in powering down and then reactivating the station, and would facilitate a long period of occupation without requiring a single crew to endure the strain of an extended flight; and, if the life support system could sustain three people, it would also facilitate partial crew exchanges and even permit a research cosmonaut to serve with successive flight crews to carry out an extended study of adaptation to weightlessness.

SALYUT 7

Salyut 7 was launched on 19 April 1982. It had been expected that the next station would incorporate a cluster of docking ports, so that it could be augmented by modules like Cosmos 1267. However, it was structurally identical to its illustrious predecessor. Although the front collar had been substantially strengthened to accommodate the extra loads expected to occur when manoeuvring with a large mass docked, Resonance tests would demonstrate that this reinforcement was unnecessary.

The new station had a revised form of the Delta automatic navigation system, considered to be sufficiently reliable to permit the station to be manoeuvred without crew supervision, and the improved Kaskad attitude control system could orientate the station to within 1 degree. The Niva video enabled observations to be recorded at any time, and then downloaded when back in communication. The long-range Mera transponder had been installed to assist in the final phase of a ferry's rendezvous, to help it manoeuvre from the closest approach of its transfer orbit to the point where the Igla could lock on, a few kilometres out. To prevent the outer surface of the large portholes from degrading, they had been fitted with covers. To overcome the degradation in the transducers of the solar panels, attachment points had been fitted along the edges of the deployment frames so that additional panels could be clamped alongside to re-establish the required output later on. The spacesuits were the same, but the new station had umbilical connectors in the forward transfer compartment. This not only enabled the cosmonauts to finish all their preparations before drawing on the life-support backpacks, facilitating a full five hours of external operations, but also guaranteed supplies in the event of difficulties in repressurising the airlock (as a faulty sensor had once suggested had been the case on Salyut 6).

It was hoped to use Salyut 7 for four years. With crews flying ever-longer missions, it was clear that there would be fewer flight opportunities. In fact, Vladimir Shatalov said that only 16 cosmonauts were training for Salyut 7. Two-person resident crews were to be visited by teams of three, so the station's environmental systems had been upgraded to be capable of supporting five people for short periods. The Rodnik water system had been modified so that a Progress tanker could pump its water directly into tanks within the unpressurised engine bay, which was more practical than having the crew unload dozens of small spherical flasks. Although the main compartment appeared to be walled off, there were spaces behind many of the panels for storage, to try to avoid the problem encountered late in Salyut 6's life where much of the apparatus ferried up over the years had to be strapped to the walls; this was to prove to be wishful thinking.

The long experience of living on Salyut 6 had prompted substantial improvement in facilities. Previously, food had been prepacked irrespective of whether that combination would appeal to a cosmonaut. Food items on Salyut 7 were packed separately, and the cosmonauts could 'pick 'n' mix' to suit their individual taste. A small refrigerator had been installed to store the fresh food which was to be delivered regularly by Progress ferries. The stove could heat tinned food for both cosmonauts at once. About 65 per cent of each meal was reconstituted freeze-dried food. Each cosmonaut was to have four meals daily; that would provide 3,150 calories. Electric toothbrushes and a new type of toothpaste were provided. The electric shaver incorporated a stubble collector, but this proved inadequate. The walls had been covered by brightly coloured washable panels, and the lighting had been strengthened and made more uniform. As before, there was both a treadmill and a veloergometer (this time mounted on the ceiling) for exercise. The multifunction Aelita biomedical test kit had been specifically designed for use in space, and it superseded the Polynom-2M. Penguin suits were still to be worn, but because the elastic strapping had been removed they were now simply comfortable work clothes. The cosmonauts were told to perform additional arm exercises to compensate. An improved Bania shower had been installed but, like its predecessors, it proved to be frustrating to use.

Most of the scientific apparatus was to be delivered by a succession of ferries, but the bulkiest items had been loaded prior to launch. This included an improved form of the MKF-6M multispectral camera, a KATE-140 mapping camera, the SKR-2M X-ray spectrometer, the RT-4M mirror-based soft X-ray astronomical telescope (an improved form of the telescope tested on Salyut 4) contained within the bulky conical mounting in the main compartment, the Spektr-15M and MKS-M spectrometers to study the ocean and atmosphere, the Yelena-F gamma-ray spectrometer for near-Earth radiation studies, the KGA-2 holographic camera and an improved Oasis plant cultivator.

Anatoli Berezovoi and Valentin Lebedev had begun training to commission the new station in early 1981. Soyuz-T 5 was launched on 13 May. The new Mera transponder was activated while still 250 km away on the final transfer orbit, but it did not detect the station until the separation was down to 30 km. The Igla locked on soon thereafter, and the Argon docked the ferry at the front port. There followed a moment of concern when the hatch would not open, and Lebedev had to stand on the roof of the orbital module with his feet around the hatch's rim and prise it open. After everything had been checked out they put their sleeping bags on the ceiling of the main compartment and retired for the first night of what they hoped would be a record-breaking mission.

As usual, for the first week, as they adapted to weightlessness, they undertook only light duties. They gave the Delta and Kaskad systems a thorough test, and then set them on automatic. To execute a given manoeuvre, they had only to enter the command in the computer. They redistributed the apparatus that had been set on frames for launch, then removed all the temporary brackets and struts. Wheat and peas were placed in the Oasis cultivator, flax in the Biogravistat and arabidopsis in the Phyton. Then on 17 May, after testing its battery and transmitters, the Iskra amateur radio satellite was ejected from the scientific airlock to mark the start of the Congress of the Young Communists' League. This was the first time that a subsatellite had been released by an orbital station.

The research programme had been organised so that the station could be maintained in a given orientation for several days at at time, to minimise propellant consumption. The most economic attitude exploited the gravity gradient to provide stability. Because much of the observational apparatus was portable, it could be set up in the forward transfer compartment, which had a ring of portholes providing panoramic coverage. The station was orientated either up or down, depending on whether the sky or the Earth was to be studied. Materials-processing experiments could be performed during these periods. Of course, the MKF-6M and KATE-140 cameras could be used only when the belly of the station was maintained facing the Earth, the manoeuvres for which precluded materials processing. Other experiments required instruments to be kept facing the horizon, and depending on the instruments this could be done whilst in the gravity-gradient mode or by placing the station in a complex rotation. All these flight regimes had to be carefully scheduled. Spending several days on one type of work was also more efficient because it eliminated the time wasted setting up and repacking apparatus. Lebedev actually spent so much time looking out of the portholes in the forward transfer compartment that he soon developed an impressive tan.

Progress 13 arrived on 25 May. As with Salyut 6, only the rear port incorporated the plumbing necessary to replenish the station's fluids. Although the cosmonauts were to have isolated the rear transfer compartment for the docking, they rigged the sensor so that the flight controllers would believe that the internal hatch was closed, and then they went to watch the approaching ferry through the tiny porthole beside the docking collar. Amongst its scientific cargo was the Magma-F furnace and 250 kg of apparatus for the forthcoming visit by a French cosmonaut. Progress 13 dropped the station to a 300-km circular orbit to accommodate the ferry which was to deliver the first visiting crew, then left on 4 June. As it withdrew, the Astra apparatus took data to observe how its engines contributed to the gaseous environment in the immediate vicinity of the station.

While assembling the Magma-F furnace it was discovered that there were errors in the instruction manual. An improved form of the Kristall furnace used on Salyut 6, this incorporated a microaccelerometer and a magnetometer to record any effects influencing the melting and crystallisation processes. It could operate at up to 900°C, and report its status to Earth as part of the telemetry stream. After experimenting, it was realised that it produced better results if it was unbolted from its mount and left to float freely, because it did not pick up the vibrations transmitted through the hull of the station.

THE FIRST GUEST: A FRENCHMAN

Soyuz-T 6 was launched on 24 June. It carried Vladimir Dzhanibekov, Alexander Ivanchenkov and Jean-Loup Chrétien, the first French cosmonaut. Dzhanibekov had been assigned in February after Yuri Malyschev had been taken ill. This, the first of the new series of international flights, was the first involving a citizen of a state not in the Intercosmos organisation. The 1979 agreement raised the profile of a close cooperation which has seen French apparatus flown on Soviet satellites since the mid-1960s. Initial enthusiasm had turned to calls for the flight to be cancelled in response to the invasion of Afghanistan in December 1979, but preparations had quietly gone ahead.

When 900 metres out, the Argon computer turned the spacecraft to fire the engine to cut its closing rate. As it turned back to face the station, prior to using the less powerful thrusters for the final phase of the approach, the strapdown inertial reference froze. The Argon aborted the manoeuvre and left the ferry in an end-over-end spin. Dzhanibekov immediately stabilised the spacecraft. This was where Soyuz 33 had been when it had encountered trouble approaching Salyut 6. But this time there was nothing wrong with the engine, so Dzhanibekov did what Rukavishnikov had longed to do: he took control, completed the approach, and docked.

Dzhanibekov and Ivanchenkov tested the Braslet. This involved wearing thin elastic rings around the thighs for up to an hour several times a day during the first few days to restrict the migration of blood from the legs into the upper torso. It complemented the Tchibis. Chrétien had a broad programme which included biomedical, biological, materials processing, and Earth and astrophysical observations. The biomedical tests focused on cardiovascular and sensory functions.

Echograph used an ultrasound scanner to provide a sectional image of the heart and a doppler sensor to measure blood flow to observe the cavity dimensions, mycocardiac thickness, arterial blood capacity and the supply of blood to the vessels. Following the chest pains suffered by Grechko upon returning to Earth, it had been realised that in the absence of gravity the heart migrated up within the chest cavity. Chrétien initially had some difficulty locating his heart because it was not where trials had led him to expect it to be. The Posture experiment studied how the body maintained specific postures in the absence of gravity. A variety of sensory organs and muscle groups (different from those used when moving) are employed to maintain posture. The experiment required Chrétien to don motion sensors and muscle monitors and place one foot in a restraint to enable spatial coordinates to be measured. With both eyes closed, he executed a series of movements to exercise sensory and locomotive functions. The apparatus took longer to install than expected, which prompted Chrétien to make recommendations for future apparatus. Once it was set up, however, the tests were performed by all of the visitors.

The Cytos-2 experiment used an incubator to study the structure and functioning of bacteria cells in weightlessness and the effect of a variety of antibiotics against bacteria. The Biobloc-3 studied the effect of cosmic rays on biological materials by sandwiching seeds between heavy-ion detectors. Materials-processing experiments were performed in the newly assembled Magma-F furnace. One measured small accelerations to develop a mathematical model of the furnace, and another studied the dissolution of a polycrystalline solid alloy in its own liquid while in a state of thermodynamic equilibrium, to develop a mathematical model of the solidification process and investigate capillary forces; a third created an alloy of aluminium and indium (which are immiscible in Earth's gravity).

The PCN low-light camera was used to photograph faint sky phenomena, including the zodiacal light, noctilucent clouds and lightning storms. Ivanchenkov who, as a result of his previous mission had become particularly adept at seeing such phenomena, assisted Chrétien. The Piramig camera could be used only in orbital darkness (currently fully 25 minutes per orbit). It was set up in the forward transfer compartment using the porthole normally used by the MKF-6M, with the hatches shut, with the lights off, and with the attitude-control thrusters off to avoid spewing glowing reactant across the field of view. Its photomultipliers could make faint ultraviolet, visible and near-infrared observations. Although the exposures lasted just a few minutes, the station was orientated for this work for up to six hours a day. A total of 350 pictures were taken of the upper atmosphere and interplanetary dust. Those of noctilucent clouds were so fascinating that the residents were later asked to use the PCN twice a week to photograph the horizon, in the hope of capturing other transient stratospheric phenomena. Chrétien's pictures were shown on the video uplink to indicate what to watch out for.

Although Chrétien started off by sleeping for the assigned period, his schedule was so heavy that within a few days he was working late and was able to snatch only a few hours rest, so towards the end he was exhausted. The visiting crew returned to Earth on 2 July. Although a long tour was planned, Soyuz-T 5 had not been exchanged; it was to be taken away by the next crew. Since the next visit was not due for several months, Salyut 7 raised its apogee to 344 km. Berezovoi and Lebedev were so worn out playing host that they were told to take several days off to recover. They spent hours searching for Dzhanibekov's watch and a roll of film from the PCN which had inadvertently been left behind. Although sad to be alone once more, they soon settled back into the familiar routine. They became slightly grumpy, however, and urged the flight controllers not to feel obliged to chat simply because they were within radio contact (this interrupted their work, which was usually at its most intensive whilst over the Soviet Union). Later they rebuked the controllers for demanding that they terminate a run with the MKF-6M even though they were nearing the target, just so that they could start an exercise period on schedule.

Progress 14 docked on 12 July. It delivered miniature onions for the cultivator, but they were so tempting that they were eaten instead. Amongst the equipment it delivered was the Korund furnace and the EFO-1 electrophotometer. The EFO-1 watched stars as they sank through the atmosphere towards the Earth's limb. By tracing the attenuation of their light it was possible to measure the optical density of the atmosphere. From the observations accumulated globally they would map the vertical structure, density and composition of the thin layers of the upper atmosphere for the Climate experiment.

SPACEWALK: EXPOSURE CASSETTES

On 22 July the cosmonauts cleared out the loose items that they had accumulated in the forward transfer compartment, retrieved the spacesuits and started preparations for a spacewalk. Because Berezovoi and Lebedev were the first crew of the new station, the excursion suits had never been tested, so they gave them a complete inspection, filled the oxygen bottle, installed the lithium hydroxide canister, and filled the water tank that fed the network of fine tubes woven into the fabric of the undergarment of the suit to keep the occupant cool. The preparations consumed most of the week. The experiments to be attached to the outside of the station (exposure cassettes delivered by Progress 14) had to be unpacked. Medusa contained biopolymers, but the others exposed samples of materials and structures to be used on future spacecraft. The 20-metre umbilicals which would provide electrical and radio links were unpacked and tested. They installed all the brackets in the airlock that would be used to hold equipment while the hatch was open, to prevent it drifting away. They recharged the battery of the video camera to be used to record the state of the transducers in the solar panels. They were keen to extend their scheduled external activities, to allow them to crawl along the entire length of the station to inspect the thermal insulation, but Kaliningrad insisted that they stay near the hatch, where the sample containers were to be attached. Finally, they donned the suits and pressurised them slightly above the ambient level. There was some concern when Lebedev discovered that his suit leaked, and close inspection revealed that a strap had caught in the rear hatch when the hinged back-pack had been closed, preventing it from forming a proper seal. The day before the big event, just in case it proved impossible to pressurise the airlock and they had to abandon the station, their experimental output was stored in Soyuz-T 5's descent module, the internal hatch was sealed, and the flasks of air which would be required to repressurise its orbital module in the event of such an emergency were checked. The preparations complete, Salyut 7 was reset for autonomous operation regime.

The spacewalk itself, on 30 July, took only 2.5 hours. Lebedev positioned himself on the foot restraint immediately outside the hatch and retrieved several packages placed prior to launch. These included Etalon, a plate that had exposed optical coatings (some of which had bubbled and peeled off), a Medusa package and a Spiral package. He then attached the new packages. He was keen to move up to another anchor point at the base of a solar panel to test the winch that a later crew would use to extend additional solar panels, but the station was approaching the Earth's shadow so he stayed where he was. Berezovoi remained in the airlock, with only his head and shoulders out of the hatch, and handed Lebedev the television camera to film the state of the solar panels. Back inside, the hatch was closed, and the airlock repressurised without incident. Lebedev was very surprised to find a 20-mm tear in the protective cover of his helmet, but he had no idea how this had happened. The air lost when depressurising the airlock was replenished by air pumped into the station's tank from the attached cargo ferry.

BACK TO WORK

After a day's rest, the cosmonauts set up the new furnace delivered by Progress 14. The Korund incorporated a revolving sample holder which could be loaded with up to a dozen ampoules. Although each was processed in sequence, the furnace could be set up to operate either automatically, or by remote control so that the retiring crew could set it for use following their departure, when it would not be disturbed by their presence. In contrast to the Splav (in which the sample was fixed and the thermal field was varied to control crystallisation) and Kristall (in which the thermal field was held constant and the sample was moved), this new furnace was capable of both moving the sample at speeds varying from a few millimetres a day to a few centimetres a minute, and of changing the temperature between 20°C and 1,270°C at rates of 0.1°C to 10°C per minute. It was hoped that this second-generation furnace technology would facilitate industrial-scale manufacturing of extremely pure semiconductors on an automated free-flying spacecraft that would be periodically visited by a crew only to reload the furnace. One of the rôles long-ago identified for the Salyut stations was refining apparatus intended to be used on automated vehicles. Unfortunately, it overheated and switched itself off, so it could be used for only a few hours at a time. It took three months for them to realise that the thermal coefficients of its heating element had been computed from trials conducted on Earth in which thermal convection had drawn off energy. In microgravity, where convection did not occur, the element had heated up so rapidly that the controller concluded that it had *over*heated. Once reconfigured with appropriate coefficients, it worked perfectly.

In early August, the arabidopsis in the Phyton produced pods; it was the first time a plant grown from seed in space had done so. A week later, the pods burst to release the seeds. The peas in the Oasis were developing well, but even more mouthwatering were the tomatoes in the Svetobloc hothouse. All of this was excellent news for the botanists.

A WOMAN VISITOR

Progress 14 departed on 11 August, and on 20 August, with the station again down to a 289 × 300 km orbit, Soyuz-T 7 took its place. It delivered Leonid Popov, Alexander Serebrov and Svetlana Savitskaya. The second woman to fly in space, Savitskaya was presented with a flowering arabidopsis to mark the visit of the first woman to the station. She was also given a floral apron and invited to play hostess for the evening meal. The males reserved the orbital module of a ferry for Savitskaya to enable her to have her own toilet, but she chose to sleep in the main compartment with them.

The main item of the visiting crew's agenda was the testing of the Tavriya electrophoresis apparatus. This was essentially a column of biological compounds which separated into homogeneous fractions when an electric current was applied. To test it, human blood protein (albumin and haemoglobin) was separated. Savitskaya filmed the process, then extracted a sample of each separated product. Later, it was used to purify urokinase (an enzyme present in human urine) and, for the first time, interferon was processed by the electrophoresis technique. The results were very encouraging and demonstrated a full order of magnitude improvement in purity over similar apparatus on Earth. The electric current acted regardless of molecular weight, so the apparatus could be used to separate a wide range of biological substances. It was hoped that improved equipment on later missions would yield extremely pure vaccines, and other pharmaceutical products, on a commercial basis.

Popov, Serebrov and Savitskaya returned to Earth in Soyuz-T 5 on 27 August. This left Soyuz-T 7 at the rear port. It had to be moved while the recovery window remained open, so it was done two days later. A fly-around had not been done with the Soyuz-T type before, and there was some concern that the narrow beam of the Igla system might not be able to lock onto a ferry only 200 metres out, but the redocking was completed without incident. As a precaution, though, the station had been returned to its automatic regime. It turned out to be much easier than in the earlier model of the spacecraft, which had not had the benefit of the Argon computer. Although they had been ordered not to re-enter the station until they re-established communications, as soon as the pressure was equalised Berezovoi and Lebedev sneaked in to get something to eat, and then retreated to their ferry to await the appointed moment to open the hatch. With the scheduled visits over, the orbit was raised.

THE LONG HAUL

There was some flexibility in the length of the mission, and the cosmonauts had been suggesting for some time that they would like to break the 185-day record. By 10 September they were once again growing irritable with the flight controllers over the merits of adhering to the daily schedule. The best days, ironically, were the rest days, because the controllers left them alone, so they spent their time off performing tricky observational experiments in uninterrupted peace. On one occasion they made a video of the ground track followed by Yuri Gagarin.

At last, on 14 September, Viktor Blagov formally invited them to extend the flight by two months. They agreed, and asked to make another spacewalk to test the apparatus that was to be used by a later crew to erect solar panels, although no commitment was made to this new request. The launch of Progress 15 had been postponed for two weeks; its arrival on 20 September was eagerly awaited, because both men were thoroughly bored with reconstituted food. As usual, the routine maintenance of the thermal regulation and environmental systems followed the unloading of replacement parts.

In late September, there was a moment of concern when the cosmonauts noticed a smell of burning insulation. They closed all the hatches to isolate the compartments and switched off the fans. With fire extinguishers at the ready, they opened a suspect panel and found a fan coated with a thick layer of dust. They decided that the dust must have clogged the mechanism and caused it to overheat, so they cleaned it. Because the entire incident had taken place when out of radio contact, they decided not to tell Kaliningrad.

By this point, a sense of wonderment had set in. One piece of apparatus had arrived that they had never seen before. When they asked what it was, they were told to eject it from the scientific airlock and track it using the sextant, so they did; they reported that it had drifted away in exactly the same way as did the rubbish which they regularly tossed out! Resident crews on very long flights would later grow used to unpacking apparatus which they had never trained with, and even to having people that they had never met appear in the hatch of a newly arrived ferry; but it was disconcerting. Another source of frustration was apparatus that did not work properly. Thus far, they had fixed the leaky gasket in the Oasis, fixed the Korund, and replaced burnt-out components in the Aelita, the EFO-1, the vacuum cleaner and the RTS transmitter in the telemetry system. But all of this paled into insignificance compared with the fault in the Delta navigation system. In late August an astrophysical run with the RT-4 telescope had been interrupted by the Delta dropping off-line. After analysis, it was decided to re-enter its programme, so in mid-September Lebedev had manually keyed a sequence of 325 six-digit instructions. In early October he had to do it again; unfortunately, he failed to key 'enter' at the end, and the entire sequence was lost. He redid it, but the checksum did not match! By the time the sequence was accepted, it was far into the night. Back in June, a test had been made to determine whether the sextant and star tracker could be used to orientate the station with an accuracy comparable with the Delta. It was possible, but it required so much attention that little other work could be done. As confidence in the Delta diminished, the station was kept orientated in the gravity gradient, even though this made some work difficult.

In October, the water reclamation system failed. When the panel was removed the cosmonauts discovered a massive blob of water floating inside. Lebedev disassembled the pump and found that the bearing had seized. By this point, they were intimately familiar with the noises onboard the station. A month earlier, he had warned that this pump had been running rough, but Kaliningrad had told him not to worry. Another irksome problem was that the STR had recently begun to let the temperature fall to its minimum level before switching on the heater. This was only its first crew and already Salyut 7 was proving to be rather less reliable than its illustrious predecessor.

At the end of October, Berezovoi and Lebedev passed the point at which they were to have returned to Earth. Ironically, having canvassed for an extension, they started to count the days to their departure. Since they had decided to stay on, it had been decided to send up the next resupply ferry to stock up the station for the next crew. Progress 15 had left on 14 October, and Progress 16 arrived on 2 November. Problems with the Delta continued, and it was put in a mode that constantly recomputed the station's position in space in the hope of catching it 'fall over', but this failed to expose the fault.

On 11 November, Lebedev awoke to find Berezovoi writhing in pain. After an hour of no improvement, he issued him with biseptol from the medikit. Once in contact with Kaliningrad, they described the symptoms; an injection of atropine was prescribed and they were told to make preparations to return to Earth the following day. Berezovoi's condition improved rapidly, however. If they had been forced to return, it would have been very bad timing, because it would have left them just two days short of the 185-day record. Time began to pass quickly. They acknowledged that it really was time to leave when they found that they were fast running out of writing paper on which to take note of observations! The Delta failed completely in early December, and it was decided to bring the cosmonauts home. They undocked on 10 December 1982, having set a 211-day record, but without making their second spacewalk. The capsule descended in darkness, in a snowstorm, was dragged along the ground by its parachute, and then rolled down a slope, which was rather disconcerting after so long weightless. The recovery window would not open for another fortnight, so they had barely begun their drug and exercise programme and were in pretty bad shape. The weather was so bad that the first helicopter broke its gear upon landing. It was deemed too dangerous to land another helicopter, so the intrepid spacefarers spent the rest of the night in the back of the disabled chopper.

Amongst the many scientific results was a monocrystal of cadmium selenide with a mass of 800 grammes, a hoard of almost 200 arabidopsis seeds produced in space, and a total of 2,500 MKF-6M pictures. In recognition of his Earth-resources work, Lebedev was declared a "distinguished prospector" by the Minister of Geology.

Progress 16 departed on 13 December, leaving Salyut 7 in the gravity gradient. The newly-repaired Korund had been left fully stocked so that smelting could continue by remote control. With the Delta navigational control system overwriting its memory, the engineers concluded that its electronics had to be replaced. The crippled station could be manoeuvred only while in radio range, when Kaliningrad could control it. Fortunately, it was orbiting somewhat above its usual 350-km altitude, so there was no danger of its orbit decaying before the next crew could be dispatched; or so it was thought.

TRIALS AND TRIBULATIONS

On 2 March 1983 a Proton placed Cosmos 1443 into orbit, and within hours it began to manoeuvre towards Salyut 7. It was reported to be an 'operational' version of Cosmos 1267, which had docked with Salyut 6 after that station's last crew had departed. It was a multi-function vehicle. Firstly, given the difficulty controlling Salyut 7, by acting as a tug it could manoeuvre and stabilise the station. As a freighter, it carried some 3,600 kg of cargo. As a power module, its pair of 20 m^2 solar panels added 3 kW to the station's supply. Finally, the attached Merkur capsule provided a 500-kg cargo-return capability. Chelomei's TKS, although now unlikely to carry a crew, had nevertheless been able to make a substantial contribution to his rival's programme. Perversely, despite having been relegated to cargo-return, the capsule had couches installed.

After an eight-day chase, Cosmos 1443 docked at the front port of the vacant Salyut 7. It would satisfy all the attitude-control and orbital manoeuvring requirements until a crew could be sent to repair the Delta. The TKS had large propellant tanks, but it could not be replenished, so it was crucial that the station's control system be repaired as soon as possible. The next favourable launch window was in mid-April. On 5 April, in preparation, Cosmos 1443 lowered the orbit to 300 km. This was a sign that the next crew would comprise three cosmonauts. An attempt to launch Soyuz-T 8 on 11 April was scrubbed due to a technical problem, as was the second attempt three days later, and it finally left the ground on 20 April, carrying Vladimir Titov, Gennadi Strekalov and Alexander Sere-brov.

Irina Pronina (Savitskaya's backup for Soyuz-T 7) had been assigned to this flight, but had been replaced by Serebrov late in the training. Just because they had left the pad did not mean that their problems were over, however. The aerodynamic shroud ripped the rendezvous antenna boom off the side of the orbital module as it was jettisoned, but since this was not instrumented there was no signal to indicate this fact; it was not detected until the crew ran through the equipment tests, whereupon it transpired that the rendezvous radar did not respond. Believing that the boom had simply failed to release, and was stuck against the orbital module, Titov made several futile attempts to free it by a series of thruster firings.

Although the rule book required the flight to be aborted as soon as possible, it was decided to permit the crew to continue. The tracking stations provided the data required to compute the transfer orbits and, the next day, they were just 10 km from Salyut 7. Without the Igla transponder, the Argon could not attempt the rendezvous, so it would have to be done by combining closure rate data from further ground tracking with what Titov could see on his Vzor periscope. Fortunately, he was approaching from the rear. Once he was close enough to resolve the shape of the complex, its apparent size on the graticule on the periscope gave him a feeling for range. All went well until he was only 300 metres out, when the Earth's shadow suddenly swallowed the station. He switched on his floodlight to show the rear docking port, and closed in to 160 metres (the point at which he had trained to take command after a computer failed), but then he sensed he was closing too fast; he aborted the approach and pulled back. By the time he emerged from the shadow he was out of radio range, and by the time contact could be re-established (on the next pass) the ferry had drifted 4 km from the station, so the mission was called off. This heroic effort was to pay a dividend later. The disappointed cosmonauts landed the following day. Cosmos 1443, having lowered Salyut 7's orbit, now boosted it back to its usual altitude.

THE SECOND EXPEDITION

In late June, Cosmos 1443 adjusted Salyut 7's orbit to 326 × 337 km in preparation for its next crew. Since this was near its nominal operating altitude, it was an indication that a two-cosmonaut crew was to be sent, and true enough, when Soyuz-T 9 was launched on 27 June it carried only Vladimir Lyakhov and Alexander Alexandrov.

Flight director Valeri Ryumin told reporters that this new team would not attempt to break the endurance record, and noted that they would be extremely busy because they had to try to follow a programme originally assigned to a crew of three. This was because Lyakhov and Alexandrov had been assigned with Viktor Savinykh as the backups for Soyuz-T 8; Savinykh had been dropped so as to carry the additional propellant required to reach the station in its high orbit.

The first task was to retrieve the new computer from a cargo rack in Cosmos 1443. This was the first time that cosmonauts had ventured into a Chelomei module in space. Apart from the Delta computer, it contained 600 other items including a pair of clip-on solar panels. To help unloading, a telescopic pole could be extended through the hatch into the station, and a sled run back and forth to transfer apparatus through the hatch. Lyakhov's immediate reaction upon seeing the racks of cargo was that it looked like a veritable warehouse. Alexandrov was a specialist in control systems, and during their first week he replaced the electronics in the Delta and reloaded and tested it, thereby restoring Salyut 7 to fully operational status.

To try to make up for their missing crewmate, Lyakhov and Alexandrov operated a 12-hour day, and a six-day week. At the end of the month, although they were working very efficiently, the doctors warned them not to burn themselves out. The heavy work of unloading Cosmos 1443 gave them "uncommonly good appetites" and they began to put on weight. The unloading of cargo continued throughout July, and was interleaved with the setting-up of experiments. The Yelena-F monitored gamma-rays, and Ryabina studied charged particles in near-Earth space. Two different instruments were available for the MKS-M spectrometer, to observe either the atmosphere or the ocean. The first phase of the scientific programme was concluded in early August. They extensively surveyed crops in the Central Asian Republics, and studied the Caspian and the Volga regions, the Sea of Azov, the Crimea, the Urals and the Caucasus. They made extensive oceanic studies, in particular the Mediterranean Sea, and also participated in an experiment in the Black Sea during which a 500-metre diameter pool of dye was released. This was part of the prepara-

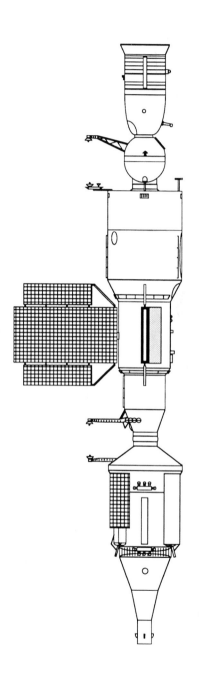

Salyut 7 with Soyuz-T 9 and Cosmos 1443.

metres

10

5

1

tions for a 20-day study of the Black Sea region, later in the month, undertaken by the Intercosmos organisation. Their observations were correlated with data from an aircraft, a research ship and from the 10-channel Fragment multispectral camera on the Meteor-Priroda satellite. In addition to observations for the Intercosmos organisation, they also imaged Brazil for UNESCO. They were so productive that in just two months they took more photographs than their predecessors had during seven months. During one 14-day period they took 3,000 pictures. They used the PCN low-light camera to record lightning bolts and aurorae. The Astra mass spectrometer had revealed that there was a tenuous stream of matter trailing the station; this was composed of air vented by the small scientific airlocks and the efflux from the attitude control thrusters.

All their results were loaded into Cosmos 1443's Merkur. If not for this capsule's cargo capacity, the extensive observational programme would have overloaded the limited capability of the Soyuz descent module. Despite their hard work, their results exploited only half of the capsule's capacity (it had been said to be capable of returning 500 kg of cargo, but they could find only 318 kg of materials to send back), so they loaded spent filters, fans and scrubbers from the environmental system so that their service life could be reassessed, together with the components of the faulty Delta. If the Soyuz-T 8 crew had managed to get aboard and complete its assigned programme, then the capsule would undoubtedly have been filled to capacity.

Cosmos 1443 left on 14 August. It could not be kept on the front port any longer because, to permit a tanker to dock to replenish the station's propellants, Soyuz-T 9 had to be transferred to that port. The departure of the entire module demonstrated that the cargo transporter version of the TKS was not meant to become a permanent part of an orbital complex. It was reasonable to jettison a cargo carrier, but it would be impractical to discard a scientific module after just a few months, so a specialised laboratory variant would need to incorporate a second docking port to facilitate continued operations. The Merkur returned to Earth on 23 August and, many years later, was sold at auction in the United States.

While Cosmos 1443 had been attached, tests had been performed to determine how well the module's thrusters could control the complex. Over the years, it had repeatedly been stated that the ultimate objective of the space station effort was to construct a large orbital complex from modular components. The simplest configuration was with all the modules in line. Following engineering trials by Cosmos 1267 docked with Salyut 6, studies had suggested that a revised control system would be able to manoeuvre a larger complex using only 20 per cent of the propellant previously consumed. Lyakhov, however, found it difficult to control the station manually, and he had ended up expending more propellant than expected. The development of a fully automatic system which could be readily adjusted to take into account the changing dynamics as a modular complex was reconfigured was a major long-term priority.

A SLIGHT LEAK

On 9 September, while Progress 17 was pumping oxidiser aboard, a pipe fractured in the unpressurised engine bay of the station. The transfer was being controlled from the ground, so, as soon as the telemetry revealed the leak, the cosmonauts were ordered to retreat to Soyuz-T 9 at the other end of the station, just in case it proved necessary to abandon the station. The telemetry convinced the engineers that this was not a problem with the pump, as the tank in the station was slowly venting. This effectively disabled the main engines and one of the station's two sets of attitude control thrusters. Even though the nitrogen tetroxide was extremely corrosive, there was no danger of an explosion, so a few hours later the crew was permitted back aboard. Although it did not totally cripple the station, it did preclude orbital manoeuvres and made attitude control difficult, which once again constrained the experiment programme. There was no official announcement of the problem. Only after the crew eventually returned to Earth did Roald Sagdeev, director of IKI, admit that "technological problems" had developed, and in his post-flight press conference Lyakhov acknowledged that there had been "a slight leak" from one of the propellant tanks. Progress 17 raised the station's orbit to 337 × 358 km, and then left on 17 September.

An attempt to launch Titov and Strekalov on 26 September almost ended in disaster. With just seconds to go, a valve failed to close and a fire broke out at the bottom of the rocket. It was fully 20 seconds before the escape tower was triggered, by which time the rocket had been engulfed by flame. As the descent module descended by parachute 4 km away, the rocket finally exploded; its wreckage continued to burn for most of the following day.

This was the first time that a crew had been forced to abandon a rocket so close to launch, and it frustrated this crew's second attempt to reach Salyut 7. They had intended to relieve Lyakhov and Alexandrov and thereby achieve the first orbital handover. While both crews were aboard, Titov and Strekalov were to have conducted a spacewalk to mount solar panels. Lyakhov and Alexandrov would have returned in early October and Titov and Strekalov would have served until either early December or early January, at which point they too would have been replaced. If this had been achieved, the programme would have taken a vital step towards permanent occupancy.

Following the launch abort, Lyakhov and Alexandrov agreed to extend their tour in order to install the solar panels. It was significant that Soyuz-T 9 was not recalled, even though it had already been in space for three months; this indicated that it was believed that it could safely stay in space for much longer than the 110 days or so (already a record) demonstrated by Soyuz-T 5 and Soyuz-T 7.

Resigned to their extended mission, Lyakhov and Alexandrov resumed work. The difficulty in orientating the station imposed "a substantial change" on their research. They left the station in the gravity gradient and concentrated on making semiconductors and alloys in the Kristall and the Magma-F, crystalline melts in the Pion-M, and biological electrophoresis in the Gel. On 22 October, Progress 18 arrived, and delivered a revised Tavriya electrophoresis apparatus and the second pair of clip-on solar panels. Taking advantage of the ferry's propulsion system, the cosmonauts took a series of MKF-6M pictures of the southern regions of the Soviet Union for a geological survey. (Dense vegetation in the summer tended to mask subtle geological structures, but pictures taken later in the year could expose this detail.)

SPACEWALK: SOLAR PANELS

On 1 November, Lyakhov and Alexandrov ventured out to mount the first pair of clip-on solar panels. The size of the main solar panels was restricted by the shape of the stepped-cylinder compartment and by the aerodynamic shroud's enclosure. The panels of a docked ferry augmented the station's power supply, but power was still limited, so the clip-on units had been developed to overcome the degradation that had been noted in the case of Salyut 6's panels towards the end of its life, and attachment points had been bolted to the frames of the main panels to accommodate a clip-on strip down each edge. As Lyakhov and Alexandrov had not trained specifically to do this, they found the fifty steps in the attachment sequence daunting. Romanenko was in the control room to offer advice, and Titov and Strekalov were duplicating the operation in the hydrotank to help resolve any problems which occurred. To extend the radio link, three communications ships were used (the Cosmonaut Yuri Gagarin was in the Mediterranean, and Cosmonauts Vladislav Volkov and Georgi Dobrovolsky were in the Atlantic).

Alexandrov took up position on the foot restraint at the base of the dorsal panel, and Lyakhov handed him the folded panel and the toolkit. The panel was fitted to the lower attachment point and the handle inserted into its winch. As he unfolded the concertina-like panel alongside the main panel, Alexandrov reported that the winch was difficult to use (this was the winch that Berezovoi and Lebedev had been eager to test). Telemetry indicated that he became most stressed during this phase of the operation. Once it was fully extended, the side strip automatically clipped into a fixture near the far end of the main panel. Finally, Lyakhov plugged in its output cables. The Sun-tracking motor had been deactivated so that the main panel could be locked in the best position to enable a cosmonaut on the anchor to access it. The other strip could not be fitted until the panel was rotated. They returned two days later. This was the first time that cosmonauts had made two spacewalks. Once the second strip was in place, Alexandrov took almost 10 minutes to tie a knot in the electrical cables to hold them in place. The new panels added 1.2 kW to the power available, which meant that full advantage could now be taken of the various materials-processing facilities aboard.

With the primary task of the flight extension over, the cosmonauts settled down to await the recovery window later in the month. Progress 18 undocked on 13 November, and Lyakhov and Alexandrov left on 23 November. Although their 149-day flight had not set a human endurance record, it had considerably extended the demonstrated orbital life of the Soyuz-T ferry.

THE THIRD CREW

Salyut 7's orbit was high enough to enable it to survive until the next crew could be sent to repair its leaking propellant system. First, however, the engineers had to develop the procedure and build the tools to enable them to undertake such an ambitious task. Meanwhile, the station was repeatedly said to be operating normally; no mention was made of the problem with the engine.

By early 1984 (with Salyut 7's orbit having decayed below 300 km), it was decided to send up a full crew, so, on 9 February, Soyuz-T 10 docked at the front port. Leonid Kizim and Vladimir Solovyov had already trained extensively to conduct spacewalks, so were well-suited to the complex repair task. If this was completed successfully, they were to try to set a new endurance record. Dr Oleg Atkov, a physician, had been added to the team to monitor his colleagues' health and to perform his own investigation of the process of adaptation to weightlessness. He had the authority to end the mission. It was announced that a medical specialist would form part of every new record-breaking crew (although in the event this proved impracticable).

Kizim's first task was to evaluate the station's remaining attitude control capability under manual control. Progress 19 arrived on 23 February. By this point, the complex's orbit was down to 282 × 286 km, so Progress 19 boosted it to 306 × 327 km. In a change to procedure, to avoid the problem of cluttering up the station with supplies, items were to be unloaded only as they were needed. To enable Kizim and Solovyov to concentrate on their work, Atkov was assigned the job of unloading cargo. Progress 19 undocked on 31 March and Soyuz-T 11 took its place on 4 April. This delivered Yuri Malyschev, Gennadi Strekalov and Rakesh Sharma, an Indian cosmonaut. It was the first time that six people had been aboard a station. Nikolai Rukavishnikov had been assigned to this flight, but when he had fallen ill Strekalov had stepped in (Rukavishnikov had suffered a run of very bad luck).

India, like France, although not a member of Intercosmos, had closely collaborated with the Soviet Union on satellite research for several years. In this case, most of the research was biomedical. Vektor used an Indian-built electrocardiograph to monitor the cardiovascular system by measuring chest movement; Anketa and Optokinez studied the relationships of the vestibular and visual systems; Membrana measured calcium loss in bone; Ballisto tested the accelerations on the body caused by the heart's pumping action and monitored changes in its shape and location as the body adapted to space. Sharma's most novel test was a Yoga experiment in which he adopted various standard positions while Atkov observed his muscular responses with the Miokomp and Briz apparatus (it was hoped that Yoga exercises might help to prevent muscular atrophy). MKF-6M and KATE-140 photographs were taken of India for the Terra project to identify oil and gas resources, to assess the biological productivity of the Nicobar, Andaman and Laccadive Islands, to assess water resources in arid regions, to evaluate sites for hydroengineering projects, and to measure ice-melt run-off rates in the Himalaya and Karakorm regions. The Earth studies were to support India's own remote sensing satellite IRS, then under development for launch on a Soviet rocket.

Malyschev, Strekalov and Sharma returned to Earth in Soyuz-T 10 on 11 April. The residents reported that their guests had worn them out. Two days later, they transferred Soyuz-T 11 to the front port, and four days after that the newly vacated port was occupied by Progress 20. It immediately raised the station's orbit back above 300 km altitude. In addition to consumables, it delivered replacement NiCd batteries, two dozen specialised tools, and specially-constructed components to facilitate the repair of the ODU.

ENGINE REPAIRS

Kizim and Solovyov ventured out on 23 April to start the complex operation. Atkov monitored their progress from Soyuz-T 11's descent module, just in case his colleagues were forced to abandon the station. A foot restraint had to be affixed to the periphery of the main compartment, alongside the engine. This proved more difficult than expected, because it involved driving pins into the hull. Curving in a 120-degree arc around the engine bay, this was to provide a workstation for one cosmonaut during the repair operation to follow. The other man was to station himself on a restraint on the cargo ferry, and a special platform on the surface of the orbital module unfolded on command from Kaliningrad. Finally, the tools and other apparatus were attached to the main workstation so that they would be able to set straight to work on the next excursion. The preparations complete, they returned to the airlock. When they returned three days later, Solovyov stood on the platform in the gap between the rear of the station and the ferry. Kizim strapped himself to the ladder that wrapped around the engine bay, and between them they could access the work site from different angles.

The first task was to replace a valve associated with the oxidiser leak. Kizim opened the access panel, then used a pneumatic punch to open the thermal blanket that protected the bay and cut a hole to access the plumbing. Nitrogen gas from a tank in Progress 20 was pumped through the stages of the replenishment system in turn until it confirmed that the length of pipe believed to be fractured did indeed leak. It had to be bypassed by replacing the nearest valves with units incorporating connectors for a bypass pipe.

The disconnection of the valve was aggravated by the discovery that some of the bolts had been sealed with glue to prevent vibrations loosening them; one took an hour to release. When the station passed into the Earth's shadow they continued under the illumination of the lamps on their helmets because, with only five hours of life support in their backpacks, and with almost an hour reserved for making their way back and forth along the station, there was just no time to waste. Finally, they managed to install the first valve. It had been hoped to fit the pipe too, but they were so far behind schedule that they had to stuff the insulating blanket into the hole to protect the engine and hurry back to the airlock. Returning three days later, they quickly attached the bypass pipe to the pump. On 4 May, they replaced the second valve and connected the pipe to provide a parallel path for propellant alongside the leaky pipe. Having done all that they could, they restored the insulation and closed the access panel. Of course, the engine would not be usable until the leaky pipe had been isolated. The tools to accomplish this were still under development, however, so the final phase of the repair operation had to be left for a later date. Its function as a foot restraint over, Progress 20 left on 6 May and the port was taken four days later by Progress 21, which delivered a third pair of clip-on solar arrays.

On 18 May, Kizim and Solovyov continued their record-breaking series of external operations by fitting the second pair of clip-on strips to either side of the lateral panel alongside the airlock hatch. This time Atkov remained in the main compartment to rotate the main panel through 180 degrees while his colleagues fetched the second clip-on from the airlock, so that they could install both panels on a single excursion. The winch used to extend the strips confirmed its reputation for being awkward to use; its handle broke. Finally, Kizim and Solovyov moved over to the other side of the station to inspect the other lateral panel and cut out a segment of its transducer array for return to Earth for its degradation to be assessed. A few days later, the station was reorientated to measure the efficiency of the new panels under a range of insolation angles. Progress 21 left on 26 May, and Progress 22 arrived on 30 May. It delivered a great deal of medical apparatus for Atkov, more replacement NiCd batteries, scientific equipment for the next visiting crew and more air to replenish that which would be lost in mounting future spacewalks.

In June and July, the cosmonauts devoted themselves to geophysical observations. They made extensive use of the Niva video, because it enabled them to offer a running commentary to the photographic work. They surveyed sites in Siberia that were being considered for hydroengineering construction, and the entire length of the Baikal–Amur railway, which was now almost finished. The Ukrainian harvest was starting, so they took multispectral imagery of this to assess its yield. In oceanic studies, they confirmed earlier reports that under certain lighting conditions shallow water becomes transparent, but extended this by reporting that they could see submerged seamounts in open ocean.

On 14 July Progress 22 raised the station's orbit to 365 × 383 km and then undocked the following day. Having decided that the efflux from the thrusters of a departing ferry degraded the efficiency of the transducers in the solar panels, this ferry used the springs of the alignment mechanism in the docking assembly to push it away from the port, and it did not manoeuvre until it was well clear (this would become standard procedure). As the ferry withdrew, the Astra apparatus monitored the gaseous environment around the station. It did so again three days later when the next spacecraft approached.

Soyuz-T 12 carried three cosmonauts, even though the station's orbit was back at its normal operating altitude. It had been launched by a new form of the Semyorka with upgraded engines. Vladimir Dzhanibekov and Svetlana Savitskaya were accompanied by Igor Volk. It had been rumoured that a woman would be included in this crew, but it had been expected to be Irina Pronina, Savitskaya's backup on Soyuz-T 7; but she had retired. Savitskaya therefore became the first female cosmonaut to make a second flight. Although it was not announced, Volk had piloted the Soviet equivalent of the DynaSoar spaceplane in atmospheric tests in the mid-1960s. He was now training to fly the Buran space shuttle,

On Salyut 7, Vladimir Dzhanibekov and Svetlana Savitskaya enjoy weightlessness.

and was flying to gain experience of weightlessness and the subsequent return to Earth. This was to assess how spaceflight degraded normal piloting skills (in effect, this question had already been answered; pilots manually landed NASA's space shuttle). Atkov paid particular attention to Volk, who was to test a medication which, it was hoped, would reduce the most unpleasant symptoms of weightlessness. In order to isolate its effects, Volk was prohibited from participating in the exercises undertaken by the others.

Savitskaya worked with an improved Tavriya electrophoresis apparatus. The small quantities of material produced by these experiments were sufficient to keep specialists busy for months. In fact, trial production of drugs refined from the results of her earlier experiments had already begun. This time she made an anti-influenza preparation for a vaccine, a genetically engineered anti-infection preparation, interferon, and an antibiotic for agricultural use which, when added to animal and poultry fodder, would increase the body weight of the animals by up to 20 per cent. These electrophoresis experiments proved so successful that an industrial-scale unit was developed (and when this eventually flew, it was named Svetlana). The Electrotopography experiment was also rerun. Because the first results had been flawed by water vapour, Kizim suggested an improved procedure, and the results were excellent. At the request of the Byurakan Observatory, Savitskaya used the Piramig to photograph the Earth's L4 and L5 Lagrangian points (the gravitationally neutral points 60 degrees ahead of and behind the Earth in its orbit around the Sun) to help to investigate the distribution of interplanetary dust.

The main event of the visiting programme, however, was the spacewalk by Dzhanibekov and Savitskaya on 25 July. Kizim, Atkov and Volk remained in the station, on one side of the airlock, and Solovyov occupied the Soyuz-T 11 ferry on the other side.

Just outside Salyut 7's airlock, Svetlana Savitskaya tests the URI welding toolkit. Note the clip-on strips on the main solar panel at her foot restraint.

The purpose of the excursion was to test the Universal Manual Toolkit (URI). This was a development of the Isparitel electron-beam apparatus built by the Institute of Electrical Welding in Kiev. Savitskaya stood on the anchor just outside the hatch and set up the workstation which carried a variety of tools to process metal in space. An electron beam was used to cut titanium and stainless steel plates. She reported this to be very easy and said that she could see the beam. A small handheld non-contacting infrared thermometer was used to measure the temperature of the material. Steel and titanium plates were then joined by tack welding. She reported producing good seams. Metal plates were joined using solder composed of tin and lead. She reported that she did not think that the seam looked very good. Finally, a small silver granule was heated in a crucible and sprayed onto an anodised aluminium plate. She complained that this was more difficult than she had expected because she could not really see the spray. This entire test was completed during a single daylight pass, and Savitskaya remained in position while the station passed through the Earth's shadow. Back in sunlight, she and Dzhanibekov exchanged places, and he repeated the test. Later, Academician Paton, director of the Institute of Electrical Welding, said that these simple tests had demonstrated that in the future, robotic systems would be able to assemble space structures using such tools. Before closing the hatch, Dzhanibekov retrieved some of the exposure cassettes and deployed others. The Astra spectrometer detected the increase in the stream of gas trailing the station resulting from venting the airlock.

The visitors departed in Soyuz-T 12 on 29 July. Their flight had lasted longer than the standard week-long international visit. The residents returned to Earth observations, mapping fault lines and ring structures (the latter are best detected from space; they had already reported two candidates near the Syrdarya and Amudarya rivers, each 10 km in diameter) to assist prospectors.

On 8 August, Kizim and Solovyov set off to finish the engine repairs. A miniature pneumatic clamp had been designed and built, and had been delivered by Soyuz-T 12. Dzhanibekov and Savitskaya had been assigned to use it during their extended visit, but their hosts had argued that the repair was their responsibility, and they had been granted permission to do it. A tape had been played on the uplink demonstrating the best way to attempt to seal the bypassed pipe.

Progress 20 had left long ago, so this time they had to work without the advantage of its anchor point. After opening the insulation blanket, they used the pneumatic press to pinch the fractured pipe. Driven by compressed air at 250 atmospheres pressure, this delivered a force of 5 tonnes, and it completely crushed the pipe, blocking it. Once they had verified the new seals using pressurised nitrogen, they reset the thermal insulation. They left the curved anchor in place, but returned with the toolkit to the airlock. Before retiring, they went to the left-hand solar panel and used another tool which fitted closely around the edge of the 20 cm^2 transducer to cut it out without risking contamination.

What Lyakhov had described as "a slight leak" had taken fully 24 hours of external activity to repair. In contrast to the spacewalks by Lyakhov's crew to attach the clip-on solar panels, this impressive series of spacewalks was hardly mentioned (*Tass* reported simply that the engine had been "serviced"), but this was undoubtedly due in part to the fact that the cosmonauts were too busy to bother with a television camera. It was later admitted that replacing valves in the engine bay had been considered to be the "absolute worst case" scenario when designing the station to be serviceable by its crew. Now that this had been done, the propellant leak which had earlier been described as "minor" was conceded to have been "very serious".

Cosmonauts had repeatedly demonstrated an ability to overcome problems which would have been deemed to be sufficient to force a previous station to be abandoned. In a very real sense, it was having a crew aboard which had enabled Salyut 6 and Salyut 7 to be sustained for so long. Although some observers criticised the amount of time that cosmonauts spent maintaining the systems, arguing that this was time lost to the science programme, this narrow viewpoint did not recognise the value of engineering knowledge derived from sustaining ongoing operations. The 'learning curve' was steep, but the cosmonauts were determined to climb it. Knowledge gained from overcoming adversity was empirical, but it was the only way to achieve the ultimate goal of establishing a permanent foothold in orbit.

Progress 23 docked on 16 August and delivered propellants to refill the ODU's tank and new scientific apparatus including the French–Soviet Sirene X-ray spectrometer and the RS-17 X-ray telescope. On 21 August, with the replenishment complete, the ODU was reactivated and found to be fully functional. Progress 23 was unloaded in record time. This was because the Sirene apparatus, which had been set up in the rear transfer compartment, could not be used until the port had been vacated. Progress 23 departed on 26 August. In early September the station contributed to the Black Sea-84 and Gyunesh-84 Earth-studies projects run by the Intercosmos organisation.

On 7 September, Kizim, Solovyov and Atkov broke the 211-day endurance record. From this point on, Atkov's assessment of their condition acquired extra significance. A crew of three had proven more efficient during the long flight than had the previous crews of two. Later in the month, they began preparations to return to Earth, then did so on 2 October, having set a new endurance record of 237 days and undertaken a record-breaking six spacewalks.

Atkov had focused his studies on the adaptation of their cardiovascular systems. He used the Aelita to make vector cardiograms, the Reograf to make rheographs and used the French Echograph which produced ultrasonic images of internal organs. He had also used the Glucometer to monitor carbohydrate exchange in cells and had found that this process operated at a greatly reduced rate in the absence of gravity, suggesting cellular functioning might be disrupted during very long flights. In the Membrane experiment he had studied the exchange of calcium in cell membranes. This had shown that if they took an anti-oxidant preparation, calcium metabolism in weightlessness was reduced. Blood samples were spun in a small centrifuge to separate the components for storage and subsequent analysis on Earth. The wisdom of including a doctor in the crew was demonstrated by the fact that he had performed a series of 30 different medical tests a total of 200 times, providing a degree of detail in monitoring the adaptation process that had not previously been possible. When he attended a press conference, Atkov said that their very presence indicated their rapid recovery. Although it was noted that their flight demonstrated that it would be possible to fly to Mars, it was acknowledged that existing technology was not capable of facilitating such a journey.

Soyuz-T 11's 182 days in orbit demonstrated that the systems of the ferry had been considerably improved. It seemed likely that six-month tours would now be standard, and that, once handovers became routine, the onerous task of pushing the endurance record would be assigned to medical specialists serving with successive resident crews.

CRISIS!

At the end of the year, with Salyut 7 in a 365 × 370 km orbit, Boris Belitski, speaking on Radio Moscow, reported that the station's systems were "in good shape", and that it still had "plenty of life" left. At the end of January 1985 its systems were reported to be "functioning normally" with its internal conditions "optimum" for the next crew.

On 11 February, however, telemetry briefly revealed a fault in one of its systems, and then contact was lost. With the station in free-drift and its Igla transponder out of action, it would be impossible for an automated ferry to dock with it to secure an independent manoeuvring capability. On 1 March, *Tass* noted that Salyut 7 had been in space for 34 months, and announced that it had "completely fulfilled" its "planned programme". This statement had a greater sense of finality than the one that had announced that Salyut 6 had completed its "primary" mission. It was hardly credible that Kizim and Solovyov would have made such a valiant effort to repair the engine if Salyut 7 was to be written off foll-owing their departure. Clearly, *something* had gone disastrously wrong. It was therefore expected that preparations would be made to launch the next station. When the next launch came, on 6 June 1985, it was Soyuz-T 13, with Vladimir Dzhanibekov and Viktor Savinykh onboard.

Dzhanibekov was making a record fifth flight. As the commander of Soyuz-T 6 he had made a manual approach to Salyut 7 from a kilometre out, and was the best choice to try to dock with an uncooperative station. Savinykh had entered training with Vladimir Vasyutin and Alexander Volkov in September 1984 for a 10-month mission from early 1985. Savinykh had been selected to accompany Dzhanibekov because he had trained (with Vasyutin) for a spacewalk to affix the last pair of clip-on solar panels. This repair mission had been organised as soon as Salyut 7 had fallen silent. Vladimir Shatalov said that their work would be "complicated". If they managed to board the station, they were to find out why it had ceased to respond to commands, and then, if possible, fix it.

Soyuz-T 13 pursued a propellant-efficient two-day rendezvous. The transition from the transfer orbit to the point at which a straight-in approach was made would have to be conducted without the benefit of the Igla radar transponder. Vladimir Titov's success in manoeuvring Soyuz-T 8 close to the station without a radar was now about to yield an unexpected dividend. Although the ground radars had been able to get him within a kilo-metre of the station, when he moved in to try to dock he had encountered difficulty deter-mining range and closure rate. If this problem could be overcome, there was every chance that Dzhanibekov would be able to repeat this process. The long rendezvous had ensured that the tracking radars had computed the orbits very accurately. Every drop of propellent had to be saved in case they had to perform extensive manoeuvres to achieve a docking. When 10 km from the station Dzhanibekov paused to assess the situation. As expected the station was completely dead. It did not respond to the command to turn on its Igla transponder and it did not orientate itself to face the approaching ferry.

Dzhanibekov closed in to 3 km, then activated the special optical sight that had been fitted to enable him to approach manually, free of the problems that dogged Titov. This incorporated both low-light optics and a laser rangefinder. The image intensifier would show him the station's position and orientation in the Earth's shadow. The rangefinder would measure the separation and compute the rate of closure. At the usual pause point, 200 metres out, Dzhanibekov studied the station. Fortunately, although it was in a very slow roll of about 0.3 degree per second, which was easily matched, it was not tumbling. He slowly flew around it to examine it from all angles, looking for any sign of catastrophic damage. It was apparent that the three solar panels had ceased to track the Sun. The two lateral panels were orientated 75 degrees apart, indicating that the station had suffered a power supply failure rather than a communications problem. Dzhanibekov aligned Soyuz-T 13 to approach the station's front port. He controlled the spacecraft using the optical sight while Savinykh called out the range and rate. Once they had matched the station's roll, they moved in and docked at the first attempt. Moments later, the station flew into the Earth's shadow. It was still in radio range with the ground, however, so the jubilant cosmonauts were able to report their success to the relieved flight controllers.

Standard procedure after docking was to establish the electrical and hydraulic links, test the hermetic seal of the tunnel, equalise pressure with the station, and then open the hatches. Unfortunately, because there was no power available they could not tell if the station was still pressurised. Opening the valve was, therefore, the moment of truth. There was air in the station. They sniffed it for toxic fumes. Satisfied, they swung back the probe and drogue assemblies and stared into the dark void (with the porthole covers closed, there was no light). Using torches, they ventured into the station. Dzhanibekov reported that the air was musty. He estimated that it was −10°C, but it was difficult to tell, as it was off the bottom of the scale of the thermometer. The water pipes had frozen, condensation had frosted the walls, and microgravity icicles had formed. Even wearing arctic clothing, fur coats and fur boots, the cosmonauts had to return to their ferry every hour or so to warm up. They had to retreat there to eat and sleep. So long as they could not power down their ferry, time acted against them. Unless they could restore Salyut 7 to life within 10 days (the limit of their powered-up ferry) they would have no option but to abandon it and return to Earth.

It was obvious that the power supply had failed and the batteries had run flat, and it did not take long to find the root of the problem. The switch which controlled the feed from the solar panels to the batteries had failed and, ironically, the batteries had drained their remaining charge running the motors which kept the panels facing the Sun. It was soon realised that the panels were still delivering power. If the batteries could be recharged, it should be feasible to restore the station to full operation. In fact, only two of the batteries were ruined; the other six should be serviceable once recharged. The culprit switch was isolated and jump leads run from the panel lines to the recharger. Soyuz-T 13 was then used to orientate the station so that its irregularly positioned solar panels collected as much energy as possible. The first battery recharged in a few hours, so the telemetry link was reactivated to enable Kaliningrad to examine the state of the other systems.

It would take several days to reactivate the environmental systems. A ventilator tube was run from Soyuz-T 13 to circulate the stagnant air, because it contained pockets of carbon dioxide that gave the cosmonauts headaches. There was soon sufficient power to run the lights and atmospheric heater. The thermal regulation system would not be able to be reactivated until the humidity (initially at about 90 per cent) had been reduced. The main fault had been overcome, but there was a long way to go before Salyut 7 would be usable as a platform for scientific research. As the vital systems were assessed, the parts and tools needed to effect repairs were loaded into the next supply ferry. Luckily, the attitude control system was found to be fully functional, so the Igla would be able to orientate the rear port towards the approaching automated ferry.

At the end of the week, the temperature finally climbed over 0°C, and the ice began to thaw. The humidity was still high, so it was crucial to ensure that the temperature did not melt the ice faster than the water vapour condenser could remove it. It was several more days before the humidity fell sufficiently for the heaters to be safely switched on. This was just in time, because Soyuz-T 13's limited resources were nearing the point at which, unless Salyut 7 had recovered, it would have had to have been abandoned. *Tass* reported that it would soon be in "normal working mode". Dzhanibekov and Savinykh had successfully achieved the most important part of their mission. Once again, human intervention had saved a crippled station. The irony of the crisis was that if cosmonauts had been aboard when the recharging switch had failed, they would have pre-empted the draining of the batteries. Clearly, although a fully operational station was a prerequisite for permanent occupancy, which was the objective of the programme, maintaining a crew aboard was likely to be crucial in keeping a station operational.

The situation improved so rapidly that, on 19 June, the MKF-6M and KATE-140 cameras were used to photograph the Kursk Oblast region for the Kursk-85 programme. It was something to do while waiting for Progress 24, which docked on 23 June. It carried propellant, water, air, clothes, three replacement NiCd batteries and a new water heater (the original had been split by the ice). Progress 24 departed on 15 July, taking away all the damaged items.

Although the next resupply ferry was successfully placed in orbit on 19 July, a fault immediately afterwards led to it being assigned the anonymous designation of Cosmos 1669. But its cargo was simply too valuable to be written off, and the flight controllers finally managed to overcome the problem. It then followed the standard rendezvous and docked without incident. It delivered two new spacesuits (the originals had been deemed unsafe, due to being frozen), the Mariya spectrometer to study high-energy particles in near-Earth space, the Rost plant chamber and a number of exposure cassettes for deployment outside.

The spacewalk was conducted on 2 August. The new semi-rigid spacesuits offered improved peripheral vision and increased mobility. The primary task was to install the third pair of clip-on solar panels (which had been delivered long ago, by Progress 21). This time, the lateral panel was commanded to rotate through 180 degrees by remote control. This done, an experimental solar transducer was deployed so that its performance could be evaluated. Well ahead of schedule, they replaced the Medusa and several technology cassettes and set up the Comet micrometeoroid detector (a French package with several chambers which could be opened from within the station; the first was to be exposed for Comet Giacobini–Zinner, the second for Halley's Comet). Although many satellites had studied meteoroids in the past, these had tended simply to measure the rate and size of the particles striking the detectors. This experiment was a step forward because not only could it focus on specific cometary streams, but the material was to be returned to Earth for analysis. In addition to being of intrinsic interest, it had been suggested that cometary dust might have played a key role in the onset of the Earth's ice ages.

Throughout August, Dzhanibekov and Savinykh concentrated on photography using the KATE-140 and MKF-6M while the Gel apparatus purified biological materials, flax grew in the Magnetobiostat and further Mariya data were taken. Cosmos 1669 undocked on 29 August, and then redocked to test an alternative control system; after a few hours it left for good.

ORBITAL HANDOVER

Soyuz-T 14 arrived on 18 September, with Vladimir Vasyutin, Alexander Volkov and Georgi Grechko. Its cargo included the EFU-Robot electrophoresis apparatus and the Skif spectrometer.

On 25 September, Dzhanibekov and Grechko left Savinykh, Vasyutin and Volkov (the original crew) aboard the station, and undocked in Soyuz-T 13. Instead of landing immediately, they withdrew and then repeated the rendezvous with the Igla deactivated, to further evaluate the transition from the final transfer orbit to the 1,000-metre point in the approach using only the optical sight. No attempt was made to redock. After some time station-keeping, they again withdrew, but they did not return to Earth until the next day. The day after that, a Proton put Cosmos 1686 in orbit, and it began a slow rendezvous with Salyut 7.

The reconstituted crew expected to stay in space until March 1986. Although by that time Savinykh would have been in space for 10 months as had originally been planned, his colleagues would have served only six months. This orbital handover was notable not only because it was the first time that one long-duration crew had relieved another, but also because it was the first time that a cosmonaut launched as part of one crew had returned as part of another. Given the earlier statement that a doctor would accompany crews on endurance missions, it would turn out to be unfortunate that this crew did not include a medical specialist.

Cosmos 1686 arrived on 2 October. This was the first time that such a module had docked with an already inhabited station. *Tass* reported that it was "similar in design" to Cosmos 1443, but did not say precisely how it differed. Despite the fact that soon after it had been launched, Salyut 7 had been said to have been designed for a four-year life, for several months there had been speculation that it might soon be replaced. Comments by Soviet spokesmen at the Swedish congress of the International Astronautics Federation indicated that it would be replaced by an orbital complex constructed from modular components. Cosmos 1686's Merkur capsule had been heavily modified. A bank of instruments had been affixed to its nose (instead of the conical re-entry package), and their control panels placed inside. The stripped-down capsule was no longer capable of returning to Earth. Vasyutin's crew had trained to use the instruments. On the original plan,

Soyuz-T 14 at Salyut 7. The picture was taken from Soyuz-T 13, during its final fly-around inspection of the newly-repaired space station.

Cosmos 1686 had been scheduled to arrive early in their tour. Its 4,500 kg of "miscellaneous cargo" included the Kristallisator semiconductor furnace and a deployable girder (a package far too bulky to fit in a Progress ferry). During October, unloading cargo was interleaved with gathering Aerosol and Mariya data, processing materials in the Pion-M furnace and evaluating the ability of the module's advanced control system to manoeuvre the complex (although it was able to perform all the desired manoeuvres without the crew's intervention).

In its 25 October report, *Tass* made its last reference to the crew being in "good health and feeling well". In fact, over the period of several days, Vasyutin had progressively succumbed to illness. This had begun with "a slight uneasiness" that had been aggravated first by an inability to sleep, then by loss of appetite. Although initially the cosmonauts had kept this to themselves, as soon as it became apparent that this was a serious illness they had reported it, and Vasyutin described his symptoms. The medics at Kaliningrad told him to relay the output from the biomedical test kit via the telemetry stream so that they could assess his condition. Meanwhile, the cosmonauts were told to prepare to return to Earth. Unfortunately, orbital dynamics would not offer a favourable recovery window until mid-November, so an emergency return would be in less than ideal conditions. *Tass*'s report on 29 October noted that the crew was implementing the flight programme, and that the systems on the complex were functioning normally, but it did not include the stock phrase indicating that the cosmonauts were in good health. Similar reports followed. By 17 November, Vasyutin's condition had deteriorated to the degree that he was suffering periods of acute pain. The doctors decided that he was too ill to continue, and ordered the crew home, which they did on 21 November. Vasyutin was examined immediately. Although his condition was "satisfactory", he "required hospital treatment" and was flown to Moscow. The report of their successful return said that the flight had been terminated by Vasyutin's "sickness". This vague reference was the first official acknowledgement that there had been a medical problem. It was suspected that he was suffering from appendicitis, but he had a prostate infection. This had manifested itself as an inflammation and a fever. This was the first time that a mission had been cut short by illness. In retrospect, considering the number of cosmonauts that spent long periods in space, it was remarkable that this had not occurred earlier.

It had been intended to send two crews to visit the residents after Cosmos 1686 had departed. One of these was an international flight, with a Syrian cosmonaut. The other was an all-female crew with Svetlana Savitskaya commanding Yekaterina Ivanova and Yelena Dobrokvashina. These flights had been postponed when Salyut 7 had suffered its power failure, and rescheduled after the Soyuz-T 13 crew had rescued it. The Syrian flight was further postponed following the recall of Vasyutin's crew. The all-female crew was disbanded when Savitskaya fell pregnant early in 1986.

The Salyut 7–Cosmos 1686 complex continued in orbit, in its automated regime. It was clearly capable of supporting another crew and there were several experiments still to be completed.

Table 4.1. Salyut 7 docking operations

Spacecraft	Docking Date	MT	Port	Undocking Date	MT	Days
Soyuz-T 5	14 May 1982	1536	front	27 Aug 1982	-	105
Progress 13	25 May 1982	1157	rear	4 Jun 1982	1031	9.94
Soyuz-T 6	25 Jun 1982	2146	rear	2 Jul 1982	1501	6.72
Progress 14	12 Jul 1982	1541	rear	11 Aug 1982	0211	29.44
Soyuz-T 7	20 Aug 1982	2232	rear	29 Aug 1982	1847	8.84
Soyuz-T 7	29 Aug 1982	-	front	10 Dec 1982	-	103
Progress 15	20 Sep 1982	1012	rear	14 Oct 1982	1646	24.27
Progress 16	2 Nov 1982	1622	rear	16 Dec 1982	-	44
Cosmos 1443	10 Mar 1983	1220	front	14 Aug 1983	1804	157.24
Soyuz-T 9	28 Jun 1983	1346	rear	16 Aug 1983	1825	49.20
Soyuz-T 9	16 Aug 1983	1845	front	23 Nov 1983	-	99
Progress 17	19 Aug 1983	1747	rear	17 Sep 1983	1444	28.87
Progress 18	22 Oct 1983	1434	rear	13 Nov 1983	1808	22.15
Soyuz-T 10	9 Feb 1984	1743	front	11 Apr 1984	1427	60.86
Progress 19	23 Feb 1984	1121	rear	31 Mar 1984	1245	36.06
Soyuz-T 11	4 Apr 1984	1831	rear	13 Apr 1984	1427	8.83
Soyuz-T 11	13 Apr 1984	-	front	2 Oct 1984	1030	172
Progress 20	17 Apr 1984	1322	rear	6 May 1984	2146	19.35
Progress 21	10 May 1984	0410	rear	26 May 1984	1341	16.40
Progress 22	30 May 1984	1947	rear	15 Jul 1984	1736	45.91
Soyuz-T 12	18 Jul 1984	2317	rear	29 Jul 1984	1326	10.59
Progress 23	16 Aug 1984	1211	rear	26 Aug 1984	1913	10.29
Soyuz-T 13	8 Jun 1985	1250	front	25 Sep 1985	0758	108.80
Progress 24	23 Jun 1985	0634	rear	15 Jul 1985	1628	22.41
Cosmos 1669	21 Jul 1985	1905	rear	29 Aug 1985	0150	38.28
Cosmos 1669	29 Aug 1985	-	rear	29 Aug 1985	-	-
Soyuz-T 14	18 Sep 1985	1814	rear	21 Nov 1985	-	56
Cosmos 1686	2 Oct 1985	1316	front	<permanently docked>		
Soyuz-T 15	6 May 1986	2058	rear	25 Jun 1986	1858	49.92

Table 4.2. Salyut 7 crewing

| Cosmonaut | Role | Spacecraft | | Duration |
		Arrive	Depart	days
Anatoli Berezovoi	CDR	Soyuz-T 5	Soyuz-T 7	211.38
Valentin Lebedev	FE	Soyuz-T 5	Soyuz-T 7	211.38
Vladimir Dzhanibekov	CDR	Soyuz-T 6	Soyuz-T 6	7.91
Alexander Ivanchenkov	FE	Soyuz-T 6	Soyuz-T 6	7.91
Jean-Loup Chrétien	CR	Soyuz-T 6	Soyuz-T 6	7.91
Leonid Popov	CDR	Soyuz-T 7	Soyuz-T 5	7.91
Alexander Serebrov	FE	Soyuz-T 7	Soyuz-T 5	7.91
Svetlana Savitskaya	FE	Soyuz-T 7	Soyuz-T 5	7.91
Vladimir Lyakhov	CDR	Soyuz-T 9	Soyuz-T 9	149.41
Alexander Alexandrov	FE	Soyuz-T 9	Soyuz-T 9	149.41
Leonid Kizim	CDR	Soyuz-T 10	Soyuz-T 11	236.95
Vladimir Solovyov	FE	Soyuz-T 10	Soyuz-T 11	236.95
Oleg Atkov	CR	Soyuz-T 10	Soyuz-T 11	236.95
Yuri Malyschev	CDR	Soyuz-T 11	Soyuz-T 10	8.02
Gennadi Strekalov	FE	Soyuz-T 11	Soyuz-T 10	8.02
Rakesh Sharma	CR	Soyuz-T 11	Soyuz-T 10	8.02
Vladimir Dzhanibekov	CDR	Soyuz-T 12	Soyuz-T 12	11.80
Svetlana Savitskaya	FE	Soyuz-T 12	Soyuz-T 12	11.80
Igor Volk	CR	Soyuz-T 12	Soyuz-T 12	11.80
Vladimir Dzhanibekov	CDR	Soyuz-T 13	Soyuz-T 13	112.13
Viktor Savinykh	FE	Soyuz-T 13	Soyuz-T 14	168.12
Vladimir Vasyutin	CDR	Soyuz-T 14	Soyuz-T 14	64.87
Georgi Grechko	FE	Soyuz-T 14	Soyuz-T 13	8.88
Alexander Volkov	FE	Soyuz-T 14	Soyuz-T 14	64.87
Leonid Kizim	CDR	Soyuz-T 15	Soyuz-T 15	49.92
Vladimir Solovyov	FE	Soyuz-T 15	Soyuz-T 15	49.92

Table 4.3. Salyut 7 spacewalks

Date	Hours	Activity
30 Jul 1982	2.5	Berezovoi and Lebedev exited the airlock, retrieved several exposure cassettes from near the hatch, deployed others and recorded the state of the solar panels.
1 Nov 1983	2.8	Lyakhov and Alexandrov attached the first clip-on solar panel to the dorsal solar panel.
3 Nov 1983	2.9	Lyakhov and Alexandrov attached the second clip-on solar panel.
23 Apr 1984	4.25	Kizim and Solovyov installed a curved ladder in a 120-degree arc around the engine compartment.
26 Apr 1984	5.0	Kizim and Solovyov opened up the engine bay, cut the thermal insulation, and replaced a valve to support a bypass around the fractured propellant pipe.
29 Apr 1984	2.75	Kizim and Solovyov fitted the bypass pipe.
4 May 1984	2.75	Kizim and Solovyov replaced a second valve to complete the bypass.
18 May 1984	3.1	Kizim and Solovyov attached the second pair of clip-on solar panels, this time to the lateral panel alongside the airlock hatch.
25 Jul 1984	3.6	Dzhanibekov and Savitskaya tested the URI toolkit, retrieved some exposure cassettes, and deployed others.
8 Aug 1984	5.0	Kizim and Solovyov employed a pneumatic press to crimp the fractured pipe, leaving the bypass to carry the propellant flow. On their way back, they cut out another sample of solar panel.
2 Aug 1985	5.0	Dzhanibekov and Savinykh attached the third pair of clip-on solar panels and an experimental transducer, retrieved the Medusa and other materials cassettes and replaced them, then deployed the French Comet experiment.
28 May 1986	3.8	Kizim and Solovyov retrieved the Comet, Medusa, Spiral, Istok and Resurs cassettes, and then erected and retracted the URS girder.
31 May 1986	5.0	Kizim and Solovyov mounted experiments on top of the girder, re-extended it, then jettisoned it. Finally, they retrieved the transducer that Dzhanibekov had installed.

Part 2: Mir

5

A base block for modular construction

Early in the morning of 20 February 1986, a Proton rocket, still the most powerful in the inventory, rose on a brilliant pillar of flame above the Kyzyl-Kum desert and then streaked north-east into the pre-dawn twilight. Within its payload shroud was the first element of what was to become the Mir space station.

Even though rumours that a new vehicle was being prepared had been leaked, the actual launch was a surprise, because Salyut 7 (which still had the Cosmos 1686 module attached) was fully operational. Actually, Salyut 7's troubled history had prompted the decision – upon the Soyuz-T 14 crew's return – to abandon it and immediately launch a new station to take its place. The illness which had necessitated Vasyutin's recall had curtailed the experimental programme, so a brief visit to complete it was required. Despite the unfinished business, the new station was launched on time in order to mark the formal opening of the Congress of the Communist Party of the Soviet Union. Unusually, video of Mir's launch was released within hours. *Tass* reported that it was fully operational.

The word 'Mir' could be translated (depending on the context) as 'new world', 'peace', or 'community'. The assignment of a name for the new spacecraft, rather than Salyut 8, combined with the fact that the two orbits were coplanar, prompted initial speculation that Mir might dock with Salyut 7 and Cosmos 1686 to further expand that complex. It was soon announced, however, that the new vehicle was a *third*-generation design that was to serve as the core of an entirely new complex, and that it was to be permanently occupied. To reinforce this point, it was reported that in addition to axial ports, it had a ring of four radial docking ports; this feature had been widely predicted, so came as no surprise. Given that its overall mass and dimensions were dictated by the configuration of the Proton rocket, Mir's basic structure was necessarily similar to its predecessors: namely, a 20-tonne stepped cylinder some 13 metres long and 4.15 metres in diameter at its widest point. When its configuration was announced, Mir was criticised in the West for having little scientific equipment, but to argue this was to fail to appreciate that the Soviet space station programme had reached the point where it was feasible to construct an orbital complex from highly specialised modules. The first element, the 'base block', was a habitat, not a laboratory. It seemed a fair bet that it would soon be expanded by modules similar to Cosmos 1686.

Upon attaining orbit, Mir had deployed a pair of solar panels to provide electrical power. These had a larger area (each 38 m^2) than those of Salyut 7 (each 20 m^2), and were deployed by a new mechanism which initially unfolded a long truss from either side of the narrower section of the main compartment, as a backbone, after which the individual panels mounted along it unfolded fore and aft. Each panel produced 4.5 kW. Although this was sufficient to operate the life support systems, more power would be required to run the equipment which was to be delivered in the specialised laboratory modules. The top panel of the second-generation Salyut design had not been installed prior to launch because the new units were so bulky that, when folded up, there was no room for it within the aerodynamic shroud. The third panel was later to be mounted on the already installed motor. Experience had shown that spacewalking cosmonauts should be able to carry out such a construction task; something else which was not fully appreciated by critics in the West.

Each panel could be rotated in only one degree of freedom, turning over a 180-degree arc (±90 degrees). The power level falls off sinusoidally with deviation from face-on illumination. An all-sky sensor noted when the station was in sunlight, and it activated the automated system. Each mount had a Sun sensor, and a controller which compared the angle of the panel with the direction of the Sun and rotated it to produce the maximum output. With both panels mounted on the same axis, the vehicle obviously had to be orientated so that they *could* face the Sun. As soon as the station flew into the Earth's shadow, the panels turned to the predicted angle for reacquiring the Sun. On a daylight pass the output from the panels ran the systems and charged the NiCd batteries. There were twelve batteries, each with a storage capacity of 60 amp-hours. The electrical system used 28 volts direct current, with 5-amp, 10-amp, 20-amp and 50-amp taps off the power distribution bus. Upon emerging from shadow, it took some time for the transducers to reach maximum output, so the batteries continued to supply power during this lag, prior to recharging. The operation of the power system was completely transparent to the crew. However, being engineers, they were familiar with how it operated, and they could intervene in an emergency.

Mir had seven computers in an integrated complex known as Strela. The main unit was the Argon 16B, which was also used by the Soyuz-T ferry's flight control system. Serving the same purpose aboard Mir, it dealt with attitude control. Frames of reference were drawn from infrared Earth-horizon sensors, a Sun sensor, and a sophisticated star sextant. Whereas the Delta navigation system of the second-generation Salyuts had required its state-vector to be periodically updated, which had imposed a supervision function on the crew, Mir's computer could be ordered to establish a given attitude, which it would then automatically maintain. It could also be programmed to run through a series of manoeuvres, but its limited memory capacity meant that it could store sequences for only a few days at a time. It would automatically keep track of when the spacecraft was in communication range of ground stations. Like the power generation and distribution system, therefore, the navigation system was highly automated, and this released the cosmonauts for productive work. Once the operating mode for a given situation had been specified, the computer eliminated the need for a cosmonaut to orientate the station, to aim instruments, and then to track the target. In fact, the computers were also able to manage resources in support of a wide range of research operations.

Although the Mir base block had six ports, only the axial ports were fitted with the antennas required to effect an automatic docking. It was not intended that more than two

Soyuz spacecraft would be docked at any given time, so the environmental systems were designed to support a maximum occupancy of six cosmonauts for short periods, with indefinite residence by crews of two or three cosmonauts. The standard crewing was nevertheless considered to be a commander and an engineer. Individual researchers would be able to make short visits during in-orbit handovers between crews, or a doctor could check out the residents and then accompany them on their return to Earth.

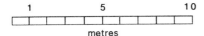

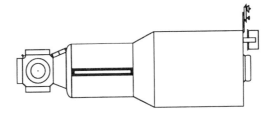

Mir base block.

AN AMBITIOUS MISSION

Soyuz-T 15 was launched in the afternoon of 13 March. With refreshing openness, the event had been announced the previous day, and it was shown live on television; this was then something of a rarity. Also, as soon as the crew reported in at the end of their first orbit to confirm that the ferry was fully functional, their mission was explained to the television audience. First, they would go to Mir, to check it out, and then transfer to Salyut 7 to complete experiments there.

Alexander Viktorenko and Alexander Alexandrov had served as the backup crew for Soyuz-T 14, and as such they would normally have been expected to move up to fly the next mission. If all that had been required was a brief visit to Salyut 7, to complete unfinished business, they might well have done so. The decision to fly a double-station mission, however, effectively ruled them out. Instead, "the most complex flight" likely to be attempted "for the next few years" (as chief cosmonaut Vladimir Shatalov put it) was assigned to Leonid Kizim and Vladimir Solovyov. They had not only trained to commission Mir, but, almost exactly two years earlier, had spent 237 days on Salyut 7, so they were ideally suited for this ambitious mission. Despite the planned workload, it had been decided not to fly a third cosmonaut, so as to be able to carry the propellant to perform the double rendezvous.

The rocket had placed the spacecraft in a 193 × 238 km orbit with its perigee over the Soviet Union. Only after the many systems had been verified was the first manoeuvre performed. This was done at apogee half-way through the second orbit (that is, whilst over the South Pacific Ocean) and it raised the perigee by 100 km to give a 238 × 290 km orbit, now with its apogee over the Soviet Union. The Earth's rotation made the orbital plane migrate about 22 degrees West per orbit. Only for a few orbits a day would the spacecraft extensively overfly Soviet territory on successive orbits. Before its track passed too far, the radars accurately determined the orbital parameters to assess the performance of the engine. At the end of the fourth pass, another burn raised the orbit to 290 × 330 km. These manoeuvres were meant not so much to regulate the orbit's altitude as its period, which was about 92 minutes, to coordinate the motions of the ferry and the Mir space station so that, some 50 hours into its flight, the ferry's final transfer orbit would take it within a few kilometres of the station, then in its 332 × 354 km orbit, whilst travelling on a parallel course with a very low relative speed. Apart from the unmanned test flight by Soyuz-T 1, and the cautious approach of Soyuz-T 13 to Salyut 7, this was the first time that a prospective station crew had made such an approach. Although it took longer, this slow rendezvous had the virtue that it was propellant-efficient. Given the ambitious plan for this mission, maintaining a high reserve of propellant was a major objective.

Mir had two rendezvous systems. The new Kurs was initially to supplement, and eventually to supersede, the Igla transponder. Igla required the station to reorientate itself continually throughout the final phase of the rendezvous to aim the appropriate docking port at the approaching spacecraft. Although this reorientation was no problem for the isolated base block, it would be impracticable once a cluster of heavy modules had been added. Using Kurs, the ferry could approach the station from any direction and then, a few hundred metres out, fly-around to begin a straight-in approach to the required port. Because Progress ferries were to resupply Mir, its rear port had transponders for both rendezvous systems. Because Soyuz-T 15 was to dock with Salyut 7, it had only Igla. Igla gave data on range, closing rate, line-of-sight angular velocity, and perpendicular deviation from the straight-in vector. At the 200-metre pause, it rolled the spacecraft to assume the correct alignment for docking. At this point, it was standard procedure to let the ferry attempt an automated docking. In this case it had been decided to leave the rear port clear so as to accommodate the first cargo ferry (which was, at that moment, being loaded with stores). Kizim took command and flew around to the front port. Since his spacecraft did not have the Kurs system, he had a laser rangefinder (similar to that used by Dzhanibekov and Savinykh in approaching the crippled Salyut 7) to generate data on separation and closing rate during the manual straight-in approach. Soyuz-T 15 docked with Mir on its first attempt. As events transpired, this was to be the only Igla-equipped Soyuz ever to dock with Mir; the new Soyuz-TM model carried the Kurs system. After pushing back Mir's Konus drogue assembly, Solovyov floated through the multiple docking adapter (which was no bigger than the ferry's orbital module) to the internal hatch that opened into the main compartment. With the lights already on, the automatic video camera caught his entry. A few minutes later, having powered down the ferry, Kizim joined him to form Mir's first crew.

A NEW HOME

Although the body of the base block was a 3-metre diameter cylinder connected to a 4-metre diameter cylinder, the compartment within was essentially a long narrow room. Despite the fact that cosmonauts could orientate themselves arbitrarily in weightlessness, space psychologists had demanded a definite sense of up-and-down, so the 'floor' was carpeted, the 'walls' were dark green and the 'ceiling' was white. A control panel with several video monitors spanned the floor immediately inside the hatch. The first thing Kizim and Solovyov did was to check the status of the systems which it displayed. The other end of the compartment had a slightly raised deck that accommodated embedded apparatus. From the viewpoint of a cosmonaut facing the controls, the multiple docking adapter was 'ahead' and (as expected) the engine was 'behind'. A system of cartesian axes had been devised to describe the layout of lockers, but this is so arbitrary that the more intuitively obvious reference frame will be used herein (a distinct sense of up-and-down may not be necessary for the cosmonauts in space, but it helps an Earthly reader). Thus, in addition to the front axial port, the multiple docking adapter had upper, lower, left and right ports.

For the first time, two small cabins had been provided. Each phonebox-sized space had a sleeping bag fastened to the wall, a small fold-out desk, and a porthole. They were located towards the rear of the compartment, one on each side wall, and were for the residents. Visitors would continue the previous practice of fastening their sleeping bags to the ceiling.

In general the cosmonauts were to follow a normal 0800 to 2300 Moscow day, five days a week. Unless there was a specific reason to disrupt their sleep cycle, their day began with two hours of personal hygiene and breakfast. Work from 1000 to 1300 was followed by an hour's vigorous exercise, an hour off for lunch, another three hours of work, and another hour's exercise. The preparation of the evening meal began at about 1900. The meals were spaced out to achieve a high degree of assimilation. The diet was designed to provide about 100 grammes of protein, 130 grammes of fat, 330 grammes of carbohydrates, and appropriate mineral and vitamin additives. In the evening, the cosmonauts were free to relax or to catch up on work as they pleased. Even in space there was paperwork to be done, and this was often done late at night. Unless a task had to be performed at a particular time, the cosmonauts were free to define their own schedule and work at their own pace. On long missions, common sense prevailed; there was little reason to hurry to get started on a task one day if it could reasonably be left to the next day. Ironically, cosmonauts had found working on days off more stimulating, because they were doing so voluntarily.

In contrast to the Salyuts, which had been dominated by the massive conical bulk of the primary instrument housing, Mir's main compartment was unobstructed. Because it provided the 'living room' for a modular complex, Mir had improved facilities for food preparation. The table had a water heater to reconstitute dehydrated food, and a stove to heat canned foods. Although it folded against the wall, the table could conveniently be left extended across the rear of the compartment (just in front of the the cabins). There was a small fridge for fresh produce, and prepackaged food was kept in lockers around the hatch to the rear transfer tunnel. The toilet (ASU) was a tiny cubicle set immediately aft of the right-hand cabin. Unlike on the Salyuts, it had a door for privacy. In addition, this compartment included a 'wash basin'. This incorporated a transparent hood which prevented water from escaping, and it had holes to accommodate the user's head and hands. Luxury indeed!

There was a comprehensive gymnasium which folded away when not in use. The veloergometer stationary bicycle was stored under the floor, at the step in level between the two parts of the compartment. Each cosmonaut had to cycle the equivalent of 10 km a day.

The treadmill was mounted behind the table, directly in front of the rear tunnel. Wearing a springy harness for traction, a cosmonaut would 'walk' the equivalent of 5 km every day. The track provided a satisfying view along the axis of the station (rather than a bare wall). The daily exercise regime was designed to minimise deterioration of the heart, bones and muscles. Although they could split the exercises into short bursts, they had to accumulate 2.5 hours a day to burn off 450 calories. They followed a four-day cycle; the first day built up speed and strength; the second day concentrated on exertion; the third emphasised endurance; the fourth was a day of rest.

The thermal regulation system (STR) employed two internal fluid loops (a cold, and a moderately-warm) and one external loop that ran through radiators surface-mounted on the narrower part of the compartment. Glycol was used as coolant because of its low freezing point. The heat exchanger was mounted in the unpressurised engine bay. The water vapour content and the temperature of the cabin air were regulated by the internal coolant loops. It could maintain a temperature in the range 16–28°C. If necessary, the regulator would increase the humidity, but usually it condensed vapour and pumped it to the main water tank.

Every effort was made to recycle water, but this was categorised by its quality and employed in different ways. Only the water vapour reclaimed from the station's air by the SRVK was potable. The water from the wash basin was pumped through a column containing ion-exchange resins and activated charcoal, filtered, remineralised, purified, and then reused only by the hygiene system. The toilet included a urine processing unit. This vapour-diffusion apparatus heated the urine with the STR's hot-loop to prompt the water molecules to diffuse through a membrane to condense on the cold-loop so that the water could be recovered. The concentrated urine was then fed into another tank. This, however, was forward planning for an electrolysis system which would release oxygen from the recovered water. Until that apparatus could be delivered, this filter could not be used and the tank was vented to space. Faecal matter was stored in a bag in a separate tank, and when this was full it was either ejected from the small airlock or dumped into a departing Progress ferry.

Oxygen and nitrogen were drawn as necessary from separate high-pressure tanks, and mixed to maintain a standard oxygen–nitrogen atmosphere at sea-level pressure. Progress ferries would pump liquid oxygen and nitrogen aboard to replenish the tanks. Later, in addition to the electrolysis unit, an apparatus was to be delivered to release a surge of oxygen by heating a solid-fuel canister. Dangerous trace constituents (carbon monoxide, hydrogen, methane and ammonia) were extracted from the air using regenerated charcoal beds and catalytic oxidisers. The carbon dioxide was extracted by passing air through a lithium hydroxide canister; with a crew of two, this canister had to be replaced every few days. On earlier Salyuts, a similarly-configured canister of potassium superoxide had been used, and this had both extracted carbon dioxide and liberated oxygen in a reaction which left a solid residue of potassium carbonate. In the lithium hydroxide scrubber, however, the surface of the crystal extracted the gas by adsorption, which did not yield oxygen. The lithium hydroxide scrubber was an interim solution though, because closed-cycle air cleansing and regeneration systems were to be incorporated in the add-on modules.

Over the first few days the cosmonauts adjusted the thermal regulation system to provide a comfortable 24°C environment, performed extensive tests of the navigation and orientation computer, and one-by-one checked out the rest of the station's systems. Only then did they set about attending to the experiment packages that had been stowed away for the launch. Progress 25 docked at the rear port on 21 March. Over the next few days, the crew unloaded cargo from its orbital module. This was a time-consuming task because it had been loaded to capacity and there was little room to move within its cramped compartment. Each container had to be unbolted and unpacked, and the food, tools, film, and assorted items stored within the base block. This job was made difficult by the fact that the quick-release bolts which held the containers in the racks had been over-tightened. In parallel with their manual labour, 200 kg of water was automatically pumped into Mir's Rodnik tanks (there were two of these, in the unpressurised engine bay). It was standard practice to leave the propellant transfer until just before the ferry was due to depart.

Like the second-generation Salyuts, Mir's ODU used UDMH and nitrogen tetroxide pumped from tanks incorporating nitrogen-inflated metallic bladders, and it employed a twin-chambered (one chamber either side of the rear docking port) orbital manoeuvring engine producing a total of 600 kg thrust. The chambers could be gimballed within a 5-degree cone to deliver their impulse through a slightly offset centre of gravity. Attitude control was by much less powerful (13-kg thrust) engines arranged in compact clusters at 90-degree intervals around the periphery of the engine bay. This system included 32 thrusters, but these were two groups of 16 with separate propellant feed pipes (forming a primary and a backup system). The pointing accuracy of the basic system was 1.5 degrees. In its precision mode it could maintain a given orientation to within 0.25 degree. When instruments requiring greater pointing accuracy were installed, so too would an improved orientation system.

Like the Salyuts, upon being placed in a low 192×238 km orbit by the upper stage of its Proton rocket, Mir had been required to manoeuvre to 332×354 km using its own engine, so its depleted tanks were replenished by Progress 25. On 25 March, and again on 18 April, the cargo ferry boosted Mir's orbit to 336×360 km. Subsequent ferries would progressively boost it towards its circular 400-km operating orbit.

Normally, direct communication was possible only when the station's ground track took it over a tracking station. On a favourable pass, the integrated land-based network spanning the Soviet Union provided at most 20 minutes coverage per orbit. With ships stationed in both the Atlantic and the Pacific, this time could be doubled on a favourable pass. A series of unfavourable passes, however, could result in the station being out of contact for periods of nine hours. The Luch relay satellites were designed to overcome this limitation. Unfortunately, there was only one satellite available; Cosmos 1700 had been launched the previous October and was now stationed over the Indian Ocean. The plan was to station three such satellites around the equator at 95° E, 200° E, and 344° E to provide a Satellite Data Relay Network (SDRN) which would provide continuous high-bandwidth communications with the scientific equipment soon to be installed on the expanded Mir complex, and enable data to be relayed to researchers immediately rather than stored on tape to be returned to Earth in descent capsules. Luch was the equivalent of the Tracking and Data Relay System (TDRS) then being developed by NASA for the shuttle and the Great Observatories which it was to deploy. As with Luch, however, there was only one TDRS satellite available at this time; the second had been lost a few months earlier, with Challenger. Mir's K$_u$-band system provided selectable channels in the 11–14 GHz range. It employed a steerable parabolic antenna mounted on a boom at the rear of the engine bay to relay voice, video and telemetry to Kaliningrad via the Luch satellite. It had been tested prior to launch, so Kizim and Solovyov transmitted a television broadcast to verify that the antenna tracked the satellite properly. Although Cosmos 1700 could relay signals for 40 minutes per pass, on every pass irrespective of whether the ground track passed over the Soviet Union, it turned out that Mir's normal work cycle often made maintaining the antenna's lock on a single satellite impracticable, so the K$_u$-band link would be of limited use until the network of relay satellites could be completed.

In the first week of April, Kizim and Solovyov performed a Resonance test (which involved determining the station's natural vibration modes in order to evaluate stresses on its structure) and continued installing apparatus unloaded from Progress 25. Their research was limited to visual Earth observations and setting up long-term experiments in plant growth. At the end of the week, it was announced that the first stage of their ambitious mission was coming to an end. Progress 25 left on 20 April, and Progress 26 took its place on 26 April. This delivered more propellant, food, water and apparatus, but the first item sought out by the cosmonauts was the mailbag. The first day of May was a holiday, and television coverage of the Red Square celebrations was relayed up to them.

In the days that followed, 500 kg of stores were transferred to Soyuz-T 15's orbital module, the ferry's batteries were recharged and its engine tested, and Mir was returned to its autonomous operating regime. It was time to move house.

AN OLD HOME

Since late April, Mir had been some 3,000 km behind the Salyut 7–Cosmos 1686 complex. On 5 May, Progress 26's engine was used to lower Mir's orbit slightly, to begin to reduce this separation. Later that day, Kizim and Solovyov undocked Soyuz-T 15 and withdrew. All three control rooms at Kaliningrad were now in use, one dealing with the Salyut 7–Cosmos 1686 complex, one with Mir, and one with the Soyuz. The next day, after a series of propellant-efficient manoeuvres, Soyuz-T 15 activated its Igla and made an automated approach to Salyut 7. The orientation of the complex was being controlled by Cosmos 1686 in response to commands from the ground. When the Igla paused 200 metres out, Kizim once again took command, and manually docked at Salyut 7's vacant rear port. This was the first time that a spacecraft had transferred *between* two orbital stations; the apparent ease with which it was achieved amply demonstrated how much the transportation system had matured. Only at this point was it announced that once the cosmonauts had finished Salyut 7's programme they were to load as much equipment as they could into Soyuz-T 15 and then, having already made history by making one inter-station transfer, they were to do so again by returning to Mir.

Upon venturing into their old haunt, the cosmonauts found it to be rather chilly, so they turned up the heater and returned to their ferry for the night. The following day, they reactivated the station's systems and replaced components such as filters, fans, scrubbers, lamps and transmitters that had exceeded their recommended service life. To round off, they performed a thorough test of the Delta and Kaskad control systems. By mid-May, the maintenance had been completed, so they turned to the research programme and settled down to making Earth observations. At the end of the month, they started preparations for a spacewalk. This involved cleaning and drying out the semi-rigid suits delivered by Cosmos 1669. On a previous visit, Kizim and Solovyov had accumulated nearly 24 hours working outside Salyut 7, so they were exceedingly well trained for the first item on the schedule for 28 May. Within 15 minutes, they had retrieved the Comet dust trap, together with all the cassettes which had been mounted outside the station by previous crews to test the way materials reacted to exposure to the space environment, including Medusa (biopolymers), Spiral (cables), Istok (connectors and bolts) and Resurs (structural metals). Once these had been secured within the airlock, they set about the main item on their agenda. The Ferma experiment involved erecting a girder. Designed and supplied by the Institute of Electrical Welding, in Kiev, the package had been delivered by Cosmos 1686.

Kizim and Solovyov set up a small work platform immediately outside the airlock hatch and affixed the 150-kg package to it. They selected the automatic option to deploy the structure. When fully erect, the 15-metre long girder comprised a set of aluminium–titanium alloy lattice-and-pin frames with a square section about half a metre on a side. Having demonstrated that the deployment mechanism worked, they retracted it. When it was back in its container, they mounted a sensor on one of the portholes, this being intended for use in a subsequent experiment. By the time they closed the airlock hatch, the cosmonauts were 45 minutes behind in the predicted three-hour schedule, but they were jubilant because they had achieved all of their objectives. The television down-link had been broadcast live by domestic television and, for the first time, shown in part by the American networks. On 31 May, having once again refurbished their suits, Kizim and Solovyov made a second spacewalk. The first task was to attach a flat plate to the top of the girder (which had been left retracted), mount a number of experiments on this, then extend the structure once again, this time manually.

One of these experiments was a low-power laser positioned to illuminate the sensor that had been deployed at the end of the first excursion. This measured the rigidity, stability and vibration modes of the structure. They used a welding kit (an improved form of the URI tested by Savitskaya) to electron-beam weld sample lattice-elements in order to evaluate this as a means of locking such joints in position. A similar girder was soon to be used on Mir, so verifying this deployment mechanism and assessing its dynamic characteristics had been the principal reason for returning to Salyut 7. Another package on the girder was Fon; this sampled the environment near the station. Previous data from the Astra mass spectrometer had demonstrated that a station created a tenuous 'tail' of gas. These experiments over, the URS girder was retracted and jettisoned. The final task was to retrieve the experimental solar transducer that Dzhanibekov had left, so that engineers could measure how it had degraded.

After a well-deserved rest, the cosmonauts spent early June making Earth observations, paying particular attention to the Kiev area following the nuclear accident at Chernobyl. Their data (together with that from the Meteor weather satellites) were used to direct the aircraft which were seeding clouds in an effort to prevent the normally heavy rains from flushing contaminated soil into the rivers and the reservoirs that supplied the Ukraine's major cities. They also performed materials-processing experiments in the Kristallisator semiconductor furnace and electrophoresis of biological materials in the Gel apparatus, and tended plants in the Biogravistat and Oasis apparatus.

In the following week though, Kizim and Solovyov began to dismantle apparatus to be salvaged. In all, they loaded 400 kg of equipment into Soyuz-T 15's orbital module. Their treasure trove included the spacesuits (there were none aboard Mir), the KATE-140 mapping camera, the EFU-Robot electrophoresis unit, the PCN low-light camera, the Echograph biomedical monitor, the Kristallisator and Pion-M furnaces, a number of spectrometers, television and video apparatus, film with 3,000 exposed images, and the sample cassettes recovered from outside Salyut 7. In mid-June it was announced that Salyut 7 was being cocooned. On 25 June, having restored the Salyut 7–Cosmos 1686 complex to its automated operating regime, Kizim and Solovyov shut the hatch for the last time and undocked. During their 50-day stay, they had successfully completed the work left outstanding by Vasyutin's crew.

MEANWHILE – ANOTHER NEW FERRY

In their absence, Soyuz-TM 1, an upgraded version of the Soyuz-T ferry, had been launched without a crew on 21 May. The intention to test a revised form of the proven Soyuz-T configuration had been announced two months earlier. The launch report said that its mission was to evaluate "the onboard systems and assemblies and the structural elements" of the revised configuration, and to conduct "joint tests" with Mir, the most obvious test being that of the Kurs rendezvous system. The new spacecraft pursued the usual two-day approach. During the rendezvous, the station remained passive; it simply provided appropriate navigational data to the approaching spacecraft, which did all the manoeuvring to line up with the front port before closing in and successfully docking with it on the first attempt. After raising the complex's orbit, Soyuz-TM 1 undocked on 29 May, then returned to Earth the following day.

The Soyuz-TM had the Soyuz-T and the Progress ODU, with the unified propellant system feeding both the main orbital manoeuvring and small attitude-control engines. It had an improved inertial unit which facilitated navigation independently of the ground tracking network. It also had an on-line display manual for the crew, and a diagnostic system that facilitated easier revision of the flight plan. This Rassvet communications link transmitted signals to Mir for relay via the Luch satellite. This meant that so long as a ferry was in line-of-sight of Mir it could maintain contact with Kaliningrad if the station was itself in contact. In addition, various modifications (notably the introduction of a lightweight parachute) had resulted in weight savings which enabled the new model to carry an additional 250 kg of payload into orbit, and to return with 150 kg (an increase in capability which had required an uprating of the retrorockets fired immediately before the descent module touched down). As would later become evident, the recovery criteria had been considerably relaxed, so that the new model could return at any time of the day.

By mid-June, Mir was only 1,000 km ahead of the Salyut 7 complex. Progress 26 replenished Mir with propellant (it had required the full loads of two tankers to make up the propellant expended by Mir's initial manoeuvres) and undocked on 22 June.

THE WANDERERS RETURN – BRIEFLY

On 26 June, Soyuz-T 15 lowered its orbit to chase Mir and rendezvoused with it for the second time. As before, the Igla made an automatic approach, and then Kizim took over and flew around to redock at the front port. On this occasion, video downlink was released. This revealed that the collar of the forward docking port was rotated compared with that of the Salyuts, with the result that the ferry docked with the plane of its solar panels at 45 degrees to (instead of coincident with) those of the station, but the reason for this new configuration was not elaborated.

On their first day back, Kizim and Solovyov unloaded the apparatus which they had salvaged from Salyut 7. Early in July, during a television report, they showed off their information system (a computerised manual for operating the station's apparatus) which they had just added to the Strela computer complex. It was hoped that this experimental unit would not only replace the many paper manuals, but also the Stroka teletype. This new system had the advantage that it could be updated by the flight controllers, to track changes in equipment and operating procedures. A significant milestone occurred on 2 July, when Kizim exceeded the 362-day accumulated endurance record previously held by Valeri Ryumin.

Most of early July was devoted to Earth photography, primarily for the Intercosmos organisation's Geoex-86 remote sensing programme. It focused on the GDR. Their results were correlated with data from other satellites, aircraft and ground teams. By this point, Kizim and Solovyov were well into the increased exercise regime which added half an hour to the daily routine to strengthen their load-bearing muscles and generally develop stamina in preparation for returning to Earth. The announcement on 14 July that they would return in two days time came as a surprise to some observers, who (accepting the statement that Mir was to be permanently inhabited) had expected this commissioning team to hand the station over to their next crew some time in September; but speculation that the 'emergency' return was prompted by a serious systems failure was unfounded. However, delays on the ground in preparing the first laboratory module for launch *had* made the intended orbital handover impracticable. Put simply, once Kizim and Solovyov had completed the limited number of experiments available, they grew bored. With no prospect of receiving the new module, they loaded their ferry with the exposed film and the sample cassettes that they had retrieved from Salyut 7, undocked and returned to Earth. Although the two men needed to be helped out of the capsule by the recovery team, and were immediately seated in reclining couches to await medical tests, they were in hearty spirits. It had been a truly remarkable mission. They had spent 50 days commissioning Mir, then 50 days finishing off research on the Salyut 7–Cosmos 1686 complex, and then another 25 days aboard Mir. Kizim had taken the accumulated endurance record to 373 days.

INTERMISSION

In late August the Salyut 7–Cosmos 1686 complex began to manoeuvre, prompting speculation that it might be about to rendezvous with Mir so that Cosmos 1686 could transfer across. Over a four-day period, however, each vehicle fired its main engine and virtually

Vladimir Solovyov and Leonid Kizim are clearly in high spirits upon completing their unprecedented multi-station mission.

expended its remaining propellant in climbing to a 480-km circular orbit. This was in marked contrast to the previous practice of deorbiting a discarded station.

With Cosmos 1686 continuing to provide attitude control, the complex was orientated in gravity-gradient mode for stability and then placed into 'suspended animation' to test how its basic systems degraded over time. Despite its problems over the years, Salyut 7 had successfully hosted ten manned ferries, 13 automated cargo ships and two of the large Cosmos vehicles. However, the key to the future was the Mir base block, then orbiting about 150 km below. The new station was working normally in its automated flight regime; nevertheless, it was reported that operations would not restart until early in the new year. The launch of Progress 27 on 16 January 1987 was covered live by domestic television, as was its docking with Mir two days later. In the six months that Mir had been dormant, its orbit had decayed to 312 × 340 km, so the ferry's first task was to boost this to 340 × 365 km.

At the end of the month, a new Luch satellite was launched, but its geostationary transfer stage malfunctioned, stranding it in a useless orbit. It had been meant to replace Cosmos 1700, which, upon running out of station-keeping propellant in October 1986, had drifted off its Indian Ocean relay station. Without a Luch relay, Mir was now in contact with Kaliningrad only for brief periods several times a day.

6

An astrophysics laboratory

In the spirit of *glasnost*, the launch of Soyuz-TM 2 was announced in advance, and the live television coverage began with the suiting-up process. In fact, Yuri Romanenko and Alexander Laveikin had started out by backing up Vladimir Titov and Alexander Serebrov, but the crews had switched priority when Serebrov had fallen ill. Ironically, because of the delay in preparing the first laboratory module, and the resultant slippage in this crew's launch, Serebrov was fully recovered by the time that Romanenko and Laveikin finally lifted off on 6 February. Soyuz-TM 2 pursued the now standard two-day rendezvous. The new Kurs system was switched on when the transfer orbit brought the spacecraft within 100 km of Mir. It initially approached the rear of the Mir–Progress 27 complex, and then automatically flew around to dock at the front. *Tass* reaffirmed that Mir was to be permanently inhabited, and added that this new crew hoped to set a new endurance record. It was Laveikin's first mission, but Romanenko had spent three months on Salyut 6 a decade earlier and had commanded a visiting Intercosmos mission. They spent the first few days readjusting to the normal Kaliningrad duty cycle because orbital dynamics had required that both the launch and the docking take place during the night.

The initial phase of adaptation to weightlessness was "rather painful" for Laveikin, so he "rested" for four days while Romanenko unloaded Progress 27. A medical check-up at the end of the week confirmed that they were adapting well. Although Laveikin had physically recovered, he later reported that it was almost a month before he felt fully at home. During the second week, they performed maintenance on expired parts of the environmental and thermal regulation systems, added elements to the electrical power distribution system, and refurbished the spacesuits. Meanwhile, Progress 27 pumped propellant and water aboard, and then undocked on 23 February.

EXPERIMENTS

Before they launched, the cosmonauts had expected the first laboratory module to follow within "a few months". Until it arrived, their primary research activity would be Earth observations. One of the first things they reported was a striking oceanic feature involving a nine-ring concentric wave form some 300 km across, centred on an area with an apparently "serene" surface. The Sever camera was "a fixed unit using a movable apparatus", and was to be used to take oblique imagery which would highlight surface relief. The new Pion-M experiment investigated heat and matter transport within a fluid (in its first test it revealed that silica aerogel in suspension formed saucer-shaped structures, fluoroplastics formed tree-shaped structures, and glassy pellets formed arbitrary but extremely robust clumps), and it was intended to assist in the design of an industrial-scale apparatus to process biological (particularly colloidal) materials in bulk. The Kolosok experiment investigated aerosol structure in microgravity.

Progress 28 arrived on 5 March and delivered a new KATE-140 mapping camera, the Gamma-1 biomedical test kit, the Korund-1M furnace, and several spectrometers. After testing the biomedical apparatus while using the veloergometer, Romanenko and Laveikin unpacked the Korund-1M (an improved form of the one tested on Salyut 7). Unfortunately, because it consumed 1 kW, its use had to be restricted until the station's power supply could be augmented. Then they installed the KATE-140 and conducted a survey of water run-off in the Caucasus, coordinating overhead KATE-140 views with oblique-imagery from the Sever camera. While Progress 28 replenished Mir's fluids on 24 March, they put accumulated rubbish into its orbital module; after it had gone they switched to general housekeeping duties while the materials experiments continued.

THE FIRST EXPANSION MODULE

Early on 31 March, a Proton placed a heavy satellite in an orbit coplanar with that of the Mir complex. The official announcement said it was a tug whose payload (Kvant 1) was an astrophysics laboratory for Mir. Being a stripped-down TKS, the tug had a pair of 400-kg thrust engines mounted peripherally for orbital manoeuvring, and clusters of 40-kg thrust engines for fine-scale manoeuvring and attitude control. At eight metres long, it comprised almost half of the 22-tonne overall mass. It was connected to its payload by a conventional probe-and-drogue docking system and, once this had been delivered, the tug was to undock and depart.

A year earlier, journalists visiting Kaliningrad had been shown a mock-up of a rear-mounted module fitted with astrophysical apparatus. To preserve the two-ended docking

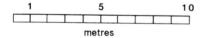

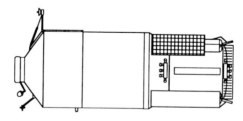

Kvant 1 with TKS tug attached.

capability, it was evident that this module would need to incorporate a rear-facing port. None of the modules which had docked with the second-generation Salyuts had had such a configuration, so clearly this was something new. In fact, it was derived from the pressure hull of the Proton high-energy physics satellites (it had been these that had given Chelomei's new rocket its popular name).

As with Cosmos 1686's slow rendezvous with Salyut 7, Kvant 1 made a series of engine firings designed to bring it alongside Mir with a low relative motion. On 5 April, when its final transfer orbit brought it within 20 km of Mir, the Igla radar transponder was switched on. Kvant 1 extended its docking probe when it was 500 metres out. At 200 metres, as it rolled to achieve the proper orientation, it lost transponder lock, began to deviate from the straight-in path, and aborted its docking attempt. Watching its video downlink, the flight controllers realised that there was every chance of a collision with the station. Romanenko and Laveikin had retreated to Soyuz-TM 2 at the far end of the base block. Despite the danger, flight director Valeri Ryumin asked them to re-enter Mir and look out of the portholes to try to locate Kvant 1. Romanenko watched it drift by in a slow roll, at a distance of only 10 metres. Later in the day, having analysed Kvant 1's telemetry, the engineers identified the source of the problem. Ryumin told reporters that they had been "unduly cautious" in defining the tolerances for permitting such a large module to approach, and that this was easily rectified. Unfortunately, since Kvant 1 had drifted far beyond Mir, it would take several days to set up another rendezvous.

On 9 April the cosmonauts again sealed themselves into Soyuz-TM 2, and Kvant 1 made its second approach. When 1 km out, its closure rate was 2.5 metres per second. By the time it was 25 metres out, this had been cut to 0.3 metres per second. This time, the final phase of the approach was uneventful. Kvant 1's probe slid straight into Mir's rear port and immediately achieved a soft docking. Unfortunately, when the command was sent to retract the probe to draw the two collars together, the main latches failed to engage. Although the sequence was rerun, the probe stopped a few centimetres short of its fully retracted position each time. When their troubled station flew out of radio range with Kaliningrad, the cosmonauts re-entered the base block and inspected the newcomer through the tiny porthole in the rear transfer compartment, but they saw nothing wrong. Kvant 1 could safely remain soft docked so long as the complex did not manoeuvre. It was eventually decided that the only option was for the crew to spacewalk, and inspect the docking mechanism.

Two days later, Romanenko and Laveikin opened one of the vacant radial ports in the forward docking adapter. Upon noticing that his suit was losing pressure, Laveikin suffered an anxious moment until he recycled a switch and rectified the fault. Then they cautiously made their way back along the 13-metre length of the base block, using a set of prepositioned hand-holds. They took a television camera with them to show the flight controllers what they found, but this did not work. As soon as Laveikin reached the engine compartment, he peered over the rim and saw "a white object" in the narrow gap between the collars. Clearly, this was what had prevented the spacecraft from drawing together. Kvant 1 was commanded to extend its probe, to open the gap, then Laveikin reached in and retrieved the obstacle, which turned out to be a cloth bag full of hygienic towels that had somehow become caught in Mir's docking assembly when the rubbish had been loaded into Progress 28.

Satisfied that nothing else was blocking the docking mechanism, the cosmonauts asked the flight controllers to command Kvant 1 to retract its probe, which it did, to achieve a hard docking. Having demonstrated yet again the ability of a human crew to troubleshoot a problem, Romanenko and Laveikin made their way back inside. It was subsequently revealed that in a medical check-up after returning from the spacewalk an anomaly was detected in Laveikin's heart rhythm, an irregularity which, if it had been detected prior to launch, would have precluded his flying. Their first task inside was to reconfigure the base block's computerised attitude-control system to take account of the altered centre of mass of the Mir–Kvant 1–Soyuz-TM 2 complex, which (with the TKS tug in place) was 33 metres long and had a mass of almost 50 tonnes. It took a long time to fine-tune the attitude control system, however, and consumed more propellant than had been anticipated.

The next day, 12 April, Cosmonaut Day, they removed the docking assemblies and entered Kvant 1 for an initial inspection. Later that evening, the tug was commanded to undock. As it withdrew, antennas for both the Igla and the Kurs rendezvous systems unfolded on the rear of the exposed payload. Unfortunately, the tug had consumed so much propellant in setting up the second rendezvous that it was unable to deorbit itself, so had to be left for its orbit to decay naturally.

Kvant 1 was a stubby cylinder 5.8 metres long and of the same diameter as the main body of the base block. It comprised three separate hermetic compartments. At the front, adjoining Mir, was the short forward transfer tunnel. A lightweight internal hatch led to the main laboratory compartment in which the controls for the scientific apparatus were located. The hatch beyond accessed the longer tunnel which led to the rear-facing docking drogue. Equipment was wrapped around this tunnel, within an unpressurised compartment. Of the 1,600 kg of scientific apparatus built into Kvant 1's structure, fully 800 kg was Svetlana, the semi-industrial-scale electrophoresis processing system for biological materials (named after Svetlana Savitskaya, who had tested an experimental form of the apparatus onboard Salyut 7 in 1984). The rest of the compartment was taken up by a variety of high-energy instruments (collectively known as the Roentgen apparatus) which had been supplied by a range of collaborating international partners.

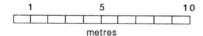

metres

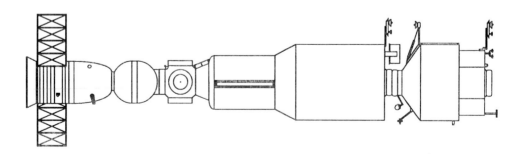

Mir with Soyuz-TM 2 and Kvant 1 (after TKS tug departure).

The Max Planck Institute and the University of Tübingen built a scintillation spectrometer, known as the High-Energy X-ray Experiment (HEXE). It incorporated four identical
Phoswich detectors sensitive to X-rays in the energy range 15–200 keV, and with a 1.6 × 1.6
degree field of view. ESA's Sirene-2 high-pressure gas-scintillation proportional-counter,
which was an improved version of apparatus flown on the EXOSAT observatory, was
sensitive to X-rays in the 2–100 keV energy range, with a 3 × 3 degree field of view. It
was to study emissions from extremely high-temperature rarefied cosmic gas. The TTM
(COMIS) wide-angle coded-mask imaging spectrometer supplied by the Netherlands
Space Research Organisation in Utrecht and Birmingham University was an improved
form of apparatus flown on the Spacelab 2 shuttle mission. Sensitive to X-rays in the
2–30 keV energy range, and with a 7.8 × 7.8 degree field of view, it was used to determine
the location of X-ray sources to an angular resolution of 2 arcminute. The Pulsar X-1 hard
X-ray spectrometer had been supplied by the Soviets. Its four identical Phoswich detectors
studied X-rays and gamma-rays in the 50–800 keV energy range, with a 3 × 3 degree field
of view. The Soviets also supplied a 180-degree wide-field scanner for survey work. This
was sensitive up to 1,300 keV, and had a one millisecond temporal resolution.

Most of these telescopes were designed to be capable of being operated remotely by
researchers on the ground. To exploit this real-time downlink, a new Luch satellite was
launched on 26 November (like its defunct predecessor, Cosmos 1897, it was placed over
the Indian Ocean), but a single relay satellite would not really be adequate.

The Glasar telescope, built by the Byurakan Astrophysical Observatory in Armenia,
was configured for ultraviolet spectrography in the wavelength range 1,150–1,350 Å. It
used an electronic image-intensifier, but exposures as long as ten minutes were required
to record faint stars. There was a small airlock in Kvant 1's transfer tunnel to enable the
cosmonauts to reload the film cassette. This was to conduct a survey of bright quasars,
active galactic nuclei and stellar associations at ultraviolet wavelengths.

A variety of other equipment had been built into Kvant 1 to assist in operating the
expanded complex. The Elektron apparatus electrolysed water reclaimed from urine and
produced oxygen (the hydrogen it released was vented into space, further adding to the
gaseous tail behind the station). It maintained the desired composition of the gaseous environment within the complex without requiring so much liquified gas to be delivered by
Progress ferries. It used long flexible air tubes that were strung through the complex by the
cosmonauts. Vozdukh contained a molecular sieve to scrub carbon dioxide from the air
(which was also vented), reducing the number of lithium hydroxide canisters which had to
be delivered by Progress ferries. These regeneration systems, together with the water
vapour condenser in the base block, were a step towards achieving a closed-cycle environment. During crew handovers, a chemical burner (Vika) topped up the oxygen.

To perform a vital engineering function, Kvant 1 incorporated six gyrodynes. The 165-kg
magnetically suspended flywheel control-moment gyros were spun in pairs at 10,000 rpm,
one pair on each of the three cartesian axes. They converted electricity from Mir's solar
panels into torque, to serve as an inertial attitude-control system which could orient the
complex without consuming propellant. Once the required position had been attained, they
damped out perturbations due to active elements (such as the crew) and held the complex
stable. It was intended that this system would be able to orient the complex to an accuracy
of 1.5 arcmin (an order of magnitude better than was possible using the base block's
thrusters), so that the telescopes could be used effectively.

In addition, the main compartment of the new module had been packed for launch with 2,500 kg of assorted cargo. The bulkiest item was a 340-kg solar panel (it was actually in two segments) to provide the power needed by the new equipment. There was pressure from astronomers to start using the telescopes to observe the supernova SN 1987A which had been detected in the Large Magellanic Cloud (LMC) two months earlier, but this was deemed impracticable until the solar panel could be erected. Unfortunately, this required spacewalking and, having discovered Laveikin's heart irregularity, the flight surgeons insisted that he not be allowed out until his condition could be properly evaluated. They requested that from now on during strenuous exercise his cardiac trace be transmitted over the downlink for analysis.

The largest of Kvant 1's four portholes was the 43-cm diameter high-fidelity optical glass in the floor. A mock-up of the module shown at the Paris Air Show had an MKF-6M multispectral camera fitted, but this had not been included in the cargo. The cosmonauts were to fit a high-precision star tracker into the next-largest (23-cm) port to provide updates for calibrating the alignment of the gyrodynes. The others were small (8-cm) observational ports for use by the crew.

With Kvant 1 mounted at the rear of Mir, it became impracticable to fire the base block's main engines. Kvant 1 had no propulsion of its own (which was why a tug had been necessary). From then on, therefore, the complex would have to rely on a docked Progress ferry to perform major orbital manoeuvres.

It was later disclosed that Kvant 1 had originally been designed to dock at the front of Salyut 7, but the station's troubled history had imposed so many delays that it had been decided to modify the module to function at the rear of Mir. This had required that pipes be installed so that a Progress tanker could pump fluids around it into the base block. It is noteworthy that if Kvant had been sent to Salyut 7 its double docking ports would have enabled it to form a permanent adjunct to that station, whereas modules like Cosmos 1443 were only temporary because they had to be jettisoned to release the front port for the next ferry. The evolutionary nature of the development of the elements for constructing a modular orbital complex is thereby revealed.

Years later, it was revealed that the Korolev Bureau had intended that Mir be built up using a number of modules of this type, each with a mass of about 8 tonnes, but had been ordered instead to use 20-tonne modules derived from Chelomei's TKS. Work on further small modules (most particularly a biotechnology laboratory, a power plant and an X-ray telescope, initially to be delivered by TKS tug, and later by the Buran shuttle) continued, but Kvant 1 turned out to be both the first and the *last* of its type. As a hybrid, therefore, Kvant 1 seems with hindsight to have been something of a 'cuckoo' in its subsequent domination of the Mir complex. It had been intended to send up Kvant 1 during the second part of the Soyuz-T 15 crew's residency, but it had not been possible to complete the modifications in time. As a result, this initial phase in the expansion of the Mir complex was far behind schedule.

After a few days spent unloading cargo, the cosmonauts updated the attitude control system's mass model to take into account the migration of the centre of gravity caused by the redistribution of so much apparatus, then powered up and tested each gyrodyne.

Progress 29 docked at the rear port of Kvant 1 on 23 April. It created a four-vehicle in-line complex 33 metres long. Although the 750 kg of propellant it delivered was less than normal, it contained 170 kg of water, 250 kg of food, 140 kg of film, 275 kg of replacement hardware, and 140 kg of miscellaneous items (some of which were for the long-delayed Syrian mission). The two men unloaded this over the next week and then, while fluids were pumped into the base block's tanks through the pipes running around the intervening module, they took a well-earned rest on 1 May to watch the Red Square celebrations on the uplink.

After a few days of exhaustively testing the reprogrammed attitude control system's ability to handle the unprecedented four-vehicle structure, the cosmonauts activated the environmental support systems in Kvant 1. Progress 29 departed on 11 May. On 21 May, while Romanenko and Laveikin were still fine-tuning the ability of the Argon to manoeuvre the complex using the gyrodynes, Progress 30 arrived. They unloaded its cargo for two days, and then set up the Yantar electron-beam spray in the scientific airlock (in the floor, between the veloergometer and the treadmill). This followed on from experiments with the Isparitel apparatus aboard Salyuts 6 and 7. It was tested by vapourising a copper–silver alloy to create a thin layer over a polymer film. After that, the newly delivered high-precision star tracker was set up in Kvant 1. With this rapid build-up, the early Western criticism that Mir lacked scientific equipment now seemed rather naive.

MEANWHILE – A NEW LAUNCHER

On the evening of 15 May the first Energiya was launched. Despite *glasnost*, there was no live coverage of the launch. This Saturn V class heavy-lift launcher carried its payload within a 38-metre long cylindrical container strapped to the side of the 60-metre core segment. Unfortunately, although the rocket functioned flawlessly, the payload did not. It was not properly orientated when its orbital insertion engine fired, and it drove itself into the atmosphere above the Pacific Ocean.

The payload was an engineering mock-up of Polyus, a platform intended to carry an industrial-scale factory for materials processing and biotechnology. It was envisaged that such platforms would either be incorporated into a future orbital complex or operate autonomously and be visited occasionally for maintenance. Nikolai Gerasimov, head of the Salyut Bureau (part of the former Chelomei Bureau), said that Polyus could be used to deliver 40 tonnes of cargo to a station in low orbit. Dr Vladimir Pallo, the platform's designer, said that it was essentially an enlarged version of the Mir base block, with a mass of 80 tonnes. Dr Boris Gubanov, chief designer of the Energiya Bureau (formerly the Korolev Bureau), responsible for the new rocket's development, said that it was to be used to launch the elements of Mir 2, which was then envisaged as a rather elaborate complex equivalent to NASA's Space Station Freedom concept.

SPACEWALKING: A NEW SOLAR PANEL

For ten days in late May, when the Sun passed through the plane of Mir's orbit, the time available for the two solar panels to generate electricity was so brief that they would be unable to recharge the storage batteries if power-hungry apparatus was used during the shadow pass. During this time, therefore, although plants continued to grow in the Rost and Phyton cultivators and electrophotometric and spectrometric atmospheric data were collected, little other scientific work could be performed. Then, on 6 June, with the power crisis over, the Roentgen telescopes were powered up for 'first light' trials. At the top of the list of X-ray sources was, of course, SN 1987A in the LMC. Testing continued for a period of several days. Unfortunately, it was found that when Mir was near the South Atlantic Anomaly the X-ray detectors suffered electron precipitation.

After a medical check in early June, Laveikin was declared fit to spacewalk to install the new solar panel. The timing of the spacewalk was dictated by a requirement that the three or four orbits during which they would be outside offer daylight passes over the Soviet Union for maximum communications. This work had initially been scheduled in April, immediately after Kvant 1's arrival, but concern regarding Laveikin's heart had prompted its postponement first to early May, then to late May and finally to mid-June.

The new solar panel had been delivered in two segments, so two spacewalks would be required. The multiple docking adapter, which was essentially a 2-metre diameter sphere, was too small to accommodate the men together with the bulky panel segment. Soyuz-TM 2 was at the front port, to provide an escape route in the event that they were unable to re-enter the base block, so its orbital module was to be used as a storage locker for the apparatus. Accordingly, on 12 June the spacecraft's internal hatch was shut to protect its descent module, then the docking adapter was depressurised with the transfer tunnel open. With the upper docking port open, one man moved out and the other eased out the panel segment.

Romanenko and Laveikin fixed a stubby cylindrical container onto the motor mount embedded in the roof of the base block, then aligned the panel with its arrays extending 'horizontally' out to either side. The container held an extensible girder similar to that

The Mir base block with Kvant 1 and a Soyuz-TM docked, showing the third solar panel deployed by Romanenko and Laveikin in June 1987.

tested by Kizim and Solovyov outside Salyut 7. This task proved much easier than they had expected. After resting and refurbishing their suits, they ventured out again on 16 June. This time they attached the second segment on top of the first, then extended first the upper and then the lower truss, to extend the panels in concertina fashion. Once the 10-metre long structure was deployed, they connected the colour-coded cables to sockets on the base block to link the new panel into the electrical system. Once again, they found themselves ahead of schedule. Having anticipated this possibility, a number of cassettes containing samples of materials to be exposed to the space environment had been placed in the makeshift airlock, so they retrieved these and affixed them to anchors on the surface of the base block. With an area of only 22 m^2, the new panel added just 2.5 kW to the 9 kW, but 11.5 kW was sufficient to run the X-ray telescopes, a furnace and the gyrodynes simultaneously, so astrophysical observations were made in parallel with smelting and additional Yantar experiments, and the first pictures were taken using the Glasar ultraviolet telescope.

In early July, a feasibility study was conducted to evaluate whether clinical analysis of blood samples could be performed in space, rather than after being returned to Earth, to provide instant results. The doctors had been "closely following" the cardiovascular data from the Gamma-1 biomedical monitor, which confirmed Laveikin's extra-systolic activity. Unfortunately, although the apparatus was sufficient to reveal this irregularity, its data were insufficient to be certain of a diagnosis, so it was decided to recall Laveikin as soon as possible. Since the problem appeared only when he exerted himself, he was told to take it easy.

July saw a very varied research programme which included photography of the Urals, Moldavia, the Crimea, the Pamir mountains, the Caspian depression, and areas in the Soviet Far East; using the Biostoykost polymer apparatus; making semiconductors in the Kristallisator furnace; using the Gel electrophoresis apparatus to process biological substances; tending the Phyton and Rost plant growth experiments; and participating in the TeleGeo-87 programme for the Intercosmos organisation. (An Earth-study programme was run each summer; this one concentrated on Poland.)

THE LONG-DELAYED SYRIAN MISSION

Progress 30 departed on 19 July, and Soyuz-TM 3 was launched three days later. It carried Alexander Viktorenko, Alexander Alexandrov and the Syrian Mohammed Faris.

Although Viktorenko was making his first flight, Alexandrov had served a 149-day tour on Salyut 7 in 1983. They had originally expected to visit Vasyutin's crew on Salyut 7 in 1985, but a series of problems had prompted their mission being reassigned to Mir. Having sat out the Soyuz-T 15 crew's commissioning of the new station, then the frustrating period during which it had remained vacant, they finally got their chance. After a flawless automatic docking, the visitors had to use a lever to crack their hatch's hermetic seal. Faris floated through the tunnel into the Kvant module and proceeded to give his hosts the traditional Arab greeting. Romanenko observed that with three people aboard called Alexander, things could get a little confused.

It was at this point that Vladimir Shatalov publically announced that Laveikin was to return home early. He emphasised that although specialists had "no serious misgivings" concerning Laveikin's heart, it was nevertheless prudent to relieve him, and Alexandrov would take his place. Although he had been given just a month's warning that his flight would be more than a brief visit, Alexandrov knew what this would involve because of his tour on Salyut 7. Deputy flight director Viktor Blagov reported that it was intended to assign cosmonauts individual missions, so in the future partial crew exchanges such as this would be commonplace.

Amongst the items delivered by Soyuz-TM 3 was the Ruchei electrophoresis unit. This was set up in Kvant 1 and used to purify interferon and an anti-influenza vaccine. The Svetlana apparatus was used to separate active microorganisms to make agricultural antibiotics to assist in stock rearing. A whole range of other typical research was carried out too. Bosra gathered data intended to improve the mathematical models of the upper layers of the atmosphere and ionosphere. Kasyun made use of the Kristallisator furnace to smelt an aluminium–nickel alloy, and Afamia used it to make a gallium–antimonide monocrystal. Palmyra mixed two substances to make a crystalline structure resembling that of human dental and bone tissue. Faris combined the standard medical checks with photography of Syria. The need for favourable daylight passes had been the primary factor in timing the visit. One objective was to use the KATE-140 to map ancient sites. The related Euphrates experiment combined visual, photographic and spectrographic observations to identify water and mineral resources and to assess local atmospheric pollution.

On the evening of 29 July, Laveikin joined Viktorenko and Faris in Soyuz-TM 2. No sooner had they settled themselves into the cramped descent module than they were told to delay undocking for two orbits, to avoid a rain storm at the primary landing site. Even so, the winds caught the parachute and carried it far off course, and it nearly came down on a small settlement. Like most guest researchers, Faris had neglected to sleep, so he was thoroughly exhausted. Laveikin, although said to be "pale" after 174 days in space, was flown to Moscow to be examined by heart specialists, who pronounced him fit to fly! Laveikin was no doubt keenly aware of the irony that if it hadn't been for the advanced biomedical equipment on Mir, the doctors would never have become aware of his cardiac irregularity, and he would still have been in space with Romanenko on what was intended to be a record-breaking mission.

A LONG SIX MONTHS

On 31 July, Romanenko and Alexandrov undocked Soyuz-TM 3 from the rear port, pulled back a hundred metres, waited half an hour while the gyrodynes rotated the Mir–Kvant 1 complex end over end, and then redocked at the front port. Progress 31 arrived on 6 August. After three days unloading cargo the cosmonauts started a run of atmospheric observations in the vicinity of a power station in Kazakhstan, for an ongoing study of the propagation of pollution. At the end of the month, they held an emergency drill by retreating to the ferry to don the Sokol pressure suits (which is what they were to do in the event of a slow air leak). The first week of September was devoted to maintenance. One particularly tricky operation involved replacing a pump in the thermal regulation system; they had to be careful not to spill glycol. While they slept, the X-ray telescopes were used by remote control; by this point, 300 observational runs had been made. As a special treat, the cosmonauts were told that they could eat the radishes and onions that they had grown. The

Biryuza apparatus was used to study the dynamics of physio-chemical processes (in particular, the formation of "spatial structures during a chemical-oscillation reaction"). Polycrylamide was made by the Gel electrophoresis unit (to make the purification of biologically
active substances on Earth more efficient). It was a very varied programme!

Progress 31 departed on 22 September. Progress 32 arrived four days later, and as it
made its final approach, flight controllers watching its downlink noted that Mir's new solar
panel flexed and shook as the Igla commanded small manoeuvres to keep the rear port
aligned.

At the end of September, Romanenko broke the 237-day single-mission endurance
record which had been set in 1984 by Kizim, Solovyov and Atkov. Romanenko was in
good physical and mental health but he was clearly growing tired, so his working day was
reduced to 5.5 hours (compared to the standard eight hours) and he was encouraged to
exploit the full sleep period (once adapted to weightlessness, cosmonauts required less
sleep). Accordingly, Alexandrov did the heavy work of unloading Progress 32's cargo
while Romanenko continued experiments. By mid-October, however, Romanenko was
rather irritable, and he suffered from fatigue and insomnia. Because he thought the tasks
monotonous, his work load was further cut (to 4.5 hours per day). Vladimir Shatalov emphasised that it was important for the morale of the cosmonauts on long missions that the
results of their experiments be examined quickly, and the conclusions relayed back so that
they could be assured that they were working effectively, otherwise they risked becoming
disillusioned. He said that when the operational life of the Soyuz ferry had been just a few
months, the frequent Intercosmos visits had meant that results could be returned to Earth
quickly, but now that a ferry could remain in orbit for six months there was no need for
such frequent flights, so unless a crew received visitors, it might never see the results of
its efforts. The obvious thing to do was to incorporate a small descent capsule into the
Progress cargo ferry (because these were dispatched frequently) so that results could be
returned incrementally; in fact, the development of such a capsule was well advanced.

On 4 October, the cosmonauts took part in a televised link-up with the participants of
an international forum in Moscow to celebrate the 30th anniversary of Sputnik 1. A stock
of commemorative envelopes had been sent up in Progress 32 to be franked in space on
the day. Ultraviolet photography with the Glasar telescope was resumed. Observations by
the Roentgen telescopes revealed a change in the X-ray spectrum of the supernova. Having increased for two months, the flux now began to decrease (this was correlated to a
brightening of the envelope of ejecta surrounding the stellar remnant); it reversed a few
weeks later and began to brighten once more. Romanenko and Alexandrov took the day
off on 7 November and watched the Red Square celebrations of the 70th anniversary of
the Revolution.

Progress 32 replenished fluids, then undocked on 10 November. It withdrew to a distance of about 3 km, then returned two hours later to evaluate a revised approach that
minimised the reorientation manoeuvres the Igla system required Mir to execute. As it
redocked, its video downlink showed that the upper panel did not flex. As a bonus, this
revised procedure had consumed about 100 kg less propellant. It departed a week later.
When Progress 33 approached on 23 November, it employed this new procedure. This
reprogramming indicated the remarkable degree of operational flexibility in the resupply
service. Alexandrov immediately set to unloading cargo. The Mirror furnace used a pair

of lamps and a mirror system to smelt at temperatures up to 1,000°C. It was to be used to test procedures for apparatus in the 'technology module' that was to be added to Mir later. An improved Mariya spectrometer was also set up to study high-energy charged particles in near-Earth space, to resume a monitoring programme started on Salyut 7, in the hope of correlating flux variations with tectonically active surface areas.

On 2 December, Romanenko passed the 300-day mark. Medical tests revealed that although the muscles in his legs had diminished by as much as 15 per cent, this atrophy was within "foreseen limits". Viktor Blagov announced that Romanenko was "very much" missing his home, his family and his friends; and just as he had towards the end of his 96-day marathon on Salyut 6 in 1978, he had become increasingly grumpy. In fact, he was now so tired that almost all the work was being done by Alexandrov. Blagov noted that although it was planned for cosmonauts to generally serve six-month tours on Mir, it was necessary to explore the process of adaptation to weightlessness over much longer periods. Romanenko later opined that the optimum length for a tour of duty would be three to four months.

HANDOVER

Progress 33 departed on 19 December. Four days later, Soyuz-TM 4 docked at the rear of the complex. It brought Vladimir Titov, Musa Manarov and Anatoli Levchenko.

When Soyuz-T 8 had been launched in 1983, the jettisoned shroud had ripped away the spacecraft's Igla antenna. Titov had nevertheless successfully performed a manual rendezvous with the Salyut 7–Cosmos 1443 complex; he had been forced to abandon the approach with only a few metres to go when the complex had flown into the Earth's shadow and he had lost his bearings. Later in that year, during the final seconds of the countdown for his second attempt to reach that station, the rocket on which Soyuz-T 10 was mounted caught fire, and he and Gennadi Strekalov had survived only because the escape tower had hauled the capsule free of the maelstrom of the exploding rocket.

Titov had originally been scheduled to fly with Alexander Serebrov (with whom he had backed up Romanenko and Laveikin), and they were to have been accompanied by Dr Valeri Poliakov, who was to have assessed Romanenko's state of health before his return to Earth. This plan had been changed initially by the substitution of Manarov for Serebrov as flight engineer, and again at the last minute when Levchenko was assigned Poliakov's seat. What had happened was that the Ministry of Aviation had insisted that a Buran pilot trainee be flown as soon as possible to gain experience of weightlessness, as Igor Volk (leader of the Buran group) had done on Soyuz-T 12 in 1984. When Levchenko floated through the hatch into Mir, Romanenko had to ask him who he was; he had never seen him before.

Throughout the week, the old residents briefed their replacements on the status of Mir's systems and reviewed their spacewalking activities. By maintaining a permanent presence aboard the complex, with one crew handing over a fully operational station to its successors, it was estimated that about a week would be saved which would otherwise have been wasted powering down and then reactivating the complex. As more and more apparatus was delivered, it also became important to have a crew show where it had stored everything. And, as the complex was expanded and serviced, a handover in orbit enabled the newcomers to be briefed on the quirks of the systems as they really were because no simulator was perfect. The programme included vegetable and animal tissue culture growth, a small aquarium, installing the Ainur electrophoresis apparatus (and then using it to grow protein crystals), and purifying interferon in the Ruchei apparatus.

On 27 December, Romanenko and Alexandrov stowed their experimental results in Soyuz-TM 3's descent module and loaded rubbish into its orbital module. Their haul included 270 ultraviolet photographs by the Glasar telescope, samples processed in the Korund-1M furnace, Earth photographs and biological samples. Their last day in space was hectic, with the two crews sorting out their various apparatus. At one point, one of them remarked that he would be able to get on better if the flight controllers would just desist from trying to remind him of everything. When Soyuz-TM 3 departed on 29 December, it carried Romanenko, Alexandrov and Levchenko. The fact that they had never trained together was remarkable confirmation of Blagov's prophetic statement that cosmonauts would soon begin flying individual missions.

The descent module touched down north-east of Arkalyk. A television camera in one of the recovery helicopters showed it descending over the frozen windswept steppe into a blizzard. The deflating parachute caught in the wind and tipped the capsule onto its side. The wind prevented the inflatable medical tent being erected, so it was decided to airlift Romanenko and Alexandrov to Arkalyk. Despite having spent 160 days in space, Alexandrov was permitted to walk to the helicopter. Although Romanenko said that he would walk, the medical team insisted that he recline on a stretcher because he had been away 326 days, which was fully three months longer than anyone else. Nevertheless, the following day he was shown on television walking with his wife in the grounds of the Cosmonaut Hotel, at the cosmodrome.

Reviewing Romanenko's rapid deterioration after about eight months in space, it was concluded that 'routine' missions should be scheduled to last four to six months. Titov and Manarov however, were hoping to break Romanenko's new record. The fact that neither had spent time on an earlier station suggested that Romanenko's cumulative record of 431 days in space would be safe for some time to come.

Levchenko died in August 1988, following surgery to remove a brain tumour. By a strange coincidence Alexander Shchukin (his backup) was killed in an accident at an air display a fortnight later, so Ural Sultanov was assigned as Volk's new co-pilot for the eagerly awaited Buran mission. When Volk was grounded, Rimantas Stankiavicus took his place, but he too died before he was able to make his familiarisation flight to Mir. All of this group were highly experienced test-pilots.

A YEAR IN SPACE?

Early on 31 December, as they became the first cosmonauts since Romanenko and Grechko a decade earlier (onboard Salyut 6) to welcome in the New Year in orbit, Titov and Manarov transferred Soyuz-TM 4 to the front. To make the event even more of a holiday, on the following day, 1 January 1988, they set up a two-way television link with their families and celebrated Titov's birthday. They had several spacewalks planned and hoped to commission the next expansion module. They were not expected to receive any visitors until well into the summer, when their ferry would be replaced.

The first half of January was devoted to astrophysical observations and to smelting semiconductors. When Progress 34 docked on 23 January it brought propellants, more consumables, the usual replacement items for the environmental systems, a replacement segment for the upper solar panel and specifically requested music cassettes. After this cargo had been unloaded, Titov and Manarov tackled a rather varied programme during the rest of the month: they studied thermocapillary processes with the Pion-M apparatus, performed the ERI experiment which tested methods of depositing galvanic coatings, and took ultraviolet pictures with the Glasar telescope. The semi-automated Korund-1M was in use throughout. The remote-control astrophysical observations concentrated on the supernova. The flux had been increasing since November, and it was now 50 per cent brighter than it had been at its peak in August. (It had been suggested that these variations might be due to the development of holes in the expanding ejecta which exposed the hot stellar remnant.) In mid-February, the video uplink was used to brief the cosmonauts on the procedures for their first spacewalk, and while Progress 34 pumped fluids aboard they refurbished their spacesuits. The excursion was scheduled for 26 February. Keen to start, Manarov floated out of the vacant radial docking ports before radio contact was established with Kaliningrad.

The first task was to attach their equipment to the anchor point near the upper solar panel. Once they were all set up, they retracted the lower array and replaced one side of it with a new panel using improved semiconductor photoelectric transducers. This was part of a long-term evaluation of the efficiency of different types of transducer and the rate at which they degraded. Actually, only six of the new array's eight segments were wired to produce electricity; the other two were instrumented to yield telemetry so that their performance could be periodically monitored by engineers. Even though the new panel segments were smaller than those they replaced, they produced 20 per cent more power (the standard silicon-based transducer converted 11 per cent of sunlight into electricity, but this new gallium arsenide transducer had an efficiency of 14 per cent) so the station's supply was actually increased slightly despite the experimental units. The downlink showed the two men working on the panels. Using the small lamps built into their suits, they worked on while the complex passed through the Earth's shadow.

The main task complete, they collected more equipment and made their way along the length of the base block, paused at the juncture with Kvant 1 to deploy the handrail built onto the conical front of the module, and then moved across this bridge to the rear of that module where they set up several experimental packages and visually inspected Progress 34. On their way back, they photographed the external surface of the complex to begin to document its exposure to the space environment. The first order of business once they were back inside was to carry out a thorough medical check; the incident with Laveikin's heart had prompted ongoing concern.

Progress 34 departed early on 2 March. During the next week, the cosmonauts used the Mariya and Pion-M apparatus, and processed materials in the Mirror furnace. Earth observations were resumed, this time concentrating on the Central Asian Republics, to assess snow and glaciers in the Pamir and Tien-Shan mountains for an ongoing study of irrigation in that region. Astrophysical observations continued under remote control (by this point in the mission, the X-ray telescopes had been used 125 times, most often aimed at the supernova). They also installed a fax machine to permit sketches of their observations to be sent to scientists in the support team. The environment within the complex was itself a subject for study. A part of this was the Akustika experiment. This measured background noise at various points (the hum from the environmental systems was typically 80 decibels). The cosmonauts reported that there was a lot of dust adrift in the station (it made them sneeze), and they complained of an irritating smell from the environmental system. It was reported in mid-March that although they were managing to follow their programme, their working efficiency was having its "ups and downs".

Progress 35 docked on 26 March. As well as the usual replacement components for the environmental and thermal regulation systems, it delivered 400 kg of food (much of which was fresh fruit and vegetables). After the cosmonauts had unloaded the cargo, routine maintenance was interleaved with research, most notably using the EFO-1 spectrometer to study the atmosphere for the Climate experiment, and the Kristallisator to investigate the formation of crystals.

On 12 April, marking Cosmonaut Day, Viktor Blagov reported that the first of the modules to be attached to the front of the complex was being prepared for launch "at the end of the year". It would incorporate an airlock, in which would be the autonomous manoeuvring unit. The second, another astrophysics module, was to be launched six months later. The third, a materials processing factory, might be operated as a free-flyer to achieve ultra-low microgravity conditions, docking only to be serviced. The final module was for life sciences, and would have a sterile compartment and a small airlock. When finished, he observed, the complex would have a mass of over 100 tonnes. For a variety of reasons, this plan for assembling the complex (which called for completion at the end of the decade) was to prove impracticable. In retrospect, considering the fast pace of this early plan, and the fact that the construction process had started, it is astonishing that the suggested rôles for the modules could be so different from those which were eventually launched.

Later in April, the cosmonauts complained that they had been receiving conflicting demands on their time from different groups of researchers. Even so, they worked with the Pion-M apparatus, made Earth-resources studies of Cuba and took astronomical photographs with the Glasar telescope (by this point, they had accumulated 20 images). Progress 35 undocked on 5 May, and Progress 36 arrived on 15 May. Radio Moscow reported that the cargo included a new detector for one of the X-ray telescopes, together with special tools needed to install it. The TTM had been suffering intermittent electrical interference since late 1987. Because the apparatus in the unpressurised compartment at the rear of Kvant 1 had not been intended to be serviced by cosmonauts, simply gaining access to the instrument would be difficult. A video had been included to show how to replace the detector. The spacewalk to attempt it was scheduled for later in the month.

During the second half of May, Titov and Manarov completed unloading cargo, and installed "a considerable amount" of apparatus which had been sent up in anticipation of the forthcoming Bulgarian mission (this was done to ensure that their visitors would be able to start their research programme immediately). The Gel apparatus was used to make more polycrylamide. Meanwhile, Progress 36 replenished fluids, refined the complex's orbit, and departed. By this point, if they had been on a six-month tour they would have been preparing to return to Earth, but they were doing so well that they had been given the go-ahead to continue.

VISITORS: BULGARIA – FINALLY

Soyuz-TM 5 was launched on 7 June with Anatoli Solovyov (not to be confused with Vladimir Solovyov) and Viktor Savinykh, and with Alexander Alexandrov (a Bulgarian research cosmonaut, not to be confused with the Soviet cosmonaut of that name). This was the second Bulgarian mission. In 1979, Alexandrov had backed up Georgi Ivanov, who had been launched in Soyuz 33 to Salyut 6 and then been prevented from reaching it when the ferry's engine misfired. Since that was the only Intercosmos mission to be aborted, this Mir visit had been arranged as a consolation. The primary objective of the mission, however, was to exchange the ferry. The flight had initially been scheduled a fortnight later to give the residents time to attempt to repair the TTM, but the Rozhen astronomical experiment had to be timed to avoid the illumination of a full Moon, so the visit had been brought forward.

When Soyuz-TM 5 rendezvoused two days later, its automatic Kurs system made a fly-around from the front to the rear of the complex and then began its approach. At this point it began to deviate from the straight-in path, so Solovyov took over and withdrew to let the flight controllers analyse the telemetry. One orbit later, after the fault had been identified and overcome, the final approach was resumed, this time successfully. In the event that Soyuz-TM 5 had been unable to dock, Titov and Manarov would have been left with a ferry nearing its in-orbit endurance limit, which might well have necessitated their return to Earth, leaving Mir vacant.

An ambitious programme was planned with 46 experiments involving remote sensing, biomedical studies, space physics and materials processing. It was supported by over a tonne of new apparatus, including the Zora portable computer that was to be used to process experimental data in orbit so that the cosmonauts could assess their results as they worked.

The main astrophysical experiment involved the Bulgarian-built Rozhen apparatus, a sophisticated system incorporating a digitally processed electro-optical telescope using the Parallax-Zagorka image intensifier and the Therma photometer. This equipment was being evaluated for an autonomous telescope. Although it was primarily intended for studying deep-sky sources, in this case it was also to be employed to observe aurorae and luminescence in the upper atmosphere. Its data was processed in orbit by the Zora computer, and then downlinked.

The medical programme addressed physiology, psycho-physiology and radio-biology. Prognoz assessed changes in the operational performance of the cosmonauts; Potential investigated the interaction of the nervous and muscular systems; Stratokinetika studied the body's movement in the absence of gravity; Pleven-87 used 15 psychological tests concentrating on locomotor functions and volition processes, the data being processed by the Zora computer; Dosug assessed the influence of music, video and games on crew morale whilst off duty; Lyulin used a microprocessor to test psychological and physiological analysis of their reactions, in particular their reaction time; and Voal used the furnaces to produce an alloy of wolfram (a form of tungsten) and aluminium. While the visitors slept, Son gathered electrophysiological data which was recorded using a long-duration cassette tape. Doza-B involved installing radiation sensors and biological samples at specific places throughout the complex to measure the exposure levels.

The Earth-observation programme involved surveying Bulgarian territory, but this part of the programme was hindered by cloud cover. The Bulgarians had built the Spektr-256 (an improved form of the Spektr-15 spectrometer) to study the atmosphere; its data was processed by the Zora computer. There were several materials-processing experiments.

Klimet–Rubidium used the Kristallisator furnace to test making extremely lightweight batteries and condensers using a mixture of rubidium, silver and iodine (because this experiment consumed so much power, it was performed while the cosmonauts slept). The Ruchei apparatus was used to separate and purify genetically engineered interferon, and an Australian experiment to study the production of vaccines was set up. Seeds of wheat, arabidopsis and ginseng were all planted in the Magnetobiostat. Savinykh noted that the work programme was "a bit too much" for a week-long visit. On 16 June, the visitors loaded 30 kg of material (much of it as data on computer disks from the Zora computer) into the descent module of Soyuz-TM 4, and the next day they departed.

REPAIRING THE TELESCOPE

On 18 June (while the landing window was open), Titov and Manarov transferred Soyuz-TM 5 from the rear port to the front, to clear the way for the repair of the X-ray telescope and to ensure that their ferry would be accessible from the multiple docking adapter in the event that they were unable to re-enter the base block after their spacewalk on 30 June.

The detector was a 40-kg cylinder, some 40 cm in diameter. They manoeuvred it, together with all the specialised tools they would require, out through one of the vacant radial docking ports, along the base block, and across the bridge to Kvant 1. A handrail on the side of the module led them to the unpressurised instrument compartment, but at this point their task was made difficult by the total absence of restraints. Whenever they grew tired, they paused to rest, so the schedule slipped. Nevertheless, they were able to cut through the 20-layer thermal blanket protecting the unpressurised compartment and gain access to the bulky telescope. Unfortunately, the 2.5-metre long instrument had not been designed to be serviced by engineers wearing space gloves, and the small bolts proved extremely difficult to release. Although the procedure recommended from the hydrotank simulations had assigned 20 minutes to this task, it actually took an hour and a half.

Hoping to catch up, the cosmonauts were astonished to discover that the apparatus before them incorporated a few embellishments that had not been on the development unit in training. Proceeding anyway, they sawed through several bolts, wiped away an unexpected resin deposit, unfastened a number of screws and cut stainless steel clips to expose the detector. Unfortunately, when a special tool was inserted to release the brass clamp which held the detector in place, the tip sheared off, rendering it useless. They placed a new blanket of thermal insulation over the hole to protect the apparatus and then returned to the airlock with the new detector. The rest of the operation would have to wait until another tool could be delivered.

When they finally closed the docking port, the two men had been out for over five hours, and were thoroughly exhausted. As soon as they flew back into communications range with Kaliningrad, they reported that they had broken the special tool and aborted the spacewalk. The British and Dutch astronomers who had built the telescope were at Kaliningrad as advisers. Although "very disappointed" by this news, they were pleased that the cosmonauts had at least been able to gain access to the instrument and carry out about 70 per cent of the operation. Engineers immediately began to make a stronger tool, to be sent up on the next cargo ferry.

BACK TO THE ROUTINE

After cleaning their spacesuits and taking a few days off to rest, Titov and Manarov celebrated their 200th day in space on 8 July. Medical tests showed them to be in good health, so it seemed likely that they would be able to achieve their objective of a year in space. The next week was spent on Earth observations to locate mineral deposits in the Soviet Union.

Progress 37 was launched on 19 July, and docked at the rear port two days later. In addition to the usual consumables and replacement equipment, it delivered a new colour television monitor, a pair of Orlan-DM spacesuits (these contained their own power and communications facilities, so did not require an umbilical link to the base block), a new computer, equipment for the impending French mission, and new tools for repairing the telescope. Unloading began immediately, and then several days were spent on routine maintenance. The rest of the month was devoted to Earth observations. The cosmonauts reported on forest fires across the Soviet Far East, the Urals and southern Siberia. They were often the first to detect such fires, and in many cases the firefighting teams were able to reach the site before an outbreak became too serious. Progress 37 replenished the complex's fluids, refined its orbit and then departed on 12 August. Three days later, Titov and Manarov exceeded the 237-day mark; Romanenko's 326-day single-mission record was in sight. Much of the rest of August was spent on Earth photography for the Tien-Shan-88 Intercosmos programme, in Tadjikistan and Kirghizistan.

AN AFGHAN VISITOR

On 31 August, Soyuz-TM 6 docked at the rear port without incident. It delivered an *ad hoc* crew, hastily thrown together by bureaucratic and political factors.

Vladimir Lyakhov had spent 175 days on Salyut 6 in 1979, and 149 days on Salyut 7 in 1983. Dr Valeri Poliakov had been assigned to Soyuz-TM 4 to perform a complete medical check on Romanenko immediately prior to his to return to Earth from his record-breaking flight, but his couch had been commandeered by Levchenko, the Buran pilot. Poliakov was the deputy head of Moscow's Institute of Medical Biology, and Dr Gherman Arzamazov, his backup, also worked there. The third member of this crew was Abdul Ahad Mohmand, an Afghan. He had taken over from Mohammad Dauran when he had lost time in training by having his appendix removed. Earlier in the year, Glavcosmos chairman Alexander Dunayev reported that the Afghan mission was to be inserted into the schedule without disturbing the French visit, which was to coincide with the crew handover later in the year. The reason for this change in the schedule was political: the flight had to be made before Soviet occupation forces completed their withdrawal from Afghanistan. As a result of his flight being brought forward, Mohmand received only six months of training (two years was more usual), but the fact that he could already speak Russian assisted the assimilation process. Despite this truncated training, he was given the flight engineer's couch in the spacecraft. In fact, when Mohmand started training, Titov and Manarov were already aboard Mir, so when he floated through the hatch they had no idea who he was. Poliakov was a different matter. They would have been well aware of the fact he was there to assess their ability to attempt to exceed Romanenko's endurance record. It was at this point that

Romanenko had begun to become noticeably weaker. The insertion of this flight into the schedule gave Poliakov the opportunity to assess Titov and Manarov just as had been intended with Romanenko (thereby effectively satisfying the previously stated goal of having a doctor oversee an attempt to break the record). Rather than make only a brief 'house call', Poliakov was to remain aboard. He had the authority to curtail the flight if the physical or mental state of health of the would-be record-breakers deteriorated. As a bonus, he was able to study his own adaptation to weightlessness. The fact that a French cosmonaut was to accompany the next handover meant that Poliakov would not be able to return to Earth with his two charges, but would have to stay on with their successors. For Poliakov, therefore, this hastily arranged flight seemed like a once-in-a-lifetime opportunity to try for the record himself.

The decision to fly the mission early had pre-empted the development of specifically Afghan apparatus, so equipment already aboard had to be used for 24 experiments. The main objective was to photograph Afghanistan using the KATE-140 camera, to map the mountainous regions, to assess water and glacial run-off, and to provide data that could be used to identify possible sources of oil and gas in lowland areas. Because most of the country was remote and inaccessible (less than 30 per cent of it had been surveyed by conventional methods) this orbital survey would be rather significant. The MKS-M and Spektr-256 spectrometers were used for ongoing Biosfera study. The EFO-1 was used for the Climate experiment. Amongst the new cargo was a small aquarium with fish to evaluate a closed ecosystem. The Ruchei apparatus was used to make more interferon; this was proving to be a very productive experiment.

The medical part of the programme involved the standard tests (namely Stratokinetika, Potential, Labrint, Prognoz, Opros and Son-K) designed to monitor the initial phase of adaptation to weightlessness. Mohmand also used the Zora computer to carry out one vestibular experiment which had not been done during the Bulgarian mission. While he and Poliakov slept, the Son apparatus recorded the electrical activity within their brains. Poliakov's own programme had involved sampling his bone marrow (reportedly a rather painful process) prior to launch; this was to be compared with another sample immediately upon his return to Earth. Realising that he had easily adapted to weightlessness, he was reportedly "somewhat disappointed" not to have been able to experience for himself the rather unpleasant symptoms which upset some cosmonauts.

On 5 September, Lyakhov prepared Soyuz-TM 5 for departure, even though it had been in space only a few months. It was loaded with accumulated film, disks and other experimental results from the resident's programme, including the ampoule from the Australian experiment which had produced monocrystals of an influenza virus antigen. (The three-dimensional structure of the crystal was analysed using an X-ray scanner; it was hoped to find much larger crystals of membrane protein than could be produced on Earth.) Compared to the hectic Bulgarian visit, Titov and Manarov had found this visit a welcome relief.

STRANDED IN ORBIT?

Lyakhov and Mohmand departed in Soyuz-TM 5 the next day. The orbital module was jettisoned according to plan; then, as they ran through the final steps of the retrofire sequence, the navigational computer received conflicting signals from the primary and the backup infrared horizon sensors, resulting from solar glare as the spacecraft passed over the terminator into sunlight. The automated control system required the sensors to confirm the appropriate alignment 30 seconds prior to firing the engine for the deorbit burn, so the conflicting signals inhibited the imminent manoeuvre. For years, recovery window constraints had required the spacecraft to cross into sunlight at least ten minutes beforehand, to provide the sensors with sufficient time to verify its orientation and stability. However, despite the increased flexibility of the latest model of the ferry, passing through orbital dawn at the critical moment had upset the system.

Seven minutes later, while the cosmonauts were interrogating the system to determine what was wrong, the sensor problem cleared up, finally asserting the signal to confirm that the spacecraft was correctly oriented, so the still-waiting computer fired the engine. The spacecraft was now far beyond the planned deorbit point. If permitted to continue, its trajectory would take the capsule thousands of kilometres downrange, which would make recovery difficult. Deciding to descend near the normal recovery area on the next orbit, Lyakhov immediately terminated the manoeuvre. The situation did not seem very dangerous. The engine was clearly functional, and whatever had confused the sensors at the critical moment was unlikely to occur again.

Once Soyuz-TM 5 had been realigned using its inertial system, the deorbit burn was rescheduled for two orbits later. The engine fired on time, but shut off six seconds into the required 230-second burn. Lyakhov immediately restarted it, but 50 seconds later the computer decided that the vehicle was out of alignment, and shut it down again. The situation had suddenly deteriorated significantly. Lyakhov was later criticised for restarting it, but he had undoubtedly been aware that his ground track had drifted too far to try again on the next orbit. Flight director Valeri Ryumin eventually announced that they would have to remain in space for another day. Overnight analysis revealed that on the second attempt the computer had tried to execute the incorrect programme (for some reason it had selected part of the rendezvous sequence that it had followed when taking the Bulgarian cosmonaut up to Mir). Having established that the engine had not really malfunctioned, the tension in Kaliningrad evaporated.

In space, however, the situation was dire. Having jettisoned the orbital module, the tiny descent module was extremely cramped for two fully-suited figures. It was little consolation knowing that this would have been much worse if the third couch had been occupied. There was no point trying to return to Mir, because the Kurs antenna and the docking unit had been discarded with the orbital module. The descent module had not been designed for prolonged independent operation. Its temperature dropped to a chilly 10°C, which made the men glad they were wearing Sokol pressure suits. Lyakhov was not concerned that they had no food (they had only dried rations for three days of basic survival following an emergency landing in a remote area), but he lamented the fact that the toilet had been in the orbital module, and he suggested that in future the module should not be jettisoned until after the deorbit burn had been completed, as in "the good old days", even though doing so would consume extra propellant. With a low perigee, the orbit would decay naturally within a few weeks, but their air supply would not last that long. To sustain the electrical systems, the spacecraft was orientated so that its tiny solar panels faced the Sun, and then put in a slow rotation for stability. Finally, the two men settled down to wait,

undoubtedly aware that if they could not complete the deorbit manoeuvre there was absolutely nothing they could do; they would die in orbit and then in all likelihood be incinerated on re-entry into the atmosphere.

When back in communication, a revised computer program was read up to Lyakhov,
and he entered this manually into the computer. Vladimir Dzhanibekov wryly noted that
this predicament was the result of a "blunting of vigilance", but Viktor Blagov blamed it
on "a combination of circumstances". Strictly speaking, however, the basic failure was the
decision to make an early morning descent rather than the late afternoon descent that had
been standard practice for so many years, because this necessitated the deorbit burn at
orbital dawn. The operational envelope had been stretched too far, and Lyakhov and
Mohmand were paying the price.

Although the Western media reported that they were "stranded" in orbit and running
out of air, the situation was actually manageable. The spacecraft had already descended
from Mir's 338 × 364 km to a 250 × 341 km orbit as a result of the various manoeuvres,
both intended and unintentional, so it was fairly well placed to finish the deorbit a day late
and land at the planned recovery site. Just to be certain however, the backup engine was
employed. Television cameras recorded its final descent through pre-dawn twilight.
Within minutes, the two cosmonauts scrambled out of the capsule and greeted their res-
cuers. Despite the apparent maturity of the technology, spaceflight was still a potentially
lethal business.

REPAIRING THE TELESCOPE – CONTINUED

The next day, 8 September, Titov, Manarov and Poliakov transferred Soyuz-TM 6 to the
front port to make way for Progress 38, which arrived four days later. In addition to stores,
it delivered a ham radio transceiver and various apparatus for the impending French visit.
For the next few weeks, materials-processing was interleaved with visual observations of
plankton and forestry surveys.

On 18 October, however, Titov and Manarov began to prepare for the spacewalk to
finish the repair of the X-ray telescope. They unpacked the new Orlan-DM spacesuits.
These comprised an aluminium alloy body section with elasticated arms and legs for
greater flexibility. They did not require a communications and power umbilical, because they incorporated a new unit in the backpack to supply power and transmit
telemetry. They would be tied to the complex by a nylon tether, however. The oxygen
tanks and the lithium hydroxide scrubber which absorbed exhaled carbon dioxide could
support external operations of up to seven hours in addition to several hours pre/post-ac-
tivity in the airlock (the previous suits had been restricted to five hours, plus a reserve). Another enhancement was a set of biomedical sensors which transmitted continuous telemetry to enable the flight surgeon to monitor the cosmonaut's health. The
total freedom of action offered by these suits was to permit spacewalkers to use the
autonomous manoeuvring unit which was to be delivered in the first of the large
modules. (Titov and Manarov had hoped to commission it, but its launch had been postponed.)

Two days later, Poliakov retreated into Soyuz-TM 6 so that he would be safe in the
event that it proved impossible to repressurise the docking adapter after the spacewalk,

and then Titov and Manarov emerged from one of the vacant radial ports. The downlink was shown live by domestic television. As they made their way with their equipment to the rear of the Kvant module, they reported that the handholds positioned specifically to assist spacewalkers were actually spaced too far apart.

At the work site, they reopened the thermal blanket and used the new tool to unseat the clamp without incident. Working rapidly, they extracted the faulty detector, inserted the new one (which had been fitted with handholds to make it easier to manipulate) and affixed a simple clamp mechanism to lock it into position. In contrast to their previous spacewalk, by the time they replaced the thermal insulation they were an hour ahead of schedule.

On their way back, they used a fine brush to clean accumulated dust off the glass of two portholes, then attached an anchor point outside the docking adaptor (to be used for an experiment on the impending French mission) and erected an antenna which would enable them to talk to radio hams (a capability which would also serve as a backup if the main communications system failed). Having achieved all their tasks, they closed the docking port with plenty of time to spare. The repair of the telescope provided a remarkable demonstration of real-time mission planning, engineering improvisation and technical prowess. In trials, the new detector proved to be reliable, so, to the delight of the astronomers, the TTM telescope was soon back in operation.

GOING FOR THE RECORD

Throughout late October, the Mariya spectrometer collected data on charged particles in the near-Earth environment. Materials processing was resumed in early November, with aluminium and copper alloys and monocrystals of zinc being produced by the Mirror and Pion-M apparatus.

By this time, Poliakov had Titov and Manarov on a "particularly rigorous" exercise regime. They were now regularly wearing the Tchibis for extended periods to increase cardiovascular capacity, and they began to consume salt water and other additives to build up body fluids. Dr Anatoli Grigoriev, director of the Institute of Medical Biology, said that they seemed to have suffered none of the deterioration that had afflicted Romanenko in his final months. On 12 November, Titov and Manarov finally broke the 326-day record, and Poliakov reported that he was happy for them to try for "a year and a day". To build up their stamina for their return, however, they cut their work cycle by half an hour, adding this to their exercise regime.

A FRENCH VISITOR

Progress 38 left on 23 November, and Soyuz-TM 7 took its place on 28 November to deliver Alexander Volkov, Sergei Krikalev and Frenchman Jean-Loup Chrétien. The launch had been put back by 5 days so that President François Mitterrand could attend.

Volkov had been a member of the Soyuz-T 14 crew, which had been recalled from Salyut 7 when Vasyutin had fallen ill. This was Krikalev's first flight. Chrétien became the first guest researcher to be granted a second flight (he had spent a week on Salyut 7 in 1982); he was first through the hatch. Titov and Manarov offered a veritable feast of jellied salmon, quail meat and candied fruit to the newcomers, who had in turn brought with them a range of specialities including vegetable soups, fish, ham and various types of cheese, paté and dressing. With six cosmonauts aboard for the first time, the base block proved to be a little cramped, especially since it was to be an extended mission. A month-long handover had been scheduled to permit Chrétien to undertake a much more ambitious visiting programme than usual. He was to return with Titov and Manarov, which meant that Poliakov would have to stay on with Volkov and Krikalev.

Chrétien's researches included ten French experiments, involving both medical and technological topics. About 580 kg of equipment had been ferried up (in Progress 37 and Progress 38) specifically for this work. Of the medical experiments, Superpocket measured the neurosensory system and the reconditioning of postural reflexes; Physalie evaluated the coordination between the body's sensory and motor systems; Kinesigraph created stereoscopic images with which to investigate the restitution of movement of the corporal segment; Circe monitored ionising radiation, gamma-ray and neutron dosages in the complex; a sophisticated X-ray scanner had been used prior to launch (and would be used again following his return) to measure calcium loss in his bones; and he was to return blood samples for analysis. In addition, an improved version of the Echograph ultrasonic scanner was to measure blood flow in deep vessels (in particular, the main truncus and venous return), and to measure the capacity of the heart and other internal organs. The Echograph's monitor screen was to be used for the Viminal experiment in which a hand-operated indicator had to be keyed in response to visual cues to test visual acuity. The Ercos experiment was a cassette to be attached to the inside of the station's wall to test the extent to which high-density computer memory chips degraded by being exposed to cosmic rays; it was to be returned after six months. The primary item on the technology programme, however, was the self-deploying Era truss.

The highlight of Chrétien's programme was the first spacewalk by a guest researcher. This had been scheduled for 9 December, then changed to 12 December by the slip in the launch. Having adapted to weightlessness so rapidly this second time around, Chrétien decided to proceed as originally planned to exploit the favourable lighting conditions that had suggested the original date (starting earlier opened up the possibility of making a second excursion if he was unable to complete his work). So, he and Volkov devoted 8 December to preparing the Orlan-DM spacesuits.

After Chrétien had exited one of the radial ports, Volkov pushed out the segmented Echantillon cassette. The Comes segment exposed small samples of paints, reflectors, adhesives, optical materials and filament-reinforced composites; Mapol exposed some polymeric materials thought suitable for creating inflatable structures in space; and MCAL exposed materials to determine the evolution of absorptivity and emissivity. The cassette also incorporated two dust traps (one active, one passive, called DIC and DMC) to collect cosmic dust for later analysis. Chrétien affixed this 0.75-metre square box to a convenient anchor point close by. Intended to be left out for six months, in the event it was to be over a year before it was retrieved.

Volkov next passed out the Era experiment. The 240-kg main package (a 0.6-metre diameter compressed stack of 1-metre long carbon fibre rods) was accompanied by a support platform, a video unit and a control panel. After the platform had been affixed to the anchor that had been mounted by Titov and Manarov, the stack was affixed to an arm on the platform set at an angle of 45 degrees, which kept it away from the surface of the complex. Once a 50-pin umbilical had been connected, Krikalev, inside the base block, triggered the controller. The articulated pin-jointed rods were supposed to spring open automatically over a four-second period to create 24 identical prisms forming a thick hexagon some 4 metres wide. Unfortunately, it did not budge. Chrétien nudged it a few times, to try to shake it loose, but it refused to deploy. At that point, the complex flew out of communications range with Kaliningrad. Rather than jettison it simply to stick to the schedule, the cosmonauts continued their efforts to release the structure. Several violent kicks by Volkov's boot did the trick, and it suddenly deployed. When fully unfolded, a set of accelerometers were attached to it, to determine its vibration modes. Upon flying back into communications range, they replayed video of the deployment process. It was later decided that the structure had been locked in its folded position by water vapour that had frozen when exposed to vacuum. An hermetically sealed container would have precluded this, but would have made deployment more complex. After the experiment was complete, the structure was jettisoned so that it would not block the radial docking port. This experiment was intended to evaluate one of the options being considered by the French Space Agency (CNES) for erecting antennas in space, as an application of its planned Hermes spaceplane.

The spacewalk had been scheduled to last for four hours, but the extra time required to finish the Era deployment stretched this to six hours, which was a new record. Despite improvements in the Orlan's environmental control system, by the time he was finished Chrétien noted that he was drenched in sweat, and there was condensation on the inside of the visor. Immediately afterwards, Poliakov gave Volkov and Chrétien a thorough medical examination.

After the trouble with the Era, Chrétien performed his next deployment experiment inside the base block. The Armadeus experiment involved erecting a tiny solar panel to test its articulated deployment mechanism. It comprised four motorised winding blades incorporating Carpentier joints to eliminate friction. He deployed the array three times, in each case executing a series of ten tests. The whole sequence was filmed by a pair of infrared television cameras, so that the deployment kinematics could be analysed.

Chrétien was required to perform the Physalie experiment, which involved collecting biomedical data while being filmed repeatedly performing a range of movements. Upon realising that it took over two hours to set up the equipment (two cameras, a cluster of monitors, and sensors on his neck, torso and legs) he vowed that if he got the chance to make a second spacewalk he would jettison it out of the airlock. Evidently the experiment, which had seemed straightforward in the laboratory, had proven awkward in space.

In mid-December, the region of Armenia which had earlier suffered an earthquake in which 100,000 people had been killed, was extensively photographed. Throughout the joint mission, the X-ray telescopes were operated remotely (the flux of hard X-rays from the supernova had decreased considerably of late). Titov and Manarov devoted an increasing amount of time to the exercise regime, and spent long periods in the Tchibis. On 15 December, they exceeded Romanenko's record by 10 per cent, which was the margin demanded by the International Astronautics Federation to officially set a new record. It was a happy pair who loaded experimental results into Soyuz-TM 6.

Titov, Manarov and Chrétien undocked on 21 December. While they waited for the deorbit burn, the spacecraft's computer became overloaded and aborted the manoeuvre.

The problem lay in a new software routine, so another attempt was scheduled for two orbits later using a backup program. Following the problems suffered by Soyuz-TM 5, the practice of retaining the orbital module until after retrofire had been reinstated, so its facilities were available to the cosmonauts while they waited. The computer functioned perfectly. About ten minutes after the completion of the deorbit burn, the orbital module was jettisoned, followed by the service module a minute later. A television camera in a recovery helicopter caught the capsule's descent into a thick layer of low cloud. For once, it settled in its upright position, and it was quite a struggle for the record-breakers to climb up to the hatch. The recovery team assisted the cosmonauts out, then down the slide. The frozen steppe was a shock after Mir's carefully regulated environment.

Despite 366 days in space, Titov and Manarov were in "good health". Titov had lost 3 kg, but Manarov had put on this amount (*gaining* weight in space was unusual). Both were in better condition than Romanenko had been. Since there was an infectious outbreak at the Cosmonaut Hotel at the cosmodrome they were flown to Kaliningrad. Although the muscles in their lower legs had atrophied through lack of use, this was found to be due to loss of intramuscular fluid rather than deterioration of the tissue, and was easily restored; the bone calcium loss proved to be no worse than that suffered by previous crews. A Tchibis configured as a lower body positive-pressure suit was used to prevent blood from draining from the lower body, and, just as immersion helped in training for weightlessness, swimming helped adapt to its absence. Within days, Titov and Manarov were strolling around the space centre. With the endurance record now set at a year, it was clear that the task of breaking it could be assigned only to an extremely motivated individual serving with successive crews, and it would be unfair to expect a crew on a routine tour to attempt it.

FRUSTRATED HOPES OF EXPANSION

Soyuz-TM 7 was moved to the front port on 22 December, and Progress 39 docked at the rear port four days later. While Poliakov continued to observe his own adaptation to weightlessness, Volkov and Krikalev anticipated a very busy plan of their own. So far, the programme had produced mixed results. Mir's expansion had fallen far behind schedule, but the cosmonauts had nevertheless been able to establish an ongoing presence, and had achieved the psychologically important year-and-a-day milestone. Now, however, it was time to switch the effort back to construction. The schedule listed by Alexander Dunayev earlier in the year had called for the docking of the first module prior to the French visit, but this had been postponed. Volkov and Krikalev now planned to receive the next *two* modules. Within weeks though, Dunayev had to announce that the launch of the first module had been postponed again.

Another change in the programme concerned the visits by foreign researchers. To date, these had been supported in the interests of fraternal cooperation, but there were to be no more 'free rides'. From this point on, guest researchers would be accommodated only on payment of a fee, this being calculated to defray the cost of their participation in the programme (both the Soyuz and any Progress cargo shipments). In the years to come, this decision was to have far-reaching consequences.

To mark the start of 1989, Volkov, Krikalev and Poliakov opened the gifts that had been delivered by Progress 39, and then resumed a full research programme that included

Earth-resources studies of Siberia and the Soviet Far East, using the Parallax-Zagorka image intensifier to study the vertical distribution of luminescence at polar, middle and equatorial latitudes, and also of the luminescence generated by the complex itself as it passed through the extremely rarefied atmosphere at orbital altitude; taking astronomical photographs with the Glasar ultraviolet telescope; and astrophysical observations using the Rozhen apparatus.

Poliakov continued to monitor the state of his colleagues' cardiovascular systems. This work was greatly assisted by the use of the Reflotron apparatus supplied by West Germany. This provided instant blood analysis, so that changes in blood chemistry could be studied as they happened; previously, samples had been collected and returned for later analysis. Given a drop of blood, it would measure a wide range of parameters, report on haemoglobin, cholesterol, uric acid and glucose, and then analyse and store the results. Poliakov periodically downlinked these data to colleagues at the Institute of Medical Biology.

On 23 January, Volkov and Krikalev began several days of routine maintenance on the environmental and thermal regulation systems and installed hydraulic components that had been delivered by Progress 39. They then used a holographic imager to record damage to the portholes caused by cosmic dust. At the end of the month, they used the Yantar apparatus to apply metallic coatings of alloys of silver–palladium and tungsten–aluminium to samples of polymer film. Mir began making favourable daylight passes of the Soviet Union in early February, so Earth observations resumed to complement data from automated satellites. Progress 39 undocked on 7 February, and Progress 40 took its place five days later. Amongst the food, the cosmonauts found treats such as pickled cucumber and sweet-smelling honey. The Diagramma experiment was carried out to measure the physical characteristics of the gaseous environment at orbital altitude. Measuring the flow around the complex would help assess the aerodynamic drag which caused the complex's orbit to decay. This was done by a magnetic discharge transducer on a short boom that was poked out of the scientific airlock, after the Yantar apparatus had been removed.

In early February it had been reported that spacewalks planned to affix a pair of star trackers to help orient the complex after it had been expanded had been cancelled due to a further delay in preparing the first new module for launch. The plan for the residents to commission two modules was now clearly impracticable, but it seemed certain that they would receive the first one. While they waited, the cosmonauts performed the work that was available. Until the new modules arrived, the power-generating capacity would be limited. This situation was aggravated by the progressive deterioration in the batteries. Experiments had therefore to be scheduled carefully to make optimum use of the power available.

The real-time nature of the planning process was again made clear in mid-February, when Alexei Leonov reported that it had not been decided whether Poliakov would return to Earth with Volkov and Krikalev when they were relieved in April. (If he did stay on, he would almost certainly break the "year-and-a-day" record so recently established.) He said that the next crew would "most probably" be commanded by Alexander Viktorenko. Since Viktorenko had trained with Alexander Serebrov, it seemed likely that he would be the flight engineer, but a few days later it became apparent that his backup, Alexander Balandin, would make the flight. The plan called for the next crew (whoever they were) to be launched on 19 April, with the old crew (with or without Poliakov) returning on 29 April. In the meantime, every effort was being made to complete the preparations for the launch of the first of the new modules.

The process of building up the complex would be complicated by the requirement to control the attitude of an asymmetric configuration. In a linear train, the centre of mass was at least on the major axis; in an L-shape it would be way outside the structure. The new modules would initially dock at the front axial port, then swing themselves around to a radial port by the appropriate Ljappa mounting. Clearly, whenever such a module was expected, the crew's ferry would need to be at the rear port. Equally obviously, it would be impossible to receive any cargo ships while the rear port was occupied. These operational limitations meant that the first new module would, sooner rather than later, have to be moved off the axis to a radial port so that the crew ferry could swap ends to enable the next cargo ship to dock. With a 20-tonne module projecting out to one side, the complex would be difficult to manoeuvre. It would be awkward to orientate, so all the experiments that required a specific orientation would be temporarily impracticable. Because an engine designed to deliver impulse axially would not be able to push through the offset centre of mass, the engine would tend to spin rather than push the complex, so it would not be able to adjust its orbit.

Two factors were therefore the key to the next phase of expansion. As soon as possible after the first module was swung out to the side, the attitude control system would have to be fine-tuned to manage the asymmetric configuration. (It would be difficult to receive incoming spacecraft until the orientation of the complex could be controlled.) Secondly, as soon as possible thereafter, the second module would have to be received and swung to the opposite side, expanding the L-shape into a T-shape, to restore the balance of the configuration. Ideally, this phase of the expansion would be completed in a period of a few months. It was therefore vital that the first module should not be launched until the second was also nearing completion.

When Volkov and Krikalev were in training, they had expected that the first module would just have been docked. When they were launched, they had expected it to arrive within a matter of weeks, and the second to arrive towards the end of their tour of duty. Within weeks of their taking over, however, it had become evident that the second module would have to be left to their successors. Now, with continued delays in preparation, it looked as if they would be lucky to receive the first. Yuri Semenov, the general manager of the Energiya Bureau, which was responsible for the preparation of the modules, later said that at this point it was decided not to send up another crew until both modules were actually finished. This decision offered two options: to keep the current residents in Mir until both modules were certified ready for launch (so that Volkov and Krikalev would be able to commission the first module, as they had trained to do), or to recall them and leave Mir unoccupied until the modules were ready. Of course, the first case would be feasible only if there was likely to be only a slight additional slippage in the schedule. If it looked like the delay would extend the tour of duty unreasonably, then there would be little choice but to take the second option, and give up, temporarily, the objective of maintaining a permanent orbital presence. As the Khrunichev factory continued to work on the new modules, the residents continued to work on the available projects, and their successors trained with the expectation of taking over in April. But nobody knew how events would turn out. This was real-time planning indeed!

Earth observations continued throughout February, then Progress 40 pumped fluids aboard, rubbish was loaded into it, and on 3 March it departed. As it pulled back, the camera mounted in the rear docking unit recorded an experiment in which a pair of large multi-link structures (each resembling an orange segment, but more sharply angled than an ellipse) unfolded from the sides of the ferry to test the ability of form-remembering materials to re-establish a defined shape. A 'unique alloy' incorporated into the material recorded a given shape and then reformed it when a current was passed through wires embedded in the material to generate the necessary thermal stimulus. It was hoped that a structure based on such technology might be useful in the creation of a large reflector in space which could be used to collect solar energy. When the experiment was finished, the discarded ferry resumed its withdrawal. Soviet spacecraft had always been designed to be radio-controlled (Yuri Gagarin had been a mere passenger). As a result, they proved to be extremely capable as automated satellites. The use of the discarded ferry to carry out an experiment alongside the station permitted a test that would otherwise have been awkward to observe. The modular approach was really paying a dividend.

The research programme during March included astronomical photography using the Glasar ultraviolet camera; making measurements of the constituents of the atmosphere using the Spektr-256 spectrometer; a study of charged particles in near-Earth space by the Mariya spectrometer; and, of course, the continuance of X-ray observations while the crew slept. Progress 41 arrived on 18 March, and it was unloaded intermittently over the following week. In addition to the usual consumables, it delivered a few replacement storage batteries and a number of other components for the electrical system. On 26 March, the cosmonauts cast votes for the new Congress of People's Deputies; amongst the successful candidates were fellow cosmonauts Valeri Ryumin, Svetlana Savitskaya and Viktor Savinykh. Also, Krikalev was told (tongue-in-cheek) that he had been reprimanded for failing to report for service in the military reserve at the appointed place at the appointed time; the fact that he was in orbit had not been accepted by the military bureaucracy as a valid reason for absence.

By late March, it had been decided to bring the cosmonauts home on schedule at the end of April. A fault had been discovered in the service module of the next ferry. Since there was not another flight-ready spacecraft available, and the fault would take time to fix, Viktor Blagov announced that the launch of Soyuz-TM 8 had been postponed for three months, until August. Vitali Sevastyanov and Viktor Afanasayev had expected to take over from the Soyuz-TM 8 crew in September, so their flight was correspondingly delayed. This was to have unfortunate consequences for Sevastyanov, because he was grounded for medical reasons before he could make his flight. Rimantas Stankiavicus (who had taken over the Buran group from Igor Volk) was to have accompanied them, but he was killed in an air crash before the flight could be reassigned.

Earth observations in early April started with environmental studies of the northern Caucasus, the Black and the Caspian Seas, and the spectacular aurorae which followed an increase in solar activity. (The cosmonauts regularly performed the Circe experiment, to monitor the radiation level within the complex; they were fairly safe, though, because with the exception of when they passed near the South Atlantic Anomaly, their orbit was far below the inner Van Allen belt.)

On 10 April, as a preliminary to leaving the Mir complex unoccupied, Progress 41 boosted it to a slightly higher-than-usual orbit. The next day, *Tass* announced that the cosmonauts had completed their research programme and had now begun the process of "mothballing" Mir. This was actually a bad time to leave the complex vacant because, after three years of operation, the base block's electrical system was in need of a major overhaul. The last thing that flight controllers wanted was for the base block to suffer a power failure while it was unoccupied. During their final few weeks aboard, therefore, the cosmonauts serviced the converters and installed the new chemical batteries which had been delivered by Progress 41 (only a few of these 74-kg cells could be ferried up at a time, so it had not been feasible to replace the entire set of twelve in the limited time available). Progress 41 departed on 21 April. Unfortunately, because it had consumed so much propellant in raising the complex's orbit, it was unable to deorbit itself, so had to be left for its orbit to decay naturally. After two years of continuous occupation, Mir was returned to autonomous flight mode on 27 April. Volkov, Krikalev and Poliakov boarded Soyuz-TM 7 and returned to Earth. Although eager to examine their colleague, who had spent almost eight months in space, the medical team's first task was to attend to Krikalev's leg, which had smashed against the control panel during the final stage of the descent; it was sore, but not seriously injured.

The return to Earth must have been a great disappointment to Poliakov because if he had stayed on after the handover, and Stankiavicus had flown in the autumn, it would have been necessary for him to have further extended his flight to 18 months, which would have given him the endurance record by a wide margin.

MIR IN AUTONOMOUS MODE

Although Mir was largely dormant, the X-ray telescopes in the Kvant module could still be operated by remote control, so astrophysical observations continued.

At the Paris Air Show, in June, Vladimir Shatalov noted that coordination between the agencies supporting the space programme was at the root of the delayed construction of the Mir complex. He pointed out that without a central coordinating agency (such as NASA) it was becoming increasingly difficult to sustain continuous operations in orbit. He added that because equipment had been ferried up to the complex as soon as it had become available, rather than when it was needed, much of the apparatus aboard was unusable without support equipment in the new modules. And with so much apparatus aboard, the base block was far too cluttered for the cosmonauts to work effectively. He optimistically forecast that the first module (which would incorporate a large airlock and deliver the autonomous manoeuvring unit) would be launched "in either September or October", with the second (which would incorporate an androgynous docking port for use by the Buran shuttle) following "at the end of the year". With regard to Buran, he said that there was no particular urgency to begin operations because it was meant to complement, rather than replace, the existing fleet of transport and resupply vehicles, and he added that improved versions of these would soon be introduced. These would have greater manoeuvrability and be able to carry up a heavier payload. In addition, Progress would incorporate a small descent capsule by which the cosmonauts would be able to return experimental results to Earth.

Within weeks though, it was announced that although the Soyuz-TM 8 crew would be launched in early September (in order to receive the first module in October) the next module would not be launched until February 1990; this was the downside to real-time planning.

Progress-M 1, the first of the new cargo ferries, was launched on 23 August. It had the combined propulsion unit and solar panels of the Soyuz-TM, together with the Kurs rendezvous system. Only now was it revealed that the front axial port incorporated the plumbing for replenishing the base block's fluids. This eliminated the need to transfer a newly arrived Soyuz ferry from the rear to the front port to make room for an incoming cargo craft. This new capability was demonstrated two days later, when Progress-M 1 docked at the front port. The output from its solar panels was fed into the main power grid, because every little helped. Despite the addition of solar panels, other savings had increased the overall cargo capacity of this revised configuration to 2,500 kg, so it was a significant step forward.

7

A microgravity laboratory for hire

Soyuz-TM 8 was launched on schedule, on 6 September 1989. A new aspect of the television coverage was the advertisements erected around the pad and emblazoned on the rocket's shroud. As expected, Alexander Viktorenko was flying as commander. He was accompanied by Alexander Serebrov after all. On their final approach, two days later, they saw the complex begin to oscillate when their ferry was only four metres off the rear port, causing the Kurs system to abort. Viktorenko took over, withdrew twenty metres, visually inspected the docking port, then closed in and docked without incident. An hour later, they opened the hatches, entered the complex and began the long process of returning it to life.

Immediate maintenance involved the replacement of another three NiCd batteries (these had been delivered by Progress-M 1) to further ameliorate the degraded power system. Then several days were spent loading and verifying new software that would enable the attitude-control system to deal with the offset centre of gravity that would result when the first module was swung off the axis. During this first week, the only active research was that using the remotely-controlled X-ray telescopes. By mid-month, however, they had set up the Gallar furnace (also delivered by Progress-M 1) in Kvant 1. This was an improved version of the Korund-1M; in its first trial it made a monocrystal of cadmium selenide semiconductor.

In mid-September it was announced that the first module would be launched on 16 October, dock on 23 October, then immediately be swung onto the upper radial port. Soyuz-TM 8 would be moved to the front port on 25 October. Only then would the cosmonauts be cleared to open up the new module. Once it had been checked out, they were to perform an "internal spacewalk" by depressurising the multiple docking adaptor in order to swap the docking drogue from the upper port to the lower port so that the next module would be able to dock. (Apart from the radial port, there was only one other Konus assembly in the tiny compartment, and it had to be transferred to the port next to be used.) Progress-M 2 would dock at the rear port on 29 October. In December, it was said, Viktorenko and Serebrov would make one spacewalk to retrieve the Echantillon cassette. On another excursion, they would install the two star trackers on Kvant 1 (this had not been done by Volkov and Krikalev). Finally, they would perform two further spacewalks to test the autonomous manoeuvring unit. On 28 January, after the second cargo craft had departed, they would fly Soyuz-TM 8 back to the rear port to free the front for the second module. Following its launch on 30 January, this would dock on 6 February and immediately be swung down to the lower port to release the axial port for Soyuz-TM 9, which would dock on 13 February. Viktorenko and Serebrov – having received both modules during their tour of duty, commissioned the first one, and tested the autonomous manoeuvring unit – were to return to Earth in late February. General Kerim Kerimov, the chairman of the State Commission overseeing operations, said that the two cosmonauts had been assigned "a colossal amount of work". This was, in fact, essentially the mission that had been given to Volkov and Krikalev 18 months earlier; and, as then, it would not be long before events on the ground rendered this ambitious schedule impracticable.

In the second half of September, Viktorenko and Serebrov concentrated on further preventive maintenance on the core systems of the base block, but additional materials were processed in the Gallar furnace and X-ray observations were made on a regular basis. A major solar flare at the end of the month caused some consternation. Its effects were detectable by the Circe apparatus (used to measure the level of radiation inside the complex twice a day) for about ten days; usually the effect of a flare lasted only a day or so. They slept in Kvant 1, because it was more heavily shielded. It was later calculated that they had suffered exposure equivalent only to adding two extra weeks to their tour.

In early October, while undergoing final checks, the first new module was declared unfit for flight following the discovery of faults in computer chips drawn from the same batch as those incorporated into its Kurs circuitry. It was decided to replace the chips in the module just in case these too were faulty. *Tass* warned that the schedule had slipped by 40 days, but Viktor Blagov announced that the launch had been rescheduled for 26 November. This forced Viktorenko and Serebrov to revise their flight plan to make use of equipment already aboard as best they could, so they devoted most of the next two months to Earth studies. The MKS-M and Spektr-256 spectrometers were used to study the upper atmosphere over the tropics to measure the amount and distribution of ozone – part of the Atlantika-89 project, carried out jointly with Cuba, to evaluate an apparent correlation between ozone levels and the formation of tropical hurricanes – and the Ukraine, Krasnodar, Stravropol, Moldavia, the Caspian Basin and Turkmenia were mapped using the KATE-140.

As they waited, Viktorenko and Serebrov undoubtedly endured similar frustrations to those felt by their predecessors. When Blagov added that the launch of the second 'technology' module had slipped to late March, they knew that they would not witness its arrival; but the highlight of the mission, the spacewalk to test the 'flying-backpack', for which they had received special training, seemed certain.

EXPANSION: KVANT 2

On 26 November a Proton rocket placed Kvant 2 into orbit. Unlike its predecessor, which had required a 'tug', this new module had its own propulsion system. This was because it was a TKS augmented by apparatus, rather than a separate payload delivered by a TKS-based tug; as such, it was a refinement of the vehicles which had temporarily docked with the second-generation Salyuts.

Unfortunately, it ran into difficulty almost immediately. One of the two solar panels did not deploy properly. The outermost three segments unfolded as intended, but the innermost did not. In fact, because the innermost panel had not locked, the outer piece of the 10-metre long panel was free to swing, which made manoeuvring awkward. After analysing the telemetry, the flight controllers put the vehicle into a slow roll, and then commanded the motor that rotated the solar panel to cycle back and forth, to cause the centrifugal force to drag the panel out. It worked. If it had proven impossible to deploy it by this means, the docking would have been attempted, and the cosmonauts subsequently sent out to try to lock the panel with tools that would have needed to be fashioned and delivered by the next cargo ferry.

On 1 December, Progress-M 1 pumped surplus propellant (that is, all except what was required to deorbit itself) from its ODU into the the complex's tanks, then departed. This new capability offered a significant bonus, because each ferry carried rather more propellant than it actually needed so that it would be able to make a second rendezvous. However, *none* of the first-generation Progress ferries had encountered any problems in docking; they had proven to be extremely reliable. The following day, as Kvant 2 made its final approach, Viktorenko and Serebrov retreated to Soyuz-TM 8. When the module was 20 km out, its Kurs system became overloaded. Concluding that it was closing too rapidly, it aborted its approach. In the event, even if it had been able to continue, a docking would have been risky because at about the scheduled time that it was to have occurred the gyrodynes dropped off-line. Four days later, after a change of procedures, Kvant 2 made another approach. This time Viktorenko locked the complex stable using the ferry's attitude-control thrusters. The chatter on the voice channel was broadcast live by Radio Moscow as Kvant 2 docked.

The downlink from the camera in the front docking port displayed the newcomer's solar panels in an 'up-and-down' alignment, perpendicular to the main panels on the base block. The reason for the 45-degree offset in the docking collar was now clear. Kvant 2 was equipped with Ljappa, a short twin-element arm with a triple-petal androgynous grapple on the end which, when it engaged an attachment point on the multiple docking adaptor, would enable the module to swing itself round onto a radial port. This twisting manoeuvre would rotate it by 45 degrees on its long axis. Since the front axial port was itself offset at 45 degrees, this supplied the 90-degree rotation needed to properly align the module on the upper port with its solar panels projecting out to either side. Despite never having been tested, the arm functioned flawlessly; the transfer took less than an hour.

Just as Kvant 1 was also often referred to as the 'astrophysics module', Kvant 2 was the 'enhancement module' (or sometimes the 'D' module, from *dusnashcheniye*, meaning 'expansion', 'enhancement' or 're-equipment'). A lot of apparatus was mounted on the outer surface, including six additional gyrodynes to assist in manoeuvring the expanded complex and two tanks to increase the capacity of the Rodnik system. Internally, it was comprised of three hermetic compartments. The main compartment contained apparatus to take the complex a step nearer a closed-cycle environment. There were two separate oxygen generators. Elektron produced oxygen on an ongoing basis by electrolysis of water, and Vika repressurised the airlock by chemical reaction without imposing a sudden demand on the main system. It also contained two cubicles, with a toilet and a shower, and had been loaded with assorted cargo for launch. The middle compartment contained built-in scientific apparatus. At the far end of the module was the airlock, whose hatch was wider than a docking port. Two Orlan-DMA spacesuits were stored in the airlock. These

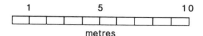

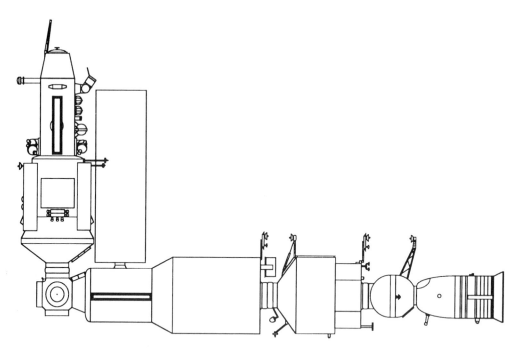

Asymmetric configuration of Mir with Kvant 2 on upper port.

incorporated changes reflecting experience with the Orlan-DM suits that had been delivered in 1988. The bulky autonomous manoeuvring unit backpack was stowed in a large box mounted on the wall of the airlock.

The scientific apparatus comprised Volna-2 to study fluid flows within capillaries to evaluate a new propellant tank design, Inkubator-2 to hatch eggs for an investigation of embryonic growth, Epsilon to assess the complex's thermal protection, the MKF-6MA multispectral camera (an improved form of the MKF-6M that had been used extensively on the second-generation Salyuts) and the KAP-350 mapping camera. In addition, the ASPG-M remote-control scan-platform was mounted outside; this carried a number of instruments, including the ITS-7D infrared spectrometer, the ARIZ X-ray spectrometer, the MKS-M2 spectrometer, and the Gamma-2 multispectral television camera cluster. Kvant 2's solar panels added 7 kW as well as six NiCd batteries to the complex's power system.

Soyuz-TM 8 was transferred to the front port on 12 December. Upon re-entering the base block, Viktorenko and Serebrov opened Kvant 2 and entered its first compartment to inspect the cargo. Two days later, they took a portable television camera with them and gave the flight controllers a tour. They spent the next week activating and checking its

Notes

1   inputs to H₂O, food, N₂, O₂ are from Progress ferries

2   CO₂, contaminants and H₂ are vented to space

3   O₂ supply can be topped up by Vika apparatus (which heats sodium chlorate to release oxygen) using replaceable canisters, and by the Elektron electrolysis units in Kvant 1 and Kvant 2

4   Rodnik water tanks in the base block are augmented by externally-mounted tanks on both Kvant 2 and Kristall

5   contaminant removal is by the base block and Kvant 1

6   CO₂ removal is by lithium hydroxide replaceable canisters in the base block, unless the Vozdukh molecular sieve in Kvant 1 is operating

7   humidity is regulated by the base block

8   urine is processed by apparatus in Kvant 1 and Kvant 2

9   air temperature is controlled by the base block

10 urine concentrate and faeces are discarded with Progress ferries

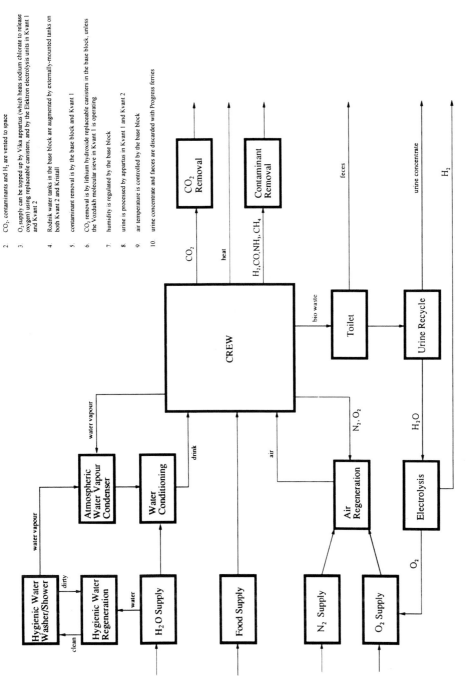

Mir atmosphere control.

systems. With Kvant 2 projecting above the upper port, whenever the orientation of the complex was changed the movement put stresses on the mating collar (if it lost hermetic integrity the entire complex could suffer sudden depressurisation), so acoustic probes were used by the Kontrol and Monitoring experiments to assess the structural stresses on the collar. No unexpected stresses were revealed.

When Progress-M 2 docked at the rear port on 22 December, it completed the first five-vehicle complex, and brought the total mass to just over 72 tonnes. In addition to more new NiCd batteries for the base block and the 256-channel Spin-6000 X-ray and gamma-ray spectrometer, it delivered an American protein-crystals experiment supplied by Payload Systems Inc. Once the batteries had been installed, the rest of the month was devoted to research. One biotechnology experiment involved long-term observation of amphibians, crustacea and molluscs to observe the reaction of their nervous systems to weightlessness in order to identify any functional changes to their vestibular apparatus. Another biological experiment grew chlorella and rice tissue cultures. Gallar was used for extended periods (often as long as six days) to process a variety of semiconductors.

Towards the end of the year, Alexei Leonov said that there would be three launches in 1990. Vladimir Dzhanibekov said that for the foreseeable future resident crews were to be formed with two rather than three cosmonauts, and that they would continue to fly tours of five to six months. Vladimir Shatalov added that five two-man crews were then training for missions. First to go would be Anatoli Solovyov and Alexander Balandin, then Gennadi Strekalov and Gennadi Manakov.

Cosmos 2054 was launched on 27 December. It took up the 344° E position in the SDRN network, over the South Atlantic. In mid-1988, Cosmos 1897 had been moved west over the South Atlantic to relay telemetry from the Buran shuttle during the critical re-entry phase of its first flight, but by early 1989 it had returned to its former position over the Indian Ocean. The introduction of this second Luch satellite greatly enhanced communications with Mir. The desired continuous network could not be realised until a satellite was positioned in the third slot (at 200° E), however. NASA was not doing that much better. Of its three TDRS satellites, only the most recent one was fully functional; a flaw in the first two satellites precluded them relaying at the maximum K_u-band data-rate planned, so they provided only a degraded service.

VENTURING OUT

To celebrate the New Year, Viktorenko and Serebrov opened the bag of gifts which they had found in Progress-M 2, and then feasted on crispy pickles, fresh lemon, canned sturgeon, blackcurrant juice and fresh fruit.

A major task in 1990 involved external activity. On 8 January the cosmonauts exited through a vacant lateral port in the multiple docking adaptor, even though Kvant 2 incorporated a purpose-designed airlock. Their ferry was at the front port, just in case they could not re-enter the base block. They got off to a poor start by inadvertently leaving open the pressure-equalisation valve for the ferry, with the result that the pressure in the orbital module dropped with the docking adaptor, and it took some time to identify and fix this problem. Undaunted, they made their way to the rear of the complex with a pair of star trackers. These bulky 80-kg packages were affixed to points on either side of Kvant 1's unpressurised instrument compartment, and would be required to assist in the task of orientating the expanded complex. On the way back, they retrieved sample cassettes from the exterior of the base block and then set up attachment points for their next spacewalk.

Three days later, the cosmonauts re-emerged. They deployed experiment cassettes (containing a variety of non-metallic test materials), then made their way back to Kvant 1 to install the Arfa apparatus to study the Earth's ionosphere and magnetosphere. On their way back, they retrieved the Echantillon cassette which had been set up by Chrétien (it had not been intended that it remain outside for so long, but this did not matter) and dismantled the anchor on which the Era experiment had been set up. Back in the docking adaptor, they swapped the Konus drogue from the upper port to the lower port, in preparation for the next module. (The earlier plan to move this as a separate "internal spacewalk" had been deemed pointless, so it had been merged with genuine external activity.)

After a rest (and a week spent unloading Kvant 2's cargo) they conducted their third spacewalk on 26 January, this time exiting the airlock. Immediately outside they affixed an open framework 'dock' incorporating magnetic clamps to mate with the autonomous manoeuvring unit. They then used handholds to move down Kvant 2 to remove its now redundant Kurs antenna. Returning to the airlock, they deployed the Danko and Ferrit exposure cassettes. Finally, they released the ASPG-M scan platform from the side of the airlock, against which it had been locked for launch.

The highlight of the mission was the testing of the autonomous manoeuvring unit. Developed by Zvezda (the manufacturer of the Orlan spacesuits), this flying backpack had been nicknamed 'Icarus' by its designer, Gai Severin. The YMK (the acronym for 'cosmonaut manoeuvring unit') was similar to the MMU backpack developed by NASA, but was somewhat larger and was covered with a thermal blanket. It had four T-shaped thruster sets, each with eight pressurised-nitrogen jets (the use of a harmless propellant meant that there would be no corrosive efflux to damage apparatus on Mir's surface). A control panel was mounted on each armrest (the left for translational and the right for rotational motions), along with a group of toggle switches and a joystick for specifying actions. It could be manoeuvred either manually or in one of two semi-automatic modes (one for optimum efficiency and the other for fast response). Unfortunately, the nitrogen bottles in the backpack had to be replaced after use, and only a few had been supplied. It was not intended for use on a regular basis, however, and was being tested for subsequent use by spacewalking Buran cosmonauts. The test could not have been carried out earlier, because the docking port apertures were too narrow for a cosmonaut wearing it to pass through; even at one metre wide, though, the airlock hatch was only just wide enough.

On 1 February, Serebrov donned his suit, followed by the YMK (fitted close around his life-support system backpack and fastened by a waist belt), and only then was the airlock depressurised.

As soon as Serebrov emerged, he 'docked' with the framework anchor which they had recently erected, so that he would not drift free. He then deployed the armrests and checked out the backpack's systems. Satisfied, he disengaged and used the thrusters to move up, away from the hatch. Although intended to facilitate free manoeuvring, in this test Serebrov was to remain linked to the anchor by a 60-metre nylon tether, just in case the backpack malfunctioned (whereupon he would be reeled back by a winch).

Viktorenko stayed in the airlock hatch to film his colleague's progress. There was a television camera built into the YMK, peering over Serebrov's shoulder. The first test involved slowly moving five metres out, coming to a halt, then returning. He repeated this three times, then moved out to about half the distance allowed by his tether to run a series of slow rotations and translations. Domestic television interrupted its schedule to show the video downlink. Once the tests were complete, he 'redocked', deactivated the YMK, and re-entered the airlock.

Four days later, Viktorenko repeated the evaluation. Before returning, he performed an experiment using the portable Spin-6000 apparatus to measure the X-ray and gamma-ray flux at different distances out to 45 metres to assess the radiation generated by the complex's movement through the plasma in the Earth's magnetic field. Once again, the YMK performed perfectly.

Next, to help recover from their exertions, the cosmonauts tested the new shower unit. A permanent cubicle in Kvant 2, the shower was an improvement over the plastic curtain devices (tested on the Salyuts) that had taken several hours to set up, use, clean and pack away again. In weightlessness, water is dominated by surface tension. Drops combine to form free-floating globules which simultaneously oscillate on a number of random axes. In the shower, a slow airflow was used to force the water from the spray head to the suction cup at the bottom of the cubicle. Despite its sophistication, it did not prove very satisfactory and the compartment was later converted into a steam room (the plumbing was finally ripped out, so that the compartment could be used for another purpose).

A DAMAGED FERRY

Progress M-2 undocked on 9 February, and as it withdrew it tested the new Luch relay. Soyuz-TM 9 arrived four days later with Anatoli Solovyov and Alexander Balandin. The residents were delighted to welcome them, as they had not had any visitors. Solovyov had been to Mir in 1988, but only for a week, and Balandin was making his first flight. The highlight of their tour was to be the commissioning of the 'technology' module; that, at least, was the plan.

The late delivery of Kvant 2 and the postponement of the next module had forced Viktorenko and Serebrov to devote more time to research than expected, and they had a fair amount of material to return to Earth. The prize in the haul was the monocrystals of various semiconductors produced by the Gallar (one of their gallium arsenide crystals was almost 300 grammes). It was subsequently calculated that this was the first flight to 'break even' in terms of balancing its cost against the 'value' of its product, raising expectations that the furnaces in the forthcoming 'technology' module would transform the complex into a miniature factory. Unfortunately, the assignment of a financial value to this semiconductor was essentially hypothetical, because it did not really feed into a commercial market. Even so, it was genuinely hoped that the complex would eventually become self-financing. The Payload Systems protein crystal package was also returned. This had processed two enzymes (hen egg white lysozyme and D-amino transferase) in 112 tests using three crystallisation processes (batch, vapour diffusion and boundary layer diffusion). There had been considerable corporate interest in sending experiments to Mir, but the US Government's ban on exporting technology to the Eastern bloc had effectively precluded it. Even now, 'exporting' the proprietary apparatus was permitted only if the package was completely sealed (it had only an on/off switch). The contract permitted the company to fly six such experiments over a period of years, so this was not just a one-off novelty. Protein crystals had been grown by similar apparatus aboard the space shuttle, but not for more than about 10 days. Crystalline protein was valuable because it enabled the normally delicate material to be studied (the structure of a crystal grown in microgravity is extremely pure and free of distortions). Protein is by far the most important substance in the body. The hope was that such crystals could shed light on genetic defects such as those producing cancers.

Viktorenko and Serebrov later reported that they had often spent as much as 80 per cent of their working day setting up apparatus, rather than using it. They had strung ropes to overcome difficulties moving around. They said that there should be foot restraints near fixed apparatus. They criticised Yuri Semenov for prohibiting the jettisoning of bags of waste from the scientific airlock, because this had resulted in 80 bags being stored in the already cramped complex. Criticism was constructive, however, so long as it improved operating procedures.

Various experiments were undertaken during the handover, including plant growth in the Magnetobiostat, production of active biological agents in the Ruchei and Biokryst electrophoresis apparatus, and production of semiconductors in the Gallar furnace. In addition, a variation of the Resonance experiment was performed to check the vibration modes of the enlarged complex (this was done periodically, to identify activities placing undesirable stresses on the structure).

On 19 February, Soyuz-TM 8 undocked. Immediately upon reaching low orbit, the Soyuz-TM 9 crew had realised that something was stuck to the outside of their vehicle, because the field of view of its Vzor optical sight was partially obscured. Nevertheless, they had continued with the rendezvous. Kvant 1's aft-facing camera had revealed that there were several loose objects projecting from one side of the ferry as it made its final approach. Before departing, Viktorenko and Serebrov performed a detour to see what these objects were, and reported that three of the descent module's eight petal-shaped thermal blankets had been torn free when the aerodynamic shroud had separated. Unfastened at the base, the blankets had peeled back "like a flower" around the collar connecting the descent and orbital modules (one projected straight out at 90 degrees, the others at 60 degrees). As they were preparing to deorbit, Viktorenko and Serebrov were told to postpone their descent for 24 hours because conditions at the landing site had deteriorated. Before they had powered down non-essential systems, they were told that a small clearing in the weather had been reported not far from the prime site, so they could return on their next pass. The half-yearly tours were pleasant for those that left in the winter and returned to glorious sunshine, but were not much fun for those that emerged from the capsule to be blasted by the chilly −30°C north wind.

The pictures taken on Soyuz-TM 8's flyby revealed that unless Soyuz-TM 9's loose blankets could be reattached they could easily block the infrared horizon sensors which orientated the spacecraft prior to its deorbit. Viktor Blagov announced that Solovyov and Balandin would have to conduct a spacewalk to reattach the blankets or to cut them off. Leonid Gorshkov of the Energiya Bureau later reported that special tools would have to be built, that these would be delivered in the next module (whose launch was scheduled for 18 April), and that the spacewalk would take place towards the end of the mission. Meanwhile, the attitude of the complex would need to be controlled to ensure that the ferry's exposed surface was neither roasted nor frozen for prolonged periods (in space the harsh thermal environment was +130°C in the Sun and –130°C in shadow). Blagov pointed out that it was also important to keep the descent module warm inside because condensation would form if it became too chilly, which risked damaging the electronics when it was powered up.

THE NEW ATTITUDE CONTROL SYSTEM

Solovyov and Balandin flew Soyuz-TM 9 around to the front port on 21 February, and Progress-M 3 arrived on 3 March. The following day, the cosmonauts installed the storage batteries it had delivered. Most of the month was spent installing the Salyut 5B computer. This had been delivered as cargo in Kvant 2, and it replaced the Argon 16B that had been initially installed in the base block. The more sophisticated Salyut 5B was to control the asymmetrical configuration by coordinating the gyrodynes in the two Kvant modules. Installing it was a tricky operation involving the integration of five processors. Then the mass distribution of the complex was accurately modelled. This involved predicting the thruster firings required for a specific manoeuvre, performing that manoeuvre, then noting the discrepancy, further refining the mass distribution model, and trying again. Iterating towards a solution not only took time, it also consumed propellant. This was just the first step, however. Once the orientation could be controlled using the thrusters, the gyrodynes were to be brought on line and the ability of the computer to control the complex without the thrusters tested. By mid-month it had become clear that controlling the asymmetric complex was more difficult than anticipated, even without involving the gyrodynes, so it was decided not to send up the next module until the problem had been resolved.

There was considerable excitement aboard Mir on 17 March when a quail egg hatched in Inkubator-2; it was the first time that an animal had been *born* in orbit. The chick flapped its tiny wings in a hopeless attempt to orientate itself in weightlessness. By early April, however, it was clear that it was not developing normally; it was extremely frail and proved unable to feed from the dispenser. Half a dozen others fared no better. Dr Ganna Maleshko of the Institute of Medical Biology reported that the hatching had demonstrated that the development of the embryo had proceeded normally. This was a significant step forward in the study of adaptation to weightlessness. Later experiments were to study the post-hatching growth phase. These embryo studies were seen as a precursor for the development of a closed-environment orbital habitat in which fowl would provide a source of food for the crew.

In early April, Solovyov and Balandin performed maintenance on the environmental and thermal regulation systems, then resumed fine-tuning the Salyut 5B's control of the complex's attitude. By this point, it was controlling the asymmetric configuration using the thrusters, so trials began to determine whether it could coordinate the gyrodynes in Kvant 1 with those in Kvant 2, which were brought on-line the first time.

The first all-up test was conducted on 19 April. Even though only eleven gyrodynes were available (one of Kvant 1's had failed), the new computer performed flawlessly, marking a significant milestone in building the complex. On a pessimistic note, though,

Viktor Blagov observed that the base block had been in space for over four years, and so was nearing the end of its nominal operational life even before the complex could be completed; it had initially been intended to have all four radial modules in place by this time. Considering the longevity of the second-generation Salyuts, however, it was clear that with careful maintenance the base block should be able to be kept operational for at least twice this time.

The extended trials of the attitude-control system had expended considerably more propellant than expected. Yuri Semenov announced that although the next module could now be launched, this would not be done until the complex had been resupplied. It had been decided to leave Soyuz-TM 9 at the front port, dock a tanker at the rear, then move the ferry to the rear port after the tanker had departed, to leave the front port free for the module which was to be launched in late May. All this was a remarkable demonstration of real-time planning. Progress-M 3 undocked on 27 April. Solovyov and Balandin loaded the Gallar and Mirror furnaces with semiconductors and set them to work.

The announcement on 6 May of the launch of Progress 42 clarified the decision to keep Soyuz-TM 9 at the front until the propellant had been replenished. The only tanker available at such short notice was an old Igla-equipped one which could dock only at the rear port; it turned out to be the last of the original model. It arrived on 8 May. It left on 27 May and Soyuz-TM 9 was transferred to the rear the following day. Three days after that, a Proton placed the new module into orbit. The hastily arranged schedule had gone like clockwork.

EXPANSION: KRISTALL

On 6 June, soon after the new module had activated its Kurs system and started its approach to Mir, one of the attitude-control thrusters malfunctioned by firing somewhat longer than required, so the automated system aborted and withdrew. A second attempt four days later (using only backup thrusters) resulted in a successful docking. The next day, the new module extended its Ljappa arm and swung itself down to the lower port. This transformed the complex into a T-shape, restoring its balance, and completed this second phase of its expansion. With things finally going their way, the two cosmonauts suggested that they be permitted to extend their tour by 10 days to allow them time to complete the commissioning of the newcomer, and their request was granted.

Kvant 3 had been dubbed Kristall because of the number of furnaces it carried (Krater, Optizon-1, Zona-2, Zona-3 and Kristallisator). The internal hatch at the far end of its main compartment led to a spherical space derived from the Soyuz orbital module. This contained the Priroda-5 Earth-observation apparatus which incorporated a pair of KFA-1000 high-resolution cameras that (with Kristall mounted on the lower port) faced forward. At the far end of the module was a docking port. This androgynous peripheral docking system (denoted APDS-89, because it was derived from the APDS-75 that had been built for Apollo–Soyuz) was intended to be used by Buran. Another such port was mounted in the side of the compartment that faced back under the base block. This was intended to take Pulsar X-2 (an X-ray telescope like Pulsar X-1 in Kvant 1) that was to be delivered by Buran and mounted using a robotic arm. Kristall also had Glasar-2 (to supplement Glasar-1 in Kvant 1); the Ainur electrophoresis unit (to augment Svetlana in Kvant 1); the Marina, Buket and Granat cosmic ray detectors; and the Svet cultivator.

Although Kristall had no gyrodynes, it had two Rodnik tanks and six NiCd batteries. The deployment mechanism of its solar panels was of a new design that allowed them to be retracted, and the mounts were detachable to facilitate their transfer to another part of the complex. Each 15-metre long panel comprised 36 in-line segments (without fore-and-after flaps to inhibit retraction) which extended concertina-like and generated 4 kW, so Kristall augmented the complex's power supply by 8 kW.

No sooner was this phase of the expansion complete than Boris Olesyuk, a worker at Kaliningrad, complained to Moscow News that the cosmonauts spent too much time maintaining the complex's systems. In addition to the need to unload the small Progress ferries, and the much heavier cargo delivered by the new modules, the installation and testing of the newly delivered apparatus took so much time that there was little left for research; their scientific product was minimal. He observed that there was a great deal of equipment in the complex which the cosmonauts did not have time to use. He noted that the Gamma-2 television package which they had set up six months earlier had not even been tested, and neither had the MKF-6MA Earth-resources camera in Kvant 2. He claimed that the difficulties encountered in upgrading the attitude-control system were attributable to the fact that there were too many computers involved, and that the problem was in making them cooperate. Further expansion was pointless, he said, without increasing the crew to permit two shifts of three cosmonauts rather than single shifts by a team of only two cosmonauts. Unfortunately for this line of argument, the life support system could sustain a crew of six for only short periods. Nevertheless, it was soon announced that the funding to finish the last two modules was to be withheld until it was demonstrated that the apparatus in Kristall really did produce economically viable results. This effectively meant that the final phase of the expansion, which at that time called for the Spektr module to be added in late 1991, and Priroda in early 1992, would be postponed for a further three years. Consequently, these two modules, which were already under construction, were mothballed in the Khrunichev factory.

SPACEWALK: LOOSE INSULATION

Even though Viktor Blagov had explained, during an interview on domestic television in February, that Soyuz-TM 9's thermal blankets were loose it was not until late May that the Western media picked up on this and reported that Solovyov and Balandin were stranded in orbit. These stories were "incomprehensible", Blagov rebutted. Reports that the crew had no "real possibility of returning to Earth" were "completely groundless", because, in an emergency, the ferry could be manually orientated for the deorbit burn. Despite efforts to minimise thermal stresses on the exposed descent module, there was still concern, however, that the wrap-around heat shield might indeed have deteriorated.

On 4 July, Soyuz-TM 9 was returned to the front port, not for the release of the rear port, but rather so that the spacecraft would be conveniently located for the spacewalk to attempt to repair the thermal insulation. Over the next two weeks, Solovyov and Balandin studied uplinked video showing procedures for the operation being tested in the hydrotank. On 17 July, they were finally set to inspect the damage.

Although exiting by a port in the docking adaptor would have reduced the distance they needed to travel to reach the work site, it would have meant making their way past the antennas mounted on the orbital module *en route* to the descent module. Given that there were no handholds, their passage could easily damage apparatus mounted on the ferry. It had therefore been decided that they should exit through the airlock and make their way to the base of Kvant 2. A kit of tools and a folded metal ladder had been provided. They were to unfold the 7-metre long ladder and angle it across the orbital module in order to reach the descent module beyond, without encroaching on the antennas. It would also serve as a work platform. Their task was to reattach or cut free the loose blankets as appropriate.

Unfortunately, they were in such a hurry to begin that they released the hatch before the airlock had fully depressurised. The other hatches on the complex opened inwards, but the airlock opened outwards so that spacewalkers could load the compartment with equipment without having to leave clearance for the hatch to swing aside. The hatch was designed to be opened in two stages. The first stage opened it a few millimetres to crack the hermetic seal and then held it close by a catch. Only after the airlock had fully depressurised was the catch to be released so that the hatch could be pushed fully open. In this case, the catch released immediately, and the pressure made the hatch fly wide violently and swing far beyond its designed limit. The cosmonauts expressed surprise at the airlock suddenly being flooded by brilliant sunlight, but set about their spacewalk oblivious to the fact that they had damaged the hatch's hinge. Free of umbilicals in their Orlan-DMA suits, Solovyov and Balandin each used two short safety lines ending with mountaineering clamps to switch between a succession of handholds. But dragging the bulky equipment with them made their progress slow. Remaining in place whilst in the Earth's shadow also slowed them down, so it took almost two hours to reach the multiple docking adaptor. The ladder proved to be more difficult to install than in the hydrotank, and this put them even further behind schedule. Upon finally reaching the descent module, they used a television camera to show Kaliningrad the exposed surface. After verifying that the thermal stress had not damaged the bonding of the heat shield, and that there was no sign of damage to the explosive bolts linking the descent module to the service module, they set about trying to reattach the loose blankets.

Clipping the blankets into position with spare pins was the preferred option. Efforts to reattach them failed because the blankets had shrunk and would not reach the locking ring. The second option was to fold each segment up, then clamp it against the orbital module, where it would not block the field of view of the sensors mounted further aft. They managed to secure two of the blankets, but the third was so badly ripped that they were unable to make as tidy a job with it as they hoped; this one was not so important, though, because it was far away from the sensors.

By this point, they had been out for almost six hours, so they did not have time to dismantle the ladder as they returned. Unhindered by tools, they made better progress. Having been on their life-support systems for so long, they could not afford to stop, so they double-checked each transfer of their safety lines in the illumination of their helmet lamps during orbital darkness. Back in the airlock, they discovered that no matter how hard they tugged the hatch it would not close the final few millimetres. They left it ajar, and passed into the middle compartment of Kvant 2, to use it as an emergency airlock. By the time this compartment had pressurised, and they were able to open their visors, they had been in their suits for almost seven and a half hours; uncomfortably close to their absolute limit. The issue was not so much the oxygen remaining in their suits, because they could plug hoses into the airlock supply, but the fact that as the lithium hydroxide neared saturation point the carbon dioxide level in the confined space would rise rapidly and carbon dioxide narcosis would set in. Although at first the cosmonauts would feel stuffy, the difficulty in breathing would accelerate their respiratory and pulse rates, and they would soon suffer impaired vision and headaches. As the body's acid-base balance was disrupted they would first become incoherent, and then lose consciousness, at which point they would be doomed to die. Leaving the airlock exposed to vacuum and retreating to the next compartment had therefore been the only option available.

Despite puzzlement concerning the hatch, there was great relief at Kaliningrad that the issue of the thermal blankets had been successfully resolved. If the cosmonauts had found that the bonding of the heatshield had cracked, or the explosive bolts had eroded, a replacement ferry would have had to have been sent up. And although this could have been launched unmanned (as had Soyuz 34) it had been decided in this case to send it with a single cosmonaut aboard so that that it would be able to dock if its Kurs malfunctioned. Anatoli Berezovoi, who had been standing by, was now stood down.

Another spacewalk was added to the schedule on 26 July. After depressurising the middle compartment, Solovyov and Balandin opened the internal hatch. They employed the television camera in the still-open airlock to show engineers the damaged hinges. A bolt had been pulled partly out, and an aluminium plate covering one of the hinges had been twisted up. This seemed to be what was preventing the hatch from closing. They moved out and, unhindered, soon reached the ladder. This could not be left extended over the orbital module because it would swing back and block the front port when the ferry undocked, but instead of bringing it back inside they had been told to break it back into two segments and fix these to the outside of Kristall on the far side of the docking adaptor. They affixed one easily, but the other proved difficult and took far longer than expected. When they returned to the airlock, Kaliningrad asked them to have another look at the hatch, and they discovered that part of the bent plate had snapped and become stuck in the hinge; once this had been removed the hatch was able to be closed. Nevertheless, it was decided to replace the damaged parts as soon as possible. It began to seem as if no mechanical task was beyond the cosmonauts.

The launch of the next crew had been put back 10 days in order to allow Solovyov and Balandin time to finish commissioning Kristall's systems. They combined this with further smelting, and the mapping of a large expanse of forest in southern Siberia that had been swept by fire the previous year. After a flawless approach, Soyuz-TM 10 automatically docked on 3 August and delivered Gennadi Manakov and Gennadi Strekalov. This was Manakov's first flight. Strekalov however, having visited both Salyut 6 and Salyut 7, suffered the frustration of an aborted rendezvous with Salyut 7, and endured the shock of a launch pad abort, was a real veteran. There had been widespread expectation that a Soviet journalist would fly with them, but the spare seat had been loaded with cargo (including a flock of newly hatched quail chicks being flown to test adaptation to weightlessness in the immediate post-embryonic phase to complement previous studies of chicks hatched in space). The residents showed their successors where everything was (as the apparatus accumulated, storing it so that it could be *retrieved* had become a major issue), and generally brought them up to date on the state of the upgraded attitude-control system.

Their flexible flight plan had resulted in a significant cargo for return to Earth which included gallium arsenide, zinc oxide, germanium oxide semiconductor and samples of epitaxial silicon, radishes and lettuces grown in the Svet apparatus, and 2,000 pictures of the Earth. Soyuz-TM 9 left on 9 August. It was now standard practice to jettison the orbital module before the service module, but in this case they were released together to preclude the loose blankets snagging at a critical moment. In the event, the re-entry was perfect. Unfortunately, Solovyov and Balandin had been so busy with spacewalks at the end of their extended mission that they had neglected their regular exercise regime, and suffered slight physical discomfort for a while after returning to Earth.

A MAINTENANCE MISSION

Manakov and Strekalov's main engineering job was to extensively rewire the base block's power supply, but they were also to attempt to repair the damaged airlock hatch before being relieved in December. Their research centred around keeping the Kristall furnaces turning out semiconductors (mainly gallium arsenide, zinc oxide and cadmium sulphide) in runs lasting up to ten days. Progress-M 4 docked on 17 August. It brought the power cables (one disadvantage in using a 28-volt line was that the cable was rather heavy) and television equipment for the forthcoming Japanese visit. A month later, just before it departed, the cosmonauts attached an experiment to its docking assembly. This was activated while the spacecraft paused about a hundred metres from the complex and the artificial plasma it created was filmed; then it withdrew.

Progress-M 5 arrived on 29 September, with more television equipment and, for the first time, a Raduga recoverable capsule. It could return 150 kg of compact material. Considering the experiments in the schedule, and the rate at which they were expected to produce results, it seemed appropriate to send a capsule up with every third or fourth resupply ferry; at 380 kg, however, this would seriously diminish the cargo capacity.

In early October, Manakov and Strekalov started preparations to repair the airlock hatch. This had been intended for 19 October, but Strekalov developed a head cold and it was postponed. On 30 October, when they finally went to replace the bent hinge plate they found that the hinge-pin was badly deformed; it was clear that the whole assembly would have to be replaced. Although the next crew was assigned this task, the urgency to repair the hatch had evaporated once it had been found that it could in fact be closed.

The Raduga capsule was a 1.4-metre long, bottle-nosed truncated cone a little under 0.8 metres across at the wide end and 0.6 metres at the narrow end. It was placed in the ferry's collar instead of its probe unit, with the narrow end projecting into the orbital module. Progress-M 5 undocked on 28 November. In contrast to its predecessors, it did not destroy itself over the Pacific Ocean; instead it flew a similar trajectory to that of a returning ferry. After it performed the deorbit manoeuvre, tracking stations computed the optimum time to command the ejection of the descent capsule. This occurred at an altitude of about 120 km, just before the ferry re-entered the atmosphere. As the parent vehicle destroyed itself over Kazakhstan, the capsule descended into the usual recovery zone. An air pressure sensor waited until it was at 15,000 metres altitude, then released the parachute.

At 4,000 metres, the radio beacon switched on to help the recovery team locate it. Despite being used for the first time, the capsule was successfully retrieved. It delivered a 115-kg haul that included the results of the ongoing smelting operation.

HANDOVER: A JAPANESE TOURIST

Soyuz-TM 11 docked at the front port on 4 December to deliver Viktor Afanasayev, Musa Manarov and Toehiro Akiyama, who was making the first 'fee-paying' visit. In general, the backups for one main expedition flew the next. Afanasayev had backed up Soyuz-TM 10 with Vitali Sevastyanov. They had previously trained for the flight in 1989 that had been cancelled following the decision to leave the complex temporarily unmanned. Two months before Soyuz-TM 10 was due for launch Sevastyanov had been grounded by the doctors, and Manarov, who had already spent a year aboard Mir, had been assigned instead. Afanasayev was making his first flight.

The fact that Akiyama was Tokyo Broadcasting System's chief foreign news editor prompted resentment in the Soviet media, which had hoped to see one of its own flown before a foreigner. The stereotypical Japanese tourist, Akiyama brought with him half a dozen cameras and a hundred rolls of film to augment the several hundred kilogrammes of television equipment which had already been ferried up. He was to make live television broadcasts whenever the complex passed over Japan, which it did for periods of up to ten minutes at a time on favourable passes. He also accumulated a great deal of video. It had been intended that he would link up with radio hams, but when Energiya demanded payment for making use of onboard equipment this was deleted from his programme.

Akiyama suffered motion sickness easily. During the two-day rendezvous, he had remained strapped in his couch. Aboard Mir, he never really adapted to weightlessness. He performed several Japanese experiments. One for Tokyo University involved him wearing a special cap equipped with sensors for thirty hours, to measure the electrical activity of his brain together with the state of his respiratory system. In addition, each night just before he retired, he performed a standard psychomotor test which required moving his hands with a predetermined pattern while his eyes remained closed.

Meanwhile, several biological experiments were set up, involving wheat and barley seeds, a ginseng tissue culture and the Vita biotechnology experiment that used animal cells to cultivate protein compounds to be used later on Earth to produce pharmaceutical preparations. Rekomb cultivated hybrid cells to be used later to make biologically active substances on Earth. Six Japanese tree frogs had been brought to record their adaptation to weightlessness. They had tiny suckers on their feet, and the objective of the test was to determine whether these helped the frogs retain normal locomotion.

Manakov, Strekalov and Akiyama undocked in Soyuz-TM 10 on 10 December, and returned to Earth. For their six-month tour, the new crew had been assigned a programme of materials-processing, Earth and astrophysical observations, and a series of spacewalks to mount a crane on the base block, and then use this to start the process of transferring Kristall's solar panels to Kvant 1.

The Gallar, Krater and Kristallisator furnaces were used throughout December. In addition, the Pion-M apparatus was used to investigate the melting and crystallisation of various materials and thermal exchange in liquids; the Mariya spectrometer was used to monitor charged particles in near-Earth space; and the Granat spectrometer was used to investigate the spatial distribution and the energy density of charged particles in the complex's orbital path.

SPACEWALKS AND TROUBLESHOOTING

After celebrating the New Year, Afanasayev and Manarov prepared their spacesuits. On 7 January 1991 they opened the airlock hatch and effortlessly unscrewed four bolts to dismantle the hinge, replaced it with a new one, and refastened the bolts to return the hatch to perfect working order. It was simply a matter of having the proper tools for the job! They then went out to affix a support bracket to a fitting on the conical skirt linking the two cylinders of the base block's main compartment (this had originally clamped the aerodynamic shroud in place) as a preliminary to mounting a crane. On their way back, they retrieved the Danko cassette that had been put out a year earlier, then extracted one of the television cameras from the Gamma-2 so that its lens could be adjusted (because the assembly was pressurised, the adjustment could not be done outside).

Progress-M 6 docked on 16 January. It delivered the crane, which had been designed by Vladimir Syromiatnikov of the Energiya Bureau. On 23 January, Afanasayev and Manarov went out to install it. It took over an hour to haul the package down the side of Kvant 2 to the support structure that they had set up previously. The 45-kg telescopic boom was just 2 metres long when stowed, but it was 12 metres long when extended. It was run out to its maximum length so that its stability could be assessed. Operated by a pair of handcranks, it could be elevated above, or lowered below the base block, and be rotated around the outside arc to reach as far back as Kvant 1. It could transfer a load of 750 kg between any two points on the left side of the complex, but care had to be taken to steer clear of all the solar panels. The crane was 'parked' against Kvant 2, so that the cosmonauts could ascend it on their way back to the airlock and then slip down it at the start of the next spacewalk. Once back at the airlock, they retrieved the Ferrit cassette.

Three days later, they went out to attach a framework mount on either side of Kvant 1 to accommodate Kristall's solar panels. They then installed a number of laser reflectors for a rangefinder to be carried by Buran on its first rendezvous (an eagerly anticipated event, then scheduled for 1992). Finally, they set up the Sprut-5 spectrometer on Kvant 2 (this measured the charged-particle flux around the complex and downloaded its data by the telemetry link).

The end of the month was devoted to routine maintenance on the environmental and thermal regulation systems while the Gallar furnace produced a monocrystal of gallium arsenide. Manarov exceeded Romanenko's 430-day record for total accumulated time in space on 6 February. Much of the next two months was devoted to Earth observations, but automated sensors continued to sample data on an ongoing basis, and the Optizon-1 furnace was used heavily. The Pion-M, used to study the processes of heat exchange in a liquid, had proved to be extremely sensitive to small vibrations from other apparatus in the complex, so it was used to *measure* the pollution of the microgravity environment.

Progress-M 6 departed on 16 March. When Progress-M 7 was 500 metres from the rear of the complex on 21 March, its Kurs system decided that it had drifted off course and aborted its approach (this was the first time that a cargo ferry had encountered any prob-

The central compartment of the Kvant 2 module, showing the ASPG-M scan platform incorporating the Gamma-2 multispectral camera.

lems). By a sad twist of fate, it carried the cake for Manarov's 40th birthday the following day. On 23 March, the ferry made another approach. At first everything went well, but when it was 20 metres out the flight controllers monitoring its downlink were alarmed to observe that it was misaligned, even though the Kurs system reported that it was on course. After hastily issuing the abort command, they watched the 7.5-tonne ferry drift within 5 metres of Kvant 1 and narrowly miss striking the base block's left solar panel. Although the diagnostics performed on the Kurs system did not detect anything wrong, it was evident that there was a major fault. But it was not clear whether this was in the ferry or the Mir complex. There was only one sure way to find out.

Afanasayev and Manarov undocked Soyuz-TM 11 on 26 March and manually flew it around to the rear of the complex where they let its Kurs system attempt an automatic approach; it drifted off course, indicating that the fault was in the Kurs system mounted on Kvant 1. Having localised the fault, they then docked Soyuz-TM 11 manually at the otherwise unusable port. Analysis of the telemetry suggested that the Kurs antenna on Kvant 1 was misaligned. Progress-M 7 docked at the front port two days later, thereby confirming that there was nothing wrong with *its* Kurs. In addition to Manarov's cake, it delivered a Raduga capsule and additional NiCd batteries for the base block. It was soon realised that the most likely cause of the problem was that Kvant 1's antenna had been disturbed by a cosmonaut during the spacewalk to affix the solar panel mountings. The possibility of replacing the damaged antenna with that at the rear of the base block (which had been redundant since Kvant 1 had blocked it) was considered, but it seemed simpler to send up a new one. There was no urgency, however, because cargo ferries could clearly dock at the front port. The replacement operation was assigned to the next crew (at about this time, it was also decided to delay their launch by a week to allow the residents time to complete their programme). While Afanasayev and Manarov undertook routine maintenance, remote-control observations by the Gamma-2 cameras were made to assess pollution levels in industrial zones in the Ukraine and Kazakhstan.

On 12 April, the 30th anniversary of Yuri Gagarin's pioneering flight, Boris Olesyuk reported in an article in *Trud* that although there were five furnaces in the Kristall module, only one worked as intended; the others were unusable because their electronic systems were faulty. This assertion was clearly at odds with the reported extensive use of the furnaces. He also claimed that even if they were all in perfect condition, power limitations prevented more than two being used at any given moment. While this was true, it would be unreasonable to expect that every piece of apparatus on the complex should be operated simultaneously on a continuous basis, so this was not a fair criticism.

Afanasayev set up an experimental thermomechanical joint near the airlock hatch on 26 April, while Manarov went to Kvant 1 to inspect the Kurs antenna. He reported that the 23-cm diameter parabolic dish was missing! He photographed the vacant enclosure to document its condition. Meanwhile, Afanasayev replaced the television camera in the Gamma-2 and retrieved exposure cassettes. The thermomechanical joint was dismantled before the hatch was finally closed.

While a cargo ferry was attached to the complex, it was responsible for manoeuvres which adjusted the orbit. In rendezvousing three times, Progress-M 7 had expended far more propellant than planned. Manoeuvring the complex had consumed most of what it had left, so now the engine in the crew's ferry had to be used to boost the complex's orbit. On 15 April the flexibility of the integrated replenishment system was exploited to feed propellant from the base block *into* Progress-M 7's own tanks, so that it would be able to deorbit itself and thereby return a Raduga capsule. When it did so, on 7 May, however, there was no sign of the capsule. If it failed to release, it would have burnt up with the rest of the craft. If it was released and returned to Earth, then its radio beacon must have failed. This means of returning valuable cargo had been criticised by veteran cosmonauts, who felt that it was preferable to send an unmanned Soyuz to resupply the complex when there was material to be returned. This would preclude replenishment of the complex's fluids, however. Evidently, different people preferred different solutions to the problem of returning cargo to Earth. With Progress-M 7's departure, the way was now clear for the next crew.

THE FIRST WOMAN VISITOR

Soyuz-TM 12 was launched on 18 May with Anatoli Artsebarski, Sergei Krikalev and Helen Sharman. As the Kurs system began its initial approach, Artsebarski realised that it was producing incorrect angular-separation data; he could *see* that Mir was not where the system believed it to be. He left the computer to close the separation to a few kilometres and then took over, closed to 500 metres, paused to line up for a straight-in approach and then docked at the front port on his first attempt. Even after the latches had engaged, he wryly reported that the computer believed it still had 100 metres to go. Given that the port had presented no problem for Progress-M 7, the fault was evidently in the ferry.

Krikalev had already spent one five-month tour on Mir. This was Artsebarski's first flight. Sharman was a British food scientist. Industrial sponsorship had been canvassed to pay the fee and to provide some experiments, but this had not been forthcoming, so the fee had been waived in return for Sharman's participation in the ongoing study of adaptation to weightlessness (unlike Akiyama, she adapted well to weightlessness). As the first woman to visit Mir, she was presented with a tiny bonsai tree upon arrival. With no special experiments to perform, she was reduced to playing the rôle of an enthusiastic tourist and she spent many hours gazing down at the Earth (as indeed did most cosmonauts in their spare moments). On a personal level, therefore, she probably benefitted more from her visit than had her often overworked predecessors.

Sharman's physical adaptation to weightlessness was regularly tested by a cardiac monitor and the Reflotron blood analyser, and her psychological adaptation was studied by the Prognoz and Pleven-87 tests. She and Krikalev set up the Electrotopograph-7K apparatus in the scientific airlock to study the surface distortions of high-temperature superconductors and advanced plastics that were exposed to space. Several biological experiments were carried out, including Vazon to cultivate ginseng, onion and chlorella, Vita to study the growth of cells producing luciferase (a biologically active albumen), and Seeds, which simply required that a bag of tomato seeds be left in the airlock during the handover so that genetic irregularities resulting from their exposure to ambient radiation could be studied when they were planted upon their return to Earth.

A recent computer failure had prevented the solar panels from continuously tracking the Sun during the sunlit part of the orbit, so the batteries were not always sufficiently charged to run all the complex's systems throughout the shadow pass; as they gradually discharged, apparatus shut down, the otherwise continuous hum from the life support system ceased, and on occasion the lights went out for several minutes. This situation was aggravated by the fact that the Sun was passing through the plane of the complex's orbit, maximising its time in shadow, a situation that transpired every two months due to the precession of Mir's orbit.

Afanasayev, Manarov and Sharman returned to Earth in Soyuz-TM 11 on 26 May. Although they had completed most of their programme, Afanasayev and Manarov had not had time to attend to Kristall's solar panels.

CONSTRUCTION WORK

Artsebarski and Krikalev had drawn a heavily construction-orientated programme calling for as many as eight spacewalks during their planned five-month tour, but the first task was to transfer Soyuz-TM 12 to the rear port on 28 May to leave the front port free for

Progress-M 8, which duly arrived on 1 June. The next week was divided between unloading its cargo (which included the replacement Kurs antenna and the tools needed to install it) and the routine maintenance which followed the arrival of replacement parts for the environmental system. Thereafter, they set the semi-automated furnaces running and made a pre-harvest survey of Kazakhstan. The MAK-1 sub-satellite was pushed out of the scientific airlock on 17 June. Carrying equipment to study the upper atmosphere, it was designed to radio its data directly to Earth, but its parabolic antenna failed to unfurl because its battery had gone flat (their predecessors had intended to release it a month earlier, but had been too busy); it was decided to send up a replacement as soon as possible.

On 25 June, Artsebarski and Krikalev ventured out to attend to the Kurs antenna on Kvant 1; it had been decided to install a new parabolic dish on the existing mounting. The mechanism was so delicate that they had to employ a small dentist's mirror to see some of the smaller components, and they had to stop work during orbital night. When they had finished, there was no way for them to test whether they had done a good job; this could only be done by an approaching spacecraft, and Soyuz-TM 12 was occupying the rear port. Back at the airlock, they deployed an experimental thermomechanical joint. A refined form of the joint tested earlier, it used a material which reverted to a predefined shape upon being heated, this thermal energy being supplied by a filament within the material to which an electrical current was applied. Such a joint was superior to a conventional mechanical hinge. The ability to test technologies on an ongoing basis greatly assisted in the development of a mechanism for erecting structures in space.

On their second spacewalk (on 28 June) their main task was to set up the University of California's Trek cosmic ray detector on Kvant 2. This was to be left outside for two years (but in the event, it remained outside for much longer). Delivered by Progress-M 8, it was a passive unit comprising layers of phosphate glass designed to track the passage of the super-heavy nuclei component of the cosmic ray flux. It was only the second American experiment to be delivered to the complex. Using the crane to move about the complex, they installed a number of small detectors to measure the spatial and energy distribution of charged particles near the complex. On the way back, they retrieved the experimental joint.

Artsebarski and Krikalev swung a work platform across to Kvant 1 on 15 July, and attached it to the 'roof' of that module. Four days later, they returned with a large box containing the Sofora girder. Its tubular rods were connected by sleeve joints made from a 'smart' titanium–nickel alloy. The first segment had been erected prior to leaving the airlock, to assist in the construction process. Once this had been affixed to the base plate, two additional segments were erected and mounted. The cosmonauts opined that it was confusing performing such assembly work in the ever-changing illumination of the daylight pass, but they had no trouble working in the illumination of their helmets during orbital darkness. More truss segments were added on 23 July, but the final (the twentieth) was not attached until 27 July. By this time the girder projected 14 metres above the platform. In order to celebrate their achievement they put a Hammer and Sickle of the Soviet Union on the top of the truss. They returned to the airlock jubilant; during 24 hours outside, they had demonstrated that orbital 'construction work' was feasible.

Throughout this entire period the automatic sensors continued to gather data for studies of space physics in the near-Earth environment (much of which furthered the study of the correlation between charged-particle flux at orbital altitude and tectonic activity on the

Earth). Astrophysical observations were conducted using the X-ray telescopes under remote control. The Vibroseismograph was used to measure micro-accelerations in the expanded complex to assess its potential for microgravity research. Additional EFO-1 data
was taken for the Climate experiment. During August, a post-harvest survey was made of
Kazakhstan, and observations were made to assess the ecology of the Aral Sea.

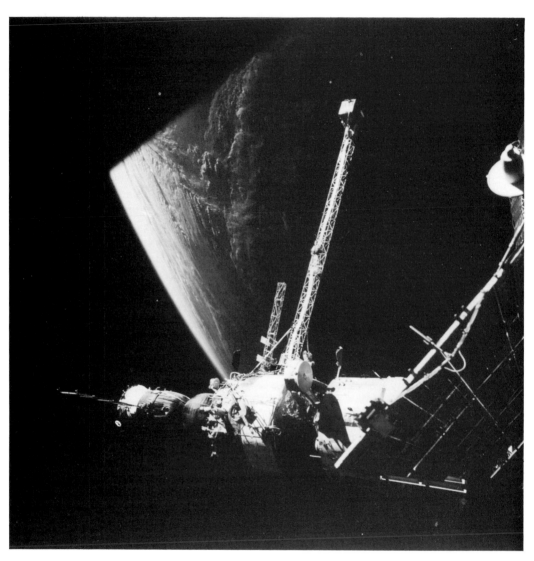

The extensive construction undertaken on Kvant 1 by cosmonauts during spacewalks is evident
in this view. The tall truss is the Sofora girder, which has the VDU thruster block mounted at its
end; umbilicals run up its length. The shorter truss is the Rapana experiment structure. The frame
on the side of the module is the motor mount for the solar panel, yet to be installed. The drum on
the short boom between the module and the base block contains the Luch antenna. A Soyuz-TM
ferry is docked at the rear of the module. Contrast the state of Kvant 1 with its pristine appearance
in 1987 (see photograph on page 148).

As Progress-M 8 withdrew on 16 August, the cosmonauts monitored an attempt to deploy an aluminium foil balloon to act as a reflector. They had attached this package to the ferry's docking assembly. Unfortunately, although the cover released properly, the balloon inflated irregularly and its fabric ripped. It had been intended to track it from the ground to measure the density of the upper atmosphere. (A lightweight structure such as a balloon would have been extremely susceptible to perturbations caused by the rarefied gaseous environment at orbital altitude.) Progress-M 9 docked at the front port on 23 August. Its cargo included a Raduga capsule and a special dispenser for Coca Cola that was flown for advertising purposes under the terms of a commercial contract with the American manufacturer.

In early September, in the aftermath of the failed coup to oust Mikhail Gorbachev as President of the Soviet Union, there was speculation in the West that Mir might be sold to NASA. When they heard this, Artsebarski and Krikalev asked whether *they* were to be included in the deal. This amusing anecdote exposed a real debate about the future of the orbital complex. In the harsh economic realism that accompanied the political chaos, it seemed certain that Buran would be cancelled; that the Energiya heavy-lift launcher, denied both this and the Polyus orbital factory module, would find itself redundant; and that with little prospect of generating a real financial return, not only would Mir not be completed, it might even be abandoned in orbit. Development for the ambitious Mir 2 complex was halted, as was work on the new ferry that would service it. Having ridden out the jokes about seeking to establish communications with NASA Mission Control in Houston, Artsebarski and Krikalev continued their programme totally oblivious to what was soon to come.

In fact, they were particularly productive throughout the second half of September, using the Priroda-5 and KAP-350 cameras to assess soil conditions and crop growth in the territory adjoining the Kara-Bogaz-Gol Bay in Turkmenia, the Golodnaya Steppe in Uzbekistan and the Aral Sea in Kazakhstan. Much of this imagery was returned to Earth in the Raduga capsule released by Progress-M 9 on 30 September.

A KAZAKH AND AN AUSTRIAN VISITOR

On 4 October, following what flight director Vladimir Solovyov described as one of the smoothest approaches he had seen, Soyuz TM-13 docked at the newly vacated front port. Commanded by Alexander Volkov, for the first time the crew contained two guest researchers, Takhtar Aubakirov and Franz Viehboeck.

Viehboeck was Austrian, and his flight was under the terms of a commercial agreement. Aubakirov, a Kazakh, was being flown by invitation. Although earlier cosmonauts had been born in Kazakhstan, this was the first time that the Intercosmos scheme had been explicitly made available to one of the internal Republics. Aubakirov had begun training only at the start of the year. He had expected to fly in November, on Soyuz-TM 14, but in July budget constraints led to the decision to merge the two visiting missions in order to save the second rocket for a later mission. Because Aubakirov displaced the assigned engineer (Alexander Kaleri), Krikalev was asked to stay aboard for another six months, and, having served a tour with Volkov, he agreed. The handover was considerably eased by the fact that he was staying on, as he already knew where everything was.

Both visitors had a very busy time. Aubakirov had two biotechnology experiments: Altyn was a genetic study of wheat, and Maskat tested the ability of certain compounds to enrich genetic material. In addition to the standard medical tests to monitor adaptation to weightlessness, he participated in several ongoing experiments, including Prognoz, Son-B and Batyr (to evaluate the effect of breathing exercises as a means of easing the initial phase of adaptation). But his primary task was Earth-resources photography of Kazakhstan in conjunction with Aral-91, the project which monitored the movement, concentration, composition, temperature and speed of dust and aerosols blowing off the recently exposed bed of the Aral Sea.

Viehboeck's programme involved fourteen experiments: one Earth-observation, three materials-processing, and ten biomedical. This work had been devised and funded by Joanneum, one of the large Austrian research organisations. About 150 kg of apparatus had been delivered by Progress-M 9. (Contrast this with Sharman's visit a few months earlier!) The Fem Earth-resources experiment involved a multispectral survey of Austrian territory using the MKF-6MA camera; Logion was to test whether an ion-emitter could cancel the electrostatic charge that builds up on spacecraft in orbit (to test a method for making a magnetospheric research satellite more sensitive by discharging it so that its own charge would not interfere with extremely fine measurements); Brillomir was to measure critical fluctuations during the decomposition of binary liquids in microgravity; and Migmas tested a mass spectrometer built for the ESA which was to be used on NASA's Freedom space station.

Medical experiments formed the core of Viehboeck's researches, however. Cogimir analysed cognitive functions during adaptation to weightlessness; Lungmon evaluated a new electrical heart-and-lung monitoring unit; Dosimir evaluated a dosimeter; Pulstrans analysed pulse transmission and heart frequency during changes of body position, and during strain; Mikrovib investigated skin sensitivity by analysing the spontaneous and stimulated microvibrations of the body's surface in the absence of gravity; Bodyfluids studied the composition and distribution of blood and bodily fluid in weightlessness by measuring the speed of sound in blood, to help determine the dynamics of transient fluid motions; Optovert measured eye movement in response to optokinetic stimulation; Mirgen used blood analysis to evaluate the effect of space radiation on genetic material; Motomir used a four-element ergometer to measure the force and velocity characteristics of the limbs to give a neurophysiological analysis of body motorics; Monimir analysed postural reflexes to assist in the development of a computerised neurological analyser by using the ergometer to investigate the movement of the subject's head and arms; and Audimir investigated changes to the auditory system. Many of these experiments would be added to the growing range of tests of adaptation to weightlessness made available to successive visitors, so that over time a substantial database of results would be built up for each test.

Towards the end of the handover, the departing crew bottled air and water samples, and collected smears from various points in the complex, just to check that there was no buildup of toxins or microflora which would pose a health risk.

Artsebarski, Viehboeck and Aubakirov departed with Soyuz-TM 12 on 10 October. It had been decided to use Soyuz-TM 13 to test the repaired Kurs antenna, because the cosmonauts would be able to complete the docking manually if it failed. On 15 October, after several successful approaches, the automated system was permitted to dock. But there was frustration four days later, when Progress-M 10 aborted its approach 150 metres from the front port. It docked successfully on 21 October. How could the *front* Kurs have been damaged?

For the rest of the year, Volkov and Krikalev combined materials-processing in the Gallar, Optizon-1, Kristallisator and Krater furnaces with Priroda-5 and KAP-350 Earth studies and an ongoing Glasar-2 ultraviolet sky survey, and the X-ray telescopes were used for astrophysical studies by remote control. The tranquillity of Mir contrasted with the shockwaves tearing apart the Soviet Union.

8

Expansion or abandonment?

On 25 December 1991, Mikhail Gorbachev resigned as President of the Soviet Union. A few days later, the Hammer and Sickle on the Kremlin was pulled down and replaced by the Russian flag, and the Soviet Union formally ceased to exist at the end of the year. It was superseded by the Commonwealth of Independent States (CIS).

Sergei Krikalev (having been launched prior to the start of his country's demise) was dubbed "the last Soviet citizen" by the Western media. Ironically, the Red Flag on the Sofora girder was now providing 'top cover' for a State which no longer existed.

January 1992 saw the research continue as if nothing had happened. By this point the X-ray telescopes were being used on a five-day cycle. Accumulated results were put into Progress-M 10's Raduga capsule, but its departure was delayed while a fault in the gyrodyne system was investigated; it departed on 20 January and returned its capsule to Earth. Progress-M 11 arrived a week later. It brought tools to enable the cosmonauts to gain access to the failed gyrodyne, the second Payload Systems package, and assorted apparatus for the forthcoming German and French visits. Although the cosmonauts had been promised a jar of honey, none was available! This seemingly trivial procurement problem in the post-Soviet chaos would soon turn into a nightmare.

Boris Yeltsin, the Russian President, established the Russian Space Agency (RSA), under Yuri Koptev, to manage civilian space operations. It assumed control of existing launchers, spacecraft and Mir, and it replaced the Ministry of Machine Building that had overseen space operations in the Soviet Union. Similar organisations were soon set up by newly independent Ukraine and Kazakhstan, and they expropriated ground stations and tracking facilities. Kazakhstan even demanded that Russia pay a fee for each launch from the Baikonur cosmodrome. In fact, the dispute over ownership of the facilities in Kazakhstan prompted the suggestion that Mir-related launches be moved to the Plesetsk Cosmodrome, north of Moscow. Berth facilities in the newly-independent Baltics were withdrawn from the ships of the tracking and communications fleet, preventing their use. Financial limitations which closed ground stations "temporarily" denied Mir use of the Luch network. For the foreseeable future, therefore, Mir would be able to communicate with Kaliningrad only while over Russian territory and the greatly diminished network of tracking stations. In the economic chaos, flight controllers displayed placards in the control room to air their dissatisfaction with the inflation that had eroded their salaries, and many of them sought secondary jobs.

It had originally been intended to complete the build-up of the Mir complex in 1990, and then start construction of the much more elaborate Mir 2 complex in 1992. Now, however, Mir 2 seemed to be a pipe dream. If Russia was to sustain orbital operations, it was evident that the existing complex would need to be kept in service until the end of the century. This was so far beyond its expected service life, that achieving it would be a tremendous engineering challenge. In fact, the first components would be technically ob-solete even before the final module was launched. The base block had been built to oper-ate for five years, but had already been in space for six years; and although the Kvants had been intended to support three years' operations, Kvant 1 had been in place for five years, and Kvant 2 for three years. In a very real sense, even if the crews' time over the next few years was devoted entirely to maintenance, the engineering expertise that this would yield would itself represent a significant result, a great step towards constructing and maintain-ing an industrial-scale orbital factory which might turn in the profit so long sought. But would the funds be provided to complete it?

In early February, Dr Valeri Poliakov denied rumours that Krikalev was seriously ill, and pointed out that on his most recent medical (conducted only a few days earlier) he had demonstrated that he was in excellent health. Some reports of Krikalev's supposed illness also claimed that he had not been returned for immediate treatment because there were insufficient funds to buy a rocket to bring him back. This assertion demonstrated a basic misunderstanding of how crews rotated. There was always a ferry *in situ*, and Soyuz-TM 13 could return at any time. Rumours of illness persisted however, and it was suggested that he might have to be evacuated by the forthcoming German mission. Once the rumour had started, it fed on itself. Meanwhile, Volkov and Krikalev prepared for external opera-tions.

The spacewalk took place on 20 February. It ran into trouble almost immediately. The heat exchanger in Volkov's suit failed. The sublimator that chilled the water which circu-lated through the suit's lining was not working properly so he overheated, and then his visor began to mist over. Volkov had to plug himself into the airlock's environmental support system using an umbilical. He stayed nearby the hatch and retrieved previously deployed exposure cassettes. Meanwhile, Krikalev made his way down Kvant 2, then along the base block, taking much longer than planned because Volkov could not swing him across on the crane. On Kvant 1, he dismantled some of the auxiliary structures which he and Artsebarski had set up as preliminaries to the construction of the Sofora structure. When this was finished, he used an ion-emitter to electrostatically sweep dust off the lens of the television camera near the rear docking port, then on his way back he retrieved the experimental solar transducers that had been mounted on the base block's upper solar panel by Titov and Manarov many years earlier.

Throughout the first quarter of 1992, the Kristallisator furnace was used for a series of experiments to investigate the process of crystallisation in silver–germanium and lead chloride–silver chloride, which are eutectic alloys (that is, they have an extremely low freezing point). The Gallar furnace was used on a continuous basis to produce a variety of semiconductors. During routine maintenance in early March the cabling that had been run throughout the complex to distribute power was replaced. Progress-M 11 departed on 13 March and Soyuz-TM 13 was flown to the front port the next day so that when it departed the front port would be left vacant. (Despite it seemingly having been repaired, the Kurs antenna at the rear had not yet been entrusted with an automated cargo ferry.)

A GERMAN VISITOR

Soyuz-TM 14 docked on 19 March, following a completely automatic approach that confirmed the integrity of the repaired Kurs system. It delivered Alexander Viktorenko, Alexander Kaleri and Klaus–Dietrich Flade. If Kaleri had not lost his seat on Soyuz-TM 13, he would now have been looking forward to returning to Earth. Flade, a German, was on a commercial visit. His 14-experiment programme included materials-processing, but his main aim was to establish baseline biomedical data in preparation for operations onboard ESA's Columbus laboratory.

The medical tests concentrated on cardiovascular measurements and investigation of hormones, plasma proteins, and redistributed body fluids in weightlessness. The CHR experiment (supplied by the Genetics Institute of Essen University) involved sampling before and after the flight to identify chromosomal aberration in the lymphocytes; ROK (Max Planck Institute) investigated responses to different orientations in weightlessness; OVI (University of Mainz) involved a special pair of goggles equipped with stimulators and sensors to study the influence on the eyes of vestibular disturbances in the absence of gravity; TON (University of Hamburg) employed a specially-built sensor to determine the interior pressure of the eye; HPM investigated hormonal changes by taking blood, saliva and urine samples; SUR studied whether the body's circadian rhythm changed in the process of adaptation to the absence of gravity; HSD (University of Berlin) used the Tchibis suit to determine tissue layer thickness and compliance; VOG investigated eye movement by video-oculography for vestibular studies; PSY used a portable computer to test perception, speech and psychomotor coordination; ISX involved wearing knee restraints while performing calf exercises to evaluate the effect of isometric exercises on muscles, blood pressure and heart rate; DOM evaluated different dosimeters as methods of measuring radiation in the complex; and KFV used the applied potential tomography (APT) apparatus and the Tchibis suit to investigate changes in the distribution and flow of body fluids (it was rather ironic that the APT apparatus had first been proposed for Sharman's flight). After assessing the results of Flade's endeavours, Germany booked a second visit.

The TES experiment used the Kristallisator furnace to measure the heat capacity of a supercooled metallic melt (in this case antimony, an alloy of silver and germanium, and two sapphire samples) and its variation with temperature. This study of the specific heat of supercooling fusions exploited the fact that in microgravity direct contact between the material and the walls of the furnace can be eliminated. Determining the heat capacities would enable other thermophysical properties of these materials to be derived. This was a basic science experiment ideally suited to microgravity.

While Flade had a very busy week, Volkov and Krikalev brought their successors up to date on where everything was located, and Viktorenko and Kaleri explained what life was like in modern Russia. With the RSA's financial crisis deepening, serious thought was being given to "temporarily" vacating Mir, so the new residents took over knowing that they might be recalled at any time.

On 25 March, Volkov, Krikalev and Flade returned to Earth in Soyuz-TM 13. This was the second time that three cosmonauts who had flown up separately returned in the same ferry. Although Krikalev had worked a 312-day double tour, he had not remotely threatened the year-and-a-day single-mission record held by Titov and Manarov. He had set a new record for spacewalking, however, having extended his accumulated time to 36 hours. Even so, the spacewalk to attend to Kristall's solar panels had been put off yet again.

Kristall had been attached to the lower port as an expedient to balance Kvant 2 until the module configured for that position in the complex arrived. Kristall was configured for a lateral port. It had been fitted with retractable solar panels so that it would not interfere with the base block's panels. Whilst it had seemed feasible to finish the construction of the complex over the period of a few years, the task of transferring Kristall's panels to Kvant 1 had been a high priority, but once financial restrictions had ruled this out, and it had become clear that the module would have to stay where it was for the foreseeable future, the preparatory construction work had become a task that was handed on from one crew to the next and achieved only as opportunity arose in the spacewalk schedule.

A QUIET TOUR

As March gave way to April, Viktorenko and Kaleri performed photographic and spectro-metric observations of agriculture around the Sea of Azov to fulfil a commercial contract with the Terra-K project which offered overhead imagery to local agricultural coopera-tives, but later in the month they divided their time more or less equally between this and renewed astrophysical observations with the X-ray telescopes. And, of course, the semi-automated furnaces kept churning out semiconductors. Progress-M 12 arrived on 22 April. After unloading the cargo, and performing the standard maintenance on the environmental and thermal regulation systems, they stripped down and overhauled the ageing base block's communications system. In May, the Vibrogal experiment was run to characterise the vibrations that polluted the complex's microgravity environment. On favourable ground passes later that month, the remote-controlled Gamma-2 cameras were used to assess the ecology of water basins, heavily forested areas and agricultural land for the Terra-K contract. Other ongoing work monitored the spread of pollution from industrial sites (in addition to the Krasnodar Kray and the industrial part of Kazakhstan, the area around Chernobyl in the Ukraine was carefully monitored). Then, in June, the Gel appara-tus was used to make more polycrylamide for use on Earth in pharmaceutical processes.

Progress-M 12 pumped fluids aboard, refined the complex's orbit, then departed on 28 June. Four days later, Progress-M 13 aborted its straight-in approach. With just 150 me-tres to go its Kurs had concluded that the roll-rate was outside permissible tolerance. Anal-ysis of its telemetry revealed that there was a fault in a new software routine, so it was reprogrammed, and the manoeuvre was completed without incident two days later. Its cargo included a pair of gyrodynes and equipment for the forthcoming French visit.

Although Kvant 1 had long exceeded its service life, five of its six gyrodynes were still functional. The gyrodynes in Kvant 2 had proven less reliable; four of its six had failed. To continue to use the gyroscopic system to orient the complex, some of the units would have to be replaced. Instead of trying to fix the failed units, however, new ones were to be installed. Unlike most of the critical systems, these bulky packages had not been intended to be serviced in orbit. Each was a pressurised spherical casing with a stubby axial projection about a metre long which supported a magnetically controlled flywheel. Installing them would not be a trivial task. On 8 July, Viktorenko and Kaleri left the airlock, moved half-way down Kvant 2, then used a pair of heavy-duty cutters to slice through the thermal blanket to mount the two new units. Back at the airlock, they tried a special pair of binoculars designed to be used by a spacesuited cosmonaut. Over the next few weeks they combined unloading cargo with on-going studies of the ecological state of rivers and lakes, using the MKF-6MA. Progress-M 13 vacated the front port on 26 July.

A FRENCH VISITOR

Three days later, Anatoli Solovyov, Sergei Avdeyev and Michel Tognini arrived in Soyuz-TM 15. Its Kurs system too, aborted on the final approach. Solovyov, who was making his third visit to Mir, took over and docked manually. This was Avdeyev's first mission. Although it was France's third flight, it was the first arranged as a commercial venture (in fact, since their earlier visits had been so productive, the French marked this docking by announcing that they would like to make further flights at roughly two-year intervals). In the post-Soviet financial crisis, these fee-paying missions not only offered an opportunity to recoup some of the cost of essential crew exchanges; they were seen as international acknowledgement of the merit of the orbital complex, and thus played a significant part in countering those who argued that Mir had become a costly irrelevance that should simply be abandoned. But the crucial issue for those in favour of continuing operations (and indeed for resuming the construction) was whether it would be feasible to sustain the base block long enough to fulfil such advance bookings.

Full opportunity had been taken of the 12-day handover to provide Tognini a varied programme of ten experiments involving medical and technological studies. Some 380 kg of apparatus had been delivered by Progress-M 13. The medical programme comprised the Orthostatism experiment to measure orthostatic resistance, by using the Echograph to monitor changes in the cardiovascular capacity and venous circulation, and invasive sampling to document hormonal changes; Illusion employed the Physalie-M apparatus to expand on an earlier study of sensory and motor adaptation; Vinimal-92 investigated sensory-motoric relationships by perception and orientation tests; Nausicca-1 measured radiation within the station (correlated with the complex's movement in its orbit); Biodose observed the biological effects of cosmic rays; Eceq measured the flux of the heavy-ion component of cosmic rays; and Immunologie-92 monitored characteristics of the immune system in space. The Alice materials experiment observed the phase-change phenomena at the critical point of a gas–liquid, and Superconductor in the Krater furnace studied the crystallisation of a high critical temperature superconductor.

The newcomers had a number of their own experiments, including the Altyn genetic experiment to investigate the transformation of plant cells; Rekomb to produce new cells and microorganisms with given properties; and Reservoir to study the development of a system for filling and emptying a capillary-tension reservoir. The residents continued to use the MKF-6MA and Priroda-5 cameras. The complex remained active even while the crew slept, because space physics sensors sent data by telemetry, and the multispectral Gamma-2 cameras and the X-ray telescopes could be used by remote control.

The handover complete, Viktorenko, Kaleri and Tognini returned to Earth in Soyuz-TM 14 on 10 August.

A NEW MANOEUVRING UNIT

Solovyov and Avdeyev set up the Krater furnace to produce epitaxial layers of silicon (that is, growing the crystalline structure on the surface of an existing crystal in such a way that the two lattices are aligned), and then reran Tognini's Alice experiment. In effect, they were waiting for Progress-M 14, which docked on 18 August. It was the first cargo ferry to use the rear port since the repair of the Kurs antenna, and it was fortunate that it managed to dock, because the most bulky item of its cargo could be unloaded only at the rear of the complex. This was because the compartment normally given over to the wet cargo had been modified to transport a package that could be unloaded only by spacewalking. As a preparation for this, on 2 September a command was issued to an angled mount in that section to open a cover which exposed this payload. This "external propulsion unit" (referred to as the VDU, because this was the acronym for the Russian phrase) was a thruster block to be mounted on top of the Sofora girder. Rather than lug the 700-kg package along the truss, a pivot had been built into a segment one-third of the way up the truss' length to enable the upper part to be folded down so that the tip came to rest directly over the hatch in the ferry from which the box would emerge.

The next day, the cosmonauts made the first of a series of excursions to install it. They folded the Sofora girder down, clamped it in position, then retrieved the Hammer and Sickle. Next, they used a special ratchet mechanism mounted in the ferry to slide the massive thruster block out of its storage compartment. Four days later, they ran an umbilical along the length of the girder, mated one end to the thruster block and the other to a plug on Kvant 1, and then attached several metal braces to the block so that it could be connected to the platform at the end of the girder. After another four-day rest, the thruster block was attached, and the girder was swung to a position 11 degrees beyond 'vertical', which placed its side-mounted thrusters in the same plane as the roll-control thrusters mounted peripherally at the rear of the base block. An 85 per cent saving in propellant would result from controlling the roll of the complex by these thrusters, because they were so much further from the complex's axis. Its introduction would reduce the frequency with which the base block's tanks needed topping up. The VDU fired jets of cold gas, and because it was a self-contained system it would have to be replaced once its tank was empty.

A fourth excursion had been set aside just in case the operation proved trickier than expected, and it was decided to use this time to carry out tasks that had been scheduled later; so on 15 September they went to the far end of the Kristall module to affix a Kurs antenna to the axial androgynous docking port, then retrieved a meteoroid collector and several exposure cassettes.

MORE ADVANCE BOOKINGS

In early October, extending its international cooperation, the Russian Space Agency signed an agreement with NASA to permit a cosmonaut to fly on the Shuttle in 1994, and for an astronaut to visit the Mir complex for an extended mission in 1995. NASA's rationale was to study adaptation to weightlessness in advance of the construction of its own Freedom space station. By inviting American cooperation, the RSA hoped to make sure that it received the funding to complete the Mir complex. Until NASA offered time on its station, Mir would be the only facility available to the international community for microgravity research. As if to confirm this case, a month later the German Space Agency signed an agreement with Russia to use Mir as a base from which to demonstrate the technology required by a highly manoeuvrable free-flying robot which would employ GPS-based navigation to rendezvous with, examine and repair satellites. A prototype of this Inspector was to be delivered by a cargo ferry sometime in 1997. Like the French, the Germans were clearly hoping that the Russians would be able to keep Mir operating for at least another five years.

Additionally, ESA contracted in November for two extended visits to Mir in 1994 and 1995, providing a total of six months aboard Mir. These flights were to yield insight into adaptation to weightlessness, and test equipment and procedures to be employed in conjunction with the Columbus laboratory. ESA flew a Spacelab on an occasional basis aboard the shuttle for the same purpose, but their duration was limited by the shuttle's inability to stay in space for more than a fortnight. With little prospect of NASA starting to assemble its Freedom space station any time soon, and the discouraging fact that its laboratory was to be the *last* module attached, ESA was keen to exploit the opportunity offered by the Russians to carry out long-term research aboard Mir.

END OF A TOUR

Their orbital construction successfully completed, Solovyov and Avdeyev resumed ongoing research. High on their list of priorities was the monitoring of the ecological state of water basins, forested areas and agricultural land for the Terra-K programme, and the spread of pollution in the industrial Krasnodar Kray and Novosibirsk Oblast areas. The furnaces were operating throughout, creating semiconductors and exotic materials (one six-day Krater run produced an alloy of barium oxide, yttrium oxide and copper oxide).

Progress-M 14 left on 21 October, with a Raduga capsule containing Earth imagery, and Progress-M 15 took its place the following week. It delivered further storage batteries, assorted scientific apparatus, and several quail eggs. These were placed in Inkubator-2 while Solovyov and Avdeyev performed routine maintenance. On 20 November the MAK-2 satellite was ejected from the scientific airlock to study the physical characteristics of the ionosphere. December was devoted entirely to research. They reran the Alice experiment, continued the Glasar-2 ultraviolet sky-survey, used the EFO-1 electrophotometer for the Climate experiment, and, at the request of meteorologists, for four days they reported the track of a tropical cyclone in the Indian Ocean. The New Year was celebrated by an all-too-brief two-way video link with their families, and then it was straight back to work. The next few weeks saw routine maintenance combined with Earth studies using the MKF-6MA. As their tour drew to an end, they at least had the satisfaction of knowing that they had completed their programme.

SPACEPORT MIR!

Soyuz-TM 16 appeared on 26 January 1993 with Gennadi Manakov and Alexander Poleshchuk. It had been fitted with an androgynous docking system so that it could test the port on the Kristall module (the Kurs antenna had been recently installed to facilitate this). The automated system made the initial approach. When it paused 200 metres out, Manakov took command and flew around to line up with Kristall. Although this was the first time that a ferry had docked off the longitudinal axis, the Mir complex now had so much mass that there was little prospect of the impulse of contact setting it rotating.

The arrival of Soyuz-TM 16 established a new construction record, because with Soyuz-TM 15 at the front and Progress-M 15 at the rear, it increased to seven the number of independent vehicles. However, it did not mark an expansion of operational capability, because the androgynous port (being different) was not intended to be used on a regular basis. Nevertheless, the Resonance experiment was rerun to assess the stresses on this unprecedented configuration.

An Israeli researcher had originally been assigned to accompany this handover, but this was cancelled, so Soyuz-TM 16 was notable for *not* delivering a fee-paying guest. Manakov had already served a tour aboard Mir, but this was Poleshchuk's first flight.

Most of the week of the brief handover was devoted to familiarising the newcomers with the state of Mir's systems, but a number of experiments were run, including using the Electrotopograph to test samples of construction materials exposed in the scientific airlock. Another experiment, supplied by NASA, monitored fluid motion in a granular material, to provide data for the development of a nutrient-delivery system in a cultivator.

Solovyov and Avdeyev returned to Earth on 1 February in Soyuz-TM 15. Manakov and Poleshchuk had a busy six-month programme scheduled, beginning with an experiment involving Progress-M 15.

On 4 February, continuing the use of cargo ferries for free-flying experiments, the Znamya package was fitted to Progress-M 15's docking assembly. The ferry withdrew 150 metres and then initiated a fast roll-manoeuvre, so that the centrifugal force dragged out eight triangular petals that formed a 20-metre diameter reflector. The orientation of the ferry had been chosen so that the mirror beamed sunlight towards the Earth, with the result that a spot of light some 4,000 metres wide scanned across Europe. To people fortunate enough to catch sight of it, it appeared like "a sparkling diamond" in the sky. This experiment was conducted to evaluate the feasibility of employing orbital mirrors to illuminate polar regions enduring extended darkness. Although frequently discussed, the reflector had only recently been built. It had initially been planned for October 1992, but Progress-M 14 had been fitted with a Raduga return-cargo capsule. The experiment over, Progress-M 15 jettisoned the reflector, and withdrew. It still had another vital test to perform, however. It returned the following day and took up station 200 metres from Mir's vacant port. A pair of controls for translational and rotational motions (identical to those used in the Soyuz) had recently been fitted to the base block's main control panel. A monitor showing the view from the ferry's docking camera enabled a cosmonaut on Mir to fly it by remote control. After commanding a series of manoeuvres to verify this link, Manakov released the ferry to Kaliningrad, which then ordered it to withdraw. Viktor Blagov reported that this important experiment demonstrated that it would be possible for a cosmonaut to manoeuvre an unmanned vehicle close alongside a complex that was festooned with projecting antennas and solar panels, because such manoeuvres could not be programmed into the automated system. It would also permit the approach of an automated ferry to be completed in the event that its Kurs system failed. (This latter capability had taken on increased significance following the difficulties encountered by three recent cargo ships.)

The base block completed its seventh year in space on 20 February. With continued preventive maintenance, it was hoped to enable it to support a full programme employing the final two modules. Progress-M 16 docked on 23 February. It delivered additional gyrodynes (to be installed inside rather than outside Kvant 2, and frames on which to mount them were included) together with new electronics for the computerised flight-control system. The next few weeks were given over to routine maintenance and rewiring the power lines of Kvant 2 and Kristall to increase the effectiveness of their solar panels, and to pool their combined output so that the resource could be used more efficiently. On 26 March, Progress-M 16 undocked under remote control, withdrew 75 metres and then redocked. (Progress-M 15 had been fitted with the Znamya package, so it had not been possible to bring that ferry all the way back in to complete the test of this system dubbed TORU.) Progress-M 16 undocked again the next day, and left. Progress-M 17 took its place on 2 April. Once this had been unloaded, Manakov and Poleshchuk prepared for two spacewalks, during which they were to mount electric motors on the frames that had been erected two years earlier on the sides of Kvant 1, in anticipation of moving Kristall's solar panels.

THE DISABLED CRANE

On 19 April, Poleshchuk made his way down the side of Kvant 2 to the base block, then used the crane to swing Manakov and the motor across to Kvant 1. It took three hours to attach it to the frame, so by the time they managed to plug it into a power outlet they were behind schedule. Worse, Poleshchuk's suit's ventilator had begun to malfunction, and he was overheating. As they made their way back, they discovered that one of the crank handles of the crane was missing. It appeared to have worked loose and drifted away! Without the use of the crane, installing the second motor would be awkward. Back inside, they reasoned that because it was a simple mechanism they should be able to fashion an alternative handle. They could not find a substitute, however, so it was decided to send up a replacement handle on the next cargo ferry. This meant that the rest of the external activities had to be postponed for at least a month. During this time, the furnaces were kept in use. Progress-M 18 arrived on 24 May. With Progress-M 17 still at the rear and Soyuz-TM 16 on the Kristall port, it was the first time that two cargo ferries had been docked. In addition to a replacement crank handle, the cargo included a Raduga capsule and apparatus for the forthcoming French visit. It was not until 18 June that Manakov and Poleshchuk ventured out. Upon repairing the crane, they used it to swing the second motor over to Kvant 1, and this time they encountered no difficulties installing it. As they rested they used their television camera to show Kaliningrad how cluttered the surface of Kvant 1 had become as a result of all the construction work.

ANOTHER FRENCH VISIT!

When Soyuz-TM 17 rendezvoused on 3 July, all the docking ports were occupied, so it paused 200 metres out, filmed the departure of Progress-M 18 from the front port, then closed in and took its place. A few hours later Progress-M 18 released its Raduga capsule,

This view of the underside of the Mir complex shows a Soyuz-TM at the front of the base block, Kvant 1 at the rear, and the Kristall module, with its docking unit, extending towards the camera. The object on the right of the base block is the crane, which runs up alongside of Kvant 2 (hidden from view). Between the base block and Kvant 1 is the Luch antenna. The projections on Kvant 1 are the motor mounts, ready to accept Kristall's solar panels. The stubby cylindeers on the unpressurised bay of Kvant 1 are the star trackers. The covers on several of the base block's portholes are open.

which returned near Orsk in the Russian Urals so as not to involve the Kazakh authorities in its retrieval; this was to become standard procedure.

The newcomers were Vasili Tsibiliev and Alexander Serebrov. Tsibiliev was on his first flight, but Serebrov had visited Salyut 7 and had served a lengthy tour aboard Mir. Because no other international participant was ready, Frenchman Jean-Pierre Haignere, who had served as Michel Tognini's backup and was therefore fully trained, was given the

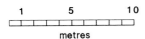

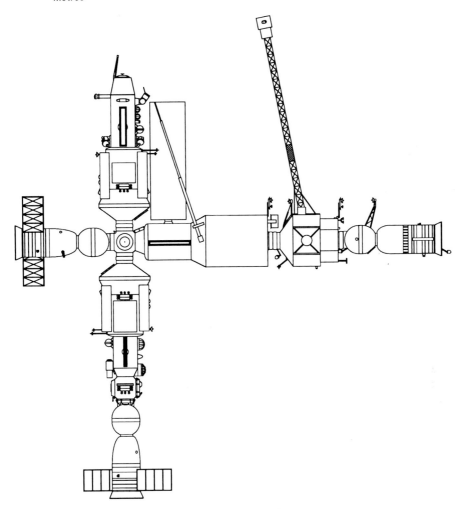

The T-shaped configuration of Mir with Kvant 2 on the upper port, Kristall on the lower port, Soyuz-TM16 on Kristall, Soyuz-TM 17 on the front, Progress-M 17 on the rear, the crane on the base block and the Sofora thruster block (with the VDU) and solar panel mount on Kvant 1.

opportunity to bring his visit forward by six months. This advancement had both an up and a down side. Although he would benefit from the longer than usual (three-week) handover, there had not been time to develop a wholly new research programme, so most of his work had to employ the apparatus which had been delivered for Tognini's visit. However, 100 kg of additional apparatus had been delivered by Progress-M 18 for new experiments (Synergies, Tissue and Teleassistance) and to permit him to adapt existing equipment to vary the research objectives.

His programme comprised the Orthostatism experiment, which innovatively combined Echograph, Diuresis and Tissue, and also used the Haut Schicht Dicke apparatus which had been used by Flade; Viminal used a miniature flight simulator to study the process of adaptation to weightlessness; Illusions studied the adaptation of the sensory-motor systems; Biodose investigated long-term effects of cosmic radiation on the body; Immunology studied the adaptation of the immune system; Synergies studied the rôle of the vestibular system in controlling dynamic equilibrium, and stabilisation of references in body synergy during complex movements; Microaccelerometer used a video camera to measure microscopic accelerations on the complex; and Teleassistance was a repeat of the Orthostatism experiment using a link-up with experts on the ground to assess how easy it would be to provide on-line technical support during complex tasks. In addition, the output from the long-running Nausicca and Eceq experiments (both of which monitored radiation) was stored away.

Manakov, Poleshchuk and Haignere returned to Earth in Soyuz-TM 16 on 22 July. The delay imposed by the need to replace the crank handle had meant that the transfer of Kristall's solar panels had not been attempted; it was passed on to the newcomers, who already had several spacewalks on their schedule.

Progress-M 17 left on 11 August, and Progress-M 19 took its place two days later. Having Soyuz-TM 16 on Kristall had provided unprecedented operational freedom, and this had been exploited to retain Progress-M 17 for 132 days. Late in 1992, NASA had approached the RSA with a view to buying a Soyuz-TM to serve as the 'lifeboat' for its own much-delayed space station. However, it needed its vehicle to remain in space for at least a year, which was twice the accepted limit of the Soyuz-TM. Because the ODU was common, it had been decided to send Progress-M 17 on a year-long test flight. After withdrawing from Mir, it was manoeuvred into a lower orbit and then powered down. When it was powered up again (on 2 March 1994), it performed several manoeuvres flawlessly. Knowing that the service module could survive an extended mission was a good start towards recertifying the Soyuz for NASA, but by that time the situation had changed.

DAMAGE ASSESSMENT

Tsibiliev and Serebrov made their first spacewalk on 16 September. They swung a platform to Kvant 1 using the crane and affixed it to the 'roof' of the module, just aft of the Sofora girder. They returned four days later to erect the 5-metre long Rapana girder on this platform. It was a scaled-down test of a structure which was intended to be used to hold the parabolic dishes of a solar dynamics power system away from the proposed Mir 2 module cluster. Once it was extended, two exposure cassettes with construction materials were mounted on it.

In early August, the Mir complex had been heavily bombarded by micrometeroids when the Earth crossed the orbit of Comet Swift–Tuttle, which had recently made its first return since 1862 and was associated with the annual Perseid meteor shower, so it was decided to add a spacewalk to the schedule to inspect the solar panels. This was on 28 September. After deploying additional sample cassettes, Tsibiliev and Serebrov made a video of the surface of the complex to assess the damage, and, to their surprise, found a 10-cm diameter hole punched straight through one of the solar panels. Upon closer inspection, they located 65 much smaller impact pits. At this point, Tsibiliev reported that he was overheating. Analysis of the telemetry from the unit in his suit revealed that the coolant was not circulating properly, so it was decided to terminate the spacewalk and add another one to the schedule to finish the survey. Viktor Blagov later reported that the complex had been reorientated to face each solar panel in turn directly towards the Sun to measure its peak output; they did not appear to have lost much power as a result of being struck by micrometeoroids.

Radio Moscow reported on 8 October that the cosmonauts had been asked to extend their tour until early in January. The Energomash factory in Samara manufacturing the uprated engines used by the crew-rated rocket would not release any engines until it was paid, and the cash was not available. This kind of production bottleneck was to become a regular feature in the new corporate-orientated rather than state-orientated economy, and it would strain the patience of programme managers and cosmonauts alike. Soyuz-TM 18 (which was to have been launched on 17 November) would be able to depart as soon as its rocket was ready, but it had been decided to postpone it so that the residents could complete a six-month tour.

Progress-M 19 undocked on 12 October and returned its Raduga capsule. Although the replacement of the crew could be postponed, replenishing the complex could not, so a rocket reserved for a military satellite was requisitioned to launch Progress-M 20, which arrived on 15 October. In addition to a Raduga capsule (the third in a row, despite the announcement that it would be used only on each second or third mission), it delivered apparatus supplied by Germany, and another two American biotechnology packages (both of which were protein crystallisation experiments, one by Boeing and the other by Payload Systems, its third).

An impromptu spacewalk was made on 22 October to deploy a new micrometeoroid detector package, and another, a week later, completed the inspection of the external surface. Apart from the large hole that they had found earlier, the solar arrays were found to be in remarkably good condition, considering that they were so old, but the thermal insulation blanket was extensively coated by soot thought to be attitude-control thruster efflux. They inspected the base of the Sofora structure to check that it had not been disturbed by stresses due to firing the thrusters at its far end, and then retrieved a sample cassette that had been outside for several years.

The downlinked video was used by engineers to assess whether the complex would support another three years of operations. Key to their assessment was the state of the solar panels. Analysis of the transducers that had been retrieved periodically had shown their efficiency to be deteriorating at a rate of 5 per cent per year. Nominally, the base block had 100 m^2 rated at 10 kW, Kvant 2 had 50 m^2 rated at 7 kW, and Kristall had 72 m^2 rated at 8 kW. Progressive degradation had reduced this ideal 25 kW by 20 per cent, and with mutual shading and losses due to off-normal insolation this could easily be cut by another 50 per cent. Despite the panels on the expansion modules, the power supply was still little better than it had been, under ideal conditions, upon the base block first attaining orbit. Mir simply *had* to be kept going, so overcoming this power crisis dominated the plans for future operations.

In early December 1993, following a remarkable announcement in September that America and Russia had agreed to integrate their plans and develop a joint space station, NASA announced that it intended to conduct a series of flights in which a shuttle would

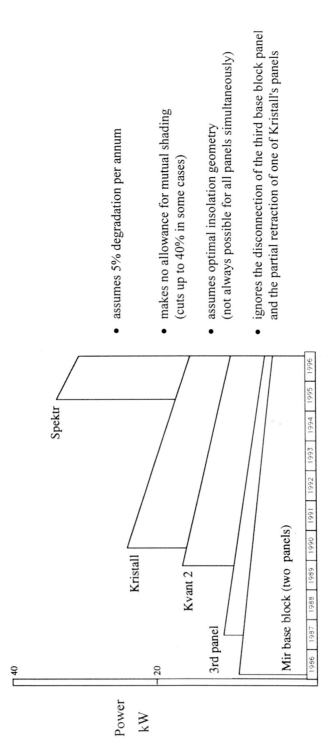

- assumes 5% degradation per annum

- makes no allowance for mutual shading
  (cuts up to 40% in some cases)

- assumes optimal insolation geometry
  (not always possible for all panels simultaneously)

- ignores the disconnection of the third base block panel
  and the partial retraction of one of Kristall's panels

Mir power generation just prior to deployment of the cooperative solar array.

visit Mir. The androgynous docking system built to enable Buran to dock with Kristall was to be installed in Atlantis, which would then be dedicated to the Shuttle–Mir flights, with the first in 1995. This was a commercial arrangement. For a fee, NASA would be able to maintain an astronaut aboard Mir on a continuous basis for two years, to carry out research and evaluate apparatus for the International Space Station. Given the shuttle's capacity to carry cargo, it was not long before it was decided to build new solar panels combining the proven Russian deployment mechanism with the most efficient American transducer cell, and drop these off during one of the shuttle's visits. Until this could be done though, the complex would remain chronically short of power.

Progress-M 20's Raduga capsule was returned on 21 November. It contained the contents of sample cassettes, a small piece of the base block's thermal blanket retrieved on the latest spacewalk, and Boeing's package. Even though the capsule used the new recovery area in the Urals, the package was handed over to a company representative in Moscow only eight hours later.

The Boeing experiment for microgravity pharmaceutical applications incorporated a dozen sample containers. It was a 'black box' that required only to be switched on, but the cosmonauts had provided a daily log of the ambient temperature and radiation level to assist the experimenters in assessing the results. The primary objective had been to grow homogeneous crystals, and the second had been to demonstrate that these could be returned intact using the Raduga capsule. The results so impressed the company that a month later it signed a contract to fly another experiment in 1994.

While they prepared themselves to return to Earth, Tsibiliev and Serebrov continued to use the Glasar-2 telescope and the Kristallisator and Optizon-1 furnaces. To assist in their successors' final training, a video was downlinked to show where everything had been stowed. Mir's cramped interior had become a storeroom, with frequently-used apparatus strapped to accessible surfaces and large bundles of other items tied in netting and dumped in little-used spaces.

As 1993 gave way to 1994, with Vladimir Titov and Sergei Krikalev training to fly as mission specialists on a shuttle, with Norman Thagard and Bonnie Dunbar training to fly a Soyuz for a three-month visit to Mir, with the Shuttle-Mir deal in place, and with the negotiations underway to agree the configuration of the International Space Station, a new sense of optimism had pervaded the programme, despite the continuing production problems that ruined flight schedules. There was no more talk of abandoning Mir as a costly irrelevance; as the only orbital complex in existence, it was a vital international resource. Riding the wave of optimism, the Khrunichev factory started to refurbish the abandoned partly-built modules which would complete the expansion of the complex.

POLIAKOV'S RETURN

Soyuz-TM 18 docked at the rear port on 8 January, but instead of the initially expected crew of Yuri Malenchenko and Alexander Kaleri, it carried Viktor Afanasayev and Yuri Usachyov, the backup crew. The third seat was occupied by Dr Valeri Poliakov, who was making his second trip. He had hoped finally to set an 18-month endurance record, but the launch postponement had eaten into his assigned time (he could not stay on after the first NASA visit began, so his return date was fixed).

On 14 January, after a short handover, Tsibiliev and Serebrov stored their results (including the latest Payload Systems package) in Soyuz-TM 17, and undocked. In this case, instead of pulling straight back, they were to manoeuvre close by Kristall and take pictures of the apparatus near the androgynous docking system to help the shuttle pilots familiarise themselves with it. Instead of withdrawing to perform a distant fly-around, it was decided to pull back a few metres and then translate down along the length of the Kristall module. Unfortunately, Tsibiliev did not realise that his translational controller was in standby mode, and he was unable to prevent a slow drift which resulted in the ferry striking Kristall a glancing blow about a metre from the docking mechanism. The gyro-dynes immediately restored stability, and the cosmonauts on Mir did not feel the impact. Soyuz-TM 18 undocked on 24 January, withdrew 150 metres, and performed a slow fly-around to look for any sign of impact damage (there was nothing dramatic), then redocked at the front port. Progress-M 21 arrived on 30 January.

Following their delayed launch, Afanasayev and Usachyov had been assigned a brief tour which would restore the year's flight schedule. It was important to be able to draw up a long-term strategy because the flight schedule would soon be integrated with flights by Atlantis. Difficulties in manufacturing spacecraft and rockets, however, made achieving this coordination increasingly difficult, and it was not long before the planned April handover became impractical.

The research programme included medical and technical experiments sponsored by the Germans, using hardware supplied by Kaiser–Threde of Munich. Some of this had been left by Flade in 1992, some more had been delivered by Progress-M 20, and all the last-minute items had been brought by Poliakov as personal luggage. When working on these experiments, the cosmonauts were to be linked by video with the researchers who had devised the equipment, both to offer instant feedback to the scientists and to receive their technical support.

History was made on 3 February 1994, when Sergei Krikalev was launched aboard STS-60 Discovery, the mission which tested the Wake Shield Facility. It was fitting that "the last Soviet citizen" had more experience living and working in space than all of his astronaut colleagues combined. Progress-M 21 left on 23 March and Progress-M 22 arrived the next day. In late March, the cosmonauts, took part in an experiment which involved aiming an electron beam at the Swedish Freya satellite while it passed by, 600 km further up the Earth's magnetic field lines. This satellite had been orbited in 1992 on a Chinese rocket to study the magnetosphere. Although this test was made to determine how the charged-particle beam was dispersed in the field, it prompted a report that Mir had been used to test a 'Star Wars' weapon. The long-running dispute between Russia and Kazakhstan over the Baikonur Cosmodrome was finally resolved on 28 March when Kazakhstan agreed to lease it to Russia in return for the equivalent of $100 million a year in trade credits. Progress-M 22 left on 23 May and Progress-M 23 moved in the next day. Upon unloading it, Afanasayev and Usachyov found that some of the food containers had been raided by opportunistic staff struggling to eke out a living from the chaotic economy. Financial restrictions had also prompted the RSA to reduce the cosmonaut corps to 50 members. At its core were 17 pilots. Headed by Alexander Volkov, this group included veterans Vladimir Lyakhov, Vladimir Titov, Alexander Viktorenko, Anatoli Solovyov and Gennadi Manakov. Many other veterans had already retired because of their age. Others now resigned because they saw little prospect of being assigned to a mission. In contrast, NASA was recruiting astronauts to gear up for the International Space Station. Meanwhile, the schedule was ruined when it transpired that the aerodynamic shroud for Soyuz-TM 19 could not be delivered on time (resulting in the already-delayed launch being further postponed). On 2 July, Progress-M 23 returned its Raduga capsule to Earth.

Soyuz-TM 19 slid into the rear port the following day to deliver Yuri Malenchenko and Talget Musabayev, neither of whom had flown in space before. Some twenty years previously, it had been decided not to launch inexperienced crews. The Kurs automated rendezvous and docking system had evidently been deemed to be reliable. Musabayev had served as Aubakirov's backup for Soyuz-TM 13. He had been assigned in place of Alexander Kaleri, in an effort to improve relations with Kazakhstan. His status on Mir became a matter of dispute: did his Kazakh citizenship make this an international visit? Musabayev was certainly no visitor; he was a career cosmonaut and was to serve a full tour as Mir's flight engineer. After another short handover, Afanasayev and Usachyov left Poliakov with his new colleagues, and returned to Earth in Soyuz-TM 18 on 9 July. Following their late launch, the newcomers had been assigned a short four-month tour in order to reinstate the schedule in October. Much of July was spent photographing Kazakhstan, particularly the area around the Aral Sea.

CRISIS AVERTED

Progress-M 24 aborted its final approach on 27 August. It was only 10 metres from the front port when it detected a misalignment, and it nearly hit one of the solar panels as it drifted by. It was not clear why it had failed to orientate itself properly, but the Kurs was reprogrammed to accept increased tolerances. Another approach was made a few days later. It was perfectly aligned as it flew straight in, but it pitched over at the last minute. This time the cosmonauts reported that they could hear it nudging the docking collar. Its video downlink confirmed that it was unable to dock because it was misaligned, so the command was sent for it to withdraw. It had propellant for only one more rendezvous. On 2 September the remote-control system that had been tested by Manakov enabled Malenchenko to dock the troubled ferry without incident. This was fortunate, because the financial restrictions that had delayed its launch had also forced the postponement of the next to November. Without Progress-M 24's consumables, the cosmonauts could well have been forced to vacate the complex in early October rather than hand it over to their successors, which would have ruined Poliakov's attempt at the endurance record. Furthermore, if the 275 kg of ESA apparatus carried by Progress-M 24 had been lost, this would have resulted in the cancellation of that mission. After a detailed analysis of the telemetry, it was decided that the fault must be in the ferry's Kurs system; it was not clear what this was, but a computer software problem seemed likely.

On 9 September, Malenchenko and Musabayev went out to examine the apparatus surrounding the docking collar to verify that it had not been damaged by the ferry when it had bumped. Then they used the crane to swing to Kristall to inspect the point where Soyuz-TM 17 had struck it a glancing blow. Although there was a tear in the thermal blanket, there was no evidence of any damage, so they resealed the blanket. This done, they attached an anchor to another shroud-fitting on the base block, in preparation for a second crane. They installed the REM experiment sent by ESA on their way back. Four days later, they inspected the mounts of Kristall's solar panels and the motors fitted on Kvant 1 to drive them once they were redeployed, and then retrieved the cassettes from the Rapana truss.

Meanwhile, to prepare for Shuttle–Mir, STS-51 Discovery evaluated the Trajectory Control System (TCS), the laser rangefinder which was to augment the shuttle's inertial navigation system in the final phase of its rendezvous. The ORFEUS-SPAS satellite, which was released and retrieved by the shuttle, acted the part of Mir during this manoeuvring test.

ESA'S FIRST VISIT

Soyuz-TM 20, carrying Alexander Viktorenko, Yelena Kondakova and Ulf Merbold was launched on 4 October, thereby re-establishing the schedule intended to synchronise with the first American visitor. As a result of his three previous visits, Viktorenko had accumulated nearly a year aboard Mir. He would have felt at home as soon as he floated in through the hatch. Kondakova, the second wife of former cosmonaut and now flight director Valeri Ryumin, was making her first flight. Merbold had already flown on two Spacelab missions as an ESA astronaut, so was the first guest cosmonaut to have prior flight experience.

Progress-M 25 undocked on 5 October. A series of manoeuvres were performed in an attempt to understand the Kurs failure, but these revealed nothing. As Soyuz-TM 20 made its final approach the next day, its Kurs system suddenly performed a yaw which threw it out of alignment. Since this was exactly what had happened to Progress-M 24, it demonstrated that, contrary to what had been thought, the problem was aboard Mir. Viktorenko took command and docked. Engineers immediately started reanalysing the telemetry, to figure out why a docking system which had worked flawlessly for years should suddenly fail.

The handover was scheduled to last almost a month so that ESA could perform an extended research programme on the first of its two commercial visits. Merbold's primary objective was 23 medical experiments, but he also had four materials-science and two technology experiments.

The medical tests monitored the adaptation of his cardiovascular, neurosensory, and muscular systems to weightlessness, the long-term objective being to develop a way to counter atrophy in muscle fibres and reduce loss of calcium from bones during long space missions. Most of the experiments reused the apparatus which had been delivered previously for the French, German and Austrian missions, thereby demonstrating that Mir had become a veritable laboratory in space. Much of it was being used by Poliakov on a regular basis. Fortunately, Progress-M 24 had delivered a compact freezer to store biological samples, a centrifuge to separate biological fluids, an IBM portable computer for data analysis, a variety of cameras and video equipment and a passive container for returning refrigerated samples to Earth, and Merbold had brought 10 kg of apparatus as luggage.

Dr Valeri Poliakov (left) during his 14-month marathon aboard Mir. Note the cluttered state of the base block.

Merbold had given samples of blood, urine and saliva prior to launch, and his musculature had been recorded by a nuclear magnetic resonance body scanner to give baseline data. The freezer would permit about 100 samples to be stored, enabling his adaptation to weightlessness to be recorded in detail. In addition, he was to monitor his fluid and electrolyte balance, fluid motion into and out of superficial tissue, any effects of changes in the central venous pressure on the erythropoietic system (that is, on the rate of production of red blood cells), and chromosomal aberrations in the peripheral lymphocytes (those in capillaries furthest from the veins into which they drain).

In contrast to his Spacelab flights, Merbold's experimental programme aboard Mir was conducted at a more relaxed pace. He worked autonomously, with far less contact with the ground; his support team was restricted to one 20-minute video conference per day. NASA, with its TDRS relay network, coordinated its shuttle flight operations far more closely. Merbold also noted the different training processes. When he had been training for a shuttle flight, he had been issued voluminous technical manuals which explained each piece of apparatus that he needed to know about, but cosmonauts relied on word-of-mouth, mutual training, and calling in the engineer that had designed a piece of apparatus if they needed additional technical information.

One week into the handover, a combination of activities inadvertently drained some of the storage batteries. Even with only essential equipment operating, the fact that there were six people aboard meant that the environmental system had to have priority on the limited power available, especially when passing through the Earth's shadow, so the gyrodynes had to be turned off for three days. During this time the complex was controlled manually with the thrusters to keep the solar panels facing the Sun in order to recharge the batteries. Recovery activities were pretty intensive for two days, then, as the power level built up, the automated systems were reactivated. Three of the batteries were subsequently isolated from the power supply. No sooner had they recovered from this crisis, however, than a small electrical fire broke out in one of the Elektron oxygen-generation systems. There are optical sensors to detect flame, small fire extinguishers are available to fight an open fire, and a suppressant system is automatically triggered in inaccessible areas. Poliakov was able to put out the fire before it became serious.

Although Merbold had to reschedule his programme as a result of the power shortage, this was not really a problem, because a fault had disabled the Kristallisator furnace (in which he was to have performed experiments to study *in situ* metal matrix composites, undercooled melts, and exotic glasses with a view to giving data of benefit to terrestrial manufacturing); this aspect of his programme was to be undertaken as soon as spare parts could be delivered, and the furnace repaired.

On 1 November, Merbold packed up his results. The frozen physiological samples (34 blood, 85 urine, and 125 saliva) were to be returned in a passive cooler, and everything else was to be returned by Atlantis. The following day, Malenchenko, Musabayev and Merbold undocked Soyuz-TM 19, withdrew 200 metres, and then let its Kurs redock to demonstrate that the problem that had affected the system was limited to the front port; this certified the rear port for the next cargo ferry. They undocked again the next day, and returned to Earth. Progress-M 25 docked on 13 November. This delivered the parts needed to fix the Kristallisator furnace.

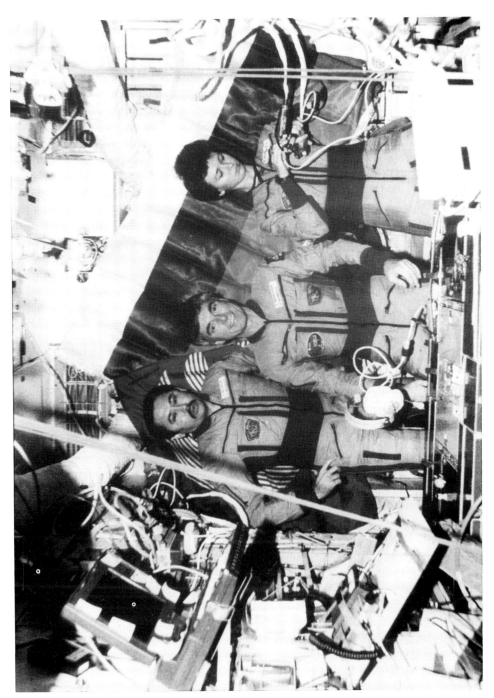

Talget Musabayev, Ulf Merbold and Yelena Kondakova in the Mir base block.

Later in the month, STS-66 Atlantis tested the unusual rendezvous procedure it was later to use to approach Mir. The opportunity of retrieving the CRISTA-SPAS atmospheric research satellite was exploited to test approaching from directly below rather than from directly ahead of a target. In fact, only once had a shuttle *not* employed this standard approach; in 1990, when STS-32 Columbia approached LDEF from overhead. Atlantis also tested the deployment of the recumbent seats to be used by Mir's residents for their return to Earth. Immediately upon landing, Atlantis was transported to Rockwell so that it could be refurbished and fitted with the equipment required to enable it to rendezvous and dock with Mir.

Viktorenko and Kondakova had been assigned a four-month tour. They were to return with Poliakov in March 1995. The launch of the Spektr module had been scheduled for December, so Viktorenko and Poliakov had inherited the task of moving Kristall's solar panels. They were to do this in late November, but even before they could unpack the spacesuits, the job was cancelled because the launch had been postponed to February (it would later slip again to April, then again to May). Crazily, a contributory cause of this delay was the late delivery of some NASA equipment which was to be delivered by the module (this delay being a combination of late shipment and customs procedures upon entry to Russia).

A new relay satellite (this time openly called Luch 1) was launched on 16 December to take up the Indian Ocean geostationary position which had been occupied by Cosmos 1897 until early 1993 (it had exhausted its propellant and begun to drift). Cosmos 2054 was still in position over the Atlantic Ocean, but its use by Mir had been restricted since early 1992. Although the two satellites would be available to improve communications with Mir at critical moments, the channels were otherwise leased commercially to relay television transmissions.

Poliakov claimed the single-mission endurance record on 9 January 1995, having spent more than a year-and-a-day in space. With a total of just over 600 days in space, so far, he also held the cumulative record by a wide margin. Given uncertainties of safe exposure levels to ambient radiation, he slept alongside the batteries in Kristall; his two colleagues used the cabins in the base block.

By early January the engineers at Kaliningrad had discovered that the computerised flight control systems of both Progress-M 24 and Soyuz-TM 20 had been improperly programmed with the centre of gravity of their vehicle, which had induced pitching and yawing deviations whilst performing the minor attitude adjustments immediately before docking. This condition would not have occurred when the vehicles were subsequently manoeuvred manually. Now that the reason for the difficulties of docking at the front port were understood, it was a simple matter to reprogram the computer. It was vital that the Kurs serving the front port be fully functional for renewed expansion of the complex to be feasible. To verify it, on 11 January 1995 the cosmonauts undocked Soyuz-TM 20, pulled back 160 metres, then let its reprogrammed Kurs redock. Another problem fixed.

9

Shuttle–Mir

A major step towards joint operations began with the launch of STS-63 Discovery, on 3 February 1995. Flown by Jim Wetherbee and Eileen Collins, its primary mission was to rendezvous with Mir. If the launch missed the five-minute slot, the additional propellant needed to complete the rendezvous would jeopardise the deployment and retrieval of the Spartan satellite. Based on previous experience, the shuttle had a one-in-three statistical chance of making it. Despite this, the launch took place at the mid-point of the window, at precisely the moment that Mir's orbital plane intersected the Kennedy Space Center, thereby eliminating the need for costly plane-adjustment manoeuvres. Mir itself was half a world away at that moment.

As soon as Discovery achieved orbit, two of its reaction control system thrusters mal-functioned, and started to leak. This led to concern that the nitrogen tetroxide might coat instrumentation mounted on the Mir complex, so the final phase of the rendezvous was made contingent on this leak being stemmed; if the leak could not be stopped, the shuttle would not be permitted closer than 125 metres. Fortunately, during the three days of ma-noeuvres leading up to the rendezvous, the leak was overcome and permission was granted to approach the androgynous port at the end of Kristall.

On 6 February, Discovery approached from below, then, with its nose high and its payload bay facing back, stationed itself directly in Mir's path. At about 2030 MT, Mir was reorientated so that Kristall's axis was aligned along the velocity vector, to aim its docking port at the newcomer. Aboard Discovery, Vladimir Titov was responsible for communications with the Mir cosmonauts; they had been talking by VHF radio virtually continuously since the shuttle had established line-of-sight contact.

Discovery paused at 300 metres to await permission to proceed with the final phase of the approach. From this point, the upward-firing thrusters were not to be used, so as not to blast Mir with their efflux (a flight mode referred to as 'low-Z' because only the thrusters aimed obliquely 'up' could be used). Mir's solar panels had been set edge-on to the shuttle to further protect the sensitive transducers. At the moment that permission was granted, Discovery started to reduce the separation. The manoeuvring was done by the shuttle; Mir was to maintain its orientation. (If it drifted, Discovery was to pause and wait for the complex to be realigned before resuming the approach.) During the final phase, Wether-bee flew the shuttle from the aft station, viewing Kristall's androgynous port through a camera set in the upper window of the Spacehab module in the payload bay (positioned near where a boresighted camera would be when Atlantis came in carrying the Russian-built docking system). Collins kept up a running commentary, providing range and closing-rate data.

As Discovery closed in, the television networks relayed the video downlinks from the two vehicles in split-screen fashion. The closing rate was so low that it required about 40 minutes – almost half an orbit, much of it during darkness – to close to the 10-metre limit. Throughout, the video showed members of the two crews waving to each other. Closest approach was achieved at 2220 MT. After 10 minutes, Wetherbee announced that he hoped their successors would be able to shake hands, then he eased Discovery back to 125 metres. At this point, the shuttle made a slow fly-around to photograph the complex using the large-format IMAX motion picture camera sitting in the payload bay. The joint exercise over, Discovery withdrew to complete the rest of its mission before returning to the Kennedy Space Center on 11 February.

From passing the 125-metre point on the approach, to finally departing, Discovery spent about three hours in close company with Mir. Since the rest of the Shuttle–Mir missions were to be carried out by Atlantis, this was to be Discovery's one and only rendezvous with the Mir complex. This "Near-Mir" flight, as it became known, set the scene for the first docking scheduled for the summer. It had demonstrated that a shuttle could approach Kristall's axial port within an 8-degree cone, and a 2-degree tolerance in orientation, then maintain its position at a point 10 metres out; and it had exercised communications and coordination procedures between Houston and Kaliningrad and between the crews. In fact, this trial had been accomplished on-the-run, because the rendezvous had been added to a long-delayed and already crowded flight plan involving a laboratory, a free-flying satellite and a spacewalk. "When all was said and done", Wetherbee reflected, "it turned out to be easy". The way was now clear for Atlantis to attempt to close that final 10 metres.

AN AMERICAN CREW MEMBER

Progress-M 25 left on 16 February, and Progress-M 26 took its place the following day. After being unloaded, it departed on 15 March. The arrival of Soyuz-TM 21 the next day was shown live by television networks in both Russia and America, because in addition to Vladimir Dezhurov and Gennadi Strekalov it delivered Dr Norman Thagard, selected by NASA to make the first visit to Mir. Although it was Dezhurov's first flight, Strekalov had been in space four times previously and had served a tour aboard Mir in 1990. Likewise, Thagard, a physician, was a veteran of four shuttle missions. The two crews displayed the Russian and American national flags in the base block to symbolise the new spirit of cooperation between the two former rival space-faring nations.

Unlike previous foreign researchers, Thagard's work was not limited to the time of the crew handover, even an extended one; he was a full member of the crew and was to serve a full tour. His programme was centred on 28 experiments, mostly biomedical. He was to study all the standard aspects of adaptation to weightlessness (the redistribution of body fluids, changes in red blood cell production and composition, muscle atrophy, and calcium loss) but was to take blood and urine samples regularly in order to track the changes on a continuous basis, in order to achieve an understanding of the processes as well as the outcomes.

When NASA first devised Thagard's programme, it had expected that the first of the two remaining modules (Spektr) would have been commissioned, and it had planned to send up his apparatus in this. As soon as it became clear that Spektr would not be ready in time, it was decided to adapt existing apparatus (such as ESA's bio-freezer) and send just sufficient equipment on regular cargo runs to ensure that Thagard would be able to per-form a core programme even if Spektr was late. Progress-M 24 and Progress-M 26 had each delivered 100 kg. The Space Acceleration Measurement System (SAMS) employed a three-axis accelerometer to make continuous measurements of ambient vibrations, and was to be used to characterise the sites where NASA's microgravity experiments were later to be set up. SAMS had been flown on STS-62 Columbia in February 1994 to calibrate microgravity experiments. It had been delivered on Progress-M 24, along with the Mir Interface to Payloads System (MIPS), which was a portable computer configured so that NASA apparatus could download data via Mir's telemetry link. Obviously the days when American apparatus could be sent up only if it was encased in a 'black box' were over.

On 22 March, Viktorenko, Kondakova and Poliakov returned to Earth. Having set a 438-day single-mission record (which is unlikely to be broken this century), Poliakov in-sisted on walking unaided to the medical tent.

The plan was for Dezhurov and Strekalov to be relieved by cosmonauts flown up in Atlantis, and then to return with Thagard on the shuttle. The timing of the two missions was planned to be such that the two resident crews would each serve a three-month tour, so that the ferry which had launched one crew could be used to return the next (the limiting factor being its six-month orbital life). This schedule was further complicated by the facts that Dezhurov and Strekalov had trained to commission Spektr, and Thagard could not complete his programme until it delivered the rest of his equipment. The final planning had to be flexible. If it appeared that Spektr's launch would be delayed by just a few weeks, the shuttle flight would be delayed by that amount to allow Thagard time to finish his work, but if the module was going to be significantly late they were to extend their tour and then return in their own ferry.

This schedule (which the Russians had recently done so much to re-establish) began to slip immediately. By the end of March, it was evident that delays in checking Spektr would push its launch from early May to mid-May, in turn pushing the Atlantis docking from mid-June to late June. It was decided that if Spektr had not been launched by early August, they were to return home at the end of that month after handing over Mir in the usual way. Dezhurov and Strekalov finished routine maintenance on the environmental and thermal regulation systems, then started upgrading the base block's core systems to further extend its operational life. (This was their primary task, and it was to consume up to 40 per cent of their working time.) Progress-M 27's arrival on 12 April confirmed that the vital front port was operational. Amongst its varied cargo was the GFZ-1 sub-satellite, and 48 quail eggs. (Thagard was to develop each egg for a specific time and then freeze it for return on Atlantis). On 18 April, Dezhurov and Strekalov cut the plumbing out of the shower cubicle in Kvant 2 (which, having been deemed a failure as a shower, had long-ago been converted into a steam room, which had been rather more successful; but the com-partment was now needed to house additional gyrodynes). The following day, the GFZ (a small sphere incorporating laser retroreflectors designed to provide geodetic data, and supplied by Germany) was ejected from the scientific airlock.

With the launch of Spektr imminent, Kristall had to be moved, but this could not be done until its solar panels had been retracted. Having been put off for so long, this task now acquired a sense of urgency. The first week of May was devoted to preparations. Unfortunately, Strekalov bruised his arm on a metal spur. Thagard applied the available medication, and the inflammation soon eased. The vital spacewalk finally took place on 12 May. The first task was to retract the panel into its container, which was then to be disconnected from the drive motor. However, the retraction proved to be trickier than expected, and Thagard had to command the mechanism to cycle a few times to complete the process. Although they had originally hoped to complete the entire transfer during a single spacewalk, they had to stop after dismounting the retracted panel. Until the panel could be redeployed, the complex's power supply would be even more restricted. They returned five days later, and swung the bulky 500-kg box on the crane across to Kvant 1, whereupon they encountered difficulties in attaching it to the new motor. The next time, they laid the necessary cables and plugged its output into the complex's power system.

The panel they moved was the one that projected out to the *left* of Kristall. This *had* to be retracted before Kristall could to be swung to the right-hand port (its assigned position in the finished complex). It had been designed to be detachable so that it *could* be redeployed; it was *able* to be moved because it could be reached by the crane, which was why the first crane had been mounted on the *left* side of the base block. It had all been carefully planned years earlier.

Once the newly mounted panel was extended, and its output verified, the two men returned to supervise the retraction of Kristall's other panel; it jammed with 25 per cent of its length still exposed, but it would not interfere with the shuttle, so it was left. This panel was not accessible to this crane, so it could not easily be dismounted and transferred (the mount for the crane on that side had been affixed but the boom had yet to be installed).

SPEKTR

Progress-M 23 departed early on 23 May to release the front port for Spektr, which had been launched on 20 May. Preparatory to reconfiguring the modules on the front of the complex to accommodate it, Kristall was powered down and two new batteries were installed to ensure that (in the absence of its panels) it would have sufficient power to operate its Ljappa arm to transfer itself onto the axial port. This was done successfully on 27 May. The following day, Dezhurov and Strekalov sealed Kvant 2, and conducted an internal spacewalk (which involved simply depressurising the multiple docking adaptor) to move the second Konus drogue from the lower port to the right radial port, and Kristall was swung onto that two days later.

On 1 June, after a long, slow, propellant-efficient rendezvous, Spektr docked at its first attempt. The next day, Dezhurov and Strekalov moved the drogue assembly from the Kristall port back to the lower port, and, soon after, Spektr was swung down. It had been designed to occupy the port opposite Kvant 2, and had similar solar panels. Late on in its manufacture, however, the far end of the module had been redesigned and it now had two additional panels mounted on an unpressurised conical compartment to boost the diminished output of the base block's panels. The cosmonauts spent two days checking out

Spektr's systems, then, on 5 June, commanded the new panels to deploy. Although one unfolded immediately, a clamp failed to release the other, and its partially unfolded segments remained near the tip of the conical compartment. The panel's motor was rocked on 8 June to attempt to release the clamp, but it remained stuck. Dezhurov and Strekalov were asked to make a spacewalk to free it, but Strekalov refused because nobody knew what the panel would do when released, and because they had no special tools. While Dezhurov and Strekalov unloaded cargo, the engineers determined how to safely release the stuck clamp. Deploying it was a high priority because, with one of the Kristall panels unavailable and a fault in one of Kvant 2's (which prevented it following the Sun), the complex was short of power despite the new arrival. Spektr, with 126 m^2 of panel, had a substantial power-generating capability. Indeed, at 16 kW, it had often been referred to as "the power module".

In fact, Spektr had originally been designated the Optizon, because its mission was remote sensing of the Earth's environment, for which it carried a range of multispectral sensors. The Balkan-1 lidar determined the altitudes of clouds; the Astra-2 spectrometer measured the constituents of the gaseous environment at orbital altitude; the Faza and Feniks spectrometers studied the Earth; and the Taurus and Grif X-ray and gamma-ray detectors measured emissions resulting from the complex's passage through the Earth's magnetic field. A variety of spectrometers (Volkov, Svet, Ryabina-4P and KR-05) and radiometers (Yusa and Neva-5) were also installed.

The French–Belgian Mir InfraRed Atmospheric Spectrometer (MIRAS), developed as a joint venture with Russia's Space Research Institute (IKI), was to scan absorption lines in the atmosphere at sunrise and sunset to identify constituents, to map their distribution and to monitor secular variations. Its data was to be stored in memory and downlinked daily. MIRAS was to contribute to an ongoing project using other satellites and aircraft to study the interaction between solar illumination and the atmosphere. An earlier form of the instrument had been used on the ATLAS-1 shuttle mission in 1992. MIRAS had been meant for installation on the Mir 2 base block, but when it was realised that this would not be constructed, the instrument had been rebuilt so that it could be sent up as cargo in Spektr, transferred internally to Kvant 2, assembled in the airlock, and finally installed on Spektr during a spacewalk.

One advanced feature of Spektr was the Pelican remote manipulator. This 2-metre long twin-segment arm, with a gripper on the end, was set between the solar panels. It was to be used to extract small experimental packages from a scientific airlock and affix them to a number of external anchor points, each of which had a socket for a power, control and telemetry umbilical. The Pelican arm was to enable exposure cassettes to be deployed and retrieved without requiring a cosmonaut to go outside (eliminating such a trivial spacewalk was important because the Orlan-DMA spacesuit was rated for just ten excursions). Energiya had developed a standardised experiment package. Once this was in place, four pallets could be unfolded individually, as required, to expose samples.

Spektr's arrival on time was welcome news to Thagard, because any further delay in delivering its 880 kg of American apparatus would have made impracticable the programme planned for the second part of his tour. His new equipment included an ergometer, a centrifuge, a thermoelectric freezer, and a number of laptops. As soon as he was certain that the freezer worked, he transferred into it the samples that had been kept in the

The Spektr module with the MIRAS atmospheric scanner mounted near its far end.

much smaller ESA cabinet. On 6 June he passed the 84-day record set 21 years earlier by the crew of the final Skylab mission, but this paled into insignificance in the light of the standard Mir tour.

The final preparation for the arrival of Atlantis involved placing Kristall on the axial port; NASA required this so that the androgynous port would be clear of the projections that might inhibit the shuttle's freedom of movement. On 10 June, the Konus assembly was moved from the lower port to the right side port, to enable Kristall to undock (with modules on both ports, this had not required depressurising the compartment), and then the transfer was made.

The Mir complex with Kvant 2 on top, Spektr beneath, and Kristall on the axis, ready for the historic docking with Atlantis. Note the partially-deployed solar panel on Spektr.

A CASE OF OVERCROWDING

After several false starts, STS-71 Atlantis was launched on 27 June. As it happened this first Shuttle–Mir mission was the 100th American space flight to carry astronauts. It was flown by Robert Gibson and Charles Precourt. Gibson had resigned as NASA's chief astronaut in order to devote himself fully to this historic mission. Greg Harbaugh, Ellen

Baker and Bonnie Dunbar flew as mission specialists. Baker and Dunbar were to perform medical tests on Mir's crew. Dr Dunbar had backed up Thagard for his Soyuz mission, and she was to handle VHF communications with Mir. Anatoli Solovyov and Nikolai Budarin were passengers on their way to the orbital complex.

The Orbiter Docking System (ODS) was mounted near the front of the payload bay. Set in a twin-triangular truss across the bay, it was basically two interconnected tubes, one leading from the mid-deck hatch to the tunnel to the Spacelab module mounted at the rear of the bay, and the other running upwards, through an airlock, to the androgynous docking system. Built long ago to enable Buran to dock with Mir, this docking system had been bought from Energiya. It had been hoped to take receipt of it in July 1994, but permission to import the pyrotechnically-equipped hardware had been delayed, so it did not arrive until September. It was shipped to Rockwell and mated with the ODS. Also in the payload bay was the TCS proven by Discovery. Of the four-shuttle fleet, Atlantis had been chosen for these Mir missions for the simple reason that it had been undergoing refurbishment at Rockwell at the end of 1993 (when the decision was taken to perform a series of Mir dockings), so the modifications necessary to accommodate the ODS had been made immediately.

Although the long Spacelab module was fitted as a life sciences laboratory (to study Thagard's adaptation to weightlessness) some scientists who were not directly involved complained that this cooperation with the Russians was being undertaken at the expense of the science programme in general, because Shuttle–Mir had commandeered life sciences and microgravity Spacelab flights. Whilst true, this criticism failed to acknowledge the potential for both areas of research likely to result from NASA being able to maintain a succession of astronauts aboard Mir.

Atlantis rendezvoused with the Mir complex on 29 June. It approached from below, as had Discovery, but it did not pass by and then ascend to dock by drawing back along Mir's velocity vector (the V-bar approach). Instead it ascended the radius vector (R-bar) that came straight up from the centre of the Earth. When Discovery had made its V-bar approach, the low-Z thrusters had consumed much more propellant than if the high-Z thrusters had been used, because the low-angled thrusters directed only a small fraction of their impulse to brake the motion, and because they had to be fired in pairs to prevent digressing from the closure axis. On the precursor flight, this penalty in propellant had been traded against the familiarity of the line of approach. By using the R-bar approach, Atlantis would genuinely save propellant. It relied on the gravity gradient of one vehicle orbiting lower than the other to brake its very slow climb during the final phase without firing its upward-directed thrusters at all. In fact, to overcome natural forces of orbital dynamics, Atlantis was required to fire its downward-directed thrusters intermittently to *maintain* its closure rate. The main attraction of the R-bar approach for NASA was that it was fail-safe; if the shuttle became disabled, gravity would draw it down, away from Mir. This precluded the possibility of a collision. Unfortunately, by some inexplicable oversight, the mission patch depicts Atlantis making a V-bar approach!

With Kristall on its front port, Mir was orientated to aim the androgynous port at the Earth, towards the ascending shuttle. At 1450 MT, Atlantis was in position 300 metres directly beneath Mir. After a pause, it closed to 100 metres, then paused again. At 1623 MT, as the two spacecraft passed north of the equator, Atlantis resumed its approach. It paused

Atlantis with the ODS near the front of the payload bay and the Spacelab module at the rear. The
Mir module nearest to the Shuttle is Kristall; its partially-retracted solar panel can be seen.

again at 10 metres. At 1655 MT, with Mir back over Russian territory, and once again in
contact with Kaliningrad, Atlantis started the final phase of the approach. It had to achieve
a docking before Mir flew out of range of the tracking network, so this gave Gibson about
15 minutes.

Stationed at the rear of the flight deck, Gibson flew the final phase of the approach,
initially by peering up through the window above his head, then at the monitor showing
the view from the camera looking up the centreline of the ODS. In the foreground was a
three-wire cross. Beyond was the target in the centre of Kristall's port. To remain in the

centre of the narrowing cone that would ensure that the two docking systems mated in alignment, he had to maintain the two marks superimposed as he slowly ascended. The video downlinks were again shown live in split-screen format.

When Atlantis paused just a few metres out, Mir was put in free-drift flight mode, to ensure that its attitude-control system would not compete with the shuttle's to control the joined mass. This done, Atlantis closed at a rate of 3 cm per second, offset by no more than 2 cm from the axis of the approach cone, and was orientated with an angular error in all three axes of less than 0.5 degree, so it was as nearly perfect as anyone could have hoped.

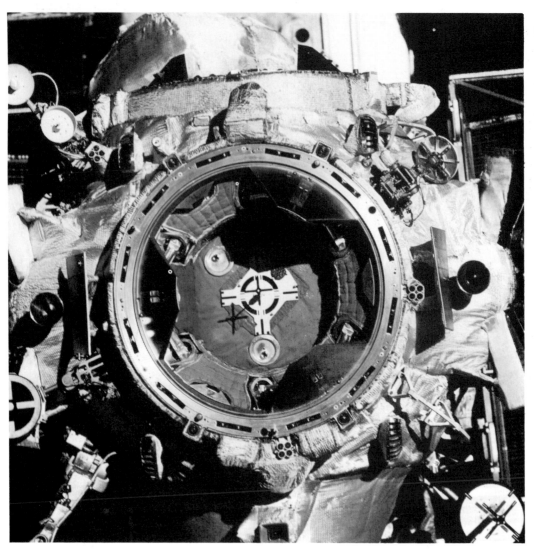

The androgynous docking port at the end of the Kristall module. The structure at the centre is the visual alignment aid. The three blades physically align the two collars for a soft docking (each has a latch on its face). The ring of the peripheral latches establishes the hard docking.

At 1700 MT, precisely on time, while over Lake Baikal, the two sets of triple-petals meshed and the capture latches they carried engaged. There was a momentary shudder as the two 100-tonne spacecraft jostled one another while springs within the mechanism damped out the residual relative motions. After about fifteen minutes in this soft-docked state, Atlantis fired its downward thrusters to force the guide-ring against Kristall, to ensure proper alignment before the twelve primary latches around the rim of two collars engaged to achieve a rigid connection. Finally, the guide-ring was retracted to establish a hermetically sealed tunnel between the two vehicles.

An hour later, after the ODS had been pressurised, Gibson made his way through its compartments. The top hatch of the ODS was opened at about 1900 MT and Gibson and Dezhurov greeted one other in the narrow tunnel. In the Kaliningrad control room, RSA chief Yuri Koptev and NASA administrator Daniel Goldin savoured the moment. The two crews then congregated in the base block for a 'photo opportunity' that amply demonstrated that Mir had not been built to accommodate *10* people. It is the procedure aboard a shuttle for the crew (when not wearing communications headsets) to share a handheld microphone to talk to Houston. Although practical within the confines of the orbiter, snaking the umbilical out through the ODS, through Kristall and on into the Mir base block proved rather awkward. One benefit of having the shuttle alongside was that it enabled continuous communications to be maintained by relaying this radio set via the TDRS network. Mir's Luch link was used only occasionally for televised conferences.

The two crews reconvened in the Spacelab module the following morning and, after a brief ceremony, got down to business. Dezhurov, Strekalov and Thagard, who were well advanced with their preparations for return to Earth, were given a thorough medical examination using the life-science equipment that included a treadmill, a bicycle, and an LBNP chamber. The scientists who had complained that the science programme had been hijacked to facilitate a political junket could hardly criticise the examinations undertaken on these three subjects who had spent over a hundred days in space. This was the first opportunity for NASA to test long-duration cases whilst still in the space environment. Even with an extended-mission pallet in its payload bay, a shuttle is limited to just two weeks in orbit. Only a permanent orbital complex could facilitate total adaptation to the space environment. Mir was the only such facility. Only by docking a shuttle with Mir could the sophisticated biomedical technology at NASA's disposal be applied to space-adapted cases. The political symbolism of the moment was rich, but this first visit provided tangible scientific results for NASA.

While the retiring Mir crew served as guinea pigs for the biomedical studies, their successors transferred cargo. In addition to 425 kg of consumables, this included a pair of heavy-duty bolt cutters for the clamp which had snagged Spektr's solar panel. As a result of the reaction in its power-generating fuel cells, Atlantis had a great deal of water to donate to Mir; this 'waste product' was usually vented to space. Kristall's port had pipes for fluid transfer, but the ODS did not, so the water had to be pumped into small containers, which were then taken aboard Mir and emptied into the Rodnik system. By the end of the visit, half a tonne of water had been transferred. Although, on the whole, Mir recycles 60 per cent of its water, only that recovered from the water vapour condenser is potable, so this top-up was very welcome. Finally, Solovyov and Budarin put their couch liners and

Vladimir Dezhurov reaches down through the Kristall hatch to greet Robert Gibson, to add the finishing touch to the historic link-up.

Norm Thagard with Gennadi Strekalov in the base block. The object floating between them is an IMAX large-format film camera.

Sokol suits into Soyuz-TM 21, and Dezhurov, Strekalov and Thagard transferred theirs to Atlantis, to formally mark the handover.

The experimental results transferred to Atlantis included frozen quail chicks, ESA's materials samples, Thagard's biomedical samples, accumulated film, semiconductors and biological materials. The Russians exploited the shuttle's payload capacity to return expired elements of the Salyut 5B computer, and items normally discarded with empty cargo ferries, so that engineers could examine them to reassess the permitted service life of each item. This aspect of the joint mission relied on quartermastering skills to track all the items going in each direction and ensure that they were properly loaded for the return to Earth (a laser bar-code reader was under development to assist future transhipment).

Late in the preparations for the mission, it had been decided that rather than keep the combined 'stack' orientated as it had been at the moment of docking (with the shuttle below Mir), Atlantis would reorientate it at the start of each day so that Mir's solar panels could generate the maximum power. Because the differential gravity field would tend to restore the most stable position, maintaining this solar-inertial attitude would involve Atlantis making frequent adjustments. The stack was to re-establish the gravity-gradient attitude each evening, and be left so overnight. When it was discovered that much more propellant was consumed in doing so than had been expected, it was realised that it was because the shuttle computer's mass model was not able to deal accurately with having such a large attached mass and, as it overcompensated using its fine-control thrusters, it oscillated back and forth, either side of the optimal alignment. The long-term solution was a better mass model, but on this occasion the orientation tolerance was relaxed (the resulting minor deviations did not seriously degrade Mir's power). The inadequacy of the computer's mass model had never been suspected. With this, and a variety of other engineering data, NASA was racing up the learning curve of how to operate a shuttle in conjunction with a massive orbital structure. Taking the technical risk out of operations planned for the International Space Station was the primary objective of the Shuttle–Mir programme. The hatches were closed on 1 July to enable Atlantis to perform manoeuvres whilst docked with Mir, to test the integrity of the ODS (it did not lose its hermetic seal) and to observe the dynamics of the solar panels (they wobbled, alarmingly), then the tunnel was reopened.

Late on 3 July, as the spacefarers bade their farewells and then congregated in their respective vehicles, Atlantis pumped up the complex's atmosphere to a pressure of 15.4 psi as a parting gift, thereby obviating the need to send up air with the next cargo ferry. Kristall's hatch was closed at 2332 MT, and the ODS hatch was closed a quarter of an hour later. This must have been a sad moment for Dunbar. Having trained for Mir as Thagard's backup, she had been assigned to the first docking mission to transfer to the complex along with Solovyov and Budarin, to continue Thagard's programme, until being retrieved by Atlantis upon its return later in the year. This plan had had to be cancelled, however, when it became clear that leaving an astronaut aboard would conflict with the 'extended mission' planned by ESA (the limiting factor being the return capacity of the Soyuz lifeboat). This was unfortunate for NASA, but since it had booked a continuous slot thereafter, the presence of an astronaut would restrict other visitors to the period of a handover. Only after NASA had used up its assigned time (mid-1998) would others be able to make extended visits. The demand for access was so great that the problem was how to fit everyone in. How fortunate it was that the Mir complex had not been abandoned in the turmoil immediately following the collapse of the Soviet Union.

Solovyov and Budarin undocked Soyuz-TM 21 from the rear port at 1454 MT on 4 July, withdrew 100 metres and then manoeuvred to the side of the complex so that they could film the departure of Atlantis 16 minutes later. The docking latches released, then the spring-loaded mechanism pushed the vehicles apart. Explosive bolts would have been fired if the latches had failed to disengage, and if the bolts failed to completely separate the two components, Atlantis carried tools to enable spacewalking astronauts to detach the docking system. In this event, Harbaugh would have unfastened the 96 bolts which held the docking system on the ODS assembly. If this had proven necessary, fouling Kristall's port in this way would almost certainly have brought the Shuttle–Mir programme to a premature end. Not all went as planned, however.

Soyuz-TM 21.

Mir had been in free-drift mode all the time that Atlantis had been docked – its Salyut 5B would not have been able to control the attitude of such an asymmetric stack – and it was disturbed when the springs in the ODS pushed the two vehicles apart. Although the attitude-control system was automatically reactivated as soon as the shuttle was well clear, and initiated a preprogrammed pitchover manoeuvre, it became confused and shut down. It was stable, however, so it was not difficult for Soyuz-TM 21 to redock.

Having withdrawn 500 metres, Atlantis now made its own photographic fly-around of the complex before departing. It returned to the Kennedy Space Center on 7 July. For the re-entry, Dezhurov, Strekalov and Thagard wore NASA pressure suits. A row of reclining frames was erected on the mid-deck for them, so that they would not have to endure sitting upright at the start of their readaptation to gravity: the time that they would be most at risk of orthostatic intolerance, and most likely to black out. Contrary to instructions, Thagard climbed out of his couch and walked away to the recovery van. The delay in launching Spektr, and the resultant delay in sending up Atlantis, had meant that instead of a 90-day tour, he had spent 115 days in space. Despite the problems, though, he had managed to carry out his entire research programme, and NASA now had detailed medical data on how the human body would adapt during the planned three-month tour on the International Space Station.

Thagard's debriefing highlighted many interesting points: Mir was roomy, and very habitable, but it definitely had the look and feel of a locker room that had been lived in for a decade; he had got on well with Dezhurov and Strekalov, but days had passed by without hearing English on the voice link, and he had suffered cultural isolation; he had yearned for his family; being a 'laboratory rat' in such a rigorous biomedical study was no joke; the requirement to log food intake was a disincentive to eating; and he had dreaded the prospect of extending his tour to six months. After four shuttle flights and a tour on Mir, Thagard retired from NASA early the following year and returned to academia.

While Atlantis and Mir had been docked, the payload which it was to deliver on its next visit (a Russian docking module) arrived at the the Kennedy Space Center. Even NASA's most ardent critics had to acknowledge that this hastily-arranged programme was off to an excellent start.

Although the primary objectives of the Shuttle–Mir programme were that America and Russia learn how to work together in space and reduce the technical risk in building and operating a joint orbital facility, the opportunity to have astronauts serve tours aboard Mir prompted NASA to review the science programme planned for the space station. This identified several dozen experiments that could be performed early, in conjunction with Mir, at little or no extra cost to the individual project budgets, so some of these projects were brought forward. Clearly, although the first docking was primarily an engineering test flight, subsequent Shuttle–Mir flights would facilitate the build-up of NASA science on Mir.

A SHORT MAINTENANCE TOUR

Safely back aboard Mir, Solovyov and Budarin ran diagnostics on the Salyut 5B to find out why it had failed. It appeared that a software oversight had misinterpreted the translation imparted by the spring-loaded docking system, an error that could easily be corrected.

Having been launched in mid-March, Soyuz-TM 21, unless it was to exceed its service life, needed to depart no later than mid-September. The newcomers were to devote most of this time to engineering work.

They started on 14 July by making a spacewalk to inspect a solar panel on Kvant 2 which had not been tracking the Sun properly for some time; they discovered that it was jammed by a piece of apparatus improperly stowed during an earlier excursion, so they released it. They then used the crane to swing down to the far end of the Spektr module to cut the clamp fouling one of its panels. Although it unfolded, the two segments at the tip stood perpendicular, degrading its output by 20 per cent. They decided to leave it like that for the moment. On their way back, they examined the right radial docking port to look for any possible cause of the slight pressure leak which had been present while Kristall had briefly been mounted there, but they could see nothing awry (when Kristall was swung back three days later, the leak did not recur).

Solovyov and Budarin were back out on 19 July. They had already connected the heliostat of the MIRAS instrument to the telescopic spectrometer in the airlock. Their task now was to move the 2.6-metre long cylindrical unit to the far end of Spektr and attach it to a fixture that had been bolted in place during the module's construction. Unfortunately, Solovyov's cooling system malfunctioned almost immediately, forcing him to plug his suit into the airlock's facilities. Since it would require both men to haul the 225-kg unit, they had to cancel the deployment. Ironically, this operation had been brought forward a month so that MIRAS could be used that much earlier, so Budarin retrieved cassettes (including the Trek cosmic ray detector which had been out since 1991) and then set up new ones. They found that as they repressurised the airlock the air leaked; the hatch had not formed a hermetic seal and it took almost an hour to overcome this problem. It had been a frustrating excursion.

Two days later, Budarin used the crane to swing Solovyov and the MIRAS package to the far end of Spektr. They attached it by three clamps to a scan platform at the tip of the unpressurised conical compartment, and connected the power and communications umbilicals without incident. There was some consternation when the unit appeared not to respond to commands. The cosmonauts reseated the connectors, but this had no effect. On their way back, they detoured to inspect Kristall's partly-extended remaining solar panel. Back inside, they went to check the internal connectors and discovered that one was not properly seated. When they reseated this, MIRAS immediately reported in to the anxious ground controllers.

Progress-M 28 arrived at the front port on 22 July. It delivered 350 kg of apparatus for the forthcoming ESA visit, and a gyrodyne. After unloading the ferry, Solovyov and Budarin settled down to maintenance tasks by replacing a gyrodyne in Kvant 2. Only in August did they find time to resume materials processing, using the Gallar furnace.

ESA'S LONG MISSION

Progress-M 28 departed on 4 September. Soyuz-TM 22 took its place the next day, and delivered Yuri Gidzenko, Sergei Avdeyev and Thomas Reiter. Matching the earlier slippage, this launch had been delayed for two weeks to allow Solovyov and Budarin time to finish their maintenance programme; they left in Soyuz-TM 21 on 11 September. In order to give ESA long-duration flight experience, Reiter, a German national, stayed aboard with Gidzenko and Avdeyev. Since the next handover was due in mid-January, Reiter expected a 135-day tour. Avdeyev had served a tour in 1992, but this was Gidzenko's first. The new residents had been assigned spacewalks to mount equipment on Spektr. They had expected to receive the Priroda module in November, but no sooner had they settled in than they were informed that its launch had been put back by four months. Then on 17 October they were advised that their tour would have to be extended by six weeks. Financial restrictions precluded paying overtime to the workers assembling the rocket which was to deliver their successors, and without a stock of rockets, such delays in production were becoming a major source of concern. ESA welcomed the extension of Reiter's mission, and announced that it wished to book a third visit (probably involving Christer Fuglesang, Reiter's Swedish backup) in 1998. This would have to fit in with a schedule which involved accommodating another French visit after the final Shuttle–Mir mission. It was intended to squeeze in all the visitors before finally mothballing Mir, so that attention could be redirected to the International Space Station.

During the first week, Reiter unpacked, installed and checked his equipment. ESA had sent a total of 400 kg of apparatus in Progress-M 28 and Spektr, Reiter had carried 10 kg with him as luggage, and another 85 kg was to be delivered by the next ferry, so this was a major enterprise for which ESA had paid dearly.

Although Reiter's programme included 18 biomedical, 10 technology, and eight materials processing tasks, the highlight of his visit was to be a spacewalk. At a more mundane level, he, like Merbold and Thagard before him, was to collect samples of saliva, urine and blood for later analysis; these were to be kept in the NASA freezer. He also wore a radiation dosimeter whose data were to be correlated with biomedical observations. A key part of his programme concerned loss of bone mass, and he tested methods of reducing it by simulating the effect of walking in a gravity field by periodically stimulating a heel and by placing an ankle under compression. He was to test his heel bones using the Ultrasonic Bone Densitometer (UBD) fortnightly to assess any benefit. One heel was the experiment, and the other was to act as the control. Most of his biomedical experiments reinforced previous studies of the cardiovascular system, visual acuity, psycho-motoric functions, posture dynamics (using the ELITE four-camera system, together with the ANBRE skintight limb-motion monitoring suit) and the respiratory system (using the RMS-2 to study lung functions). Progress-M 29, which docked at the rear port on 10 October, delivered the samples for the furnace experiments and the ESEF exposure cassettes that he was to deploy outside. He had concentrated on biomedical monitoring to ensure that he followed his adaptation to weightlessness, but in mid-October he began to prepare to go outside.

The Orlan-DMA spacesuits were unpacked on 18 October, adjusted to fit their new users, tested, and then replenished. On 20 October, Avdeyev and Reiter made a five-hour spacewalk – the first by an ESA cosmonaut. Avdeyev used the crane to swing Reiter to Spektr so that he could insert the ESEF cassettes into fixtures mounted on the conical compartment near the MIRAS platform. There were four of these cassettes: three were passive traps to accumulate ambient particulate debris, and the fourth was an active system that recorded the speed, mass and trajectory of each impact. Umbilicals were connected so that the clam-shell covers of the traps could be opened by remote control from within Mir. Each cassette was to be opened for a specific period. They were to be closed while spacecraft were manoeuvring nearby, in order to prevent contamination (the Astra mass spectrometer had long ago shown that this was a significant source of local pollution). On their way back they replaced the cassette in the Komza experiment. The next day it was announced that because his mission was to be substantially extended, Reiter would make a second excursion early in the new year. One of the ESEF cassettes was opened immediately, and another was opened specifically to sample dust from the Draconid meteoroid stream, which is associated with Comet Giacobini–Zinner.

A problem was discovered on 1 November. One of the atmospheric processors had become ineffective. Located in Kvant 1, the Vozdukh apparatus prevented the buildup of carbon dioxide. As a short-term measure, the lithium hydroxide scrubber (which had served as the base block's primary system prior to the delivery of the regenerative unit) was started up. Its saturated lithium hydroxide canisters had to be replaced on a regular basis, and the stock of canisters was sufficient only for 30 days. During the next few days, the problem was traced to a loss of pressure in the primary coolant loop, which circulated in pipes running from the base block to Kvant 1. Preparatory to removing the carbon dioxide, the Vozdukh apparatus first dehumidified the air. This released a great deal of heat that had to be removed by the base block's coolant loop. Without this active cooling, the Vozdukh apparatus was ineffective. Once the cover had been taken off the hydraulic pump in Kvant 1, a 2-litre free-floating blob of glycol was exposed. This leak was traced to a pipe feeding the main pump. The outflow was stemmed by applying a quick-setting putty-like material. After a pressurisation test, the system was reactivated. A permanent repair would be necessary though, so new tools and materials were added to the manifest of the next cargo ferry. To increase the margin of safety provided by the lithium hydroxide system, it was decided to add canisters to the next shuttle manifest.

THE DOCKING MODULE

STS-74 Atlantis was launched on 12 November. This second Shuttle–Mir mission was flown by Kenneth Cameron and James Halsell, together with mission specialists Jerry Ross, William McArthur and Chris Hadfield.

This time the rear of the payload bay held a special Docking Module (DM) supplied by Energiya. There was no need for the sophisticated biomedical apparatus provided by the life-sciences Spacelab because there were no long-stay Mir residents to be retrieved. The 4.6-metre long, 2.2-metre diameter, 4.2-tonne 'module' was really an extremely stretched Soyuz orbital module with an androgynous port at each end and, as such, it could have been transported by the service module of a Progress-M ferry. The docking of such a novel payload was probably best handled by Atlantis however.

On 14 November, Hadfield used the remote manipulator to unstow the DM and position it directly over the ODS. Opportunity was taken to collect configuration data for the computerised Space Vision System (SVS). Reference dots on the two units were tracked by oblique-angle video cameras while Hadfield manoeuvred the DM above the ODS. The movement and relative displacement of these references were compared with a computer model which, when refined, produced an animated boresighted-viewpoint to assist in mating modules on the ports of the International Space Station (on which direct viewing will often be impracticable). With the DM in position, the arm was placed in "limp" mode and

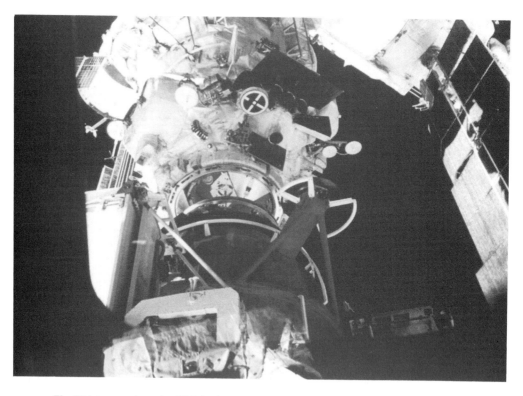

The DM, mounted on the ODS in the payload bay, is eased slowly towards the Kristall module. The final approach was conducted 'in the blind', because Ken Cameron did not have a direct view of the docking mechanism; leaving the DM in place on Mir would make subsequent visits simpler.

Atlantis fired its downward thrusters to nudge the extended guide ring of the ODS up against the DM to soft-dock. Once the residual relative motions had been damped, the mechanism was retracted to hard-dock; only then was the arm withdrawn. If there had been a problem, Ross and McArthur were prepared to make a spacewalk to try to overcome it.

Atlantis rendezvoused with Mir on 15 November. It flew the same R-bar approach as before, and, also as before, the shuttle was responsible for all the manoeuvres. This time, though, Kristall was mounted on the radial port. With the DM on top of the ODS (providing 4 metres of additional clearance) the solar panels projecting from Kvant 2, Spektr and the base block were not a problem; it had been the lack of clearance between the panels and the shuttle's cabin that had required Kristall to be transferred to the axial port for the first docking. The complex was orientated so that Kristall aimed its port at the shuttle. Cameron did not have a direct view of the DM's docking system, so the remote manipulator arm was fixed in position above and to the side of it to give a side view of the final few metres of closure. Once the DM was in place, no further 'blind dockings' would be needed. To provide a little extra clearance for the DM, the base block's panels had been rotated face-on to the Kristall module. Atlantis had never manoeuvred in such a confined space before, but Cameron had it fully under control and made it look easy.

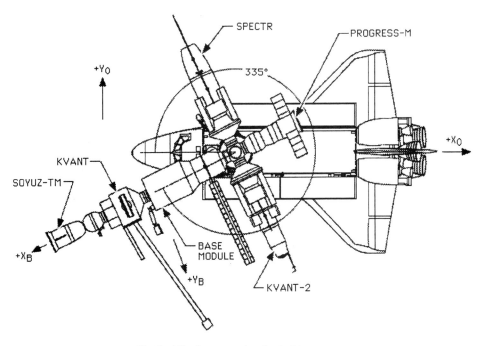

Shuttle–Mir alignment using the docking module.

Atlantis docked at Kristall's axial port at 0928 MT. Atlantis' crew opened the ODS hatch, disassembled its centreline camera and moved into the DM for the first time. The hatch to Kristall was opened at noon. Cameron and Gidzenko shook hands through the tunnel, as had Gibson and Dezhurov before them. Then everybody gathered in the base block for the ritual photo opportunity. With the national flags of Russia, America, Canada (for Hadfield) and Germany (for Reiter) on display, it was evident that Mir had indeed become a major international resource. In fact, Yuri Koptev noted that because Mir had been underfunded by 180 billion roubles in the current financial year, if it were not for the 350 billion roubles from the fee-paying missions, continued operations would not have been possible.

For the first time, the shuttle really was serving as a space truck delivering a module and cargo. In all, the astronauts delivered 275 items and collected 195 items. One tonne of cargo for Mir comprised 250 kg of food, 450 kg of water, some 300 kg of assorted items for NASA's research, and various items (including a guitar) for the residents. In case the coolant loop failed before a permanent repair could be carried out (disabling the Vozdukh air regenerator again), 20 lithium hydroxide canisters were off-loaded to give an extra margin of safety without encroaching on the manifest of the next cargo ferry.

The 375 kg of cargo retrieved included processed materials, computer disks full of experimental data, biomedical samples, and components of expired and broken hardware that had been retained so that the engineers who had built it could examine them. The Russians had been able to return such amounts of cargo only once before (from Salyut 7, using a Merkur capsule). By this simple act, therefore, Atlantis made a real contribution to Mir operations.

An international crew. Left to right, in two-tone suits, are STS-74 crewmembers Ken Cameron,
Jim Halsell, Chris Hadfield, Bill McArthur and Jerry Ross. Mir's residents were Sergei Avdeyev,
Thomas Reiter and Yuri Gidzenko. It is evident that Mir had not been meant to house such a large
crew!

Atlantis performed manoeuvres designed to re-evaluate its ability to control such an
offset centre of gravity. This time it was not necessary to orientate Mir in solar-inertial
attitude because the Sun angle was better (so the power level was higher). The hatches
were closed at about 2030 MT on 17 November, but Atlantis did not undock until
1116 MT the following day. The docking module was left on Kristall. The shuttle withdrew to perform a photographic fly-around before departing; it landed at the Kennedy
Space Center two days later. Docking with Mir had begun to seem routine! The fact that
the DM had been attached without problem boosted NASA's confidence that it would be
able to mount modules 'in the blind' on the International Space Station. This tricky flight
had "far exceeded expectation", said shuttle manager Tommy Holloway.

After the flight, Dan Goldin said that Mir was "proving to be an ideal test site for vital
engineering research", and that the first two dockings were "already paying back benefits",
by providing "proximity and docking operations" and by "simulating an early construction
flight".

Atlantis left behind more than the DM. Two bulky boxes containing solar panels were
affixed to it. One was a retractable panel like those developed for Kristall (in fact, it was
the panel which was to have been carried by the forthcoming Priroda module, the launch
of which had again been postponed, this time from the first to the second quarter of 1996).
The other, the Cooperative Solar Array (CSA), integrated this proven deployment mecha-

nism with the most powerful American transducer. It had an area of 42 m^2, and could generate 6 kW. At last, the solution to Mir's chronic power problem was in hand. These new panels were simply cargo on the DM, though; they would have to be erected by cosmonauts making spacewalks. Furthermore, since Kristall's retracted panel was now clear of the shuttle's operating envelope, it could be re-extended to augment the power supply.

BACK TO THE ROUTINE

After the departure of Atlantis, Gidzenko and Avdeyev resumed engineering work. The gyroscopic attitude-control system was taken off-line for several days so that they could carry out preventive maintenance on the gyrodynes in Kvant 2. In November, Reiter carried out a number of materials experiments by making semiconductors, alloys, and glasses in ESA's Titus furnace. An internal spacewalk was performed on 8 December to move the Konus drogue from the right to the left side of the multiple docking adaptor in preparation for Priroda. Progress-M 29 left on 19 December, and Progress-M 30 arrived the next day. In addition to the usual consumables, it contained 62 kg of cargo to support Reiter's extended tour, and a selection of Christmas presents. After a snort of brandy to see in the New Year, Reiter set up another Titus experiment, on this occasion to study the thermal properties of undercooled melts. Gidzenko and Avdeyev repaired the coolant pipe. (Progress-M 30 had delivered glycol and a by-pass.)

Reiter's chance at a second spacewalk came on 8 February. First, Gidzenko pushed the YMK of the airlock and strapped it to the framework immediately outside, so that it would no longer clutter up the airlock. Then he and Reiter retrieved two of the ESEF cassettes from Spektr. Gidzenko tried to dismantle Kristall's redundant Kurs antenna, but its bolts were too tight and he had to leave it. (It had been intended to send it back so that the engineers could examine it for signs of exposure.)

By 20 February 1996, the base block had been in orbit for ten years. Soyuz-TM 23 was launched the next day, and then, in daily succession, Progress-M 30 departed, and Soyuz-TM 23 docked. It delivered Yuri Onufrienko and Yuri Usachyov. Although this was Onufrienko's first flight, Usachyov had served a tour on Mir in 1994. Most of the handover was devoted to the formidable task of bringing the newcomers up to date on where equipment was within the complex. (As an illustration of this problem, Reiter had been unable to find his centrifuge for two months.)

Gidzenko, Avdeyev and Reiter returned to Earth in Soyuz-TM 22 on 29 February. The 'long' visit that ESA booked had turned out to be a standard six-month tour. Unlike Thagard, Reiter appeared to welcome the news of his extension. He left most of his frozen bio-samples and the ESEF cassettes to be returned by Atlantis. Analysis of the active ESEF experiment revealed that the complex passed through a stream of debris twice a day, at which time it suffered 5,000 microscopic impacts in an interval of only one minute. This was useful data for the designers of the micrometeoroid blanket which was to protect the International Space Station. By the end of 1995, the RSA had decided to postpone its contribution to that complex, so that it could continue operating Mir through to the next century. NASA had agreed to lay on more Shuttle–Mir flights to help sustain the ageing complex. The prospect of Mir's retention prompted Germany to book a visit (although it was a member of ESA, it, like France, desired to pursue its own research), Japan signed a contract to have a series of life-sciences and radiation experiments conducted aboard Mir, and China expressed an interest in an acclimatisation visit because it planned to expand its own programme to include human spaceflight. There was clearly nothing like a successful track record to encourage investment.

On their five-month tour, Onufrienko and Usachyov were to perform five spacewalks and to commission Priroda (the launch of which was now due in April). On 15 March, on their first excursion, they mounted a crane on the previously-fitted anchor in order to provide coverage on the right side of the base block, so that the solar panels on the DM could be accessed.

10

Fully booked

Flown by Kevin Chilton and Richard Searfoss, STS-76 Atlantis docked at the DM on 24 March 1996. In its payload bay this time was Spacehab, a 3-metre long, 4-metre wide cylinder trimmed flat at the top in order not to block the view of the astronauts on the flight deck. In effect, this module doubled the habitable volume of the shuttle, and it accommodated 50 racks of the type on the mid-deck, which could readily be used to carry either scientific apparatus or cargo.

Although Spacehab was developed as a private venture to offer a cheap facility with a fast turnaround on the ground for commercial microgravity research, it was NASA's commitment to use it which guaranteed private funding. Unfortunately, it did not attract many independent contracts, and its continued use was in doubt when the shuttle began to make resupply visits to the Mir complex. The original plan had been to use Spacelab (which had been assigned to most of the flights anyway) to carry cargo. Spacelab was a laboratory module, though, and its internal configuration did not lend itself well to bulk haulage. Spacehab turned out to be "a solution in search of a problem". Having been devised as a research facility, its niche turned out to be cargo transport. A contract was signed in July 1995 to fly one of the existing pair of Spacehab modules on this visit, and to expand the other by mating it to the pressure hull of the engineering test article to make a double-length module to be carried on later flights. The single-length Spacehab, which occupied a quarter of the shuttle's payload bay, was mounted well aft behind the ODS. Because its hatch was on the same axis as the mid-deck hatch (rather than on the axis of the module, as with the Spacelab) it needed just a straight tunnel to link to the ODS, so it was more convenient. A Spacehab mounted up near the flight deck was restricted to 1,300 kg of payload, but when set further back it could carry 2,000 kg. It delivered 980 kg of materials for Mir (including food, a transformer, and a gyrodyne), and 740 kg of apparatus for NASA's programme (including television cameras and lighting kit for the Glovebox). In addition, Spacehab contained a freezer for returning the rest of Reiter's biomedical samples and a Biorack of microgravity experiments. Further experiments were set up on the mid-deck and there was a Get Away Special (GAS) pallet in the payload bay. This time, a total of 500 kg of scientific results and assorted apparatus was off-loaded from Mir, for return to Earth.

Linda Godwin and Michael R. (Rich) Clifford spacewalked on 27 March. Since this was the first time that astronauts had worked outside a shuttle which could not chase after

This side view of Mir shows the docking module (at the end of the Kristall module) facing the camera. The second crane has been installed and 'parked' against Kvant 2. The object strapped to the framework anchor on top of Kvant 2 is the obsolete YMK backpack. The concertina-like object projecting forward is Kristall's partially retracted solar panel.

them if the tether snapped, their backpacks had SAFER manoeuvring units fitted. Because it was the first time that astronauts had worked in the immediate vicinity of the Mir complex, the surface of which was unfamiliar, they were not to stray beyond the upper end of the DM. After dismantling the DM's now-redundant camera, they connected clamps to its handrails and then mounted the Mir Environmental Effects Package (MEEP) cassettes on these. Two of the cassettes were to study ambient particulate matter (one was passive, but the active one recorded the time, size and trajectory of the particles striking it). The other

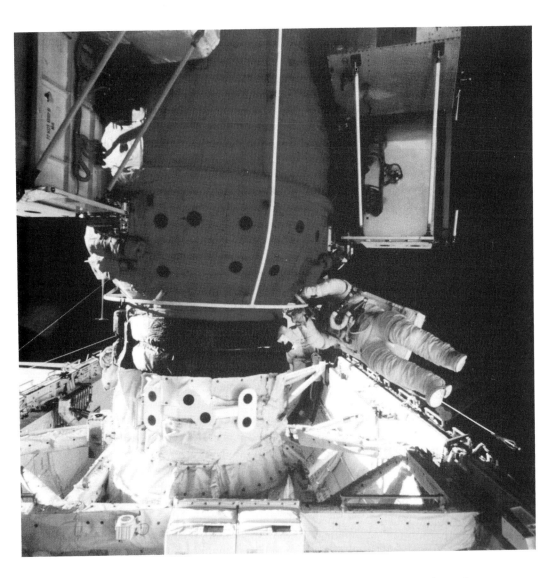

Astronaut Michael R. (Rich) Clifford, STS-76 mission specialist, works at a restraint bar on the docking module (DM) of the Mir space station during the 27 March 1996 spacewalk. Clifford teamed up with fellow astronaut Linda M. Godwin to mount the Mir Environmental Effects Payload (MEEP) experiments on the DM. The extravehicular activity (EVA) of astronauts Clifford and Godwin marked the first spacewalk while Mir and Atlantis were docked.

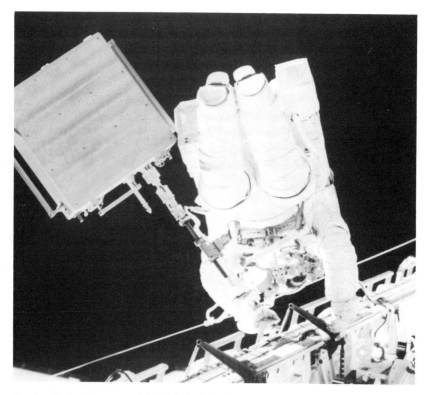

Carrying the MEEP cassette, Linda M. Godwin, STS-76 mission specialist, moves along the
longeron of Atlantis' cargo bay starboard side towards the docking module (DM) on Mir during
the 27 March 1996 spacewalk.

two cassettes exposed construction materials (including insulation, paints and optical coatings) to be used with the International Space Station. They then tested a new portable
foot-restraint and a tether system intended for use in assembling that structure. Dr Ron
Sega (Bonnie Dunbar's husband) coordinated external activities from the flight deck.

When the DM was sealed at 1338 MT on 28 March, Dr Shannon Lucid, veteran of four
previous shuttle flights, stayed aboard Mir. Atlantis left the next day and – after two
waive-offs for bad weather in Florida – landed at Edwards Air Force Base in California.
The shuttle was due back in early August, but Lucid had her own Sokol suit and couch
liner, just in case she had to return to Earth in Soyuz-TM 23.

PRIRODA

Lucid began by documenting the NASA equipment aboard the complex — an onerous job
with which her hosts could sympathise. Following Thagard's example, she slept in Spektr,
where it was relatively quiet.

Priroda was finally launched on 23 April. Unlike its predecessors, it followed a fast
rendezvous to minimise the period of independent flight because it relied upon chemical
batteries for power (the solar panel that it was to have carried had been delivered on the

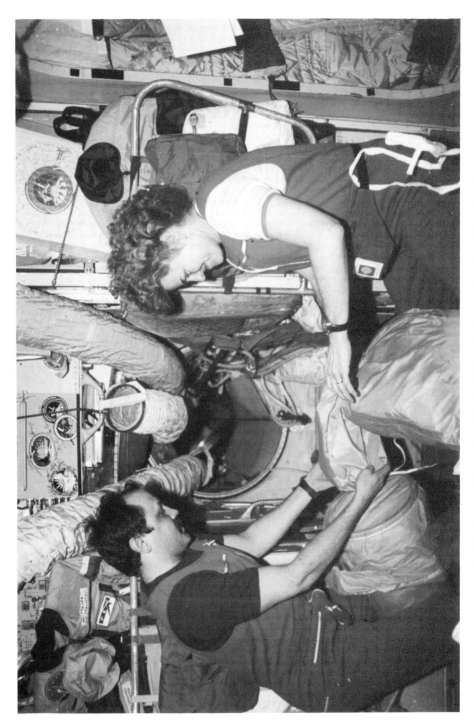

Yuri Usachyov and Shannon Lucid, in the Mir base block, unpack supplies delivered by Atlantis.

DM). Because previous modules had had to make several approaches to Mir before finally managing to dock, it had extra-large propellant tanks, and in case it encountered difficulties in the final phase it had been fitted with the remote-control system which had proven itself when docking errant ferries. Two days into the rendezvous, one of its battery buses dropped off-line, robbing it of half its power. Nothing further happened, however, and, on 26 April, Priroda docked at the axial port on its first attempt. It was swung onto the left radial port the next day, thereby completing the Mir complex.

Priroda incorporated a broad array of remote-sensing instruments for ecological and environmental investigations, and its capabilities included monitoring the propagation of industrial pollutants, mapping thermal variations across the ocean, studying energy and mass-exchange processes at sea level, measuring the height of ocean waves, measuring the mean temperature and the vertical structure of clouds, measuring wind direction and speed, multispectral Earth-resources studies, and operating as a relay station for Project Centaur by uploading information from automated geophysical stations at remote sites.

Most of its built-in apparatus had been developed collaboratively by the members of the Intercosmos organisation. Ikar comprised three sets of microwave radiometers. Ikar-N was aimed straight down at the ground track, the Ikar-D scanning radiometer sampled obliquely, and Ikar-P offered a panoramic view; this sampled 11 wavelengths overall. Istok-1 was a 64-channel multispectral infrared radiometer to study the oceans. Greben was an altimeter to measure mean sea level along the ground track; the Ozon-M spectrometer measured the concentrations of ozone and aerosols in the upper atmosphere; the Moz multichannel spectrometer analysed reflected insolation to study the oceans; medium-resolution (MSU-KS) and high-resolution (MSU-E) optical scanners provided views of clouds, and the Travers synthetic aperture radar provided a medium-resolution ground-imaging capability for looking through cloud.

In addition, Germany supplied the MOMS-2P electro-optical multispectral imager. This had previously been flown on the Spacelab-D2 shuttle mission. It could be used in various modes, including high-resolution mapping of surface relief. It carried its own GPS-based navigation, so that the position of the complex could be recorded with each image. Because such a navigation system could also trace Mir's trajectory, it potentially rendered obsolete the network of ground tracking stations. France developed a lidar for atmospheric studies, which in addition to measuring the vertical structure of clouds, could detect tropospheric aerosols (and so complement the stratospheric capability of Ozon-M). Since it had originally been proposed for Salyut 7, it was named Alissa (l'Atmosphere, Lidar Sur Salyut).

Between them, the Spektr and Priroda modules provided a formidable capability for remote sensing. Particularly in the case of the oceans, simultaneous observation by the various complementary sensors would be synergetic. The narrow sensors would detect plankton, ocean currents and processes operating near the ocean–atmosphere interface, and the lidar would see the vertical structure of the atmosphere and the distribution of tropospheric aerosols. Over land, the microwave radiometers detected soil moisture and measured the energy balance of the surface, and the imaging radar saw through cloud to map terrain. Knowing any of these parameters in a narrow time-frame on a global scale was valuable; the ability to correlate them was far more so. Spektr and Priroda had been designed as a pair, specifically to provide the broadly-based correlated data required to achieve an understanding of global hydrological cycles.

Complete at last! The Priroda module is mounted on the port opposite Kristall (which has the DM projecting from its end). The 'cooperative solar array' had been installed on Kvant 1. This view is from below, so Kvant 2 is hidden; the Sofora structure can just be seen behind the docked ferry.

The NASA equipment carried by Priroda was stowed as cargo. It comprised 284 kg of apparatus for biomedical studies, and 400 kg for several microgravity experiments, including the Mir Electric-Field Characterisation Experiment (to record ambient emissions in the 400 MHz to 18 GHz radio range), the High-Temperature Liquid-Phase Experiment; Canada's Queen's University Experiment in Liquid Diffusion (QUELD) experiment, and the samples for the Optizon LIquid-Phase Sintering Experiment (OLIPSE) to be run in the Optizon furnace in the Kristall module. Before Priroda's remote-sensing apparatus could be commissioned, its cargo had to be unstowed. But before that, the 168 chemical storage batteries which had sustained it during its rendezvous had to be stripped out. (In the process, it was discovered that the reason that one set had been lost was that a faulty relay had tripped and isolated it.) Each battery was unbolted and sealed in a plastic bag to contain any gaseous emissions. On 7 May, Progress-M 31 arrived at the front port; it was unloaded in record time so that the batteries could be dumped in its orbital module, and on 12 May Priroda was declared fit for scientific work. As they inspected their new toys, the cosmonauts noted that it felt like Christmas had come early that year.

Lucid's first task was to set up SAMS in the new module to assess its microgravity environment. The centrepiece of NASA's apparatus was the Glovebox for procedures requiring physical isolation. She installed the television cameras and lighting apparatus that she had brought with her in Atlantis. The Microgravity Isolation Mount (MIM) was a Canadian-built apparatus designed to isolate microgravity experiments from ambient vibrations in the 0.01–100 Hz range using a stabilising magnetic field. Lucid set it up in the Glovebox so that residual accelerations transmitted onto the surface of a liquid could be videotaped; SAMS recorded the ambient vibrations so that the MIM's ability to filter them could be assessed. It had once been feared that only a free-flying platform would be able to perform sensitive experiments, but MIM offered the prospect of being able to do so within the 'noisy' environment of an inhabited complex; if it worked, NASA's microgravity programme was to be expanded by successive astronauts.

Lucid's other research tasks involved processing OLIPSE samples to study melting, using the QUELD isothermal furnace to study boundary-layer processes to measure the diffusion coefficients of semiconductors, binary-metals and glasses, ongoing embryo studies of quail egg development, monitoring ambient radiation, air and water quality, and a variety of biomedical tests. Her favourite task, however, was Earth photography for geological, ecological and environmental studies, since this gave her an opportunity to use the high-fidelity porthole in Kvant 1.

Unlike on a shuttle mission, in which every activity is carefully assigned a slot on a timeline, Lucid was free to set her own pace. She worked through a four-day task list that was updated daily by Houston, and she checked off items as and when she managed to achieve them. The only significant problem she encountered during the early part of her tour was when a card in the MIPS system failed, denying her this downlink capability until a replacement could be sent up on the next cargo ferry. MIPS was proving to be extremely useful. Not only did this enable data on computer disks to be dumped on an ongoing basis (instead of being saved for return at the end of the flight), it also enabled Lucid to keep in touch with her colleagues using e-mail, which ameliorated the sense of cultural isolation felt by Thagard. In fact, the support system had been improved both at Kaliningrad and Houston, so Lucid was more effectively integrated into this distributed scientific team. On

Shannon Lucid with Yuri Usachyov and Yuri Onufrienko ('the two Yuris') alongside the Glovebox inside the Priroda module.

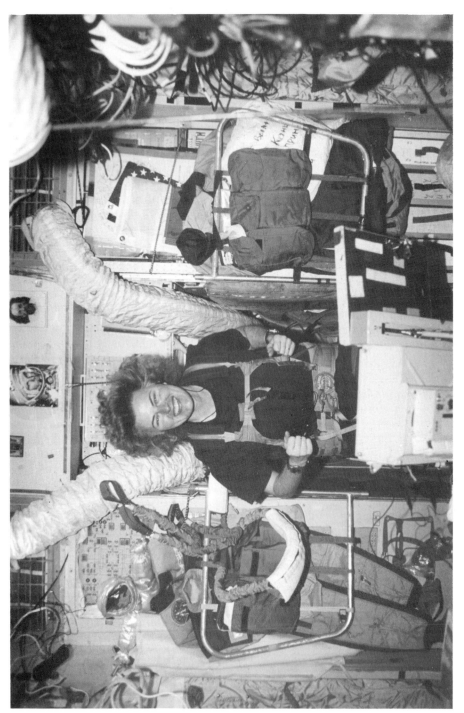

Shannon Lucid exercising on the treadmill in the Mir base block.

the other hand, she came to value the autonomy of having to operate on her own during the periods that the complex was out of radio communication. Lucid later urged the International Space Station's mission planners to remember that although it was standard to design experiments to require very little crew involvement because on a shuttle time was at a premium, an astronaut on a long tour would need to feel far more involved in the work in order to sustain interest, and she recommended two-way video conferencing with the designers of the experiment.

While Lucid was setting up the NASA experiments, Onufrienko and Usachyov had an extremely busy time spacewalking. On their first walk (early in the morning of 21 May) they used the newly installed crane to transfer the cooperative solar array from the DM to Kvant 1, and mounted it on the motor that had been in place for some time on the right side of the module. The excursion ended with a piece of theatre: the inflation of a 1.2-metre long replica of a soft drink can that was filmed against an Earth backdrop, for subsequent use in a Pepsi advertisement. Four days later, they returned to Kvant 1 and manually extended the solar panel with a hand crank. NASA was keen to have this new solar transducer tested, so that its output and degradation could be compared with computer predictions. Once the redeployable panel already on Kvant 1 degraded to the degree that it was ineffective, it was to be jettisoned and replaced by Priroda's panel which, until then, would be kept stowed on the DM. On 30 May, they mounted the MOMS-2P on a scan platform at the end of Priroda. On 6 June, they replaced cassettes in the Komza on Spektr, deployed the SKK-11 cassette on Kvant 2, and mounted two micrometeoroid packages (PIE and MSRE) on the anchor of the vacant Trek experiment for NASA. The Travers radar on Priroda required the deployment of a large framework dish on a boom projecting out from the side of the module (similar antennas had been used by the automated Almaz satellites) but it snagged, so Onufrienko and Usachyov deployed it by hand on 13 June, after which they dismantled Kvant 1's Rapana truss (it was strapped to the base of Sofora) and set up Strombus in its place, inclined aft at 11 degrees.

Onufrienko and Usachyov were probably not too surprised on 21 June to hear that they would have to extend their tour by 40 days because financial problems had delayed the fabrication of the launcher which was to have delivered their successors in late July. The RSA took this opportunity to announce that in order to save on rocket costs, tours of duty would be standardised at six months, the maximum permitted by the Soyuz-TM's service life.

On 12 July, Lucid was told that she too would have to extend her stay aboard Mir. Inspection of Columbia's solid-rocket boosters after STS-75 in June had revealed that the new sealant used in the field-joints had suffered significant hot-gas penetration, so it had been decided to refit Atlantis with boosters employing the original sealant. It would be late September before Lucid could be retrieved, which meant that she would after all be able to assist in the French visit, which was to accompany the next handover. With things going well, on 15 July she exceeded Thagard's 115-day record for an American in space.

The postponement of Atlantis had serious ramifications. Without the consumables it was to deliver, Mir would not be able to sustain the expanded crew of the forthcoming handover, so another ferry had to be sent. On 24 July, its launch was aborted less than a minute before ignition, when a propellant sensor failed. Progress-M 32 finally got off the ground on 31 July. Progress-M 31 undocked on 1 August, and Progress-M 32 took its

In addition to the ferry docked at the front port, this view shows the Travers antenna after it had been deployed by Usachyov and Onufrienko.

place the following day. It had to be unloaded immediately, because (on the revised schedule) the next crew would need its port. Meanwhile, on 9 August Gennadi Manakov was found to have a heart irregularity, so he and Pavel Vinogradov were dropped and the backup crew was assigned to the forthcoming flight. It was decided that the French researcher would not be swapped, however, because only a short visit was intended.

In mid-August, Lucid performed an extensive series of combustion experiments in the Glovebox. The Candle Flame Experiment (CFE) was subsequently augmented by the Forced Flow Flamespread Test (FFFT), which involved forced airflow across the flame to study the flammability of samples of cellulose and polyethylene solid fuel, and the ignition process. She burned eight types of fuel and 80 candles with various characteristics, and the tests were recorded by the Glovebox's cameras for analysis. The study of microgravity combustion was basic science, but if it yielded insight applicable on Earth, it would have significant commercial consequences. Lucid's other results included 40 QUELD samples, 70 OLIPSE samples and a large number of computer disks, video tapes and rolls of film.

Progress-M 32 undocked on 18 August, and Soyuz-TM 24 took the vacated port the next day to deliver Valeri Korzun, Alexander Kaleri and the French female cosmonaut researcher Claudie Andre-Deshays. Mir was now sufficiently large to offer a fair degree of privacy; Lucid continued to sleep in Spektr, and Andre-Deshays settled into Priroda.

By this time, Lucid had finished her core research, so was using her extended stay to set up the BioTechnology System (BTS) to be used by her successor (John Blaha) in protein crystal and cell culture experiments. She planted wheat in the Svet cultivator for the Greenhouse experiment to study plant reproduction, metabolism and biochemistry, and gathered data for the Anticipatory Postural Activity (APA) experiment to study posture whilst undertaking various activities. Despite the daily exercise routine, Lucid was looking forward to cycling with the wind in her hair and the Sun beating down; that and a shower. As part of a medical study, she regularly used the Metabolic-Gas Analyser System (MGAS) to analyse her expiration during exercise on NASA's ergometer, and donned the Belt-Pack Amplifier System (BPAS), which used sensors to measure muscle stimulation, for comparison with pre/post-flight data. Meanwhile, priority on the power supply was assigned to Andre-Deshays.

Being a specialist in rheumatology familiar with neurological medicine in aerospace, Andre-Deshays was ideally qualified for the cardiovascular and neurosensory tests that formed the basis of her research. She conducted experiments designed to investigate the rapid changes to the cardiovascular system that occur in the early phase of adaptation to weightlessness, particularly how the body senses blood pressure and regulates its flow; this involved donning the Physiolab harness, which supported the various sensors. The neurosensory research involved strapping into the Cognilab instrumented chair, which measured the body's response to muscular stimulation under different conditions. Her programme also included a number of technology experiments. Dynalab (a variation on the Resonance theme) measured the propagation of vibrations through the complex, and Alice-2 used a furnace which accurately controlled the temperature of a fluid and a CCD camera to record fine-scale phenomena at the critical-point phase transition (this followed up on the original Alice experiment conducted by Tognini).

Onufrienko, Usachyov and Andre-Deshays left in Soyuz-TM 23 on 2 September, and returned to Earth. Andre-Deshays' main conclusion upon her return was that two weeks was not enough to settle down to life aboard Mir and fulfil an experiment programme. Within a month, France had booked two more visits to Mir (each of which is to be a full tour) in 1999. On 3 September, Progress-M 32 (which had been station-keeping a few kilometres behind Mir) moved back in and docked at the rear; it was the first time that a cargo ferry had moved from one end of the complex to the other. Although its dry cargo had been unloaded, its propellant had not, so it had to return to replenish the base block's tanks.

Lucid spent her brief time with the new residents preparing to return to Earth, but on 7 September she claimed Kondakova's 169-day woman's record, and on 17 September, with

Shannon Lucid examines the fast-growing wheat in the Svet cultivator in the Kristall module.

Shannon Lucid inspects the Sokol pressure suit that she would have used if she were to return to Earth in a Soyuz. Alexander Kaleri is visible in the hatch.

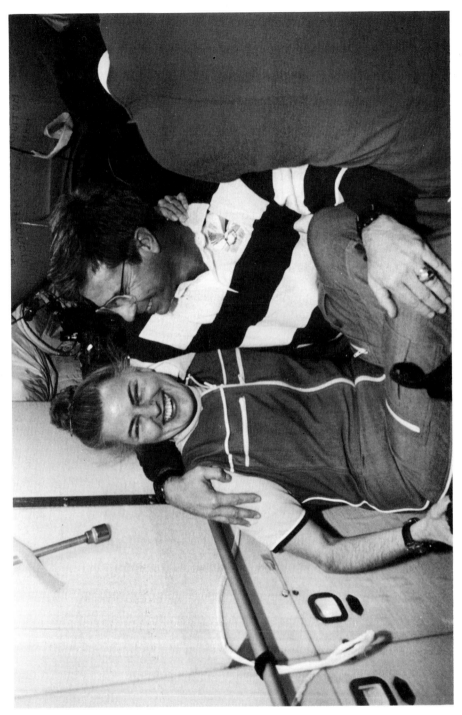

Shannon Lucid greets John Blaha, who had come to relieve her as NASA's first crew exchange on Mir.

STS-79 Atlantis already on its way, also claimed Reiter's 179-day visitor's record. The shuttle carried the double Spacehab, loaded with the long-overdue supplies, within its payload bay; one of the especially-requested items was a 2-litre bottle of nitrogen for the Elektron oxygen-production system. (The original apparatus in Kvant 1 had been out of service for some time, and the purge tank of the backup in Kvant 2 had mysteriously depleted.) Atlantis, flown by Bill Readdy and Terry Wilcutt, rendezvoused with Mir on 19 September. Although the docking with the DM was not conducted 'in the blind', the final approach was complicated by the fact that the recently-deployed cooperative solar array projected within 3 metres of the shuttle's nose. In addition to John Blaha, Atlantis carried mission specialists Jay Apt, Carl Walz and Tom Akers. The moment that Lucid and Blaha transferred their Sokol suits and couch liners marked the official exchange; it was the first time that astronauts had handed over to one another in space. Just as crews of residents did, Lucid briefed her successor on where everything was and how best to work aboard the complex. Lucid and Blaha had flown together twice, in 1991 and 1993, so they were old friends.

Akers, the 'loadmaster', evaluated a new method of handling the cargo. Instead of trans-ferring individual items from the lockers, as before, items had been pre-loaded into locker-sized canvas bags that were easily handled, and the transfer was tracked with the bar-code reader. Of the 2,250 kg transferred to Mir, the largest item was a gyrodyne. A total of 1,000 kg was retrieved. In addition to Lucid's results (in 20 such canvas bags), a gyrodyne was being returned for refurbishment together with an expired Orlan suit so that its degra-dation could be assessed. Stowing the return cargo turned out to be trickier than expected, because it had to be weighed and located so as to uphold centre-of-mass requirements; there was nothing like doing something for real for learning how to do it.

Atlantis departed on 24 September, and went on to conduct an important test in the Spacehab. The Boeing Active-Rack Isolation System (ARIS) had been developed to facil-itate microgravity experiments in the 'noisy' environment of the International Space Sta-tion. In contrast to the MIM magnetic system, ARIS relied on a mechanical suspension system to damp out micro-accelerations. Immediately upon landing at the Kennedy Space Center, Lucid was imaged with a nuclear magnetic-resonance scanner to record the den-sity of her skeleton and key musculature. To her surprise, she rapidly readapted to gravity. Shuttle–Mir project manager Frank Culbertson said that Lucid had "set the standard" for NASA's work aboard Mir. She was subsequently welcomed into the select group of astro-nauts to have been awarded the Congressional Space Medal of Honor. Upon taking over, Blaha expressed his view that there was no better way for NASA to prepare for the future than to have a cadre of astronauts learn to live and work in space by spending tours on Mir, because everything from the exercise regime to the logistics system was new to the agency.

Blaha's programme built upon Lucid's, but included several new experiments. One of the packages delivered by Atlantis was a 'powered transfer' with mammalian cartilage cells for the tissue-growth experiment in the BTS. He began this even before the shuttle de-parted, and it was to extend throughout his tour. The Diffusion-controlled Crystallisation Apparatus for Microgravity (DCAM) used a semi-permeable membrane to grow protein over a long period, with the growth being filmed. The Binary-Colloid Alloy Test (BCAT) was to study crystallisation of alloys of colloids. In the first month, Blaha processed sev-

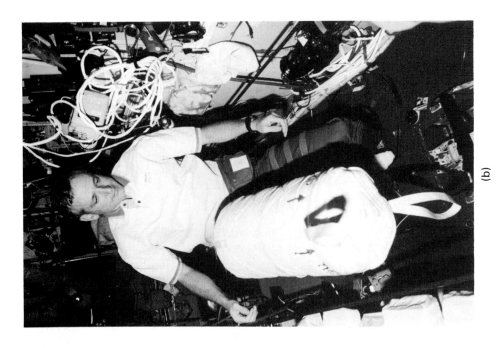

(b)

(a)

(a) Carl Walz transports a bagged Orlan spacesuit to Atlantis, for return to Earth, so that it could be examined to reassess the service life of the remaining suits. (b) Tom Akers offloads bagged experiment apparatus to Mir to facilitate Shannon Lucid's research.

(a)

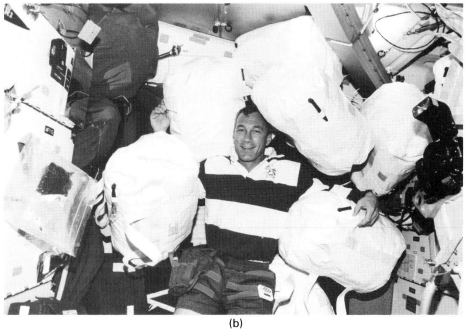

(b)

(a) Tom Akers, the STS-79 loadmaster, checks the inventory of stores in the Spacehab module. Note that much of the cargo is identified by Cyrillic coding. (b) Terry Wilcutt surrounded by water cannisters. The fuel cells that provide power for the shuttle produce water as waste, and it is usually vented. On Mir, however, water is a crucial resource that provides oxygen for the crew. These cannisters were emptied into Mir's Rodnik tanks.

(a)

(b)

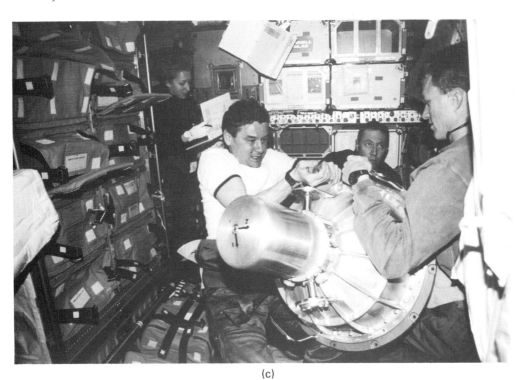

(c)

(a) Atlantis' prodigious cargo capacity enabled the Russians to return faulty gyrodynes of the attitude control system to Earth for refurbishment. Rich Clifford unpacks a new unit in the Spacehab module on STS-76. (b) Bill Readdy transfers one on STS-79. (c) Valeri Korzun, Mike Baker and Brent Jett work on one of the bulky 165-kg units on STS-81. Marsha Ivins, as the loadmaster for the double Spacehab module, checks the inventory.

eral rapid-growth samples using different relative concentrations, each of which took a day, and then set up a three-month run. He had a number of technological tests to perform to provide information important for fitting out the modules of the International Space Station. One involved an instrumented foot-restraint which monitored how much his body moved whilst he worked 'in place', and he used a push-off pad that measured the force imparted in moving around in the complex. The Passive Accelerometer System (PAS) measured low-intensity continuous effects due to air drag and differential gravity in order to further characterise the microgravity environment. This augmented data from a set of accelerometers and strain gauges that he installed to measure the sharp transient stresses on the complex resulting from manoeuvres and docking operations, and thermal effects due to flying into and out of the Earth's shadow, for the MIr Structural Dynamics Experiment (MISDE); once set up, this apparatus was to gather data throughout the time that NASA had microgravity experiments running.

Progress-M 33's resupply had been scheduled for mid-October, but its rocket was late. When it arrived on 22 November, it brought heavy-duty power cables (which were to be put in place on a forthcoming spacewalk), and the first Japanese experiments. One required collecting the bacteria and mould commonplace on Mir, with the intention of being able to eliminate it from the International Space Station. The other was a study of the long-term effects of exposure to cosmic rays, and involved exposing human DNA, silk-worms, soil bacteria and yeast to determine the effects on cell structure and to reveal any resulting genetic damage. Blaha was already analysing the bacteria and fungi on the complex, regularly sampling air, water, human skin and smears from exposed surfaces and incubating them. Although the base block had been lived in for a decade, it was in remarkably good shape in this respect; the rich green mould that had broken out aboard Salyut 4 was clearly just an unpleasant memory. Another of Blaha's experiments was a radiation monitor using a Tissue-Equivalent Proportional Counter (TEPC) that stored its data and downloaded them at regular intervals.

ONE THING AFTER ANOTHER

Five Progress resupply flights had been made in each of the years leading up to the first Atlantis visit, but there had been just three in 1996. Mir operations accounted for most of the RSA's budget, which had been declining since 1989 (it was now worth just 20 per cent in real terms of what it had been at that time). It had already had to abandon the modules which it had hoped to add to the International Space Station, and now it had come to rely on the shuttle to sustain its existing complex. Yuri Koptev warned that unless the slashing of his agency's budget ceased, it would no longer be able to make up the shortfall with Mir's limited capacity to accommodate fee-paying guests, so the complex might have to be vacated once the contract with NASA expired. The situation was clearly too serious for this to be an idle threat.

On 2 December, Korzun and Kaleri set out to complete the connection of the output from the cooperative solar array into the power system. To date, only half of its capacity had been available via the socket on Kvant 1, so that the degradation of its transducers could be monitored. This involved running a 22-metre cable across to the base block, to a socket by its top panel (this had degraded to the extent that it could now barely deliver 1 kW even under ideal insolation; it had been decided to disconnect it so that this socket could be used by the more powerful panel). The job was finished a week later and then, after the base block's power distribution system had been reconfigured, the CSA was brought on line for the first time on 11 December. Whilst outside, the cosmonauts also retrieved the Rapana and mounted it as an extension of Strombus (so that the sensors at its far end would be as far as possible from the body of the complex), and relocated the Kurs antenna on Kristall to the DM to restore the system's angular coverage. By this point, the surface of Kvant 1 was covered by a variety of installations, and care had to be taken not to disturb anything when spacewalking. Unfortunately, during their first excursion the cosmonauts accidentally yanked out the plug for the ham-radio antenna, so on the second they traced the cable and reseated its connector to restore the system to life. If it had not been for the late launch of Progress-M 33, these activities would have taken place earlier.

Meanwhile, on 6 December, the wheat that Lucid had planted in the Svet in August completed its cycle and produced grain, so Blaha 'harvested' it. This was the first time that a staple had grown to maturity, and it had important implications for the prospects of making a station self-sufficient in food production. The wheat and most of the grain was frozen for return to Earth, although a few of the seeds were immediately planted to start a second crop, but these were to be uprooted and taken back to Earth at the end of Blaha's tour so that they could be subjected to a comprehensive biochemical analysis. During a press conference at the end of the year, Blaha was asked whether he was eager for the shuttle to retrieve him on schedule; he replied laconically, "if it gets here, it gets here". Clearly, like Lucid, he was enjoying his tour. In fact, the shuttle was once again launched on its first attempt; despite the early fears of the programme's critics, no launch attempt had been scrubbed during the final moments of the countdown.

Mike Baker and Brent Jett tested upgraded software as they docked STS-81 Atlantis on 15 January 1997, and the MISDE sensors measured the resultant stresses on the Mir complex.

The double Spacehab module was loaded with the now routine cargo mix for Mir. Marsha Ivins was the loadmaster in control of the cargo transfer activities performed by mission specialists John Grunsfeld and Jeff Wisoff. Meanwhile, a treadmill that did not transmit vibrations was tested, taking another step towards minimising the pollution of the microgravity environment by the crew's presence. Having handed over to Dr Jerry Linenger, Blaha left aboard Atlantis on 20 January. Upon landing in Florida two days later, he was "absolutely stunned" at the strength of the Earth's gravity; unlike his predecessors, therefore, he obliged the doctors and let himself be carried out of the shuttle.

As a physician, Linenger's programme was primarily biomedical, but he had a number of fluid physics and materials-processing assignments too, and, by making a spacewalk to retrieve two dust traps set up a year earlier and deploy the Advanced Materials Exposure Experiment (AMEE), he was to become the first astronaut to wear the Orlan suit.

Reinhold Ewald arrived a month later in Soyuz-TM 25, to accompany the handover by which Vasili Tsibliev and Alexander Lazutkin took over from Korzun and Kaleri. Since Ewald had been the backup for the ESA mission in 1992, the biomedical tests to be carried out during his two-week visit built directly upon that work; this though, was the first German-sponsored visit. It turned out to be rather more eventful than intended. On 24 February, thick smoke suddenly filled the complex. During handovers, when there were more than three people aboard, a chemical burner was used to supplement the oxygen output of the Elektron regenerative system. The canister in the Vika unit in Kvant 1 had split, released oxygen into the electronics, and started a fire. Although the cosmonauts were on the scene within seconds with portable extinguishers and smothered the system with foam, the combustion was sustained by the direct injection of oxygen for almost 10 minutes. The entire unit was reduced to soot-blackened scrap. For the first hour or so, until the air filtration system could extract the worst of the smoke, everybody wore full-face masks and portable oxygen bottles. Although eye irritation was reported, Linenger decided that there was no lung damage due to smoke inhalation. Nevertheless, for the next few days they wore small filter masks so as not to breathe in particulates. This was the worst fire yet. On an earlier station, it would probably have resulted in evacuation, but Mir could not simply be abandoned, so the mess was tidied up and work resumed. The split canister was extracted from the ruined Vika and stowed in Soyuz-TM 24, and then returned to Earth on 2 March.

Mir's problems were not over, however. Firstly, on 4 March, when Progress-M 33 made its return, the TORU remote control system failed, so it was ordered to withdraw and deorbit itself. The next day, the Elektron unit in Kvant 2 had to be shut down because a bubble of air had blocked the flow of water in the electrolysis canal. This was a serious problem, because the other Elektron had been dead for some time, and now, following the

Marsha Ivins struggles to stow the cargo in the double Spacehab module for return to Earth. Vital lessons were learnt concerning cargo handling procedures during the Mir resupply flights; these will be applied to optimise operations with the International Space Station.

John Blaha welcomes Jerry Linenger, who had come to relieve him aboard Mir.

fire in the Vika in Kvant 1, only one method of oxygen production remained – the Vika in Kvant 2. If that failed, there was only a few days' worth of bottled oxygen (this was an emergency reserve). There was understandable reluctance to rely on the Vika, but there was no other option. The real issue was the supply of canisters; in the absence of the regenerative Elektron system, one canister would have to be burned per person per day. Since it was clear that this supply would not last until the next visit by Atlantis, unless the next cargo ferry arrived on time and brought additional canisters and parts for the Elektron, Mir would *have* to be evacuated.

With a ferry docked at either end, the Mir complex orbits orientated in the gravity gradient for stability.

One clear lesson from earlier stations was that an empty station could be disabled by a fault that could easily have been fixed if a crew had been aboard. To leave Mir vacant, even for a short period, would be to invite trouble. As if to underline this danger, on 19 March an orientation sensor failed and the attitude control system tried to correct what it took to be an unexpected motion. By the time the cosmonauts were able to intervene, the gyrodynes had built up a runaway three-axis roll. The computerised gyrodynes had to be shut down, the rotation cancelled manually by the thrusters, then the complex stabilised in the gravity gradient while the fault was analysed. The automated system was finally reactivated using a backup sensor. It is doubtful that control could have been reasserted from the ground, because problems with the communications link meant that Mir was currently in direct communication with Kaliningrad for only 10 minutes on favourable passes.

The situation began to improve however, when on 8 April Progress-M 34 slipped into the rear port without incident. In addition to the usual spares for the atmospheric and environmental systems, it brought components to restore communications and to fix the least damaged of the two Elektrons, and a stock of 60 canisters for the surviving Vika.

With the short-term future of the complex assured, Tsibliev, Lazutkin and Linenger settled back down to work. Although Linenger had only a month to go before being relieved by Michael Foale, the residents were advised that they would need to extend their tour by six weeks because the rocket which would launch their successors was late; Frenchman Leopold Eyharts was, at this stage, to accompany that handover.

On 29 April, Tsibliev and Linenger went out to retrieve the PIE and MSRE and to mount the AMEE package. Despite having been drummed into helping with maintenance chores, Linenger had made progress with his science assignments. Indeed, by processing 50 samples for the QUELD experiment, he had already completed his own programme and was working his way through the samples intended for his successor.

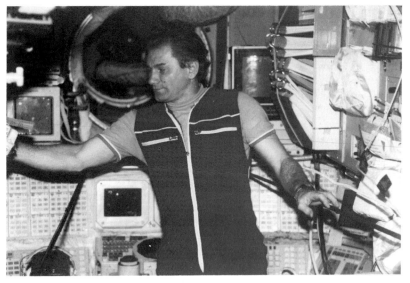

Cosmonaut Vasili V. Tsibliev, Mir-23 mission commander working in the base block of the space station.

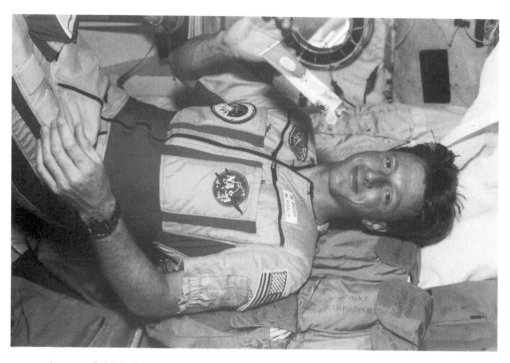

Astronaut C. Michael Foale, cosmonaut researcher for Mir-23, enjoys some off-duty time in his new quarters aboard the space station. Foale, who replaced Jerry M. Linenger aboard Mir, is holding a beverage tube.

STS-84 Atlantis arrived on schedule, on 17 May. Its logistics load, which included 1,200 kg of Russian material and 500 kg of assorted apparatus for NASA's science programme, was the heaviest yet. The Russian cargo included a stock of lithium hydroxide canisters to build up a reserve, a gyrodyne and a new 120-kg Elektron unit to replace the one in Kvant 1 that had failed long ago; this old Elektron, and the remains of the Vika unit that had caught fire, were returned for examination. In addition to retrieving Linenger, Atlantis collected 400 kg of accumulated NASA research results. A particularly welcome visitor for the residents was Yelena Kondakova.

In addition to a busy programme of protein crystallisation, Earth observations, life sciences, materials processing and engineering tests, Linenger's replacement, Michael Foale, was to help with maintenance tasks, help unload the cargo ferry scheduled for early June, look after Mir while Tsibliev and Lazutkin made two spacewalks in early July, and finally tag along while Soyuz-TM 25 was flown to the rear port preparatory to the August handover. The repair of the environmental systems got off to an excellent start, but then, completely unexpectedly, disaster struck in the shape of an errant ferry. Having been filled with rubbish, Progress-M 34 undocked on 24 June. As the redocking of Progress-M 33 had had to be abandoned, a recertification trial of the TORU system had been laid on. When the ferry made its new approach the next day, 25 June, Tsibliev lost control; the 7-tonne ferry came in too low, ran beneath the complex, and smashed into the Spektr module, badly mangling a solar panel and puncturing a coolant radiator. The twisting of the panel's motor installation, and the force on the hull transmitted through the struts supporting the conformal radiator, combined to break the hermetic seal of that module, and the air inside began to vent to space. After disengaging, the ferry drifted away from the complex.

Foale had been on his way to Kvant 1 to observe the approaching ferry through the small porthole, to use a laser rangefinder to verify the TORU's closing-rate data, when Tsibliev suddenly ordered him to retreat to the Soyuz at the opposite end of the complex. As Foale made his way across the cramped docking adaptor, he heard "a loud bang." Lazutkin, who was still in the base block with his feet anchored to the floor, felt the shockwave propagate through the structure. Within seconds, there was a faint hiss; the inner ear is sensitive to a drop in pressure, and it was immediately obvious that air was escaping to space. Tsibliev reported that the Spektr module, which projected downwards from the front of the docking adaptor, had to be the source of the leak; its hatch had to be closed, and fast.

In addition to the air tubes that snaked through the hatches, cables had been laid between the modules to form an integrated power grid linking the various solar panels to the storage batteries distributed throughout the complex. There was no time to de-install the cables, they would have to be *cut*. If the hatch could not be closed, the three men would have no choice but to retreat to the Soyuz and leave the air vent, which would certainly

The Mir space station backdropped against a massive array of clouds over the South Pacific Ocean and the Tasman Sea. The Spektr module with its four solar panels is uppermost in this view.

mean abandoning the stricken station. A heavy cutting tool was stored in the docking adaptor for precisely such an emergency. Cutting the cables was in itself a risky task, because the short circuits could easily result in blow-outs, so recovery would involve a long process of progressive verification of the state of the systems. In the event, it took several minutes to clear the aperture and fit the hatch. By that time, however, it was evident that it was a *slow* leak, because when the pressure was checked immediately after the hatch had been closed it was found to be only about 15 per cent below normal. Further monitoring showed that the integrity of the docking collar had not been breached as it had absorbed the energy of the impact.

With the station secure, it was time to tackle the immediate consequences. The quartet of solar panels on Spektr had contributed almost half of Mir's power, but, with the cables cut, this power was no longer being fed into the power distribution system. The complex had survived the decompression, but it now faced a power crisis. It was vital that the batteries not be drained, so each man dived into one of the other radial modules, and switched off all the apparatus. Nevertheless, a crisis was in the making because the complex needed power to be able to *continue* making power. The power output from the solar panel is related to the angle of insolation by a sinusoidal function; it produces its peak output only when it is face on. As Mir flew around its orbit, it generally kept the same orientation with respect to the Earth below, so the panels had to be turned to track the Sun, and it took power to turn the motor. Each panel had its own motor. If the batteries were depleted too far, there would be insufficient power to turn the panels, which in turn meant that no power was available to recharge the batteries. It was a runaway process; a spiral to oblivion. The vital thing was to switch off *everything*, to conserve the batteries, and then, with half the generating capacity denied, endeavour to remain on the safe side of the cusp in the power curve. The situation was complicated by the fact that the orientation of the complex was being controlled by the gyrodynes. These consumed electrical power, and were sensitive to fluctuations. They had to be switched off; this left the complex in free-drift mode. It was at about this point that Mir flew back into communications range with Kaliningrad. The first the flight controllers knew that there was a problem was when the telemetry stream failed to materialise; the transmitter had been turned off along with everything else. In any case, there was nothing the ground team could do, the cosmonauts were effectively on their own.

Soyuz-TM 25 had been powered up immediately, just in case it proved necessary to make a hasty escape, and now it was the only part of the complex not crippled by the power loss. Its thrusters were used to stabilise the complex and then to reorientate it so that the majority of its remaining solar panels faced the Sun. It took several hours of continuous adjustments to fully charge the batteries; only then could Mir's control systems be reactivated. For the next few days, as they inspected the station, the cosmonauts worked by torchlight, and they set up a sleep roster so that they would not be taken by surprise by a sudden deterioration. And there were still the slow-to-develop problems which could, in the end, force abandonment. With power low, the cooling system was ineffective. The soaring temperature was not just uncomfortable for the crew; the Vozdukh carbon dioxide scrubber overheated and had to be turned off, which forced reliance on the limited supply of lithium hydroxide canisters. The Elektron suffered, so waste water could not be electrolysed to make oxygen, which forced reliance on the sole-remaining Vika, and this

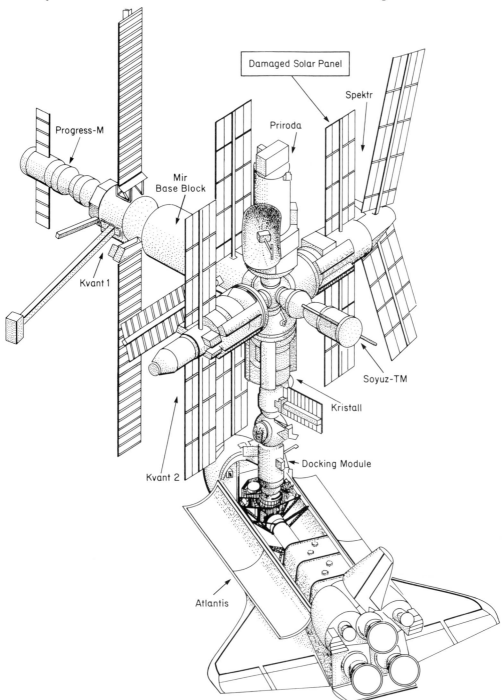

The complete Mir complex, with the space shuttle Atlantis docked, as it was in May 1997. The solar panel on Spektr damaged in the Progress-M 34 collision is indicated.

meant having masks and fire extinguishers to hand. The toilet could be used, but the reprocessing system could not be used, because the tank for the Elektron was full. And so it went. Nevertheless, the situation *was* improving with each day. It was by no means clear, however, that they would *not* be forced out. One thing was sure. If Mir was vacated, it was unlikely to be reoccupied. The spacecraft and its crew were in a symbiotic relationship; just as the crew relied on it for their survival, it relied on their continued presence. But basic survival was not the objective, Mir was a laboratory, and if it was to have a long-term future, then, at the very least, the undamaged solar panels on the Spektr module would have to be brought back on line.

In the original scheme, after Progress-M 34 had redocked, verifying the TORU system, it was to have been discarded to clear the aft port for its successor, which was already on the pad. This launch, set for 27 June, was postponed to give time to work out a way to restore the power, and to fabricate the necessary material and tools. It was concluded that although it might be feasible to run external cabling from Spektr's panels to sockets outside the base block, it would be simpler to try to reconnect them internally. With Spektr exposed to vacuum, however, the hatch would have to remain closed. How could new cables be run through a sealed hatch? An ingenious scheme was conceived. One of the Konus drogues in the docking adaptor would be modified to serve as a air-tight electrical junction. There were two of these drogues. One was kept permanently on the axial port for dockings. The other, which was detachable, had been moved around the radial ports as necessary to facilitate the movements of the add-on modules, which swung themselves around on their short Ljappa arms. This Konus had last been used to accept the Priroda module. It comprised the hollow drogue, which was essentially a conical guide plate, and the bulbous end cap that contained the clamp for the mechanism at the tip of the probe. The two parts were connected by a ring of bolts. The plan was to remove the end cap and the clamp, and to bolt on a new unit with a set of electrical sockets on either side. If this could be mounted on Spektr's hatch, which, like the other modules, was currently sealed by a simple disk-like cover, then the situation should be retrievable. It would require an *internal spacewalk*, however, to install this hatch and lay the necessary cables within the stricken Spektr module. Trials in the hydrotank by Anatoli Solovyov and Pavel Vinogradov, Mir's next crew, demonstrated that the procedure was manageable, so the apparatus was loaded into Progress-M 35. This was launched on 5 July. When this closed in to dock two days later, the view from its camera clearly revealed the extent to which one of Spektr's solar panels had been twisted by the collision, but the other three looked to be undamaged.

Kaliningrad had hoped that Tsibliev and Lazutkin would be able to effect repairs within a week, but when the cosmonauts reviewed the proposed procedure they requested additional time to prepare, so the internal spacewalk was pushed back a week. Fortunately, the Konus was not already on Spektr's hatch, so the preparatory work of swapping the junction plate for the probe mechanism could be done in a shirt-sleeved environment, which would make the job of working on the two dozen bolts more manageable. It was important that Spektr's power be restored before the August handover, because the new crew's arrival would place an extra demand on the environmental system, and because power would be needed for the experiments to be performed by the visitor. A spacesuited rehearsal without depressurising the docking adaptor was scheduled for 15 July, and the real thing was set

The damage to one of the Spektr's four solar panels is clearly shown in these two views obtained by the camera on Progress-M 35 as it closed into dock with the Mir space station on 7 July.

for 18 July. With power restored to the maximum available, the next crew would be able to concentrate on external activities designed to locate and plug Spektr's leak, so that the module could be repressurised and, with a little luck, much of its apparatus salvaged. The extent to which this proved practicable would determine NASA's future aboard the complex. The module housed 50 per cent of its science apparatus. It was questionable whether there would be sufficient scientific yield to justify continuing the Shuttle–Mir programme if Spektr had to be written off. If the errant ferry had hit and depressurised Kristall, with its androgynous docking system, the programme, would, of necessity, have been curtailed, and Foale would have been forced to return with Tsibliev and Lazutkin in the Soyuz, because Atlantis would be not returning in September. A great deal, therefore, was riding on the outcome of this makeshift repair.

Although the preparations progressed well, the mounting stress evidently took its toll on the commander. On 12 July, Tsibliev reported feeling a heart arrhythmia. When an EKG confirmed this the following day, he was ordered to take a combination of heart medication and tranquillisers to make him relax. It was also decided that Foale should support Lazutkin in the tricky repair, which was put back another week to allow Foale time to prepare. Foale had backed up Linenger in the preparations for Linenger's spacewalk, so he was familiar with the basic procedures for using the Orlan suit. On 16 July, while rehearsing the process of disconnecting the cabling within the docking adaptor, Lazutkin unplugged the cable that distributed data from the main orientation sensors to the gyrodynes. Unfortunately, he had neglected to switch over to the backup first, so the system immediately began to drift, with the result that the power output from the solar panels fell. The crew scrambled to switch off apparatus, but the complex was rapidly overwhelmed by another power crisis. In a rerun of the earlier process, the cosmonauts were obliged to retreat to the Soyuz and use its thrusters to reorientate the complex so that the solar panels could feed power and recharge the batteries.

Undeterred, the cosmonauts argued to be allowed to proceed with the repair on 25 July, but they were informed that it had been decided to reassign the job to their successors, who had been rehearsing the necessary procedures in the hydrotank. Accordingly, the Soyuz-TM 26 crew, Anatoli Solovyov and Pavel Vinogradov, arrived on the station on 7 August. Following the departure of Tsibliev and Lazutkin, the plan called for Vinogradov to lead the repair efforts to reconnect the undamaged solar panels on Spektr. He and Solovyov were due to don spacesuits for the internal spacewalk on 20 August; Foale would wait in the Soyuz during the repair. This operation, if successful, would restore power to much of the station. The next major task for the new crew would be to locate and fix the hole in Spektr, and up to six spacewalks had been scheduled for this operation. Foale was due to venture outside with Solovyov in a spacewalk on 3 September in a first attempt to pinpoint the puncture. Once Spektr had been retrieved, it might even be possible to jettison the damaged solar panel and replace it with a new one.

Without a guarantee of power for his experiments, Leopold Eyharts was reassigned to the next flight, scheduled for early 1998. This slippage would propagate on down through the visiting programme. Mir was fully booked with commercial research to the end of the century, but it was far from clear that the complex would last long enough to sustain the demand for its services.

Table 10.1. Mir docking operations

Spacecraft	Docking		Port	Undocking		Days
	Date	MT		Date	MT	
Soyuz-T 15	14 Mar 1986	1638	front	5 May 1986	1612	51.98
Progress 25	21 Mar 1986	1416	rear	20 Apr 1986	2324	30.38
Progress 26	27 Apr 1986	0126	rear	22 Jun 1986	2225	56.87
Soyuz-TM 1	23 May 1986	1412	front	29 May 1986	1323	5.96
Soyuz-T 15	26 Jun 1986	2346	front	16 Jul 1986	1307	19.55
Progress 27	18 Jan 1987	1027	rear	23 Feb 1987	1429	36.17
Soyuz-TM 2	8 Feb 1987	0228	front	30 Jul 1987	0034	171.92
Progress 28	5 Mar 1987	1543	rear	27 Mar 1987	0807	21.68
Kvant 1	9 Apr 1987	0436	rear	<permanently docked>		
Progress 29	23 Apr 1987	2105	rear	11 May 1987	0711	17.42
Progress 30	21 May 1987	0953	rear	19 Jul 1987	0420	58.77
Soyuz-TM 3	24 Jul 1987	0731	rear	31 Jul 1987	0328	6.83
Soyuz-TM 3	31 Jul 1987	0348	front	29 Dec 1987	0855	151.21
Progress 31	6 Aug 1987	0228	rear	22 Sep 1987	0358	47.06
Progress 32	26 Sep 1987	0408	rear	10 Nov 1987	0609	45.08
Progress 32	10 Nov 1987	0847	rear	17 Nov 1987	2225	7.57
Progress 33	23 Nov 1987	0439	rear	19 Dec 1987	1116	26.27
Soyuz-TM 4	23 Dec 1987	1551	rear	31 Dec 1987	1210	7.85
Soyuz-TM 4	31 Dec 1987	1229	front	17 Jun 1988	1018	168.90
Progress 34	23 Jan 1988	0309	rear	2 Mar 1988	0640	38.14
Progress 35	26 Mar 1988	0122	rear	5 May 1988	0536	40.17
Progress 36	15 May 1988	0613	rear	5 Jun 1988	1512	21.37
Soyuz-TM 5	9 Jun 1988	1957	rear	18 Jun 1988	1411	8.76
Soyuz-TM 5	18 Jun 1988	1427	front	6 Sep 1988	0255	79.52
Progress 37	21 Jul 1988	0234	rear	12 Aug 1988	0832	22.25
Soyuz-TM 6	31 Aug 1988	0941	rear	8 Sep 1988	0505	7.81
Soyuz-TM 6	8 Sep 1988	0525	front	21 Dec 1988	0633	104.04
Progress 38	12 Sep 1988	0522	rear	23 Nov 1988	1513	72.41
Soyuz-TM 7	28 Nov 1988	2016	rear	22 Dec 1988	0945	23.56
Soyuz-TM 7	22 Dec 1988	0959	front	27 Apr 1989	0328	125.73
Progress 39	27 Dec 1988	0840	rear	7 Feb 1989	0946	42.04
Progress 40	12 Feb 1989	1330	rear	3 Mar 1989	0446	18.64
Progress 41	18 Mar 1989	2245	rear	21 Apr 1989	0546	33.29
Progress-M 1	25 Aug 1989	0919	front	1 Dec 1989	1202	98.11
Soyuz-TM 8	8 Sep 1989	0225	rear	11 Dec 1989	1123	94.37
Kvant 2	6 Dec 1989	1521	front	8 Dec 1989	-	2

Table 10.1. (continued)

| Spacecraft | Docking | | Port | Undocking | | Days |
	Date	MT		Date	MT	
Kvant 2	8 Dec 1989	-	upper	<permanently docked>		
Soyuz-TM 8	11 Dec 1989	1143	front	19 Feb 1990	-	60
Progress-M 2	22 Dec 1989	0841	rear	9 Feb 1990	0533	48.87
Soyuz-TM 9	13 Feb 1990	0938	rear	21 Feb 1990	0656	7.88
Soyuz-TM 9	21 Feb 1990	0715	front	28 May 1990	1548	96.36
Progress-M 3	3 Mar 1990	0305	rear	27 Apr 1990	-	55
Progress 42	8 May 1990	0245	rear	27 May 1990	1109	19.35
Soyuz-TM 9	28 May 1990	1612	rear	4 Jul 1990	0208	36.41
Kristall	10 Jun 1990	1447	front	11 Jun 1990	-	1
Kristall	11 Jun 1990	-	lower	27 May 1995	0328	1810
Soyuz-TM 9	4 Jul 1990	0234	front	9 Aug 1990	0809	36.23
Soyuz-TM 10	3 Aug 1990	1546	rear	10 Dec 1990	0548	128.58
Progress-M 4	17 Aug 1990	0926	front	17 Sep 1990	1643	31.30
Progress-M 5	29 Sep 1990	1627	front	28 Nov 1990	0915	59.70
Soyuz-TM 11	4 Dec 1990	1257	front	26 Mar 1991	-	111
Progress-M 6	16 Jan 1991	1935	rear	16 Mar 1991	1547	58.84
Soyuz-TM 11	26 Mar 1991	-	rear	26 May 1991	1013	61
Progress-M 7	28 Mar 1991	1603	front	7 May 1991	0300	39.46
Soyuz-TM 12	20 May 1991	1830	front	28 May 1991	1410	7.82
Soyuz-TM 12	28 May 1991	1452	rear	10 Oct 1991	0352	134.54
Progress-M 8	1 Jun 1991	1345	front	16 Aug 1991	0217	75.52
Progress-M 9	23 Aug 1991	0454	front	30 Sep 1991	0454	38.00
Soyuz-TM 13	4 Oct 1991	1038	front	15 Oct 1991	0410	10.73
Soyuz-TM 13	15 Oct 1991	0554	rear	14 Mar 1992	1443	151.36
Progress-M 10	21 Oct 1991	0641	front	20 Jan 1992	1014	91.15
Progress-M 11	27 Jan 1992	1231	front	13 Mar 1992	1144	45.97
Soyuz-TM 13	14 Mar 1992	1510	front	25 Mar 1992	-	11
Soyuz-TM 14	19 Mar 1992	1532	rear	9 Aug 1992	-	143
Progress-M 12	22 Apr 1992	0320	front	28 Jun 1992	-	67
Progress-M 13	4 Jul 1992	2055	front	24 Jul 1992	1249	19.66
Soyuz-TM 15	29 Jul 1992	1151	front	1 Feb 1993	0600	186.76
Progress-M 14	18 Aug 1992	0421	rear	21 Oct 1992	1946	64.64
Progress-M 15	29 Oct 1992	2206	rear	4 Feb 1993	0345	97.23
Soyuz-TM 16	26 Jan 1993	1032	Kristall	22 Jul 1993	-	177
Progress-M 16	23 Feb 1993	2318	rear	26 Mar 1993	0950	30.44
Progress-M 16	26 Mar 1993	1007	rear	27 Mar 1993	0721	0.88

Table 10.1. (continued)

Spacecraft	Docking Date	MT	Port	Undocking Date	MT	Days
Progress-M 17	2 Apr 1993	0912	rear	11 Aug 1993	1936	131.43
Progress-M 18	24 May 1993	1225	front	3 Jul 1993	1958	40.31
Soyuz-TM 17	3 Jul 1993	2024	front	14 Jan 1994	0737	194.47
Progress-M 19	13 Aug 1993	0400	rear	12 Oct 1993	2059	60.71
Progress-M 20	15 Oct 1993	0225	rear	21 Nov 1993	0538	37.13
Soyuz-TM 18	10 Jan 1994	1450	rear	24 Jan 1994	0612	13.64
Soyuz-TM 18	24 Jan 1994	0801	front	9 Jul 1994	-	166
Progress-M 21	30 Jan 1994	0656	rear	23 Mar 1994	0420	51.89
Progress-M 22	24 Mar 1994	0940	rear	23 May 1994	0457	59.80
Progress-M 23	24 May 1994	1000	rear	2 Jul 1994	1247	39.12
Soyuz-TM 19	3 Jul 1994	1755	rear	2 Nov 1994	1330	121.82
Progress-M 24	2 Sep 1994	1745	front	4 Oct 1994	2156	32.17
Soyuz-TM 20	6 Oct 1994	0328	front	11 Jan 1995	1157	97.35
Soyuz-TM 19	2 Nov 1994	1405	rear	4 Nov 1994	1025	1.85
Progress-M 25	13 Nov 1994	1204	rear	16 Feb 1995	1605	95.17
Soyuz-TM 20	11 Jan 1995	1223	front	22 Mar 1995	(0400)	69.65
Progress-M 26	17 Feb 1995	2122	rear	15 Mar 1995	0527	25.34
Soyuz-TM 21	16 Mar 1995	1046	rear	4 Jul 1995	1454	110.17
Progress-M 27	12 Apr 1995	0101	front	23 May 1995	0340	41.11
Kristall	27 May 1995	(0500)	front	30 May 1995	-	3
Kristall	30 May 1995	-	right	10 Jun 1995	-	11
Spektr	1 Jun 1995	0458	front	2 Jun 1995	(2100)	1.67
Spektr	2 Jun 1995	(2300)	lower	<permanently docked>		
Kristall	10 Jun 1995	-	front	17 Jul 1995	(0630)	37
STS-71	29 Jun 1995	1700	Kristall	4 Jul 1995	1510	4.92
Soyuz-TM 21	4 Jul 1995	1538	rear	11 Sep 1995	0727	68.66
Kristall	17 Jul 1995	(0800)	right	<permanently docked>		
Progress-M 28	22 Jul 1995	0840	front	4 Sep 1995	0910	44.02
Soyuz-TM 22	5 Sep 1995	1428	front	29 Feb 1996	1020	176.82
Progress-M 29	10 Oct 1995	2332	rear	19 Dec 1995	1215	69.53
STS-74	15 Nov 1995	0928	Kristall	18 Nov 1995	1116	3.07
Progress-M 30	20 Dec 1995	1910	rear	22 Feb 1996	1026	63.63
Soyuz-TM 23	23 Feb 1996	1723	rear	2 Sep 1996	0820	190.63
STS-76	24 Mar 1996	0534	DM	29 Mar 1996	0408	4.94
Priroda	26 Apr 1996	1543	front	27 Apr 1996	-	1
Priroda	27 Apr 1996	-	left	<permanently docked>		

Table 10.1. (continued)

Spacecraft	Docking		Port	Undocking		Days
	Date	MT		Date	MT	
Progress-M 31	7 May 1996	1254	front	1 Aug 1996	2047	86.33
Progress-M 32	3 Aug 1996	0203	front	18 Aug 1996	1334	15.48
Soyuz-TM 24	19 Aug 1996	1852	front	7 Feb 1997	-	173
Progress-M 32	3 Sep 1996	1335	rear	21 Nov 1996	-	79
STS-79	19 Sep 1996	0615	DM	24 Sep 1996	0533	4.97
Progress-M 33	22 Nov 1996	0358	rear	6 Feb 1997	-	75
STS-81	15 Jan 1997	0655	DM	20 Jan 1997	0515	4.93
Soyuz-TM 24	7 Feb 1997	-	rear	2 Mar 1997	-	23
Soyuz-TM 25	12 Feb 1997	1851	front	- Aug 1997	-	-
Progress-M 34	8 Apr 1997	2131	rear	24 Jun 1997	-	-
STS-84	17 May 1997	0434	DM	22 May 1997	0505	5.02
Progress-M 35	7 Jul 1997	1000	rear	- - 1997	-	-
Soyuz-TM 26	7 Aug 1997	2002	-	- - -	-	-

Table 10.2. Mir crewing

Cosmonaut	Role	Spacecraft		Duration
		Arrive	Depart	days
Leonid Kizim	CDR	Soyuz-T 15	Soyuz-T 15	125.00
Vladimir Solovyov	FE	Soyuz-T 15	Soyuz-T 15	125.00
Yuri Romanenko	CDR	Soyuz-TM 2	Soyuz-TM 3	326.48
Alexander Laveikin	FE	Soyuz-TM 2	Soyuz-TM 2	174.14
Alexander Viktorenko	CDR	Soyuz-TM 3	Soyuz-TM 2	7.96
Alexander Alexandrov	FE	Soyuz-TM 3	Soyuz-TM 3	160.26
Mohammed Faris	CR	Soyuz-TM 3	Soyuz-TM 2	7.96
Vladimir Titov	CDR	Soyuz-TM 4	Soyuz-TM 6	365.94
Musa Manarov	FE	Soyuz-TM 4	Soyuz-TM 6	365.94
Anatoli Levchenko	CR	Soyuz-TM 4	Soyuz-TM 3	7.91
Anatoli Solovyov	CDR	Soyuz-TM 5	Soyuz-TM 4	9.84
Viktor Savinykh	FE	Soyuz-TM 5	Soyuz-TM 4	9.84
Alexander Alexandrov (B)	CR	Soyuz-TM 5	Soyuz-TM 4	9.84
Vladimir Lyakhov	CDR	Soyuz-TM 6	Soyuz-TM 5	8.85
Valeri Poliakov	CR	Soyuz-TM 6	Soyuz-TM 7	240.90
Abdul Mohmand	CR	Soyuz-TM 6	Soyuz-TM 5	8.85

Table 10.2. (continued)

Cosmonaut	Role	Spacecraft		Duration
		Arrive	Depart	days
Alexander Volkov	CDR	Soyuz-TM 7	Soyuz-TM 7	151.46
Sergei Krikalev	FE	Soyuz-TM 7	Soyuz-TM 7	151.46
Jean-Loup Chrétien	CR	Soyuz-TM 7	Soyuz-TM 6	24.75
Alexander Viktorenko	CDR	Soyuz-TM 8	Soyuz-TM 8	166.30
Alexander Serebrov	FE	Soyuz-TM 8	Soyuz-TM 8	166.30
Anatoli Solovyov	CDR	Soyuz-TM 9	Soyuz-TM 9	179.06
Alexander Balandin	FE	Soyuz-TM 9	Soyuz-TM 9	179.06
Gennadi Manakov	CDR	Soyuz-TM 10	Soyuz-TM 10	130.86
Gennadi Strekalov	FE	Soyuz-TM 10	Soyuz-TM 10	130.86
Viktor Afanasayev	CDR	Soyuz-TM 11	Soyuz-TM 11	175.08
Musa Manarov	FE	Soyuz-TM 11	Soyuz-TM 11	175.08
Toehiro Akiyama	CR	Soyuz-TM 11	Soyuz-TM 10	7.92
Anatoli Artsebarski	CDR	Soyuz-TM 12	Soyuz-TM 12	144.64
Sergei Krikalev	FE	Soyuz-TM 12	Soyuz-TM 13	311.84
Helen Sharman	CR	Soyuz-TM 12	Soyuz-TM 11	7.88
Alexander Volkov	CDR	Soyuz-TM 13	Soyuz-TM 13	175.12
Takhtar Aubakirov	CR	Soyuz-TM 13	Soyuz-TM 12	7.92
Franz Viehboeck	CR	Soyuz-TM 13	Soyuz-TM 12	7.92
Alexander Viktorenko	CDR	Soyuz-TM 14	Soyuz-TM 14	145.60
Alexander Kaleri	FE	Soyuz-TM 14	Soyuz-TM 14	145.60
Klaus-Dietrich Flade	CR	Soyuz-TM 14	Soyuz-TM 13	7.92
Anatoli Solovyov	CDR	Soyuz-TM 15	Soyuz-TM 15	188.90
Sergei Avdeyev	FE	Soyuz-TM 15	Soyuz-TM 15	188.90
Michel Tognini	CR	Soyuz-TM 15	Soyuz-TM 14	13.90
Gennadi Manakov	CDR	Soyuz-TM 16	Soyuz-TM 16	179.03
Alexander Poleshchuk	FE	Soyuz-TM 16	Soyuz-TM 16	179.03
Vasili Tsibliev	CDR	Soyuz-TM 17	Soyuz-TM 17	196.74
Alexander Serebrov	FE	Soyuz-TM 17	Soyuz-TM 17	196.74
Jean-Paul Haignere	CR	Soyuz-TM 17	Soyuz-TM 16	20.67
Viktor Afanasayev	CDR	Soyuz-TM 18	Soyuz-TM 18	182.02
Yuri Usachyov	FE	Soyuz-TM 18	Soyuz-TM 18	182.02
Valeri Poliakov	CR	Soyuz-TM 18	Soyuz-TM 20	437.75

Table 10.2. (continued)

| Cosmonaut | Role | Spacecraft | | Duration |
		Arrive	Depart	days
Yuri Malenchenko	CDR	Soyuz-TM 19	Soyuz-TM 19	125.96
Talget Musabayev	FE	Soyuz-TM 19	Soyuz-TM 19	125.96
Alexander Viktorenko	CDR	Soyuz-TM 20	Soyuz-TM 20	169.22
Yelena Kondakova	FE	Soyuz-TM 20	Soyuz-TM 20	169.22
Ulf Merbold	CR	Soyuz-TM 20	Soyuz-TM 19	31.52
Vladimir Dezhurov	CDR	Soyuz-TM 21	STS-71	115.4
Gennadi Strekalov	FE	Soyuz-TM 21	STS-71	115.4
Norman Thagard	CR	Soyuz-TM 21	STS-71	115.4
Anatoli Solovyov	CDR	STS-71	Soyuz-TM 21	74
Nikolai Budarin	FE	STS-71	Soyuz-TM 21	74
Yuri Gidzenko	CDR	Soyuz-TM 22	Soyuz-TM 22	179.03
Sergei Avdeyev	FE	Soyuz-TM 22	Soyuz-TM 22	179.03
Thomas Reiter	CR	Soyuz-TM 22	Soyuz-TM 22	179.03
Yuri Onufrienko	CDR	Soyuz-TM 23	Soyuz-TM 23	192.84
Yuri Usachyov	FE	Soyuz-TM 23	Soyuz-TM 23	192.84
Shannon Lucid	CR	STS-76	STS-79	188.20
Valeri Korzun	CDR	Soyuz-TM 24	Soyuz-TM 24	196
Alexander Kaleri	FE	Soyuz-TM 24	Soyuz-TM 24	196
Claudie Andre-Deshays	CR	Soyuz-TM 24	Soyuz-TM 23	15.76
John Blaha	CR	STS-79	STS-81	-
Jerry Linenger	CR	STS-81	STS-84	-
Vasili Tsibliev	CDR	Soyuz-TM 25	Soyuz-TM 25	-
Alexander Lazutkin	FE	Soyuz-TM 25	Soyuz-TM 25	-
Reinhold Ewald	CR	Soyuz-TM 25	Soyuz-TM 24	20
Michael Foale	CR	STS-84	STS-86	
Anatoli Solovyov	CDR	Soyuz-TM 26	-	-
Pavel Vinogradov	FE	Soyuz-TM 26	-	-

Only genuine Mir crew members are listed; the members of shuttle crews who visit for a few days are not included.

Table 10.3. Mir spacewalks

Date		Hours	Activity
11 Apr	1987	3.6	Romanenko and Laveikin opened a radial docking port, inspected the rear docking system to discover why Kvant 1 could not achieve hard-dock, found a bag of rubbish within its drogue, tugged this out, then watched the probe retract to complete the docking.
12 Jun	1987	1.9	Romanenko and Laveikin opened a radial docking port, began to erect a solar panel on the motor built into the roof of the base block.
16 Jun	1987	3.25	Romanenko and Laveikin opened a radial docking port, completed erecting the solar panel, plugged in its cables, then affixed sample cassettes.
26 Feb	1988	4.4	Titov and Manarov opened a radial docking port cover and retracted one of the solar panel segments to replace it with a new one which included a section which produced telemetry to monitor degradation in performance.
30 June	1988	5.2	Titov and Manarov opened a radial docking port, opened the thermal protection blanket at the rear of Kvant 1 to gain access to a malfunctioning X-ray telescope, but the tool supplied snapped when they tried to use it to release the detector.
20 Oct	1988	4.2	Titov and Manarov opened a radial docking port, used a new tool to release and replace the X-ray detector, to repair the telescope, then affixed an anchor near the docking adaptor in preparation for a French apparatus; they wore the new Orlan-DMA suits for the first time.
9 Dec	1988	5.9	Volkov and Chrétien opened a radial docking port, set up the Echantillon cassette, attached the ERA apparatus to the anchor, deployed it, tested its vibration modes, then finally jettisoned it.
8 Jan	1990	2.9	Viktorenko and Serebrov opened a radial docking port, installed two star trackers on Kvant 1 in order to help to orientate the to-be-expanded complex.
11 Jan	1990	2.9	Viktorenko and Serebrov opened a radial docking port, retrieved the Echantillon cassette, dismantled the anchor used by the ERA apparatus, set up exposure cassettes and then swapped the drogue from the upper to the lower port in the multiple docking adaptor.
26 Jan	1990	3.1	Viktorenko and Serebrov exited Kvant 1's airlock, affixed an anchor just outside, dismantled the Kurs antenna at the other end of the module, set up exposure cassettes, and then erected the scan platform.

Table 10.3. (continued)

Date	Hours	Activity
1 Feb 1990	5.0	Viktorenko and Serebrov exited the airlock and Serebrov tested the YMK autonomous manoeuvring unit.
5 Feb 1990	3.75	Viktorenko and Serebrov exited the airlock and Viktorenko reran the YMK test.
17 July 1990	7.25	Solovyov and Balandin exited the airlock, set up a ladder so that they could inspect the loose thermal blankets on Soyuz-TM 9's descent module, pinned them back, left the ladder and returned to the airlock, but then found that the hatch would not shut, so they had to retreat to Kvant 1's central compartment to use it as an emergency airlock.
26 July 1990	3.5	Solovyov and Balandin inspected the airlock hatch to discover why it would not shut, then went to remove the ladder off their ferry, attached it to Kristall for storage, then returned and were able to close the airlock hatch only by extreme effort.
30 Oct 1990	3.75	Manakov and Strekalov attempted to repair the airlock hatch, but realised that the hinge needed to be replaced.
7 Jan 1991	5.3	Afanasayev and Manarov replaced the hinge to repair the airlock hatch, affixed an anchor to the base block, retrieved sample cassettes and removed a camera from the scan platform.
23 Jan 1991	5.6	Afanasayev and Manarov exited the airlock, installed a crane on the new anchor, then retrieved sample cassettes.
26 Jan 1991	6.4	Afanasayev and Manarov exited the airlock, installed framework mounts on either side of Kvant 1 (preparatory to transferring Kristall's solar panels), fitted laser-reflectors (for a rangefinder that was to be used by Buran during its rendezvous), and then installed the Sprut-5 spectrometer.
26 Apr 1991	3.6	Afanasayev and Manarov exited the airlock, inspected a faulty Kurs antenna on Kvant 1, retrieved sample cassettes, replaced the camera in the scan platform, then tested a mechanical joint.
25 Jun 1991	5.0	Artsebarski and Krikalev exited the airlock, replaced the broken Kurs antenna, and set up a mechanical joint.
28 Jun 1991	3.5	Artsebarski and Krikalev exited the airlock, set up a charged-particle spectrometer and the Trek cosmic ray detector, then retrieved the mechanical joint.
15 Jul 1991	5.8	Artsebarski and Krikalev exited the airlock and mounted a work platform in Kvant 1.
19 Jul 1991	5.5	Artsebarski and Krikalev exited the airlock, affixed the Sofora package to the mount, then started to erect the girder.

Table 10.3. (continued)

Date		Hours	Activity
23 Jul	1991	5.6	Artsebarski and Krikalev exited the airlock and built more of the Sofora girder.
27 Jul	1991	6.8	Artsebarski and Krikalev exited the airlock, finished erecting the Sofora girder and placed a Hammer & Sickle on the end of it.
20 Feb	1991	4.2	Volkov had to remain near the airlock when the cooling unit in his suit failed, so Krikalev dismantled obsolete fixtures from the Sofora girder site, then retrieved an experimental solar cell.
8 Jul	1992	2.1	Viktorenko and Kaleri exited the airlock and cut through thermal insulation on Kvant 2 to install two new gyrodynes.
3 Sep	1992	3.9	Solovyov and Avdeyev exited the airlock, tilted and then locked the Sofora girder down across the cargo ferry at the rear of the complex, then cranked the thruster pack up out of the ferry.
7 Sep	1992	5.2	Solovyov and Avdeyev exited the airlock, ran an umbilical along the Sofora girder, affixed metal straps to the thruster pack, and retrieved the Hammer & Sickle.
11 Sep	1992	5.8	Solovyov and Avdeyev exited the airlock, attached the thruster pack to the Sofora girder, then swung it back up.
15 Sep	1992	3.6	Solovyov and Avdeyev exited the airlock, put a Kurs antenna on Kristall's axial androgynous port, and then retrieved sample cassettes.
19 Apr	1993	5.5	Manakov and Poleshchuk exited the airlock, attached a motor to the (previously installed) framework on the side of Kvant 1, plugged it into the power supply, then found that one of the handles of the crane was no longer there!
18 June	1993	4.6	Manakov and Poleshchuk exited the airlock, fitted a replacement handle, then installed the second motor and plugged it in.
16 Sep	1993	4.3	Tsibliev and Serebrov exited the airlock, affixed a platform on Kvant 1 and then mounted the Rapana container on it.
20 Sep	1993	3.25	Tsibliev and Serebrov exited the airlock, erected the Rapana girder and set up sample cassettes.
28 Sep	1993	1.9	Tsibliev and Serebrov exited the airlock, set up some sample cassettes and retrieved others, and then began to video the state of the complex's surface.
22 Oct	1993	0.6	Tsibliev and Serebrov exited the airlock, deployed a meteoroid package, and continued recording the complex's state.
29 Oct	1993	4.2	Tsibliev and Serebrov exited the airlock, completed recording the complex's state, checked the base of the Sofora girder, and retrieved sample cassettes.

Table 10.3. (continued)

Date	Hours	Activity
9 Sep 1994	5.1	Malenchenko and Musabayev exited the airlock, inspected the front port for damage by the impact of Progress-M 24, then Kristall for damage following the collision with Soyuz-TM 17, affixed an anchor on the base block (for a second crane), then set up several sample cassettes.
13 Sep 1994	6.0	Malenchenko and Musabayev exited the airlock, inspected the fittings of Kristall's solar panels, swung to Kvant 1 to inspect the motors fitted to receive them, inspected the Sofora girder, then retrieved experiments previously mounted on the Rapana girder.
12 May 1995	6.25	Dezhurov and Strekalov exited the airlock and retracted the left-side solar panel on Kristall.
17 May 1995	6.8	Dezhurov and Strekalov exited the airlock, and released, transferred and attached the panel to the motor on the left-side of Kvant 1.
22 May 1995	5.25	Dezhurov and Strekalov exited the airlock, finished stringing cabling from the solar panel to the base block, then partially retracted the remaining panel.
28 May 1995	0.4	Dezhurov and Strekalov depressurised the docking adaptor, then moved the radial drogue from the lower to the right port (for Kristall).
2 Jun 1995	0.5	Dezhurov and Strekalov depressurised the docking adaptor, then moved the radial drogue from the right to the lower port (for Spektr; it was moved back a few days after Spektr docked, but with both modules on, doing so did not require depressurising the docking adaptor).
14 Jul 1995	5.6	Solovyov and Budarin exited the airlock, inspected a faulty solar panel on Kvant 2, released a clamp to deploy a stuck panel on Spektr, then inspected the right-side radial port for any sign of damage.
19 Jul 1995	3.2	Solovyov had to remain near the airlock when the cooler in his suit failed (they had intended to deploy the MIRAS package), so Budarin retrieved sample cassettes (including Trek), and set up new ones.
21 Jul 1995	5.6	Solovyov and Budarin exited the airlock and affixed the MIRAS package to a scan platform on the end of Spektr.
20 Oct 1995	5.25	Avdeyev and Reiter installed the ESEF cassettes on Spektr (near MIRAS) and exchanged the cassette in the Komza experiment.

Table 10.3. (continued)

Date	Hours	Activity
8 Dec 1995	0.6	Gidzenko and Avdeyev depressurised the docking adaptor, then moved the radial drogue from the right to the left port (for Priroda).
8 Feb 1996	3.1	Gidzenko and Reiter retrieved two of the ESEF cassettes, put in a new one, then tried (and failed) to retrieve the mounting of a redundant antenna for examination.
15 Mar 1996	5.8	Onufrienko and Usachyov installed a second crane (on the right side of the base block).
27 Mar 1996	6.1	Godwin and Clifford left Atlantis, retrieved a camera from the DM and deployed the four-cassette MEEP.
21 May 1996	5.3	Onufrienko and Usachyov transferred the cooperative solar array to Kvant 1 and installed it on the waiting motor.
25 May 1996	5.7	Onufrienko and Usachyov extended the cooperative solar array.
30 May 1996	4.3	Onufrienko and Usachyov attached the MOMS multispectral imager outside Priroda.
6 Jun 1996	3.6	Onufrienko and Usachyov swapped the cassettes in the Komza package, and deployed a materials cassette and two micrometeoroid traps.
13 Jun 1996	3.5	Onufrienko and Usachyov dismantled the Rapana truss and left it alongside the Sofora girder, then erected the Strombus truss in its place. On the way back, they completed the deployment of the stuck Travers radar dish on Priroda (its automated mechanism having failed).
2 Dec 1996	5.9	Korzun and Kaleri started to string cables to the base block for the cooperative solar array into the power system, then retrieved the Rapana and remounted it on top of the Strombus so that the sensors at its end would be as far from the station as possible.
9 Dec 1996	6.6	Korzun and Kaleri completed laying the power cable to the base block and then transferred the Kristall Kurs antenna to the DM.
29 Apr 1997	5.0	Tsibliev and Linenger retrieved PIE and MSRE, and then deployed AMEE.

Part 3: International Space Station

11

Origins

After Werner von Braun was captured in Germany, on his way to surrender to the allies, towards the end of the Second World War, the US Army transported him, his team of rocket engineers, and as much V-2 hardware as it could capture, to America. It then set him to work assembling and firing his rockets on the test range at White Sands, New Mexico, so that his American counterparts could copy the technology. Although Hitler had seen the V-2 as a long-range ballistic missile, to von Braun it was the key to space exploration. In 1952, *Collier's Magazine* began a series of articles by von Braun in which he advanced his vision of the human colonisation of space, and his lucid prose was accompanied by Chesley Bonestell's magnificent paintings of winged spaceplanes, enormous rotating space stations and ungainly interplanetary vessels.

It must all have appeared fantastic at the time. The sound barrier had only just been broken, and the idea of spaceships was widely considered to be so far beyond what was technologically feasible that it was the realm of science fiction, if not sheer fantasy. Even so, von Braun's logic seemed impeccable; the winged shuttle would build a space station which would serve as the way-station to the Moon and the planets. At some time in the future, possibly by the end of the century, such things should be feasible.

The furious pace of aerospace research in the 1950s gave rise to ever-faster higher-flying aircraft. By the end of that decade, despite rapid progress in developing ballistic missiles, the high-performance X-15 rocket plane seemed to offer the promise of orbital flight. Combining technologies, the US Air Force began to design the X-20 DynaSoar as a vertically-launched spaceplane for performing orbital reconnaissance and delivering nuclear weapons. In reacting to the shock of Sputnik, the newly-established NASA set out to send a man into orbit before the Soviets. Since spaceplanes required considerable development before they would be practicable, it had no option but to use a rocket and a ballistic-return capsule. NASA regarded this Mercury man-in-space-soonest project as a 'crash' response to a political contest, and the inelegant capsule as an expedient which would serve until a spaceplane became operational. The way to the Moon and planets still seemed to be by building von Braun's 'wheel in space', and, left to itself (after the knee-jerk response which was Mercury) NASA might have followed this strategy.

But, following Gagarin's historic orbit and Kennedy's challenge to reach the Moon before the Soviet Union did, and to do this as rapidly as possible, it became evident that not only was there not time to develop the reusable spaceplane, there would not be time to construct the orbital way-station. A more direct means of reaching the Moon had to be devised; a massive throw-away rocket and a ballistic capsule that used quick-to-develop technologies was the only practical option. Even though his space station now seemed redundant, von Braun rose to the challenge, and threw himself into the development of the Saturn V, the enormously powerful rocket which would send Apollo to the Moon.

Even while NASA was working flat out to reach the Moon within the decade of the 1960s, it was looking beyond the achievement of Kennedy's objective, to use Apollo to extend and broaden its operations. It proposed installing modular instrument bays in the Apollo service module to support a scientific programme involving astronomical, solar and Earth observations. The intention was that this be undertaken in the 1970s, in parallel with the later lunar flights. To emphasise its evolutionary nature, this programme was called Apollo Applications. A space station was designed, but it called for the uppermost stage of the Saturn V to be converted to an 'orbital workshop' that could by visited by a succession of Apollo crews

Unfortunately, as soon as NASA had met Kennedy's challenge, America turned its back on spaceflight, which it now judged to be a luxury that the nation could no longer afford. The Nixon Administration, noting this change in the public mood, and eager to reduce federal spending, swung its budgetary axe at Apollo Applications. The only part of the programme to escape was the Skylab space station project. Furthermore, because the Saturn V production line was ordered to be shut down, NASA was ordered to make use of the hardware already built for the Moon programme, which meant that this had to be cut back too.

Cost-conscious Nixon did have a strategy for space, however. In energy terms, the most costly part of any spaceflight is achieving low Earth orbit. Most of the Saturn V's potential energy was expended in climbing up out of the Earth's deep 'gravity well'. All the subsequent manoeuvres were relatively inexpensive. Reducing the cost of achieving Earth orbit was, in the long term, the key to an expanded space programme. In developing its post-Apollo plan for Nixon's consideration, NASA proposed a 'modular' approach involving, not surprisingly, a reusable spaceplane whose primary mission would be the assembly of a space station in low orbit. In the fullness of time, this 12-person station was to serve as the base from which to colonise the Moon and to launch a flight to Mars. All NASA got was the 'Space Shuttle', but without a space station to assemble there was precious little need for it. Therefore, to ensure that its services would be in heavy demand, it was designated the 'national space transportation system' and, like the mighty Saturn V, all other 'expendable' rockets were ordered to be phased out. This was a bold policy which risked crippling the entire satellite programme if ever it became necessary to ground the shuttles for a significant period. Apart from its Skylab, Nixon grounded NASA for the rest of the 1970s to ensure that it would concentrate on the development of the reusable shuttle which would provide cheap access to low orbit.

As events transpired, Skylab ran into trouble even before it escaped the atmosphere: hypersonic aerodynamic forces ripped off one of its two solar panels, together with part of the meteoroid and thermal protection shield; it was critically short of power and soon overheated. It could not be abandoned, however, so the first crew of veteran astronaut and moonwalker Pete Conrad and rookies Paul Weitz and Joe Kerwin were dispatched with the mission of rescuing it, which they did by dragging out the remaining solar panel (then jammed closed by debris from the missing thermal shield) and by erecting a 'parasol' to protect the bare metal from the Sun. Their spacewalking was a magnificent demonstration of the value of a human presence in space. A series of three crews spent almost six of the 10 months from May 1973 aboard Skylab. Compared to Salyut, Skylab was huge; in fact, Salyut's stepped cylinder could have fitted into Skylab's main compartment with plenty of room to spare. Skylab was so voluminous that astronauts tested a flying-backpack similar to that which was to have been tested during Gemini spacewalks (and was later used by shuttle astronauts) without even going outside. Skylab's main instrument was a powerful solar telescope. The project had proven that the Apollo hardware could indeed have been used to mount an extensive science programme in Earth orbit. Ironically, all the Skylab flights were made while Salyut was in the doldrums and suffering failure after failure. For NASA's astronauts, however, there seemed to be no prospect of work in the lean years until the shuttle was ready to fly.

The shuttle took longer to develop than expected, but its debut, on Cosmonaut Day, 12 April 1981, was spectacular. In July 1982, after four engineering test flights, it was declared operational. A second orbiter, Challenger, joined the first, Columbia, in April 1983. On the first four operational flights, half a dozen communications satellites were deployed. On the fifth, in November 1983, Spacelab was flown. Because this had been developed jointly with the European Space Agency (ESA), that agency supplied its own astronaut, which made STS-9 the first international shuttle mission.

12

Competition

As soon as the shuttle was declared operational, NASA dusted off its space station plan and resumed lobbying for support, and, to its surprise, found key members of the Reagan Administration to be very receptive. Significantly, NASA proposed developing the space station as an international collaboration. This would not only share the cost, it would also serve to insulate its funding from the vagaries of the annual Congressional review process. Reagan was taken with the idea, and particularly liked the international aspect, so in his State Of The Union Address to Congress in January 1984 he directed NASA, within a decade, to build a space station that would be permanently occupied by eight astronauts. A few months later, ESA decided to build a laboratory which could be added to the space station. In 1989, Japan announced that it too would add a module to the complex.

Although the structure shown to Reagan consisted of a compact cluster of modules centred on a pair of solar panels, the design selected was Grumman's Power Tower concept. This was a 120-metre keel which was to be aligned vertically in orbit (to exploit the gravity gradient for stability) and a cluster of modules at either end to facilitate Earth and astronomical observation simultaneously. A horizontal cross-beam halfway up the keel would carry solar panels. As the design was reviewed, however, doubts began to arise. Would such a long, narrow keel be sufficiently rigid? How would the astronauts move from one cluster of modules to the other? Other issues concerned the specification of the modules themselves, which were linked together in such a way that it would be very difficult to isolate a specific compartment in an emergency. As a result, the Power Tower was rejected in late 1985.

A year later, having evaluated submissions, NASA accepted the Dual Keel concept by Lockheed and McDonnell Douglas. It had two 150-metre vertical trusses linked top and bottom by 45-metre spars for increased structural strength. Like the Power Tower, it would exploit the gravity gradient to maintain a vertical orientation and it would have Earth-observational and astronomical apparatus at the ends, but in this case these would be remotely operated from a single cluster of modules halfway up and between the two keels, beside the horizontal truss supporting the solar panels. This eliminated the need for a frequent-use transportation system along the keel, and the modules themselves incorporated key safety features. As before, once the station was complete, astronauts would be delivered and retrieved by a succession of space shuttles.

Within months of this revision, however, the enormous Dual Keel had been dropped from the configuration. With it went the dedicated Earth and astronomical observational sites. This left only the module cluster on the truss with the solar panels. Stripped of its vertical structure, gravity-gradient stability was impracticable. External apparatus would have to be mounted on the truss; the heavy keel (it was argued) could be added later. To give the project a sense of identity, it was decided to name the structure 'Space Station Freedom'. It reflected Reagan's hostility towards what he had so recently dubbed "the evil empire" of the Soviet Union. So, just as Apollo had been two decades earlier, the space station became a symbol of the Cold War.

Despite having selected the design, the funding to begin construction work was not forthcoming. Instead, the annual Congressional budget review demanded a succession of redesigns to reduce costs. During much of this hiatus, the shuttle fleet was grounded following the loss of Challenger. In 1990, Congress sent NASA back to the drawing board and told it to come up with a design that would cut $6 billion from the total cost. Ninety days later, NASA proposed a configuration that could be crew-tended within five years, and then gradually be upgraded to allow permanent occupancy. To cut the number of shuttle flights required for the assembly sequence, the truss had been shortened by 30 per cent, and the modules had been scaled down by 40 per cent to ensure that they could be delivered fully outfitted. The occupancy level of the habitat had been cut from eight to four astronauts. Already frustrated by NASA's funding delays, ESA now faced not only the prospect that its laboratory would not be launched until the end of the century, but also that there would not be sufficient crew capacity to staff it permanently.

Daniel Goldin took over from Richard Truly as NASA Administrator in April 1992, and one of his first acts was to instigate yet another redesign of Freedom, to further slash costs. At the same time, he ordered that a 'lifeboat' be added to ensure that astronauts would be able to return to Earth in an emergency, or if the shuttle fleet became grounded for some reason. By the end of the year, the design had been so 'descoped' that NASA began to consider using a Soyuz as the lifeboat in order to save the cost of developing its own 'assured crew-return vehicle', as it had so prosaically designated it. This was a remarkable turnaround. In July 1991, NASA had agreed to fly a cosmonaut on a space shuttle in return for an astronaut visiting Mir. Now NASA was proposing to operate its own fleet of Soviet-designed spacecraft. This reflected the progressive thaw in the Cold War since Mikhail Gorbachev's rise to power in the Soviet Union, and its end upon the collapse of the 'evil empire'. Negotiations to define the modifications NASA would require to the standard Soyuz were immediately set in motion.

When the Clinton Administration took over in January 1993, the provocative name 'Freedom' was quietly dropped. In March, NASA was sent back to the drawing board, to come up with a more cost-effective plan which would be compatible with Clinton's promise to reduce the deficit. NASA submitted three options on 7 June. Option-A was to use proven hardware – including a propulsion, guidance and navigational system used on spy satellites – and incorporate systems intended for Freedom which would clearly be cost-effective. Option-B was essentially what had already been proposed previously, and would maximally exploit the work done on Freedom. Option-C was a 'minimalist' facility that would exploit existing hardware: Columbia would be decommissioned and its engine block mated to a specially-modified external tank, and then used to put a large one-piece

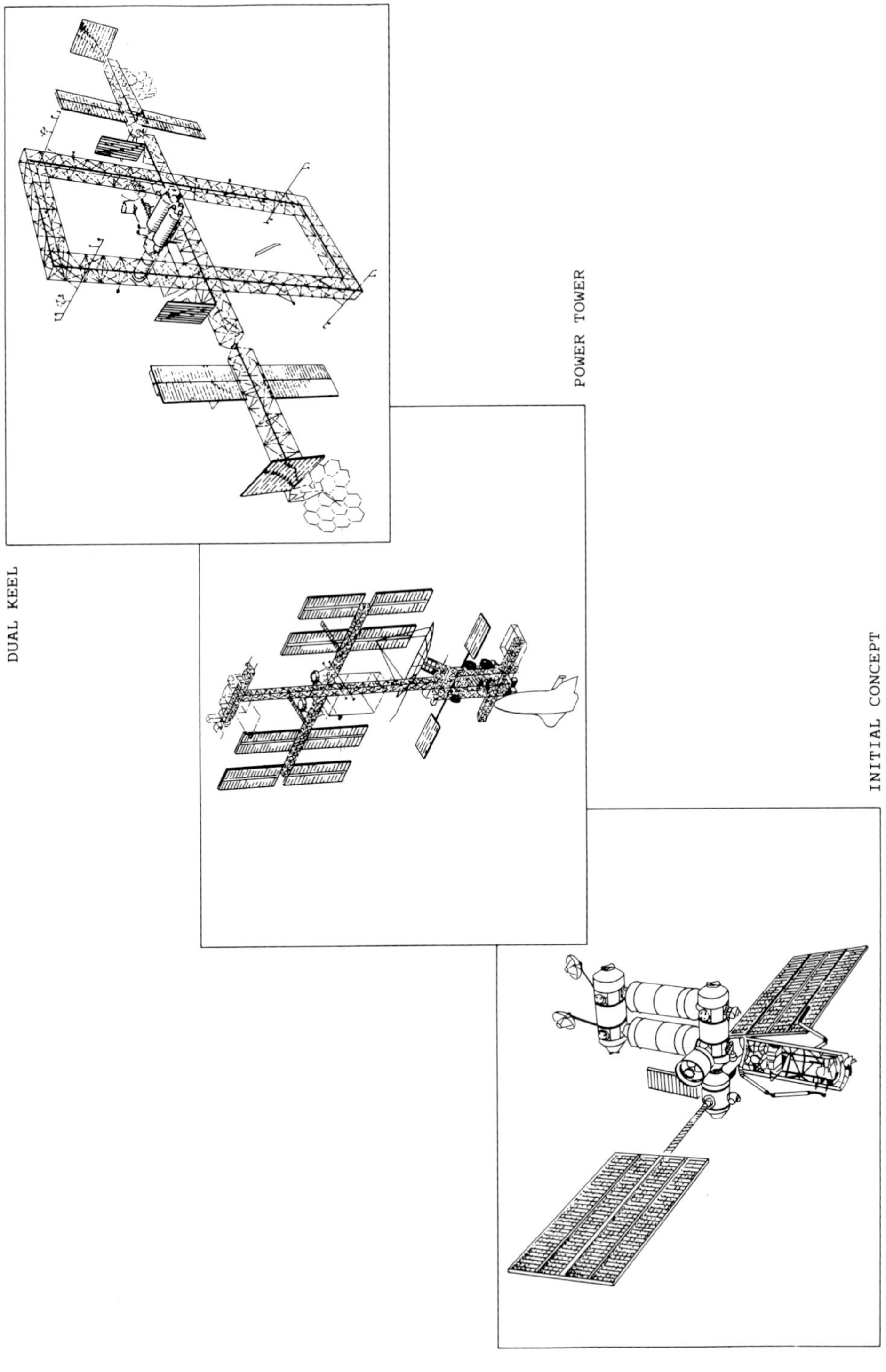

DUAL KEEL

POWER TOWER

INITIAL CONCEPT

Evolution of space station Freedom.

space station into orbit more-or-less completely fitted out (as a kind of latter-day Skylab). On 17 June, the White House told NASA to develop Option-A, and asked Congress to award the necessary funding, which it did on 23 June, with a majority of a *single* vote. Boeing was nominated as prime contractor on 17 August, and the detailed project definition for 'Alpha' was sent to the White House on 7 September 1993.

MEANWHILE – IN THE SOVIET UNION

In the late 1980s Russia expected the Salyut-based Mir complex to have at most a three-year life following completion in 1990, and the intention was to start construction of its replacement in 1992. It envisaged constructing an elaborate structure resembling the contemporary Freedom concept, using Buran to deliver modules and assemblies to be erected by spacewalking cosmonauts. But this ambitious plan was cancelled in 1991 when it was decided to delay the final phase of the expansion of Mir by at least three years. In the aftermath of the collapse of the Soviet Union, Russia decided to cancel Buran. Soon afterwards the production line for the Energiya launcher was shut down, which killed any chance of launching the ambitious Mir 2. A far more conservative design for Mir 2 was proposed using Mir technology. Construction would start in 1996. The new base block would dock with Mir, the Spektr and Priroda modules would be transferred to it, and the rest of the obsolete complex would then be discarded. Additional modules would be added to complete the new complex.

A revision of this plan in 1993 argued that it would be a waste to add to the existing complex; the last two modules should be grounded until the new base block was ready, so that they could dock with it immediately. Subsequent expansion would be by smaller modules, each of which would be devoted to a particular function and be replaced once it had served its purpose. This would permit the expansion to proceed rapidly and introduce a degree of flexibility that would enable the complex's programme to be varied. One advanced feature of the more elaborate concept was retained, however. In addition to the standard transducer-based solar panels, a pair of solar-dynamics mirrors (which would be somewhat more efficient than panels) were to be mounted at opposite ends of a long truss.

Despite such planning, it was clear that it would be far cheaper to keep the existing complex operating than it would be to replace it, so, unless something disabled it, every effort would be made to keep it operating.

13

Unification

On 2 September 1993, US Vice President Al Gore and Russian Prime Minister Viktor Chernomyrdin signed an accord declaring that the two governments had agreed to merge their space programmes; in effect to integrate Freedom/Alpha with Mir 2 to create a single space station. It would accommodate the European and Japanese modules that had been planned for the original international station. One significant change however, was that the new structure would be constructed in the same orbital plane as the existing Mir complex. This announcement took NASA's international partners completely by surprise. Although seemingly spontaneous, the intergovernmental accord arose directly from NASA's negotiations with the RSA to adapt the Soyuz to serve as a lifeboat for its Alpha space station. Although Russia had eagerly embraced Western-style democracy, its transitioning economy was in turmoil. It was clearly in America's strategic interest to offer support. As competition in space had symbolised the Cold War, collaboration in space seemed to express the new spirit of cooperation perfectly.

It was ironic, though, that spaceflight – the technology which had been developed by the two nations primarily as a means of illustrating the superiority of their individual ideological systems – should now become a symbol of the end of the Cold War.

As with going to the Moon, the real motivation to build a space station was political, not scientific. This was made clear by Senator Tom Harkin (D-Iowa), a long term critic of a NASA space station, who changed his mind and spoke in favour of it during the Senate debate immediately following the announcement of the joint effort: "I believe the agreement ... is so monumental, and so important to the future of our relations with Russia, and to the future of all space exploration ... that we must vote for the station". Whilst giving its support to the project, the House space policy committee insisted that NASA develop an independent capability to complete and operate the station just in case the Russians proved unable to deliver. The resulting vote was two-to-one in favour of proceeding with the joint venture. At last, it appeared, the political commitment to build the space station was in place.

NASA set out to evaluate how it could best integrate planned Russian hardware into its Alpha configuration. Its outline plan, with three phases, the first involving visits by the shuttle to the Mir complex, was delivered to the White House on 4 November. This was eagerly accepted on 29 November. The formal invitation for Russia to join NASA, ESA,

NASDA and the CSA in the development of the international facility was issued on 6 December. Russia accepted, and the RSA signed with NASA on 16 December. ESA, having met, decided in October to cut its Columbus laboratory to something little better than a permanent Spacelab module in order to be able to afford to develop a 'tug' and an Apollo-style spacecraft, and it formally affirmed its support in January 1994. The first meeting of the full international team (which included prime contractor Boeing and its subcontractors Rockwell, Lockheed and McDonnell Douglas) was held on 24 March to review the overall systems-design of the new proposal. In June, even as Congress continued to snipe at NASA's budget, Gore and Chernomyrdin signed the interim cooperation agreement to build what was then informally dubbed 'International Space Station Alpha.' In responding to concerns expressed by some in Congress that the Russians would not be able to deliver on the deal, Chernomyrdin assured, tongue-in-cheek, that Russia would continue with the project even if NASA had to pull out!

A MATTER OF DEGREE

The placement of the new orbital complex at an inclination of 51 degrees to the equator was a major compromise for NASA. It had planned to construct its space station at 28 degrees (the latitude of the Kennedy Space Center, Florida) because, by flying due east immediately after launch, the shuttle would receive the greatest benefit from the Earth's rotation, and so would be able to carry the maximum payload. Almost doubling the inclination would significantly reduce the shuttle's payload, which would necessitate redesigning some of the larger structural elements of its planned structure. On the other hand, some elements could be eliminated because the functionality would be provided by Russian hardware. It soon became clear that the majority of the hardware which NASA would contribute to this new structure would be more-or-less the same as proposed originally for Freedom.

For the Russians too, the 51-degree inclination involved a compromise, because some had wanted Mir 2 to be placed into a more northerly orbit so as to be able to overfly all of Russian territory (the traditional 51-degree orbit used by Russian space stations does not even reach Moscow). Like NASA, the RSA hoped to use most of the structure intended for Mir 2, without modification.

TOGETHER WE STAND

Far from paying for itself by producing semiconductor crystals, exotic metal alloys, active biological agents or Earth-resources imagery which could be sold commercially in an economy in which the rouble was on a downward inflationary spiral, it transpired that the Mir complex could attract foreign currency (that is, *real* money) by selling time to international partners, and there was no partner more significant than America. For Phase One of this joint programme, which would run for the four years leading up to the start of the new station's construction, NASA would pay $100 million per annum, and for the last two years it would be able to maintain an astronaut aboard Mir. The immediate priority for RSA was, therefore, to sustain Mir.

Just as the fee-paying international visitors had rebutted criticism of the original Mir complex as a costly irrelevance, and had prevented its abandonment, the series of visits by the shuttle called for by Phase One ensured that its life would be further extended. The agreement to develop the new station with NASA appeared to have guaranteed that the design effort put into Mir 2 would not be wasted.

Similarly, International Space Station Alpha was NASA's lifeline. The concept first proposed in the early 1980s had been for a $5 billion space station that, by this point, would have been in the final stage of its construction. But Congress had repeatedly demanded redesigns to produce a smaller station that would be cheaper to construct and to operate. NASA had worked through a series of successively more limited designs and, despite having spent $10 billion, still had no hardware in orbit. And with each passing year, space station funding became an easier target for those in Congress who wished to see the project killed. The critics that viewed the space shuttle as an expensive irrelevance saw shooting down the space station as a way to deny the shuttle one of its primary *raisons d'être*. A commitment to work with the Russians offered NASA exactly what it needed most—secure financial commitment to build, assemble and operate a specific structure in orbit. Assembly would start in mid-1997 (immediately after Phase One, the Shuttle–Mir flights) and would conclude in 2002 with the installation of the European and Japanese modules. Although it was to have a minimum lifetime of fifteen years, experience with Mir suggested that with due care and attention the International Space Station should be able to be kept operational until the 2020s.

TWO STATIONS IN ONE

The political agreement in place, the two agencies got together to define the detailed configuration and the assembly sequence.

NASA started with its Option-A (Alpha) configuration, restored the node to link the habitat module to the laboratory module, reinstated the laboratory's capacity, and also reintroduced the universal joint which would enable the solar panels carried on the truss to track the Sun, rather than requiring that the entire complex be reorientated. In the end, about 75 per cent of the hardware planned for the 1992 variant of Freedom was incorporated. Viewed in retrospect, the decision to use the Soyuz-TM as a lifeboat appeared prescient indeed.

One immediate decision was to replace the Lockheed Bus-1 manoeuvring unit by a Russian tug. This was to be built by Khrunichev, which manufactured the TKS ferry and the Mir expansion modules derived from it. The design was actually owned by the Salyut Bureau (formerly the Chelomei Bureau) but, earlier in that year, this had merged with Khrunichev. Apart from the fact that a 'Salyut tug' would cost less than half of a Bus-1, the Russian vehicle had the key advantage that it could be replenished in orbit. The simplest way forward was to order the off-the-shelf vehicle, and somehow mount this on the Alpha structure. But to do just this would be to miss the point of cooperating with the Russians.

In its own concept, NASA could not leave a crew aboard its station until it was virtually complete. During construction, it would be 'tended' only whilst a shuttle was docked. Residence would not be feasible until a habitat module had been put in place. If the Mir 2 base block was incorporated immediately, however, it would eliminate this unproductive phase.

The introduction of these Russian modules immediately advanced the 'permanent habita-tion capability' of the Alpha schedule by two years. It would take as long to assemble, but would be occupied throughout.

It then became clear that the key to integrating the two technologies was the Salyut tug. With the design of the complex maturing, this tug became the interface between the two technologies. The basic TKS propulsion unit was given a multiple docking adaptor with an androgynous axial port to mate with the first of NASA's modules, and a pair of standard radial ports for Soyuz-TM lifeboats. It would be linked at the other end to the Mir 2 base block. As the functionality of the 'tug' was refined, it acquired logistics and electrical sup-ply capabilities, and was soon renamed the 'functional energy block' (and, by using the Cyrillic acronym, the FGB). NASA's desire to eliminate the crew-tended phase of the construction was therefore achieved by forming a twin-module 'core' of Mir technology providing orbital manoeuvring, attitude control, power supply, logistics and crew facili-ties. While NASA built up one end of the complex using much of what it had designed for Alpha, the RSA was free to construct Mir 2 around the base block.

Russian practice would have been to launch and check out the new base block, then dock the FGB to this before NASA delivered its first module (an interconnecting node), but NASA wanted to launch the FGB and connect its node *before* sending up the base block and its commissioning crew. The construction sequence thereafter had yet to be agreed, and the schedule might easily slip as its details were worked out, but at least the process of construction had finally begun. As NASA Administrator Dan Goldin put it following substantial votes in favour of the station in both parts of the Congress in July 1994, the space station was "no longer just a design ... we are building hardware". To answer Congress's doubts about Russian participation, NASA agreed to buy the Salyut tug outright.

CONTRACTS AND COMMITMENTS

After resolving details of configuration, schedule and cost, NASA finally signed the pro-duction contract for US space station hardware with Boeing, as prime contractor, in Jan-uary 1995. Boeing then began negotiations with its major subcontractors. Although Boe-ing would not play a direct role in building European and Japanese hardware, it had to ensure that whatever was built was compatible and would properly integrate into the over-all system. Boeing had subcontracted the FGB 'core module' to Lockheed because it had already linked up with Khrunichev to commercialise the Proton rocket. Lockheed signed a contract with Khrunichev on 8 February, and, a few months later, agreed a fee of $200 million for the construction and 'on-orbit delivery' of the FGB. This contract was a signifi-cant departure for NASA, which had previously operated only on a 'no exchange of funds' basis with its international partners. Khrunichev was to build the base block for Energiya under a separate contract which would be funded by the RSA.

Although Khrunichev had not made a TKS vehicle with the actual configuration of the FGB, it was a highly modular design (of its 30 principal systems, 22 were basically the same as those of Mir's Kristall module, and the other 8 were present on Kvant 2), so project definition was rapid. Fabrication would be a slow process, however, because it was Russian practice to build half a dozen test articles before producing the real thing. By

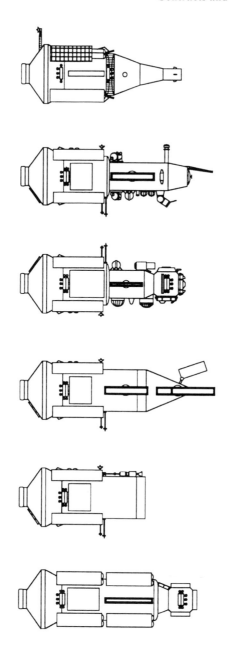

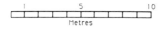

Comparison of all TKS-based spacecraft (top to bottom): Cosmos 1443, Kvant 2, Kristall, Spektr, Priroda, FGB.

relying on computer analysis instead of testing, NASA hoped to build flight articles straight away. But at least the contract had been signed, and work could begin.

Not everyone was in favour of the new spirit of cooperation. If it was decided to control the orbital complex from Houston, Kaliningrad would be redundant. Some in Russia worried that NASA might overwhelm them. Others wondered whether Russian technology would be subsumed as the station grew. Not everyone at NASA was happy either, and the assembly sequence was changed so that NASA could deliver one of its own solar panels earlier rather than later, to ensure that its laboratory would have sufficient power.

Some in Congress expressed concern that relying on Russian technology for the 'core modules' represented an exposed 'critical path', and demanded that NASA make contingency plans. To achieve a financial saving by involving the Russians, there had to be a significant Russian commitment in terms of hardware. It so happened that the Russians could provide precisely what NASA most needed—day-one support for propulsion, attitude control, guidance, navigation, power generation and a habitat (and they could do so using *proven* systems). Perhaps the most balanced contemporary assessment was that of Philip Culbertson, NASA's space station programme director, who acknowledged that "Russian technology is clearly as reliable as ours. But anybody who thinks it's going to save us a lot of money, or schedule, or complexity just doesn't understand".

To ensure that the crew of its proposed Alpha configuration would be able to return to Earth in an emergency, NASA had planned to buy a pair of specially modified Soyuz spacecraft and deliver them within the shuttle's payload bay. Following the decision to cooperate with the Russians on a joint space station, it was deemed more appropriate to send up a *standard* Soyuz-TM ferry, and integrate this into the ongoing crew-rotation cycle, so that there was always an up-to-date spare *in situ* just in case shuttle-delivered astronauts had to be evacuated.

On 28 September 1995, in an unprecedented move, Congress voted NASA a multi-year budget of $2.1 billion per annum through 1996-2002, thereby guaranteeing the financial commitment required to manufacture and assemble the space station. Other components of the agency's budget would continue to be subject to annual review; this was a special deal for the space station. As overall expenditure diminished, some expressed concern that the space station would be bought at the expense of the independent science plans. To put it into perspective, space station funding is just 15 per cent of NASA's overall budget; or to put it another way, at 0.015 per cent of the annual federal budget, the space station will cost each American less than $10 per annum. The Department of Defense spent $250 billion in 1995 alone.

Unfortunately, despite a government pledge, Energiya did not immediately receive the financial backing required to construct the 'service module' (as the new base block had become). Funding to build the proposed new research modules seemed unlikely. It began to look as if the RSA's hope of resurrecting Mir 2 was wishful thinking.

In November 1995 the RSA announced its intention to transfer Spektr and Priroda to the new station. For this to be feasible, the two structures would not only have to be orbiting at the same inclination to the equator, but also be coplanar. NASA's response was that it had no objection to anything which would increase the science capability of the complex, so long as achieving it did not delay the assembly schedule or increase its cost. On 11 December, a team of RSA engineers arrived in Houston to discuss how the transfer could be performed. A radical change in plan was then tabled by the RSA. It would make more sense (it was suggested) to attach the FGB to the front of Mir 1. By making the maximum possible short-term use of their assets, this would let the Russians postpone building new modules until the space station was operational. With ongoing maintenance, it was confidently argued, it should be possible to keep the original base block operational until then.

Although the Russians argued that using Mir 1 as a basis for assembly would allow faster commissioning of the new station because it could support continuous occupancy by larger teams, NASA rejected this idea because it would involve a significant change. Not least amongst NASA's concerns was that it intended to begin assembly in low orbit because, at the Russian inclination, the shuttle did not have the capacity to lift a full load to Mir's orbit. After the new station had been assembled, it would be boosted to its 400 km operating altitude.

In refusing to start with Mir 1, however, NASA was tempting disaster; the RSA might decide that there was little point in participating in the assembly of a new facility if doing so meant that it had to give up its existing station. If Russia withdrew, support in the Congress (which had far more to do with the need to support the fragile Russian democracy than it did the need for a laboratory in space) might evaporate as rapidly as it had solidified, which would very likely *kill* the project. As a compromise, therefore, NASA reaffirmed that it had no objection to the transfer of modules from Mir once the space station was complete, and it offered to assist in the transfer. However, it insisted that the RSA deliver the service module as promised. To help the Russians keep Mir operating while the space station was assembled, NASA extended Phase One by adding two flights in 1998 to the Shuttle–Mir joint programme.

As the negotiations continued into the new year, NASA agreed to add two shuttles to its schedule in 1999 to deliver Russia's Solar Power Platform (SPP). This truss for solar panels, gyrodynes and a roll-control thruster block was to be attached to the upper port of the service module's multiple docking adaptor. Eliminating these Proton rockets greatly relieved the pressure on the RSA's finances. All these agreements were formally ratified in July 1996. Another contentious issue resolved during this time concerned the Soyuz lifeboat. NASA had been surprised by the RSA's revelation that the couch of the ferry accommodated only a narrow range of body sizes, a restriction that would exclude half the astronaut corps. Energiya said that it could modify the couch, but only if it was paid to do so; NASA agreed.

As a package, these accommodations satisfied both sides. By continuing operations with Mir, the RSA ameliorated the consequences of its insurmountable funding crisis. At the cost of an extension to Shuttle–Mir, and an acceptable adjustment to the middle of the assembly sequence, NASA was able to keep Russian long-term support without the need to change the already agreed configuration. This revision had one other significant result. By insisting on retaining Mir, Russia ended the debate between Kaliningrad and Houston over which facility should assume primary responsibility for commanding the new station. It would be Houston.

With Russian participation declining, it became clear that the space shuttle would be the workhorse of the assembly phase. Of two long-planned upgrades, the aluminium–lithium Lightweight External Tank (LET) was on track for use in 1998, but after starts and stops the Advanced Solid Rocket Motor (ASRM) had been finally cancelled in late 1993. The LET will make a significant contribution to improving the shuttle's delivery capability at the 51-degree inclination. Atlantis, Discovery and Endeavour are to be devoted entirely to space station assembly work. Columbia will play a minor rôle; the first of the fleet, it is four tonnes heavier, and could not deliver the larger elements to the station's altitude. Assembly is therefore dependent on operating the three shuttles; the schedule will obviously slip if a shuttle is grounded.

The naming of the first crew in early 1996 marked a significant milestone. William Shepherd, then serving as NASA's deputy manager for the space station programme, was to be the station's commander. Anatoli Solovyov would command the Soyuz in which they would be launched. Sergei Krikalev was named as flight engineer. Both Solovyov and Krikalev had previously flown on the shuttle. Their assignments balanced NASA's insistence on overall American command authority with the fact that the hardware at that time would be predominantly Russian. After a behind-the-scenes dispute on this issue, Solovyov was replaced by Yuri Gidzenko. Meanwhile, chief astronaut Robert Cabana had been named to command STS-88 Endeavour, the mission which would mount the NASA node on the FGB. By this point, NASA had quietly dropped the name 'Alpha'; it was now simply the 'International Space Station'.

The switch to a 51-degree inclination orbit meant that the shuttle would not be able to carry the fully outfitted NASA laboratory. Most of its equipment racks would have to be delivered later and installed in space. NASA had originally intended to use a laboratory-sized logistics carrier to resupply Freedom, but this was not in the Option-A configuration. Rather than try to reinstate it, NASA had decided to accept an Italian offer to supply a Mini Pressurised Logistics Module (MPLM). It struck a deal whereby in return for donating this cargo module, a refurbished test article fitted with a large centrifuge would be delivered by the shuttle free of charge to serve as a life-sciences module; this was NASA's 'no exchange of funds' principle at work.

It was to be the RSA's responsibility to replenish the service module's propellant. The RSA already had a new ferry on the drawing board to supersede the Progress-M, because its 'wet cargo' capacity was just 1,000 kg. Dubbed Progress-X, this was to be launched using the Ukrainian Zenit, which was a spin-off from the Energiya rocket. In 1996, however, it was decided to cancel this and revert to a straightforward cargo form of the TKS because, if expanded tanks were fitted, it would require only one launch per year.

Perceiving an opportunity to supersede the Soyuz-TM as the station's lifeboat, ESA proposed an Apollo-style descent-only capsule. This Crew-Return Vehicle (CRV) was actually a derivative of the Crew-Transfer Vehicle (CTV) that it was considering as an application of the Automated Transfer Vehicle (ATV) being developed as an upper stage for the Ariane-V heavy-lift rocket. ESA was eternally torn between its perceived need to be independent and its clear inability to finance this luxury. It had hoped to use its own state-of-the-art spaceplane, the Hermes mini-shuttle (which is why it had ordered the Ariane-V heavy-lift launcher), but when this was cancelled in 1993 for financial reasons the conventional capsule seemed to be the only way that it could achieve its long-sought objective of an independent human spaceflight capability. Both the CTV and CRV were

(b)

(a)

(c)

Elements of the International Space Station. (a) The Functional Cargo Block (FGB) in Khrunichev's manufacturing facility in Moscow. (b) Node 1, the first element of the station to be manufactured in the USA, and which is scheduled to lift off aboard Endeavour in July 1998, along with Pressurised Mating Adaptors (PMAs) 1 and 2. (c) An electrical analogue of the Service Module undergoing tests.   The Service Module is the first fully Russian contribution to the station, and is planned for a December 1998 launch.

cancelled in 1995 in order to safeguard the ATV tug, which was now to be used to send cargo to Columbus.

Although by this time it was clear the way forward was to cooperate and to integrate, ESA still sought a measure of independence. Columbus had been 'descoped' for financial reasons, and, being only half the size of what had originally been designed for Freedom, was now officially a 'facility' rather than a 'laboratory'. Now, with Ariane-V nearing flight test, ESA began to consider whether it should further cut Columbus so that it could be launched by its own rocket and delivered using an ATV. When the new rocket crashed into the Atlantic Ocean on its maiden flight in June 1996, they had second thoughts.

NASA, in contrast, after so many years of pushing paper, was forging ahead with its part of the project and refused to be distracted. Boeing finished the aluminium pressure hull of a test article of the node in April 1995, and the first flight article followed in June. A problem was encountered during pressure tests, however, and in May 1996 eight spars had to be installed in the nodes to strengthen their structure around the four radial ports. By the end of the year, the hulls of the laboratory and the virtually identical habitat were finished. Rockwell, meanwhile, had built the solar panels and the truss to carry them, and Alenia, in Italy, finished the hull of the MPLM.

Although NASA was on schedule, the refabrication of the nodes to meet pressurisation requirements had been costly ($100 million). Accumulating overruns cut into the reserves of its capped budget, so it started to consider postponing expensive aspects of the science schedule in order to balance the books in the long term. NASA's problem was basically an accounting issue, but the RSA was facing a real cash crisis; the federal funds it had been promised had not been forthcoming, so it could not pay Energiya, its prime contractor, and so Khrunichev was not being paid. By September 1996 it was clear that work on the service module was behind schedule, with little prospect of time being recovered. By the turn of the year it was acknowledged that it would not be available for launch before the end of 1998. Since the FGB was being paid for by NASA, it was on schedule, despite a setback when its internal bulkhead had burst during a pressure test.

As a contingency – in the event that the service module could not be added before the FGB expended its propellant, leaving the complex without any attitude control – NASA decided to reconfigure a manoeuvrable stage developed to deploy military satellites after launch by a Titan rocket. Since this was already certified for use with shuttle payloads, it was expected to be a fairly straightforward process to install a docking system so that it could act as an 'interim control module'.

In April 1997, NASA decided to put back the launch of the Node, which had been set for December, by at least six months, in order to accommodate the slippage in the fabrication of the Russian service module. It hoped, however, to make up much of this delay by pursuing a more aggressive schedule thereafter. To launch the FGB and the node on time just to stick to the schedule had been deemed pointless. The Russians hoped to have the service module ready by the end of 1998. Although construction of the FGB was already finished, NASA, which owned it, asked Khrunichev to modify it so that its large storage tanks could be replenished with propellants directly from a docked Progress tanker, rather than by way of the service module. The revised schedule called for the FGB to be launched in June 1998, the Node in July, and, after more or less the originally envisaged interval, the service module will be added in December so that the first crew can be launched on a Soyuz rocket in January 1999.

14

ISS

The 'core' of the International Space Station is the Russian service module and the FGB. Both 20-tonne elements are directly derived from Mir technology. The service module has the same overall configuration as the Mir base block, with a five-port docking adaptor on the front, and it will provide environmental support and facilities for a crew of three during the assembly. The solar power platform truss to be erected on its upper port incorporates a thermal radiator, a set of gyrodynes for attitude control, and a roll-control thruster block, as well as mounts for up to eight solar panels of a similar design to the 'cooperative solar array' on Mir, each of which is capable of generating 6 kW for NiCd batteries. Rails running along the length of the truss will carry ESA's robotic arm.

The FGB is structurally similar to Kristall in that it has a long cylindrical main body with a small spherical docking adaptor at the far end. In this case, however, the adaptor provides only three ports: one axial, one upper and one lower. The first of NASA's nodes will be attached to the axial port. Soyuz-TM ferries will dock at the radial ports, and the service module will be docked at the far end. It has a pair of 420-kg thrust engines for orbital manoeuvring, clusters of 40-kg thrust and 1.3-kg thrust engines for attitude control, and solar panels providing 4 kW of power. The main compartment is fitted with lockers for dry cargo. The service module will use the same ODU as the Mir base block (burning UDMH with nitrogen tetroxide). This will perform all orbital manoeuvres and periodically reboost the station's orbit. The FGB has externally-mounted tanks to store six tonnes of propellant for the service module, which has pipes to carry propellant from the rear port, where replenishment will occur, to these tanks. It is intended to maintain a year's supply of propellant at all times to safeguard against orbital decay in the event that a resupply ferry is lost. Three or four Progress-M replenishments will be required annually.

Unlike the Russian modules, which are autonomous and are required to rendezvous and dock, the NASA modules are delivered by shuttle, so they contain no navigation, propulsion or power generation systems. Being totally passive, they are reliant on other resources for manoeuvring and for electrical power, both of which are initially provided by the FGB.

The pressurised modules and the internal environmental systems were developed by Boeing. All the modules are cylinders with an external diameter of 4.42 metres and an internal diameter of 4.25 metres. Unlike Russia's vehicles, which comprise variously-sized and variously-shaped compartments, these modules form 2-metre square-section rooms.

Internally, each module offers a definite sense of 'up and down'. All the lighting strips are in the ceiling, and the dark floor contains utilities and storage lockers. All internal equipment is mounted on the walls, in a bank of standardised racks that can be swung out so that the equipment can be accessed and replaced.

Each module has a shallow conical cap at each end incorporating a 2-metre diameter attachment port that contains a 1.3-metre square hatch engaged by eight latches, and has an axially-mounted 20-cm diameter porthole. Connectivity is offered at both ends – even though it is not required for the basic configuration – in order to maximise the potential for future expansion of the complex.

The surface of the welded-aluminium pressure hull is waffle-patterned for structural strength, and is covered by thermal and micrometeoroid blankets. It has attachments for mounting the module within the shuttle's payload bay, and handrails for spacewalking astronauts. There are no holes in the hull for instruments, and there is no provision for mounting equipment externally.

NASA has two basic module types: the shorter modules of the nodes and the longer modules of the laboratory and the habitat. Each node is 5.5 metres long, and serves as a nexus for attaching other modules. It incorporates four radial attachment ports, all offset to one end of the module to facilitate racks in the adjacent 'walls'. The first node provides command-and-control and power-distribution functions, and the second provides storage.

An adaptor is required to mate the first node to the FGB. This is a semi-conical tube with a 2-metre diameter NASA collar on one end and a narrower APDS docking system on the other end (the detailed shape of a 'pressurised mating adaptor' resembles a stack of successively narrower rings, with their axes progressivly offset slightly towards one side). Two other cones are configured to mate with the shuttle's orbiter docking system.

The 8.5-metre long laboratory, mounted between the nodes, is built from three annular rings, with a 0.5-metre wide porthole offset to one side of the central section. It contains a dozen racks for apparatus mounted in international standard payload racks or the payload-integration racks (EXPRESS, to EXpedite the PRocessing of Experiments to Space Station) that NASA has developed so that the drawer-class and mid-deck-class shuttle payloads and Spacehab payloads can be used in a standard rack. The structurally identical habitat contains a galley (an oven, a freezer, a hand washer, a water dispenser and a rubbish compactor), a gymnasium, a shower, a toilet, a wardroom (by the porthole) and individual cabins for six crew members.

The final two pressurised elements of the NASA part of the structure are the airlock and the cupola, both mounted on the nodes. The 2.2-metre wide cupola forms a blister 1.4 metres deep which provides a ring of peripheral windows for panoramic viewing of operations by the remote manipulators.

An oxygen–nitrogen atmosphere at 14 psi is maintained and regulated by drawing from supercritical cryogenic nitrogen and oxygen tanks. As on Mir, air temperature and humidity is regulated by condensing heat exchangers. Carbon dioxide is removed by a molecular sieve and, as with Mir's Vozdukh, the in-flow is dehumidified by desiccant, passed through zeolite to extract the CO_2, then reheated and released; the cycle switches every four hours for regeneration. There is no equivalent of Mir's Elektron electrolysis system for making oxygen, so the cryogenics will have to be topped up periodically. A mass spectrometer will analyse the constituents of the air and monitor dangerous trace components, and a combination of charcoal and sorbent beds will remove contaminants such as carbon monoxide (CO), hydrogen (H_2), methane (CH_4) and ammonia (NH_3). Unlike Mir, the ventilation ducting is contained in the structure of the module interfaces: a fan with 4 m^3/minute flow circulates the air through the pressurised structure. Flame detectors in open areas and smoke detectors in confined areas detect fires which will be dealt with by an automatic CO_2 suppressant system.

Although Mir passes air through a condenser to extract the water vapour, this is the only recovered water which is returned to the primary (potable) water tanks; all other recovered water is deemed suitable only for electrolysis and personal hygiene. NASA, however, has developed a multi-filtration system using heaters and filters which is able to yield potable water from the shower and galley as well (this apparatus will need to be serviced fortnightly), and a separate vacuum filtration unit which can extract water from urine. Solid waste, however, is to be off-loaded onto the shuttle and returned to Earth.

The truss, thermal control, command, communications and navigation systems were designed by McDonnell Douglas. Signals are relayed via the TDRS network to maintain continuous communication by S-band uplink and K_u-band downlink. Spacecraft in the immediate neighbourhood employ UHF. The GPS network is used for navigation so as not to tie up ground-based tracking systems. A gyrodyne-like system provides full three-axis stability.

The 94-metre long truss, which is aligned perpendicular to the complex's axis, is an open aluminium frame connected to a spur mounted on top of the laboratory. It is made up of a central part, which comprises half the total length, with a hexagonal section 4.3 metres wide, and 2-metre wide square-section 'port' and 'starboard' extensions. The 13-metre long spur is linked to the laboratory by the module truss structure. It holds the mobile transporter, the portable work-platform for spacewalking astronauts, and the GPS antenna. The port hexagonal-section truss contains within it a UHF antenna and a component of the crew and equipment translation assembly, and has a thermal radiator mounted on its trailing edge. The square-section segment has the inner solar panel mount and associated storage batteries, a short segment, then the outer solar panel mount. This substructure is mirrored on the starboard side, except that the antenna within is for the S-band uplink.

The station travels with its truss horizontal and with its principal axis on the velocity vector, so that shuttles can use the standard V-Bar approach. Flight operations are to be managed by the Space Station Control Center at the Johnson Space Center in Houston.

The mobile transporter is a 2.5 × 1.5-metre flat pallet which supports the remote manipulator's mobile servicing system. It runs on twin rails on the leading edge of the hexagonal truss. Tiny carts will be able to transfer crew and equipment along the hexagonal truss to work sites faster than would be possible by manual translation.

The thermal regulation system uses internal coolant loops which circulate water and transfer heat to external loops which pump ammonia coolant through the 23-metre long radiators on the trailing edge of the truss (three radiators are mounted in-line on pallets on either side), shaded from direct sunlight.

Rockwell supplied the solar panels, batteries, and power distribution system. Each unit uses an 11.8-metre long square-segment truss. This contains six high-capacity NiH batteries, and mounts a rotating joint which orientates the two panels and (perpendicular to and in their shade) a local thermal radiator 13.6-metres long and 3.3 metres wide. Each 35.8-metre long by 11.5-metre wide solar panel frame has 287 m² of transducers, and provides a peak of 23 kW. The panels were located on the truss because they were far too big to mount on the modules. Placing them so far away, however, introduced rather long cabling. Since power loss is proportional to current as well as to cable length, the output from the transducers is converted to high-voltage–low-current for transmission, then converted back by direct-to-direct converter units to low voltage for use; this is the same principle as used on commercial power transmission lines. All NASA-supported modules on the International Space Station use 160 volts dc, therefore. The shuttle fuel cells produce 30 volts dc. Mir technology operates on 28 volts dc. Fully half of the 100 kW will be made available to the three laboratories so, unlike in the case of Mir, science operations will not be starved of power.

Having supplied the space shuttle's remote manipulator system, the Canadian Space Agency designed the space station's arm. This 17.7-metre long heavy-duty arm will be used to lift modules from the shuttle's cargo bay and position them for attachment. The Space Vision System (SVS) (tested on STS-74 Atlantis, when mating the DM to the ODS) will enable the operator to work 'in the blind'. As with the shuttle's arm, the station's arm will be used to move spacewalking astronauts. The Mobile Remote-Servicing Platform carries a tool kit for the arm's end-effector, and mountings, power and communications sockets for items to be mounted externally. The Special-Purpose Dextrous Manipulator is a pair of 2-metre long multi-jointed arms set on the mobile platform, for manipulating external payloads with a dexterity comparable to that of the human hand.

The international modules have the same 4.42-metre diameter in order to fit within the shuttle's payload bay.

Italy's 6.6-metre long Mini Pressurised Logistics Module (MPLM) has a cargo capacity of 9 tonnes, mounted in the equivalent of 16 standard racks in a wraparound configuration. Its life-sciences module is the same, but with a 2-5-metre tilting centrifuge at its far end. ESA's Columbus module is likely to use an MPLM pressure hull, and offer a capability for life sciences, fluid physics and materials research comparable with that of the eight-rack Spacelab. ESA's ATV resupply ferry will dock at a pressurised mating adaptor.

The 10-metre long Japanese Experiment Module (JEM) is a far more elaborate facility, with two small arms to manipulate apparatus on an unpressurised pallet mounted at its end, and a collar on its roof for a 4-metre long pressurised experiment logistics module. The H-2 rocket is eventually to be used to launch these resupply modules.

The sprawling football-field-sized structure with modules clustered in the centre and massive solar panels mounted at each end of a framework truss may be a far cry from the elegant wheel in space imagined by von Braun, and the magnificent double wheel depicted by Stanley Kubrick in his classic science fiction film *2001: A Space Odyssey*, but it nevertheless has an elegance of its own.

STATION ASSEMBLY

Although the detailed manifesting might be revised again to meet operational needs, the broad assembly sequence is straightforward. The process will start with the Proton rocket that places the FGB into orbit. STS-88 Endeavour, flown by Robert Cabana and Fred

Sturckow, will rendezvous, and Nancy Currie will use the remote-manipulator arm to at-tach the node, with its axial pressurised mating adaptors, to the FGB. Then Jerry Ross and James Newman will connect power and communications cables during a spacewalk. After the shuttle departs, the FGB will look after the dormant node.

At this point, NASA will place a short square-section truss on the upper port of the node. This truss will incorporate communications apparatus and gyrodynes (in NASA par-lance, 'control moment gyros') for attitude control. The third, and final, Pressurised Mating Adaptor (PMA) will be placed on the bottom of the node. A solar panel assembly will be mounted on this truss. With its power supply guaranteed, the laboratory will then be added. The main NASA modules are to be connected in-line, so adding the laboratory will involve docking at the lower PMA, dismounting the axial PMA, replacing it with the labo-ratory, then mounting the PMA on that. Because the fully-fitted laboratory would be too heavy, it will be delivered half empty; the rest of the racks and the scientific apparatus will be ferried up on the next few flights using the MPLM, along with the airlock, the remote-manipulator arm, and unpressurised pallets carrying batteries and stocks of bottled gas. The service module will then be added, and the first crew launched by Soyuz. This ends Phase Two, and marks the point at which NASA will be able to run a full science pro-

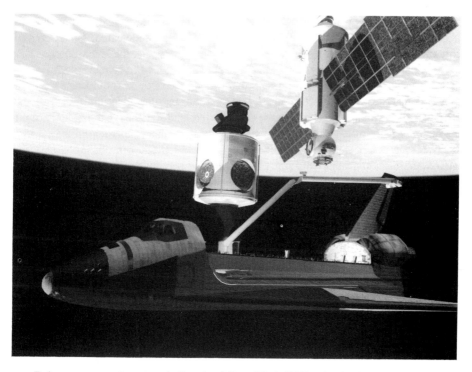

Endeavour prepares to capture the Functional Cargo Block (FGB) using the shuttle's mechanical arm in this artist's impression of the first space shuttle assembly flight for the International Space Station, mission STS-88 scheduled to launch in July 1998. The shuttle will carry the first US-built component for the station, a connecting module called Node 1, and attach it to the already orbiting FGB, which supplies early electrical power and propulsion. The FGB will have been launched about two weeks earlier on a Russian Proton rocket.

gramme in its laboratory during the 'utilisation flights' that will combine assembly with experiments.

One of the early tasks of Phase Three will be to mount the Solar Power Platform (SPP) on the upper port of the service module. The first step in assembling NASA's own truss will be to affix the spur and the hexagonal-section core segment to the top of the laboratory. This will be extended at each end to accommodate the triple-panel thermal radiator sets. It will be at this point that the second node will be mated to the laboratory (repeating the procedure for relocating the axial PMA). The observational cupola will be mounted on the first node's left port to enable astronauts to oversee the operations of the station's robotic arm. Following this, a solar panel will be placed on each end of the truss: the first panel will be moved to the end of the port truss and the final panel will be mounted at the far end to balance up the structure. At this point, the Japanese and ESA laboratories will be added. As soon as NASA's habitat is in place and the lifeboat (currently stated to be a Soyuz) docks at the FGB, the crew can be increased. This completes Phase Three.

The Russians intend to keep the Mir complex operational until late in this sequence, and only then start to add the modules to the service module to build up that end of the complex. ESA plans to affix a remote manipulator on the SPP truss to assist in docking these modules. The configuration of this part of the complex is yet to be defined, but is likely to exploit modules similar to those on Mir.

Any assembly task which cannot be undertaken with a remote-manipulator arm is to be performed manually. A total of 600 hours of spacewalking is anticipated, for which training is expected to be as highly choreographed as it was for fixing the Hubble Space Telescope. At the time, that mission was expected either to make, or to break, NASA's plan to rely on spacewalks to complete the assembly of the space station. To build up a cadre of experienced astronauts, NASA decided to put astronauts on scheduled shuttles specifically to conduct spacewalks designed to test assembly procedures. Since external activities always involve pairs of astronauts, this meant that two members of each crew had to be spacewalk-qualified. Shuttle flight crews were not allowed to work outside in case they suffered an accident which would jeopardise their ability to fly the spacecraft back to Earth. And because a shuttle docked with the space station would not be able to chase after an astronaut whose tether had broken, an emergency manoeuvring unit built into the standard backpack (appropriately called SAFER) was introduced.

In January 1996 STS-72 Endeavour first evaluated the portable work-platform, then the rigid-umbilical (to run coolant fluid and power to externally-mounted equipment) which was connected to mock-up avionics boxes. In March, the STS-76 Atlantis spacewalks on the DM provided high-fidelity training near Mir. Unfortunately the jammed airlock hatch on STS-80 Columbia in November prevented testing of the Orbital Replacement-Unit Transfer-Device (the 4-metre crane which was to have been tested by transferring a mock-up of the station's 250-kg battery).

A GUIDED TOUR

Adopting the frame of reference introduced earlier for the Mir complex ('up' being above a cosmonaut at the service module's main control panel, and with the engines 'to the rear') we find that this has the happy outcome that the International Space Station's velocity vector is 'forward', and that with the truss set horizontally, 'down' is towards the Earth, 'up' is towards the zenith and the terms 'port' and 'starboard' (as defined by NASA) do indeed correspond to left and right. As the axis of the pressurised 'stack' is aligned with the velocity vector, when the service module's engine is fired it boosts the orbit, thereby obviating the need to reorientate the complex to perform this manoeuvre.

From the point of view of the commander sitting at the control panel therefore, the internal hatch in front leads into the multiple docking adaptor. This is a 2-metre diameter sphere with five docking APDS ports, each 0.8-metre wide. It contains no apparatus; it is a nexus. There is a Soyuz docked at the lower port. This is powered down, and so is dark. A thin plastic tube snakes from a vent in the service module through the compartment, straight through the ferry's 2-metre wide orbital module to the descent module, to keep the air circulating. The solar power truss is on the top port, so this is shut. Since Mir 1 is still operating, the Russians have yet to start to expand this new facility, so the other radial ports are unoccupied.

Ahead is the FGB, and just inside, in a tiny recess in the floor, is its control panel. The main compartment is a long narrow corridor, and lockers cover its every surface. At the far end is the hatch to its docking adaptor, a 2-metre diameter sphere with 3 APDS ports. Another Soyuz is docked at the lower port; the top one is unoccupied. Beyond the axial port the architecture changes dramatically.

On the other side of the narrow hatch is a NASA pressurised mating adaptor. This 2-metre long tunnel flares out to a 2-metre wide connector ring with a 1.3-metre square hatch in its centre leading to node-1. This houses the control systems in racks on either side. At the far end is a ring of square hatches. The one on the left opens straight into the blister-like cupola; its panoramic view is breathtaking! The lower one leads into the habitat (at its far end is a mating adaptor, and sometimes a shuttle docks at it). The upper hatch is closed (a short box-like truss containing gas bottles is on it). To the right is the spacewalk airlock (there is a shuttle docking system at its far end). In contrast to the Russian part of the complex, which gives the impression of a number of rather isolated compartments linked by narrow hatches, this part of the complex is wide open.

Straight ahead is the laboratory, the heart of the NASA part of the complex. Racks of scientific apparatus line its walls, utilities run beneath the floor, and the ceiling contains storage racks. Node-2 is beyond. The lower port is unoccupied, the Columbus module is to the right, the Japanese module is to the left, and Italy's life sciences laboratory with its centrifuge is above. Straight ahead is another mating adaptor, the one most often used by shuttles, and there is a shuttle docked now. The adaptor's port is connected to the orbiter's docking system, the far end of which is the 'floor' in terms of the shuttle's local frame of reference. (In its docked position, the orbiter is nose up, with its belly forward).

Mounted on the roof of the NASA laboratory is the central section of the truss. This projects far out to either side, each having a set of thermal radiators on a rotating mount and a pair of solar panel mounts on universal joints. On the leading edge of the truss, running along the hexagonal segment, a pair of rails carry the pallet holding the remote manipulator arm assembly, and within the core of the truss are the rails for the carts that are used by spacewalkers to transport equipment along its length to the external-payload sites above and below its main section. There are all manner of communications boxes in there too. Short cranes are used by spacewalkers to swing the heavy battery packs from the cart out to their stations within the outer parts of the truss, by the solar panels.

The International Space Station is pretty big. The straight module-stack is 50 metres in length, and the truss projects about as far on either side. Even with the international modules clustered around the leading node, there is still plenty of potential for adding to the complex, so this really is just the start!

STATION OPERATIONS

A crew will typically have a commander, a flight engineer and four research scientists, and they will serve tours of three or six months. Each will have a personal laptop that will serve as the main interface to the station's systems. It will be plugged in wherever convenient to undertake a given task, and will retrieve operating manuals from the main computer as required. (The use of this 'soft' interface eliminated the need to provide a physical link to each and every system from a large bank of infrequently-used switches and dials.)

With a permanent staff of six, the sense of isolation on the International Space Station should be considerably less than on the two-seat Salyut. With a shuttle docked at the same time as a Soyuz handover, there could be as many as 16 people aboard for periods of a week or so. Furthermore, with a mix of nationalities, the *cultural* isolation should also be diminished, especially if communication with family and friends by videolink and by e-mail is easy.

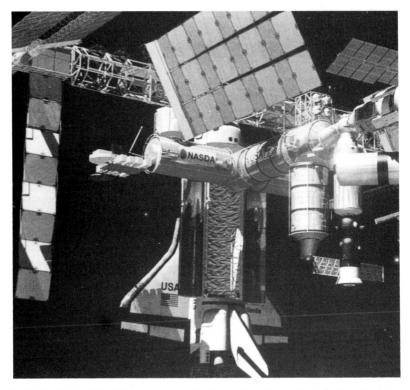

The NASA, ESA and NASDA components of the International Space Station as they should appear at the end of Phase Three.

A 90-day service cycle will be operated, each cycle providing a 30-day microgravity environment period, 10 days of maintenance, a second 30-day slot, then a rendezvous, resupply, reconfiguration and reboost period, which means that arrivals and departures must be scheduled to fit in with experiment runs as well as with orbital dynamics. The station's orbit will be boosted following a shuttle visit and immediately prior to the next resupply tanker's arrival. A shuttle will dock at the node at the front of the complex.

Although every crew member will have to be sufficiently knowledgeable to operate a Soyuz lifeboat, and the Russian crew (who will arrive and depart by Soyuz) will have to be capable of performing rendezvous and docking manoeuvres, there will be no need for the other personnel (who will arrive and depart on shuttles) to be pilots; they will be the station's equivalent of what NASA has termed 'payload specialists' on the shuttle. The crewing mix will reflect the contributions to development of the International Space Station made by the various international partners.

ROUND-THE-CLOCK SCIENCE

With a full crew, the International Space Station will be able to conduct science on a 24-hour basis. The commander and flight engineer will manage station operations and conduct routine maintenance so, in contrast to Mir, these essential activities will not be undertaken at the expense of the science programme.

Astronauts performing science on the space station will not be working in isolation; they will work with ground-based experimenters using video and data communications. NASA has variously referred to this as 'interactive science' or 'telescience'. Hardware for this has already been tested. In July 1994 STS-65 Columbia flew the International Microgravity Laboratory for the second time (IML-2), and groups of specialists from 15 nations at various sites observed the life-sciences and materials-processing experiments by video and operated apparatus directly from the ground. When two of the four video links failed, real-time scheduling had to be introduced to allocate time to experiments on a dynamic priority basis; although frustrating for the researchers, this was undoubtedly a valuable bonus for the mission planners. In October 1995, STS-73 Columbia flew the US Microgravity Laboratory (USML-2). Scientists and astronauts constantly revised the experiments to refine results, just as would be done in a normal laboratory. As well as a video downlink, an uplink enabled the scientists to show the astronauts how to conduct supplementary procedures (astronauts on tours in space will not be able to rehearse all experimental contingencies). STS-78 Columbia flew the Life and Microgravity Spacelab (LMS) in June 1996, and further tested procedures.

MICROGRAVITY RESEARCH

In contrast to serving as a way-station to deep space, the International Space Station is to be an orbital research laboratory. Having drastically 'descoped' its design to make it affordable, NASA was forced to dedicate its laboratory to microgravity research. The Europeans and Japanese followed suit, with the result that the programme will concentrate on life-science and materials-science. In contrast to Mir, the International Space Station will not be short of power.

In the absence of commercially funded research, most scientific work on the station will be government funded, at least until its commerciality can be demonstrated and cost of exploitation evaluated. Perhaps the Wake Shield Facility (WSF), flown three times on the shuttle, represents a realistic model. WSF was a remarkably simple idea: a spacecraft bores a hole in the rarefied upper atmosphere, leaving a semi-vacuum in its wake. The saucer-shaped Wake Shield created a 10^{-14} torr vacuum. This was sufficient to permit a semiconductor to be built up layer by layer (by epitaxal growth, a process investigated earlier, in Mir's furnaces). After initial stabilisation problems which limited the duration of the experiment and the purity of the product, trials in 1995 and 1996 demonstrated that the crystal produced offered the prospect of chips with a significant improvement in performance, for which a premium rate could reasonably be charged. This project had been funded jointly by NASA and the University of Houston's Space Vacuum Epitaxy Center, and Space Industries Inc. constructed the test vehicle. Development and testing cost $30 million over a six-year period. If commercial funding is forthcoming, it is planned to build a large shield to operate as a free-flyer alongside the International Space Station. Unlike Mir's crystal output, this would feed into a highly competitive semiconductor industry. An output of just 200 kg per year would provide an estimated turnover of $50 million. It is to be hoped that other microgravity research already tested on Mir and on the shuttle – protein-crystal growth and electrophoresis of biological agents for the pharmaceutical industry – will follow a similar route to commercialisation aboard the International Space Station. Over the years, the Salyut and Mir space stations proved to be effective test beds for a wide range of space-based research.

With new equipment constantly being ferried up, as well as new research modules, Mir facilitated incremental development and trial of experimental apparatus (such as for materials-processing in furnaces and electrophoresis) so that it could later be installed in automated satellites. The resupply ferries served as vehicles for testing apparatus which needed external deployment, a notable example being the spectacular space mirror. The cosmonauts observed all of these experiments. Mir's existence encouraged impromptu projects, many of which were simple improvements of earlier apparatus; the Magnetobiostat, for example, resulted from the cosmonauts making a television appeal for ideas. But other apparatus proved baffling. One small package arrived with instructions that it be ejected from the airlock, and its departure filmed. On the other hand, they became very familiar with the cultivators and furnaces, because they needed frequent servicing. Yet to satisfy US technology transfer laws, the first American apparati sent were 'black boxes' that the cosmonauts were required simply to switch on and off.

To fully exploit its laboratory on the International Space Station, NASA will have to encourage and facilitate a similar mix of continuity and spontaneity in its research. This will require shuttles to deliver apparatus on an ongoing basis. Experiments will need to be devised, fabricated and flown in under a year for such a 'try-it-and-see-what-happens' culture to thrive. These will be the station's equivalent of the 'Get Away Specials' flown on the shuttle. Ideally, a budding PhD will be able to devise an experiment and have it flown in time for the result to form the core of a thesis submission. It would be unwise to restrict the programme to applied science that is near the point of commercial viability; it must support pure science too.

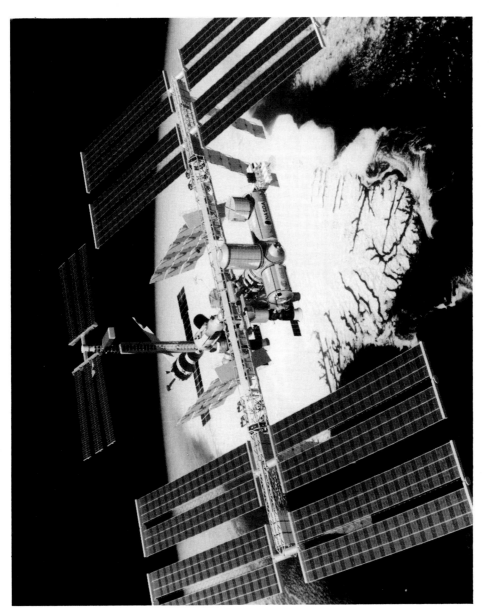

A striking depiction of how the entire International Space Station should appear when it is complete.

OBSERVATIONAL PROGRAMMES

The cosmonauts of the Salyuts and Mir have used a number of telescopes to make solar, astronomical and astrophysical observations at a wide range of wavelengths, and have used a variety of remote-sensing apparatus to observe the Earth, its atmosphere, its oceans, and its magnetosphere.

One area of science that is conspicuous by its absence from the International Space Station is astronomy. Earlier station configurations with vertical keels provided mounts for telescopes, and would have had the necessary pointing accuracy. There are, though, significant disadvantages to putting telescopes on an inhabited space station. Firstly, the microvibration from the personnel and the systems which sustain them will degrade the imagery. Secondly, even if the station uses a gyroscopic attitude-control system, efflux from the thrusters of visiting spacecraft will contaminate the optics. Maybe, later, some wide-field or omni-directional instruments – perhaps detectors for the enigmatic transient burst sources – will be installed on the horizontal truss, but the most effective way that the International Space Station could support the astronomical community would be by servicing a fleet of free-flyers like the Spartan satellites that are deployed and retrieved by the shuttle.

As yet, there are no plans to put solar telescopes on the International Space Station, but a solar telescope could be mounted on a rotating joint similar to that designed for the transducer and solar-dynamics power units. Maybe, eventually, a telescope comparable with that on Skylab could be installed.

Although the planned service life of the International Space Station's basic structure is a decade, there is every expectation that it will remain operational for a quarter of a century, so it is highly likely that a great deal of equipment that has not yet been thought of will have been designed, built and installed long before it enters its second decade.

MIR'S LEGACY

It is indeed a pity that the science activities envisaged for the original Power Tower and Dual Keel concepts were dropped during the 'descoping' to make the configuration more affordable. Without remote-sensing facilities, the International Space Station will be *less* capable than Mir. Before that complex is decommissioned, therefore, its most useful equipment should be transferred to the International Space Station.

Of course, it would have been better if – as the Russians suggested – the FGB docked at Mir's front port, so that the existing complex could act as the 'core' of the structure. Maybe common sense will eventually prevail, and Mir will be attached to the rear of the International Space Station, finally to combine the best of both worlds.

Part 4: Conclusions

15

The learning curve

The key point to acknowledge in assessing Soviet human spaceflight is that since its lunar programme faltered in the late 1960s it has doggedly pursued the development of a space station capable of continuous habitation. To properly assess this *quarter of a century* of effort, it is first necessary to review the origin and evolution of the Soyuz spacecraft.

THE INCREDIBLE SOYUZ

The Soyuz was designed in the early 1960s as a general purpose vehicle that could be launched on the Semyorka rocket, used for both Sputnik and Vostok. In a sense, it was Korolev's first 'spaceship,' because the Vostok had simply drifted in the orbit into which it had been put by its carrier rocket until it initiated a simple ballistic descent. The Soyuz was to be able to *manoeuvre*, both to rendezvous with other spacecraft in space and to refine its trajectory during atmospheric re-entry. It was to have a modular design, so that it would be capable of undertaking a *variety* of missions in both Earth and lunar orbit. In contrast to its American counterpart (Apollo), it had three basic elements rather than two: in addition to the descent capsule and the service module, it carried an 'orbital module' which provided living space for the crew, and which was needed because the descent capsule was so cramped.

Soyuz's modularity made it extremely flexible, and specialised forms of each of its basic elements could be developed to suit different mission requirements. For example, one version of the descent module had a strengthened heat shield in order to survive the higher speed atmospheric re-entry imposed by a direct return from the Moon. The orbital module could carry different docking systems and, depending on its specific mission, be fitted as a residence or a laboratory. Similarly, the power of the engine in the service module could be tailored to given mission requirements. This meant that the lightest form of each module could be employed for each specific vehicle. The Earth-orbital version could be launched by Korolev's own Semyorka, but Chelomei's more powerful Proton, fitted with one of Korolev's own stages (borrowed from the rocket he was developing for the full lunar landing mission) was needed to boost the lunar version from 'parking orbit' towards the Moon. In fact, two forms of the spacecraft were developed to fly in cislunar space, a fact which further demonstrates the flexibility of the modular design.

The circumlunar form was not meant to enter lunar orbit, so it had a stripped-down service module. This was required only to perform midcourse corrections to refine the return trajectory, and because this was computed from tracking by Earth-based radar, it did not need to be able to navigate independently in deep space. The orbital module was deleted from this variant because the Proton did not have the capacity to lift it. Although it had been intended to fly a cosmonaut around the Moon before the Americans did, this particular race was narrowly lost in 1968 (in fact, Apollo 8 went into lunar orbit, which was even more impressive) and with its propaganda value lost the circumlunar option was dropped. Without an orbital module, it would have been a very tiring journey, even for a single cosmonaut.

The full lunar mission, which would have had a lander based on Soyuz technology, had a service module with a more powerful engine to boost out of lunar orbit and head for home, and a descent module incorporating onboard navigational capability for use in lunar orbit. This form was tested without a crew several times in the early 1970s – as indeed was the lander – but all the manoeuvring trials were performed in Earth orbit. To test other aspects of the lunar mission, the 'standard' Soyuz flew several missions in the late 1960s to prove first-orbit rendezvous (to simulate rendezvous with the returning lander), and cosmonauts spacewalked from one vehicle to another (as was to be done in lunar orbit to transfer to and from the lander).

Soyuz rendezvous in Earth orbit relied on ground radar to track the orbits and on the control centre to compute the necessary manoeuvres; onboard radar was used only in the final few kilometres of the approach. In contrast, American Gemini astronauts took pride in using onboard inertial and computer systems to derive their own manoeuvres. Apollo's system (supplemented by sextant sightings) was able to navigate to the Moon and back on its own. In this respect, therefore, the Soyuz was relatively primitive. On the other hand, it was standard practice to provide each form of the Soyuz with only the facilities to fulfil its mission; the Earth-orbital variant did not require this independence. Similarly, since the lunar plan called for an external crew transfer, the docking unit was required only to engage the two craft for a short period, and a simple mechanism was bolted to the front of the orbital module. With the decision to build a space station in low Earth orbit, a docking system with a collar to establish a hermetic seal was developed so that cosmonauts could transfer directly.

Following the deaths of the Soyuz 11 crew due to a pressure leak, both the descent and service modules were heavily modified. With the mission of such a 'ferry' reduced to a single rendezvous and deorbit, the propellant tanks fitted into the service module to facilitate extensive orbital manoeuvring were eliminated. Because it would take a day to reach the station, and only several hours to return to Earth, the heavy solar panels were replaced by rechargeable chemical batteries. To protect the crew from the kind of decompression accident which had killed the Soyuz 11 crew, cosmonauts would now wear pressure suits. To accommodate the necessary independent life support system the crew had to be cut to two. This new Soyuz variant, with its two-day endurance, was, therefore, explicitly configured to serve as a space station ferry.

The 'standard' form which had the improved safety features and simplified engine of the ferry, but retained the solar panels, would fly independent missions. In addition to the Apollo–Soyuz link-up (when an androgynous docking system was carried), this has flown carrying a telescope (on Soyuz 13) and an Earth-resources camera (on Soyuz 22) mounted on the docking system.

The modularity of the Soyuz was exploited to create the Progress ferry. The descent module's avionics were moved to the service module, the descent module was replaced by a simple casing containing tanks for 'wet' cargo, and storage racks were installed in the orbital module to carry 'dry' cargo.

The Soyuz-T was introduced in the late 1970s. It restored the solar panels so that a vehicle which experienced difficulty in rendezvousing would not be forced to break off and return to Earth. For improved manoeuvrability, it was given the Argon computer of Salyut 4 and the unified propulsion unit of Salyut 6. The propellant saved by jettisoning the orbital module prior to deorbit enabled it to carry a heavier payload, which restored the third seat. Soyuz-T was evidently intended to be used with Salyut 7, but Salyut 6's unexpected longevity led to its being introduced early. The Argon offered a navigational capability comparable with Gemini, in the sense that it was limited to manoeuvring in low Earth orbit, but it was sufficient for space station operations. The Soyuz-TM introduced with Mir had a more sophisticated rendezvous system, improved communications, and even greater payload, but some of this was lost when it was decided to revert to the old procedure of disposing of the orbital module only after the deorbit burn, in order not to deprive the crew of its facilities in the event that they encountered difficulties with this manoeuvre and had to stay in space another day. The orbital module itself was modified several years later to incorporate a 'blister' porthole for improved forward vision whilst manoeuvring in close.

Although Sergei Korolev designed his spaceship to be able to perform a variety of roles, it is unlikely that even he would have expected it still to be in front-line service at the millennium. Even though the Soyuz killed two crews and nearly stranded another in orbit, Chelomei's Merkur spacecraft, which was more capable but needed the Proton to launch it, was never adopted; Korolev's is the only rocket to have been trusted to carry cosmonauts. The Semyorka had been designed to propel a heavy nuclear warhead on a ballistic arc. Its lifting power defined the dimensions and mass of the Soyuz spacecraft, which defined the crew capacity, which defined the requirements of the environmental system of the space station, whose dimensions and mass were set by the Proton rocket. Thus, it is evident, the structure and operation of the Mir complex derive from technical compromises made long ago. The new Energiya rocket offered the prospect of building an orbital village with a crew of dozens but, upon its cancellation, there was no role for the 13-tonne Zenit-launched six-person spacecraft that was under development, so this was cancelled too, which left only the venerable Soyuz.

Although it is natural to consider the Soyuz to be 'old technology', especially when it is compared with the American space shuttle which, more than any other spacecraft, looks the part, it was the simplicity of the Soyuz that made it attractive as a lifeboat (at least in the short term) for the International Space Station.

AN EVOLUTIONARY DEVELOPMENT PROGRAMME

We have seen how the space station began as a reconnaissance platform, but because there was an urgency to introduce it, and because Chelomei's spacecraft was at an early stage of development, the project was split so that Korolev could rebuild a variant using proven systems. The first station, therefore, appeared to be a straightforward extension of Soyuz technology. Although the process of step-by-step *evolutionary* development predomi-nated, it is evident with hindsight that there was a bold strategic plan at work to develop a permanently inhabited station and, had it not been for an unfortunate series of accidents early on, this goal *might* have been achieved many years before it finally was. (The Soyuz 11 deaths were unrelated to the station, after all; the next station was lost in a rocket malfunc-tion, and the first military variant was crippled by a system not aboard Korolev's configu-ration.) The stages in this development process are now clear.

The designers of the first Salyut stations, which had only a single docking port, had had to fit everything needed for the mission into the vehicle prior to its launch. Whereas Che-lomei had envisaged a large resupply ferry, a Soyuz could not transport much extra equip-ment or consumables. So, to maximise the scientific programme, the crew's facilities were basic and their tenancy was limited by the station's consumables.

The operational life of a station was limited not only by the food and water available, but also by the propellant supply, which was consumed both in overcoming orbital decay and in altering the orientation to perform experiments and to optimise the output of the solar panels. The habitable life of a station was also limited by the ability of its environ-mental system to regulate the temperature and maintain a breathable atmosphere. If, there-fore, the crew ran out of air, food or water, or the station ran out of propellant, or the thermal control system broke down, or toxins accumulated, it would then become unin-habitable.

As it happened, though, the ability of the first Salyut stations to support prolonged habitation was never an issue, because the stripped-down ferry introduced following the Soyuz 11 accident carried chemical batteries instead of solar panels. These limited the time available to effect the rendezvous, so it had to abort and return to Earth if a problem developed. Simply reaching the station represented a significant milestone in a mission; all too often this was not achieved. A station continued to consume propellant while the fault was investigated. All of the early stations were intended to sustain two periods of inhabitation. Salyut 1 would have been revisited if the Soyuz 11 accident had not taken place. Upon the failure to revisit Salyut 3, there was insufficient time to mount another mission within the station's operational life. For the first three years, lost opportunities abounded. It was not until Salyut 4 that a station was re-entered, and by the time that the second crew finally departed, the environmental system, which had sustained a total of three months inhabitation, was so worn out that the humidity had risen to the degree that rich mould had coated the walls. Even so, Salyut 4 *more* than made up for all the problems that had afflicted its predecessors.

Meanwhile, a variant of the Soyuz configured to carry cargo had been devised. The first dockings between two Soyuz spacecraft (in the late 1960s, in preparation for the lunar programme) had been performed automatically, but dockings with Salyut stations were undertaken manually after the automated system had delivered the spacecraft to the 200-metre point. Soyuz 15 tested an automated system. Although its repeated failure to dock with Salyut 3 frustrated the first attempt to place a second crew aboard a station, it is noteworthy that even at this early stage in the programme a resupply ferry was under de-velopment. Indeed, Vladimir Shatalov reported as much in explaining why the crew had not been allowed to take over and dock manually. The significance of this trial is that even if this resupply ferry *had* been available, it could *not* have serviced Salyut 4, because it required a crew aboard the station to unload its cargo, and there was only one docking port. It was, however, a clear indication of what was to come. It was also typical of the Soviet evolutionary approach to technology: refining proven hardware in order to build up the required operational capability.

The addition of the second docking port and the introduction of the automated cargo ferry for the second-generation stations (Salyut 6 and Salyut 7) transformed the state-of-the-art in just a few years. A dozen Progress cargo ferries replenished consumables and brought components to overhaul the environmental system, and enabled Salyut 6 to sustain five crews. It had been designed to support occupation for 50 per cent of an 18-month to two-year period (it was hoped to achieve a 90-day, a 120-day and a 175-day mission), but in the event it lasted twice as long.

Table 15.1. Space station occupancy

Station	Main expeditions (days)	Total
Salyut 1	23	23
Salyut 3	16	16
Salyut 4	30, 63	93
Salyut 5	49, 18	67
Salyut 6	96, 140, 175, 185, 13, 75	684
Salyut 7	211, 149, 237, 168, 50	815
	Total	1,698

The Progress had a capacity of just over 2 tonnes, split more or less evenly between 'wet' and 'dry' cargo. Experience with Salyut 6 revealed that only about 15 per cent of the mass was for the science programme; the rest was to replenish consumables for the station and its crew. With a total daily mass requirement of 16 kg, this meant that a ferry was required at two-monthly intervals. Sometimes the dry cargo was unloaded immediately, and this took several days, but often (because the vehicle was retained until the docking port was needed) the job could be done incrementally; this had the advantage that the ferry's orbital module could serve as a warehouse, which greatly relieved the cramped situation in the station. Before the ferry was jettisoned, it would replenish the station's propellant tanks and then raise its orbit. Finally, its orbital module would be loaded with discarded food packets, dirty clothes and expired components from the environmental system, all of which were destroyed when the ferry was deorbited and burnt up in the atmosphere. Despite earlier difficulties with automated dockings, the Progress ferries came and went like clockwork.

As soon as the endurance record exceeded the in-orbit life of the Soyuz, the second port permitted the ferry to be replaced. Rather than send up a new ferry empty and dock it automatically, like the Progress, visiting crews were dispatched and guest researchers worked aboard the already inhabited station for periods of a week or so before taking away the old ferry; one Salyut 6 crew, on a six-month flight, hosted four visiting crews.

Essential to the second-generation Salyut's success was the unification of the orbital manoeuvring and attitude-control systems and the installation of plumbing to enable the propellant tanks to be replenished. Unfortunately, the refuelling operation could only be undertaken at the rear port. This imposed the operational requirement that a new Soyuz left at the rear had to be flown around to the front once the old one departed, to keep the rear clear for cargo ferries. This was unfortunate, but, despite the potential for failure, every fly-around was completed without incident. Salyut 6, therefore, was a triumphant success in every respect. Salyut 7 followed after a year's intermission, and although it was structurally identical, more of its systems had been made accessible to the cosmonauts so that they could be serviced.

The new Soyuz-T had a longer service life, so for Salyut 7 the timetable for visiting missions could be relaxed. The goals of Salyut 7 were twofold: first, to further extend the endurance limit, and second, to have a succession of crews hand the station over to one another as a step towards permanent occupancy. Although the record was extended (but only by two months), a series of technical problems prevented a handover until just before the station was decommissioned.

Salyut 7 lasted six months longer than its illustrious predecessor, and was inhabited for a slightly higher percentage of its time, so it was by no means a failure, but it was dogged by problems. In addition to having to have a bypass installed on its engine (the prospect of requiring spacewalking cosmonauts to perform engine repairs had been considered to be impracticable, until a leak developed), at one point it actually had to be salvaged from deep freeze when the power failed whilst it was temporarily unoccupied. In overcoming all these problems, however, the programme reached a level of maturity far beyond the state-of-the-art bestowed by Salyut 6, which, by-and-large, had suffered no serious failures. In hindsight, instead of the stated objectives of pushing the endurance record and demonstrating ongoing habitation, Salyut 7's most significant contribution was in showing that spacewalking cosmonauts could perform intricate engineering tasks. Previously, a station leaking propellant would have been reluctantly abandoned. Now it seemed that no task was too difficult.

Table 15.2. Number of spacewalks

Station	Number	Hours
Salyut 6	3	5
Salyut 7	13	50
Mir	60	260
	76	315

Each spacewalk involved two cosmonauts, so the total experience is actually twice this duration.

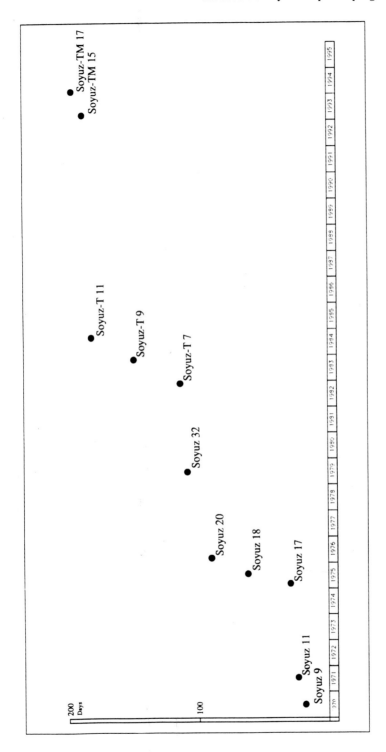

Soyuz in-orbit time.

Also, exploiting the capabilities of the Soyuz-T, every effort was made to overcome rendezvous mishaps which previously would have led to immediate aborts. Even when the Igla transponder was ripped off Soyuz-T 8, fast work by the radar tracking network and the Kaliningrad flight controllers enabled the ferry to fly in from its transfer orbit to the 200-metre point. Although all commanders were trained to take over if the automated system failed, it was expected that the Igla would supply separation and closing rate. Although denied these cues, Vladimir Titov began a manual approach. He was forced to abort when a few metres out because the station flew into the Earth's shadow; there is no doubt that, given another minute, he would have been able to dock. The disruption to Salyut 7's programme was severe, but the lessons learnt from this fully-manual approach set the basis for Vladimir Dzhanibekov's successful docking with the station when it was derelict. The early 1980s, therefore, offered many opportunities for overcoming adversity, and confidence was running at an all-time high when Mir was launched in 1986.

COMMISSIONING MIR

Mir got off to an extremely impressive start. Its commissioning crew checked it out, then flew to Salyut 7 to finish their predecessors' programme before returning once more to Mir with as much of the now mothballed station's apparatus as their diminutive ferry could carry.

Given the prior use of Cosmos 1443 to supply Salyut 7, it is surprising that such a cargo carrier was not sent up to Mir early on, rather than the second Progress ferry (the first had been needed to replenish Mir's propellant after its climb to operating altitude, a service that the big module was not equipped to offer). A full load of standard portable science apparatus would have facilitated a viable work-programme while the cosmonauts awaited the dispatch of the first of the specialised research laboratories. In the event, once they had exhausted the potential of the equipment that they had salvaged from Salyut 7, Kizim and Solovyov were obliged to return to Earth, leaving the station vacant. A cargo module could have been sent up to Mir whilst they were on Salyut 7, and they could have unloaded it upon their return. The wherewithal to carry out productive work on the base block would at least have made feasible an orbital handover, and started the permanency of occupation so eagerly sought. An MKF-6 multispectral camera would have provided the basis for continuing the Earth-resources programme, for example.

CONSTRUCTING THE MIR COMPLEX

The process of assembling Mir, although evidently delayed by almost a year, began with Kvant 1. This was docked at the rear of the base block. It *had* to be placed there, because to have put it on the front axis would have blocked access to the radial ports on the multiple docking adaptor, inhibiting further build-up. Placing it at the rear, however, blocked the base block's main engines and, since Kvant 1 did not have any engines of its own its addition made the complex reliant on a docked cargo ferry for orbital manoeuvring. Kvant 1's presence, therefore, seems rather awkward. It was, in fact, something of an interloper.

Kvant 1 had been built to sit at the front of Salyut 7, and, if that station had proven more reliable it undoubtedly would have been launched following the departure of the Cosmos 1686 laboratory. It incorporated a set of X-ray telescopes, and gyrodynes with which to orientate its host with sufficient pointing accuracy. Adapting it to sit at the rear of Mir had involved the installation of piping to transfer fluid from a resupply ferry to the base block, because otherwise its arrival would have restricted Mir to resupply only via the front port (a facility which had not been available on Salyut 7). Sending Kvant 1 to Mir was, therefore, simply a case of exploiting 'off the shelf' hardware. The Korolev Bureau, which had designed Kvant 1, had originally planned to build more such modules, but it had been ordered to develop a suite of modules similar to Cosmos 1686 (which was an adaptation of Chelomei's TKS) by replacing the Merkur capsule with additional built-in apparatus. The development of the first of these big modules was running late.

There were ports for four modules on the front of the complex. Although the base block had been specialised as a habitat, the first of the new modules was to enhance the crew facilities rather than to serve as a specialised laboratory. Kvant 2, therefore, added environmental and waste management systems, a shower and a second toilet, as well as a purpose-designed airlock to facilitate spacewalking, after all the docking ports had been occupied.

Since it was to be standard procedure to dock a new module at the axial port prior to using its Ljappa arm to swing it to the proper radial port, at such times the resident ferry had to be at the rear. Because this prevented resupply, it was clear that the new module could not be left on the axis for long. Kvant 2 was to sit on the upper port. Once it was in place, the complex would be L-shaped until the next module arrived to convert it into a T-shape. While asymmetric, the complex's orientation would be extremely difficult to control and it would be impractical to adjust its orbit to overcome atmospheric drag. It was vital, therefore, that this awkward period be kept as short as possible. This phase of the assembly could not start until the first *two* modules were ready. Kvant 2 could not be launched until its partner was ready. They formed a matched pair, with the same overall configuration, and they were developed in parallel. Work on the second pair of modules which, although similar to each other, were of a different design, followed at a relatively relaxed pace.

At this point (around mid-1988), it appears that there was a change in the plan, and it was decided to bring forward the 'technology' module in the hope that its five furnaces would produce sufficient semiconductor material to ameliorate the cost of continued operations, and this meant that Kvant 2 could not be launched until Kristall (as the second module became known) was ready. A succession of crews flew in the expectation of commissioning these new modules, but it was December 1989 before Kvant 2 was sent up, and it then proved more difficult than expected to upgrade the computer to manage the asymmetric configuration. Calibrating the computer's mass model consumed a prodigious quantity of propellant, so a resupply tanker had to be hastily sent up. Like Kvant 1, Kvant 2 had a set of gyrodynes; in this case, however, they were to facilitate reorientating the growing complex without consuming propellant. Being the first to be added, these two modules *had* to carry the gyrodynes, and since these provided *sufficient* manoeuvring capability, none of the later modules needed to have them. Furthermore, Kvant 1 *had* to be in place before Kvant 2, so that its gyrodynes would be available. Although modular, the complex was a closely integrated structure. Bringing forward Kristall meant that it had to be temporarily stationed on the lower port to balance the complex; it was to be swung onto the side port for which it had been designed once the module that it had displaced became available.

Each module extended its solar panels as soon as it reached orbit in order to provide power during the rendezvous. The panels on the modules for the upper and lower ports had frames similar to those on the base block. The modules for the side ports had a type of panel which could be retracted, so that they would not interfere with the base block's solar panels when the module was swung to the side. These retractable panels had been made detachable so that they could be *transferred* to another part of the complex. Since Kvant 1 had none, Kristall's panels were to be remounted on it. During spacewalks the cosmonauts affixed frames to the sides of Kvant 1, then mounted drive motors on these to accommodate the panels, and installed a crane on one side of the base block to swing the first of the bulky packages to the rear of the complex. However, it had been decided that no more modules would be added until Kristall's furnaces proved their commercial viability; in effect, this indefinitely postponed further expansion. With Kristall stuck on the lower port for the foreseeable future, it was decided to leave its panels in place. The cosmonauts were not short of external work to do, however. During a marathon series of spacewalks they erected the Sofora girder on Kvant 1 and set the roll-control thruster block on top of it; a very impressive piece of construction work. Although this thruster block (which was carefully positioned so as to fire in the same plane as the base block's own roll-control thrusters) would save propellant, once its tank ran dry it would need to be replaced. This structure's existence indicated not only considerable forethought, but also the extent to which constructing the Mir complex was dependent on spacewalking. And it was ironic that this construction centred on Kvant 1, the interloper.

It was not until the fortunes of the programme began to improve that the go-ahead was finally given to complete the construction of the final two modules. Even then, the first of Kristall's panels was not transferred to Kvant 1 until immediately before Kristall was moved to facilitate the first docking with the space shuttle. The other panel was simply retracted because it had been decided to fit a new one, delivered by Atlantis, on Kvant 1.

Thus it took *five years* to finish the process of installing solar panels on Kvant 1. This exercise demonstrates the interaction between long-term and short-term planning in the programme. The configuration of the modules was closely allied to the locations that they were to occupy in the complex; a redeployable solar panel was developed, a crane to transfer it was devised, supporting frames were fabricated, motors were installed and cables were strung along the station to deliver the resulting power supply, all of which indicated long-term planning. Yet the job was carried out rather haphazardly, with the individual tasks being passed from one crew to the next as the schedule for the modules repeatedly slipped. By assembling the orbital complex far more slowly than originally planned, it was left perpetually short of power. Nevertheless, the cosmonauts showed remarkable ingenuity in adapting their programme to both of these frustrations, in order to make productive use of their time. This indicated one important difference between Mir and its Salyut predecessors. Whereas previously a crew would not be sent up until their research module was ready, keeping Mir inhabited permanently meant that its crew had to absorb delays, fix broken equipment, and do their best with what was available. Life on Mir was, therefore, much more Earth-like in this respect than before. And frustrating though it must

have been for all concerned, relentlessly pursuing long-term objectives within the constraints of short-term expedients must have produced a valuable database for operational
planning. It must be borne in mind, however, that we would have known nothing of this if
the Soviet Union had not 'opened up' following Mikhail Gorbachev's *glasnost* initiative,
for otherwise everything would have been presented as having gone according to plan.

The notion of completing the Mir complex within two years, using it for three years,
and then replacing it with something much more elaborate, was clearly over-optimistic.
Kvant 1 was not launched until one year after the base block. There was then a wait of
nearly three years for Kvant 2, with Kristall following in six months. Then there was a
five-year gap to Spektr, and another year for Priroda. As a result, the base block was 10
years old when the assembly process was completed.

Table 15.3. Space station launches and re-entries

Spacecraft	Launched	MT	Re-entered
Salyut 1	19 Apr 1971	0340	11 Oct 1971
Salyut 2	3 Apr 1973	(0300)	28 May 1973
Cosmos 557	11 May 1973	(0300)	22 May 1973
Salyut 3	25 Jun 1974	0138	24 Jan 1975
Salyut 4	26 Dec 1974	0715	3 Feb 1977
Salyut 5	22 Jun 1976	2014	8 Aug 1977
Salyut 6	29 Sep 1977	0950	29 Jul 1982
Salyut 7	19 Apr 1982	2245	2 Feb 1991
Mir	20 Feb 1986	0028	-
Kvant 1	31 Mar 1987	0406	-
Kvant 2	26 Nov 1989	1601	-
Kristall	31 May 1990	1433	-
Spektr	20 May 1995	0733	-
Priroda	23 Apr 1996	-	-

CREWING

The training programme for guest cosmonauts lasts about two years and includes aspects
of flying the Soyuz ferry relevant to handling emergencies; for anyone already fluent in
Russian, this course can be compressed. Commanders and flight engineers each receive
comprehensive training prior to being assigned to a crew. The system has changed over
the years, but cosmonauts might start out supporting a flight without actually being part of
the crew, and then be assigned to a specific mission without any real expectation that they
will fly it, because it is a backup role, but with the understanding that they will be first in
line for the next compatible mission. Early on, all-rookie crews were sometimes formed,
but after a number of docking failures a rule was introduced that each crew had to include
a veteran. This rule has recently been relaxed.

Although crews are generally formed a year or so in advance, changes in the flight assignments are common, and substitutions for illness and accidents are not unknown. Valeri Ryumin, for example, although he had recently completed six months on Salyut 6, stood in for Valentin Lebedev when the latter injured a knee a few weeks before he was due to fly. Some crews were formed at very short notice to meet operational exigencies, an excellent example being the pairing of Vladimir Dzhanibekov and Viktor Savinykh at four months notice to rescue Salyut 7. Crew switches tend to ripple on. For example, after Soyuz 25's failure to dock with Salyut 6, Alexander Ivanchenkov stood down from the next crew so that Georgi Grechko could inspect the docking system. As a result, when Vladimir Kovalyonok (the Soyuz 25 commander) was moved up to take the next flight, he was teamed up with Ivanchenkov. The Romanenko–Grechko team, formed after the problem developed, therefore had just two months to train for the flight that finally broke the Skylab endurance record. The final expedition to Salyut 6 was mounted only after a maintenance crew refurbished the station, which gave the Soyuz-T 4 crew three months to prepare.

In some cases, however, even a fully-trained visitor could wait years to make his flight. Mohammed Faris of Syria, for example, was repeatedly rescheduled because of Salyut 7's problems, and was eventually reassigned to Mir. On the other hand, political factors meant that Abdul Ahad Mohmand, the Afghan trainee, had to fly after six months of training, which was feasible only because he was fluent in Russian; in fact, his hosts were already aboard Mir when he entered training. The ripple effect from this was that flight engineer Alexander Kaleri lost his seat and Sergei Krikalev was obliged to serve a double tour.

Occasionally fate has intervened at the preflight medical, and the prime and backup crews have swapped. Valeri Kubasov had symptoms of a lung infection, so Soyuz 11 was flown by its backup crew. Recently, Gennadi Manakov was discovered to have a heart irregularity, so at just 10 days notice Valeri Korzun and Alexander Kaleri found themselves in line for a six-month tour. Surprisingly, few cosmonauts have fallen ill in space. Vladimir Vasyutin developed a prostate infection and the crew was recalled from Salyut 7, but, in the case of Mir, when Alexander Laveikin was diagnosed as having a heart irregularity, he was replaced in space, and his commander (Romanenko) went on to break the endurance record.

Although it is important that the members of a crew be psychologically compatible, too much can be made of this. Whilst it is true that fighter pilots tend to be competitive, they are also highly professional. Cosmonauts understand their systems, know what is expected of them, and accept the deprivations in return for the chance to fly. Excluding antagonistic competitors, the process of selecting crews inevitably reduces to choosing either 'likes' or 'dislikes'. It turned out that people with similar personality traits were able to substitute for one another in their interrelationships without effort. They worked well together for short periods, but often became irritated with one another in the longer term. Anatoli Berezovoi and Valentin Lebedev, during 211 days on Salyut 7, gradually diverged. They did not actually argue; they just avoided one another. They were simply *too* alike to get on well. It turned out to be better to choose people with complementary traits because, being aware that they had different backgrounds, expertise and outlook, they tended to coordinate, acknowledge one another's point of view, and come up with a wider range of solutions to any problem. Once such a team had harmonised, it proved extremely stable. Ironically, therefore, the hasty pairing of Romanenko with Grechko, and Kovalyonok with Ivanchenkov, each of which was a team of disparates, was very successful. One factor that assisted in making disparates effective was matching single-mindedness with analysis of alternative options. It was also often the case that the flight engineer was considerably older than the commander, who was often on his first flight. So far, all commanders have been male.

Salyut had been designed to support three cosmonauts, but the loss of the Soyuz 11 crew forced modifications to the ferry that reduced the capacity to two. The introduction of the two-ended second-generation station meant that crews could receive visitors. The Intercosmos missions were flown to meet Salyut 6's operational requirements, rather than for their own sake. The inability of Soyuz 33 to dock tore the schedule into shreds for a year. As a result, some guest cosmonauts had to wait a long time to fly, and often trained with several commanders. The visits also offered opportunities for veterans like Valeri Bykovsky, Viktor Gorbatko, Valeri Kubasov and Nikolai Rukavishnikov to fly again. Although it would have been feasible, no in-orbit handovers were attempted with Salyut 6. The Soyuz-T reinstated the third seat, and Salyut 7's environmental system was able to accommodate a crew of three for extended periods and six for short periods.

Psychologists initially feared these expanded crews would fragment – two excluding the third – but in practice this did not occur. The first crew to make an extended mission included Dr Oleg Atkov, but he was already an outsider because his rôle was to observe his colleagues. In the next case, the commander (Vasyutin) fell ill, and when his comrades rallied to him it reinforced the team spirit. The first three-person resident crew on Mir again involved a doctor (Poliakov). The restored ferry seat has been used to enable visitors to accompany the handover rather than to expand resident crews. It is only recently that fee-paying visitors have begun to remain for full tours and, as foreigners, the visitors are once again naturally outsiders. Despite the efforts of his colleagues to make him a full member of the crew, Norman Thagard felt culturally isolated. Shannon Lucid though, integrated much better; in fact, she made the point that the crucial thing is not the technology but a crew having a common purpose, and always working as a team.

The task of forming compatible crews was complicated when cosmonauts started to fly individual missions, coming and going opportunistically, and on several occasions a ferry returned with three cosmonauts who had been launched separately. When Alexandrov replaced Laveikin, for example, he returned with Romanenko, whom he had joined, and Levchenko, who had been making a short visit. Similarly, when Krikalev was obliged to extend his tour, he returned with Volkov, his new commander, and Flade, the visitor.

To illustrate the sometimes *ad hoc* nature of the crewing strategy, consider the case of Dr Valeri Poliakov. Although Dr Oleg Atkov had flown a complete 236-day mission on Salyut 7 with two colleagues to monitor their medical state (with authority to order them home if he thought it expedient), it was clearly impractical to keep a doctor aboard Mir full time. Instead, it was decided to send up a doctor towards the end of a record-breaking flight, so that the cosmonauts could be examined prior to their return.

Poliakov had planned to assess Yuri Romanenko as he entered the final stage of his 326-day mission (he had been suffering progressive exhaustion) but Poliakov's seat on Soyuz-TM 4 was taken by a Buran pilot. Vladimir Titov and Musa Manarov, who took over, hoped to extend the record to a year. Towards the end of their mission, therefore, Poliakov was assigned a seat on Soyuz-TM 6, which carried an Afghan on a brief visit. By remaining aboard with the residents, Poliakov committed himself to staying on with the *next* resident crew, because a French researcher was to accompany the forthcoming handover. Furthermore, because his new colleagues (Alexander Volkov and Sergei Krikalev) were *not* to host any visitors, and another Buran pilot was to accompany the next handover, Poliakov had committed himself to serving with Vitali Sevastyanov and Viktor Afanasayev *as well*. Unfortunately for Poliakov's hope of setting an 18-month record, the decision to vacate Mir until the new modules were ready meant that he had to return to Earth after 240 days. Undeterred, he subsequently spent 14 months on Mir, setting a record which is likely to stand into the next century.

Table 15.4. International space endurance records

Cosmonaut/astronaut	Launch spacecraft	Launch date	Days in space
Yuri Gagarin	Vostok 1	Apr 1961	0.07
Gherman Titov	Vostok 2	Aug 1961	1.05
Andrian Nikolayev	Vostok 3	Aug 1962	3.93
Valeri Bykovsky	Vostok 5	June 1963	4.97
Gordon Cooper Pete Conrad	Gemini 5	Aug 1965	7.92
Frank Borman Jim Lovell	Gemini 7	Dec 1965	13.75
Andrian Nikolayev Vitali Sevastyanov	Soyuz 9	June 1970	17.71
Georgi Dobrovolsky Viktor Patsayev Vladislav Volkov	Salyut 1	June 1971	23.76
Pete Conrad Paul Weitz Joe Kerwin	Skylab	May 1973	28.04
Al Bean Jack Lousma Owen Garriott	Skylab	July 1973	59.49
Jerry Carr Bill Pogue Ed Gibson	Skylab	Nov 1973	84.04
Yuri Romanenko Georgi Grechko	Salyut 6	Dec 1977	96.42

Table 15.4. (continued)

Cosmonaut/astronaut	Launch spacecraft	Launch date	Days in space
Vladimir Kovalyonok Alexander Ivanchenkov	Salyut 6	June 1978	139.60
Vladimir Lyakhov Valeri Ryumin	Salyut 6	Feb 1979	175.06
Leonid Popov Valeri Ryumin	Salyut 6	Apr 1980	184.84
Anatoli Berezovoi Valentin Lebedev	Salyut 7	May 1982	211.38
Leonid Kizim Vladimir Solovyov Oleg Atkov	Salyut 7	Feb 1984	236.95
Yuri Romanenko	Mir	Feb 1987	326.48
Vladimir Titov Musa Manarov	Mir	Dec 1987	365.95
Valeri Poliakov	Mir	Jan 1994	437.75

Given the importance of the biomedical research, it was not surprising that doctors were inducted into the cosmonaut corps. In 1964, Voskhod 1 was the first spacecraft to carry a multiple crew. Pilot Vladimir Komarov was accompanied by Dr Boris Yegorov and Dr Konstantin Feoktistov. Yegorov was a medical specialist, and Feoktistov was one of Korolev's senior engineers. From this point, all multiple-person crews (other than the Almaz crews, which were drawn exclusively from military pilots) included a military commander and a civilian flight engineer. In contrast to NASA, which advertised for its astronauts, the Korolev Bureau selected flight engineers from within its own ranks. By the same token, the Buran pilots formed a completely separate group. As a result, when Anatoli Levchenko (the first of the group to fly) visited Mir, the residents had to ask him who he was.

Although a number of women were selected as cosmonauts, few actually flew. The plan to send a three-woman crew to Salyut 7 just before decommissioning it was scrubbed when Svetlana Savitskaya fell pregnant (at that time, each crew had to have an experienced member, and she was the only woman available). A number of scientists and journalists trained, but never flew. As a result, the predominantly male corps of pilots and engineers had to perform work which, on Skylab, NASA assigned to scientists, a split that became even more obvious when the shuttle entered service, and six or seven astronauts flew as a matter of routine. On the other hand, the second-generation Salyuts and Mir hosted a succession of visitors. Admittedly some were essentially sight-seeing jaunts, but the majority (and nowadays all) had serious scientific objectives.

Another difference between the cosmonauts and their American counterparts is that, as spacecraft designers, the flight engineers migrate back and forth between passive and active roles, and serve as flight directors at Kaliningrad. The only direct employment of astronauts in NASA's control room is as 'Capcom', the main communications link with a crew in space.

The two training regimes are very different too. Whereas shuttle training now relies on state-of-the-art computer-based self-tuition systems, Mir crews employ a traditional lecture format with classroom exercises and formal examinations. The Russians rely on word of mouth rather than voluminous documentation, preferring to call in the designer of a particular piece of apparatus if a cosmonaut needs to know about it. In the Russian system, the knowledge is informal. Both make heavy use of simulators and underwater training, of course, so there is also much which is common to both systems. At first it was only ESA astronauts that cross-trained for Mir who experienced this culture shock, but then, following the agreement to cooperate, astronauts and cosmonauts are now training together. Ironically, whereas NASA has recruited new astronauts to work with the International Space Station, the cosmonaut corps has been cut back to a small cadre, many of whom have already served several tours of duty. Of the Buran group, who were all highly experienced test pilots, the leader, Igor Volk, was grounded for medical reasons, and several others have died or were killed in crashes (including Anatoli Levchenko, Alexander Shchukin and Rimantas Stankiavicus, all of whom were assigned familiarisation flights on Soyuz). Upon the Buran project's cancellation in the post-Soviet rationalisation, the group, which dubbed itself the 'Wolfpack' (Volk meant 'wolf' in Russian), was disbanded. Some members of the main cosmonaut corps, in contrast, have made many flights; Gennadi Strekalov has set a record of six flights, equalling the American record held by John Young.

Mir has passed through several phases of crewing strategy. The first crew could not take a guest with them because there was no returning crew to take the visitor home. In any case, they had an extremely demanding mission involving both commissioning Mir and mothballing Salyut 7, so Soyuz-T 15 needed all the propellant it could carry. If the planned handover had taken place, it might well have been accompanied by the Syrian cosmonaut who had originally been meant to visit Salyut 7. In the event, even though it had been announced when Mir was launched that it was to be *permanently* inhabited, once it became clear that the first of the science modules could not be launched on time, it was decided to vacate the station. Consequently, the second crew could not take up a visitor either. However, in addition to achieving two handovers, the following five flights each carried a guest cosmonaut. During this time, after the first crew had commissioned the Kvant 1 module, the emphasis switched to extending the endurance record to a year in space. Thereafter, the complex was left vacant once more until the next module was readied for launch. No visits were hosted during Soyuz-TM 8–10, while the front of the complex was expanded with the Kvant 2 and Kristall modules, so these crews had long and lonely tours. With two exceptions (Soyuz-TM 16 and Soyuz-TM 19), a visitor has since accompanied every flight. It is also noteworthy that this latter series of visits was by fee-paying researchers rather than by invited guests. During this time, the endurance record was decoupled from routine operations (Poliakov flew with successive residents who flew regular tours). Mid-tour visits were abandoned half-way through this series, so visitors had to accompany crew exchanges. By way of compensation, the handover period was lengthened to allow visitors up to a month aboard, rather than a week. Only recently have fee-paying researchers been granted extended missions. NASA's Norman Thagard was closely followed by ESA's 'long' mission with Thomas Reiter, which, in effect, was extended into a full tour. Until the continuous two-year accommodation slot bought by NASA for a succession of astronauts delivered and retrieved by the space shuttle expires, other visitors are restricted to handover periods. The RSA's decision to exploit the service life of the Soyuz-TM ferry, by making tours last six months, reduced opportunities for these visits to two per year. Mir crewing, therefore, is now a dynamic process involving a multitude of organisations from different countries.

WEIGHTLESSNESS

The basic objective of the space station programme was to discover whether the human body could survive long-term exposure to weightlessness, and the subsequent return to Earth. It is difficult now to appreciate the uncertainty which faced the first Salyut crew, and the shock of discovering them dead in their couches upon returning to Earth. It was clear that they had survived three weeks of weightlessness without any debilitating effects, but it was far from certain that, if the technical malfunction had not claimed their lives, they would have survived the return to Earth. Ironically, much longer flights later revealed that a three-week flight was just about the worst duration to choose, because no sooner had the body adapted to the space environment than it was forced to readapt to gravity, and do so from a state of relative weakness.

In general it takes a few days for the vestibular disorientation to abate, and about a week for the pooling of blood in the upper torso and puffiness in the face to wear off, during which time the body increases urination to shed what it takes to be excess body fluid. For the first week, cosmonauts spend an hour a day in a lower-body negative-pressure unit, to assist the heart in coming to terms with the fact that it is no longer pumping blood against the tug of gravity, and take plenty of water to prevent dehydration. As the body adjusts, the capacity of the cardiovascular system reduces, and the heart migrates further up into the chest cavity. Physical exertion is avoided during this period of physical adaptation, but thereafter, to prevent the muscles from atrophying, a rigorous daily exercise programme is followed. By the end of the month, the body has settled into a state compatible with the weightless environment.

With cosmonauts making repeated flights, it turned out that adaptation was easier the second time around; it was as if the body remembered what to do, and adapted faster. A short flight for familiarisation prior to embarking on a long mission seemed to have real benefit. After the rule was introduced that at least one member of each crew had to have flown in space before, it was found that an experienced cosmonaut could teach a rookie how to adapt to weightlessness, thereby speeding up the process and making for more productive missions.

As mission duration increased beyond about four months, it was found that there is an initial period during which every crew is highly motivated and extremely productive. In some cases there was then a lull during which motivation faltered until the flight neared its end, at which point the crew had a 'second wind' and renewed its effort. Of course, not every crew suffered this relatively unproductive phase, and often only one member of a crew did, whilst the other worked extra hard to make up. This 'gap' depended on the flight duration. For example, whilst Yuri Romanenko was very productive throughout his three months on Salyut 6, he became increasingly depressed during the latter half of an 11-month stay aboard Mir. Vladimir Titov and Musa Manarov, on the other hand, who followed him, spent a year aboard without ill effects. Once in-orbit handover from one crew to the next was routine, the task of extending the endurance record was assigned to medical specialists, and routine crewing became a succession of relatively short 'tours of duty'. Romanenko had recommended four months, but it was later reasoned that it was best to exploit the in-orbit service life of the Soyuz ferry, so six-month tours became the norm.

One factor which greatly affected a crew's long-term performance was the length of their day. Although the familiar idea of a 'solar day' means very little in the context of a 92-minute orbit, the human body is linked by dint of its evolution on the planet to such a daily rhythm. However, the precession of their orbit imposed by the Earth's equatorial bulge meant that crews of the early Salyuts tried to follow a 'day' which made optimal use of the time that they were over the tracking network, and so were in communication with the control centre. Unfortunately, week after week of rising half an hour earlier each day proved debilitating, and the cosmonauts became progressively more exhausted. The switch to a standard 24-hour day, synchronised with the control centre in Moscow, greatly eased the situation; this is broken only for events dictated by orbital mechanics, such as a spacecraft docking, and each 'late night' is followed by a 'lie-in' to catch up on lost sleep. In fact, it has been found that whereas newly-arrived resident crews sleep a full eight hours, their requirement gradually diminishes to about five hours. Crews on brief visits inevitably skip sleep in order to concentrate on their work, with the result that they burn out after a week. Resident crews, on the other hand, stick to a normal five-day week as much as possible. They can work on through their evenings and at weekends should they choose, but books and videos are available for relaxation. (After so many years of habitation by so many crews with different tastes, Mir has a substantial library of each; John Blaha, for example, took a set of *Star Trek* tapes with him).

In the days when the in-orbit life of the Soyuz was only three months, a ferry had to be replaced several times during a record-breaking flight, and this operational requirement allowed the Intercosmos Organisation to invite its members to send researchers on brief visits to Salyut 6. As a result, each short-stay crew expanded the database on the initial phase of the process of adaptation to weightlessness. In addition, a whole range of tests was devised to take data (both subjective and objective) on every aspect of space flight. These studies were further developed by the fee-paying Mir visitors. The tests focused on vestibular, hormonal, chromosomal and immunological changes, the capacity of the cardiovascular system, the heart's rhythm, structure and migration within the chest, the composition and distribution of body fluid, the capacity of the respiratory system, and bone loss and muscle degradation. Other topics included posture, skin sensitivity, sources of physical irritation, changes to the senses of hearing, taste and visual acuity, aspects of brain activity and cognitive function, and a wide range of psychological tests to monitor self-assessment of working efficiency and the crew's relationships amongst themselves and with the controllers. (Behavioural psychologists also used the audio/video downlink to independently assess the state of mind of each member of a crew.)

The cosmonauts were, therefore, 'laboratory rats' as well as pioneering engineers and research technicians, and they endured a great deal in the interests of biomedical science. For example, the procedure for sampling bone marrow taken pre/post-flight from some visitors was unpleasant. At a more mundane level, Norman Thagard noted that the need to log his food intake proved to be a disincentive to eating. Although cosmonauts accept the exercise regime (and on Earth regard it as a tonic) it is not a very pleasant experience in space because, in the absence of gravity, sweat clings to the body by surface tension and pools in deep puddles in the body's concavities, particularly over the sternum in the chest. Furthermore, in the cabin's artificial environment there is no breeze to cool the body and dry it off. An all-over body rub afterwards with a towel is hardly adequate; *ergo* the re-peated efforts over the years to devise an effective shower.

In addition to the strict daily exercise regime during the flight to keep the muscles in trim, certain extra measures are taken in the final few weeks of a flight. A drug is taken to stimulate the body and to develop stamina. The lower-body negative-pressure garment is used to draw blood into the legs, and thereby increase the capacity of the cardiovascular system and impose an increased load on the heart, and, as the body accumulates fluids in response, inordinate quantities of saline solution are drunk to overcome dehydration. A cocktail of vitamins and mineral supplements to balance the electrolytes rounds off the pre-return procedure.

Contrary to initial concerns, the 3-4 g load imposed by re-entry does not represent a serious threat even after a prolonged period of weightlessness. The fact that the crew of a Soyuz descent module takes this load flat against their backs, with their feet above the torso, helps greatly (similar couches are employed on Atlantis for those returning from Mir by that means). Although most cosmonauts have to be assisted out of their capsule, after a few minutes resting in the recliners nearby they are able to stagger to the medical tent.

After a six-month tour it generally takes a couple of days for the vestibular system to recover, but once the sense of balance and coordination is restored it is possible to walk, albeit with a distinctive gait, for as long as stamina lasts. During the first week or so, it is standard procedure to apply positive pressure to the lower body (by wearing a pair of pilot's g-suit leggings) to prevent blood draining into the legs at the expense of the head, and the resulting blackout from orthostatic intolerance, and to ease the adaptation of the heart to the increased load of pumping against gravity. Usually, despite exercises on the treadmill and ergonometer, the leg muscles are weak. In contrast, the arm muscles tend to be strong due to moving cargo about. Although most cosmonauts tend to lose weight in space, a few have actually gained several kilogrammes. One effective way of easing the readaptation to Earth is to swim, because, just as it simulates training for spacewalking, immersion masks the force of gravity. In fact, it is not uncommon for a newly-returned cosmonaut to try to float out of bed upon awakening, and to set objects in mid-air in the expectation that they will remain there. These minor irritations apart, even the weakest individual was fully recovered after a few weeks. After his first long mission, Georgi Grechko complained of chest pains for several days, but was fine the second time; it seemed that the first time it took several days for his internal organs to 'drop' back into place. Valeri Ryumin recovered much faster from his second six-month tour, which he volunteered for within months of returning from his first. The vital point though, is that the recovery time has been *decoupled* from the duration of the mission. It is clear that a six-month tour is medically and psychologically practical. Furthermore, the fact that none of the cosmonauts to have served such tours has yet shown any permanent side-effects indicates that exposure to weightlessness does not trigger any runaway life-threatening biological processes.

Given this certainty, it might legitimately be asked why NASA is devoting so much effort to biomedical research on its shorter periods aboard Mir. Put simply, it hopes to follow the adaptation process on a daily basis, in order to study it in detail. Dr Poliakov had made a significant start on this during his 14-month marathon, but NASA intends to employ more sophisticated apparatus to measure bone density, muscle mass and blood chemistry throughout, and to dovetail this into pre/post-flight studies.

The worst effect of exposure to prolonged weightlessness is bone demineralisation; this is because it seems not to reach a plateau. The calcium leaches into the bloodstream and accumulates in the renal glands, so kidney stones could ultimately be an issue. The effect on the skeleton is similar to osteoporosis, but is much more pronounced. On Earth a sufferer typically loses bone mass at 2 per cent per annum; in space the loss rate is an order of magnitude greater. In fact, the effect is not uniform; large bones such as the femur tend to lose most. Norman Thagard developed what cosmonauts refer to as 'chicken legs' after losing nearly 10 per cent of the mass of his femur in five months. Bone loss is not a simple correlation with duration. One cosmonaut lost 8 per cent after six months, which was the same as was lost by the final Skylab crew in three months. To complicate the situation further, whereas one cosmonaut lost fully 20 per cent in five months, another lost 15 per cent in seven months. The effect also depends upon the initial state of the bone. Shannon Lucid, for example, had relatively dense bones so, even after six months she was within acceptable bounds. The effect *is* reversible, but the recovery rate is slow. One side-effect of this investigation is likely to be the increased understanding that could eventually benefit Earthly osteoporosis sufferers.

THE FRUSTRATIONS

One of the frustrations of living in a weightless environment was that unless an item was firmly strapped down it floated freely. Elasticated straps were run across every flat surface to grip small items in place temporarily; unused items were wrapped in nets and bundled in spare nooks. Although the fan-driven air flow tended to carry loose items to the grilles, small things were often lost. Yuri Romanenko lost his watch, and Jean-Loup Chrétien lost a roll of exposed film. Thomas Reiter could not find his centrifuge blood separator for two months! One night Anatoli Berezovoi and Valentin Lebedev stayed up late looking for the locking mechanism for the docking drogue, and after several hours they found it behind the cone, just out of sight. This illustrates the folly of trying to impose a rigid schedule on activities in space.

Improvements to the telemetry system meant that the doctors could monitor the data from biomedical sensors during spacewalks. This was a two-edged sword, however. It was sufficient to reveal that Alexander Laveikin suffered a heart irregularity while under stress, but was inadequate to form a diagnosis, so he was recalled. Follow-up tests revealed that his condition was not life-threatening, but he never flew again. No doubt he wishes the suit monitors had not been fitted.

Although the cosmonauts put up with the regular biomedical tests, they found them frustrating because it took a long time to unpack the sensors, pastes, cables, cuffs and belts and plug in and verify the monitoring units. Even then, they had to wait until they were within radio range to be able to transmit data. Furthermore, unless the tests could be completed within that pass, they had to break off and wait for the next orbit. Finally, the apparatus had to be disconnected and packed away again. It was a major item on the schedule.

Whilst everybody involved recognised the need for teamwork, there was occasional friction between the cosmonauts and the flight controllers. At times the consequences of different motivating factors surfaced. Cosmonauts became particularly irritated at being hustled to stick to the exercise schedule at the expense of an experiment then underway. On one occasion, having set up a camera, they were told to break off, even though this would have meant not being able to take the intended pictures.

On several occasions, cosmonauts urged their colleagues on the ground not to feel obligated to chatter simply because the radio link was established. All of this interaction was valuable data for the psychologists, of course. Recently, while one crew packed to return home, a cosmonaut observed that he would get on faster without being reminded of what needed to be done. There are times, however, when the crew is not disturbed unnecessarily, such as during their ablutions and at meal times. Overall, the fact that the radio link was not continuous was probably welcomed by most crews. Though Shannon Lucid acknowledged the need for teamwork, she also relished the self-reliance being out of contact for the long periods imposed. Often, the only sound on the downlink is laughter, as the crew enjoy their own company. The two-way video link proved crucial on long flights, because it enabled crews to both see and hear their families.

The arrival of a resupply ferry was always a highlight because of the fresh food that it brought. No matter how nourishing the prepackaged food was, the cosmonauts soon grew bored with it (in part, this was due to the change in the sense of taste in space). The switch from complete-meal packs to shared variety packs from which individuals could pick-and-mix was a welcome development. Much of the food was pretty normal; it just had to resist flaking (items such as bread and cake were cut into bite-sized chunks and moistened so as not to crumble). Nevertheless, everyone craved fresh food. It was not unheard of for cosmonauts to eat the onion bulbs sent for planting in a cultivator for an experiment. To make amends, one cosmonaut took a big plastic cucumber with him and claimed that it had grown whilst the station had been vacant! It turned out that watching a plant grow in space was therapeutic, even for cosmonauts with no previous interest in horticulture. Simply gawking out of the porthole at the Earth rolling by beneath was always a popular pastime, and it was not unheard of for a cosmonaut to find a companion *asleep* at the window. The view out of the porthole is surpassed only by the view out of a suit visor during a spacewalk. On the other hand, hand-pumping the waste out of the toilet must surely be one of the least appealing daily assignments.

Imagine the frustration of riding a rocket into orbit in the expectation of spending several months aboard a space station, and then having to return to Earth 48 hours later, because of a failure to dock! Even successfully boarding a station was no guarantee of a smooth flight. Laveikin suffered the disappointment of being recalled, and Valeri Poliakov had his first attempt to set a record frustrated when Mir was temporarily vacated. On the other hand, Sergei Krikalev was obliged to stay on to serve a double tour. It must have been frustrating to train specifically to commission a module, only to have it not appear;

to have to cancel a spacewalk because the crane handle worked loose and drifted away; to order a piece of apparatus and, upon unpacking it, discover that it was not what was requested; to find that the food packs had been rifled by hungry members of the launch team; or worst of all, to collect data day after day, and not know if anything worthwhile was being achieved.

Despite the danger and the frustrations, however, there is no shortage of volunteers to fly. Indeed, quite the reverse. Now that six-month tours on Mir are the norm, and mid-tour visits have been terminated, there are precious few flight opportunities.

THE WORK

With so many automated satellites in use, it is fair to ask the value of maintaining a human presence in low orbit. What could cosmonauts see that the sophisticated sensors could not?

Although vision was initially degraded upon entry to weightlessness, this recovered and, eventually, cosmonauts found that they could distinguish extremely fine variations of colour which, to their disappointment upon seeing the results of their picture-taking, standard panchromatic film was unable to capture. Thus, visual observations of oceanic currents, thin oil slicks above shoaling fish, and blooming plankton proved particularly valuable. It turned out that the recovery of visual acuity was hastened if an experienced cosmonaut was able to offer tuition, to draw the newcomer's attention to the fine detail.

In addition, the cosmonauts were rather amazed to find that, under suitable lighting, they could see the details of shallow sea floors, and seamounts in open ocean. It turned out that sun-glint was very useful, because it enabled inundation to be studied, even in heavily vegetated areas. Even though there were meteorological satellites far above, the cosmonauts were often asked to make visual observations of hurricanes. They assessed glaciation and snow coverage, melt-water run-off and river capacity. Cosmonauts on long missions made particular note of seasonal variations. It turned out that topographic detail was best studied in the autumn, after the vegetation thinned out and before snow obscured the terrain, so the search for faults likely to contain valuable natural resources was best done in the autumn. Cosmonauts on later tours noted that prospectors had set up at sites identified by their predecessors. Some communicated directly with fishing fleets to steer the skippers onto the most promising-looking shoals. The cosmonauts were often the first to report forest fires. More can be seen from low orbit than might be imagined.

It is not generally appreciated that from low orbit the Earth does not simply look like a map; topographic relief is discernible. In fact, the perspective from Mir is comparable to viewing vertical structure within a 4-cm band from a height of two metres. Because the rate of change of perspective is faster (due to the speed of orbital flight) the eye is easily drawn to the vertical detail, and since the view is much wider the overall impression is actually more striking than from an aircraft. Flying across a tropical storm for example, the depth of the hollow eye is pronounced. Similarly, not only is the vertical segment of a volcanic plume noticeable as it rises to the stratosphere, if it is caught by the jet-stream it can be followed downwind for thousands of kilometres. And even from an altitude of 400 km, the volcano itself can be seen poking up from the surrounding terrain. All this sense of depth is completely missing in photographs, however. The binocular vision of the human eye is surpassed only by a stereoscopic camera.

Crew visual observations were of course augmented by apparatus designed to make specialised studies. Multispectral cameras made possible comprehensive surveys of the mineral resources of the unexplored wilderness of Siberia, and assessment of agriculture in the Ukraine, the Soviet Union's 'bread basket'. And of course topographic mapping was performed. Undoubtedly, this overhead imagery from the later Salyuts contributed to the national economy by selecting routes for hydrological and transportation projects, but later attempts to sell high-resolution output from the Almaz radar-imaging satellites on the world market did not prove as lucrative a business as hoped.

Atmospheric studies using spectrometers made feasible the first global survey of the processes involved in heat transfer to space, and water vapour and aerosol distribution in the upper atmosphere. The data revealed a steady-state glow above the equator that had not been suspected. Bright transient flashes high above thunderstorms were also noted. These are believed to be a form of lightning discharge to the ionosphere, rather than to the surface. Measurements of charged particles at orbital altitude revealed a correlation with tectonic activity. It now seems that rock under tectonic stress generates a localised intensification of the Earth's magnetic field. From the vantage point of a high-inclination orbit, cosmonauts could peer obliquely down on aurorae, follow the glowing bands for thousands of kilometres across the surface, and correlate activities at each pole. Another upper-atmospheric phenomenon which crews were able to make unique studies of was noctilucent clouds. These turned out to be more common, and more extensive, than had been believed.

Another beneficial effect of having cosmonauts present was the testing of apparatus intended for later installation on the automated satellites. They were able to troubleshoot problems with prototype equipment ranging from furnaces to plant cultivators, and to work around design flaws.

The materials-processing research covered organic as well as inorganic materials. A wide variety of alloys of substances which are immiscible on Earth were made and their crystalline products returned for study, as were many different types of semiconductor. Despite the hope that installing a bank of improved furnaces would transform Mir into a commercial 'factory', it was difficult to show economic viability because the output did not feed an enterprise economy. The biological materials-processing, on the other hand, was of immediate benefit to the pharmaceutical industry. Electrophoresis proved to be a highly effective way of separating out active biological substances. Polycrylamide gel could be used to refine further products on Earth, and this was manufactured in space, as were interferon and a variety of vaccines. Automated versions of both the furnaces and the electrophoresis units were subsequently installed on automated satellites. American experiments have grown crystals of proteins and 3-D tissue cultures for later study. The advantage of doing this aboard Mir was that the crystal could grow for months rather than a fortnight, which is the longest that a space shuttle can remain in orbit.

Mir's capability to support biomedical, Earth-observational and materials-processing research has been demonstrated many times, a capability that expanded with each new module and each visiting crew, as apparatus accumulated. Later visitors could rerun the experiments to broaden the database. Although storing it all became an issue, having a range of apparatus in-house also meant that later visitors could adapt experiments to test for related factors in order to expand on earlier work. The residents too, could continue experiments set up by their visitors, and could fill in gaps in Earth-studies coverage lost due to adverse weather. A large part of the value of Mir, therefore, was that its research was ongoing and long-term.

It was a long struggle, but cultivators were perfected which enabled higher plants to grow in space. Arabidopsis (a weed) was first to complete its cycle and yield seed, but wheat recently became the first *staple* to be 'harvested' in space. Experiments revealed that chlorella thrived in space, and so could be used in a closed-cycle aquarium. Both of these results are encouraging for food production on future stations, and in the long run for O'Neill's colonies. Another key element of this ultimate habitat has yet to be studied though; fish, frogs and quail chicks have all been successfully hatched in space, but, as yet, as far as is known, no human being has yet been conceived in space. If procreation proves impossible, then it will kill stone dead all dreams of claiming the Solar System as the human race's own private back yard.

COLONISING SPACE?

Space is an inherently radiation-intensive environment. Cosmonauts on long flights build up significant doses. The exposure from living in low orbit (within the protection afforded by the Earth's magnetic field) is far less than it would be in deep space, but the biomedical effect of cosmic rays is yet to be seen. Once the transportation infrastructure is available, it should indeed be feasible to build the colonies in high orbit envisaged by O'Neill. Since these will be spun to create *artificial gravity*, surviving short periods in weightlessness will not be an issue, rather it will be long-term exposure to radiation that will pose the biomedical threat.

THE BURAN SHUTTLE

No discussion of the Mir station can ignore the Buran space shuttle. On the other hand, not *too* much should be made of it because, in the final analysis, it proved to be unnecessary. The VKK (the acronym for atmospheric spacecraft) was ordered by the military in the late 1970s to counter the American space shuttle, the development of which was backed by the Department of Defense. As such, it was developed independently of the space station programme.

Surprisingly, considering the secrecy surrounding its development, Buran's maiden flight in 1988 was shown live on television. Even more remarkably, it flew completely automatically to evaluate the avionics. It could not carry a crew until the development of the environmental systems was completed but, thereafter, it was to be introduced into service.

The prospect of operating Buran in conjunction with Mir prompted the installation of the androgynous ports on the Kristall module. It is known that it was to deliver an X-ray telescope for attachment to the side port. A module resembling Kvant 1 was once displayed in a mock-up of its payload bay, so it is possible that it would have been used to deliver such modules to further extend the front of the complex once the cluster there was complete. It is also possible that the need to have the androgynous port in place for the shuttle was a contributory factor in bringing Kristall forward. Its construction duties apart, Buran prompted the development of the YMK cosmonaut manoeuvring backpack sent up in Kvant 2 (for trials prior to being used by cosmonauts working in Buran's payload bay).

Although conceived independently, the two programmes were clearly seen as being mutually supportive. Certainly, apart from being used by Soyuz-TM 16, and then by the American shuttle for its visits, Kristall's androgynous ports have been surplus to requirements.

Several mission profiles were considered for testing Buran. One involved launching it without a crew so that it could automatically rendezvous with Mir. After a fly-around to enable the residents to inspect its thermal tiles, Buran would dock with Kristall. Two pilots on the station would then transfer to Buran, undock, put it through its paces, and then, depending on how things had gone, either redock or return directly to Earth.

In preparation, two members of the Buran pilot group flew on Soyuz in 1984 and 1987, and another flight was planned for 1989. This was to provide experience of weightlessness and to assess piloting skills immediately following return. As events transpired, these were the *only* members of the group to fly in space. Buran was cancelled, for financial reasons, by post-Soviet Russia.

With hindsight, it is clear that Buran consumed enormous resources which could probably have been better spent. It left an important legacy, however, in the form of the Energiya rocket.

THE ENERGIYA ROCKET

The cancellation of both Buran and the Freedom-style Mir 2 complex led inevitably to the termination of the Energiya rocket production line. Since the Proton would be able to launch the Mir-style components of the scaled-down Mir 2, there simply did not seem to be any 100-tonne payloads requiring this massive rocket.

Without a shadow of doubt, in retrospect, Energiya's cancellation will be seen to be a lost opportunity as great as America's termination of the Saturn V rocket two decades earlier. The absence of *both* these heavy-lift launchers limited the International Space Station to offering full-time occupancy to only a few people. This is because the configuration of its modules is constrained by the delivery vehicle; that is, either by the lifting capacity of the Proton rocket and the dimensions of its aerodynamic shroud, or by the lifting capacity of the American shuttle and the dimensions of its payload bay.

If *either* of these superboosters had been in service, the International Space Station would probably have been designed as a cluster of 100-tonne Skylab-sized modules, and could have accommodated several dozen people. The irony is that far from denying the shuttle its *raison d'etre* in *assembling* a small station, if a heavy-lift rocket had been retained, the shuttle would have been ideally suited to servicing a *large* station, providing logistical support and carrying crew a dozen at a time in a passenger module in the payload bay. Now that would have been something!

THE BOTTOM LINE

Once a space-based system becomes operational, it is taken for granted. This is particularly so with communications, meteorological and remote-sensing satellites. As soon as consumers forget what it was like *not* to have the benefit of such systems, it becomes increasingly difficult to compute the 'bottom line' of the cost-benefit analysis. This is because the benefit of a space-based system was originally computed in terms of the cost of the best alternative, or, indeed, the cost of not having any viable alternative.

The advent of communications satellites revolutionised the world. Despite their high development cost, communications satellites are, over time, more cost-effective than the best alternative which, in the early-1960s, comprised a grid of line-of-sight transmitters and a few sea-floor cables, all of which represented a tremendous investment in familiar technology. Just as the piston-engined airliners were rendered all but obsolete by the introduction of the Boeing 707 jet-engined stratocruiser, the Atlantic cable was rendered obsolete by Telstar and the Relay satellites that followed it. The television industry is now *completely* dependent on satellite links, the business world would grind to a halt without cheap communication, and the world-wide-web of the internet users would be impossible without instant linkage. The 'bottom line' for communications satellites is a big plus.

The case for meteorological satellites is similarly certain. In this case, though, it is necessary to recall that until the introduction of the Tiros cloud-coverage satellites in the 1960s, severe weather often struck without warning. Nowadays, it is not unusual for a local television station to offer a weather forecast that covers an entire continent. There is now almost always sufficient warning to prepare for the arrival of severe weather. In fact, there is invariably condemnation if the weather manages to spring a nasty surprise on the forecasters. A major item in the benefit column of weather satellites, as espoused immediately following their introduction, when the cost of not having their forewarning was still fresh in popular memory, was the real *saving* resulting from knowledge of an approaching storm. So, just as the world cannot now operate without communications satellites, it is, in a very real sense, reliant on meteorological satellites.

A similarly solid case can be made for the Earth-resources remote-sensing satellites, whose multispectral imagery can be processed to reveal a wide variety of distinctions in incredible detail, across wide areas. This technology was developed in the early 1970s. NASA chose to deploy it on automated satellites in Sun-synchronous high polar orbits. The Soviet Union assigned the testing to cosmonauts, and then deployed it operationally on its later Salyut stations. For the Soviet Union, therefore, the assessment of the value of remote-sensing technology was inextricably linked to the cost of operating its wider space programme.

In the late 1970s the cosmonauts took detailed pictures of the proposed route of the 3,500-km long railway which was then being laid in Eastern Siberia, from Lake Baikal to the River Amur on the Manchurian border. This imagery revealed many unsuspected hydrological and geological features. These had to be taken into account in planning the route and constructing the necessary bridges and tunnels. In the case of just one tunnel, "many millions of roubles" could be saved by forewarning of unsuspected hazards. All of these savings contributed to the cost of operating the Salyut space station. In 1980, *Tass* reported Vladimir Shatalov as saying that the space programme was "an organic part" of the economy, and that "every rouble invested in space exploration ensures a ten-fold return". In 1983, *Tass* reported that pictures from Salyut 7, during its first year of operation, had led to a saving of 30 million roubles in topographic mapping, and 100 million roubles in the cost of comprehensive prospecting for oil and gas resources. Clearly, the Soviets regarded their Salyut stations as a vital component of their national economy.

It is a little more difficult to quantify the benefit of performing a task in space that is either very difficult or impossible to do on Earth, such as physical processes that exploit microgravity. Materials-processing has long been a core element of the work performed on Soviet space stations. The processing capacity of the furnaces has improved over the years, and a wide variety of crystals for semiconductors and exotic alloys – many totally impractical in a gravity field – were returned. But what was their value? It was calculated that the Soyuz-TM 8 mission to Mir was the first to break even, in terms of balancing its operating costs against the value of its output. Unfortunately, the financial value of the semiconductor returned was largely hypothetical, because the material did not feed into a commercial market. This, therefore, presented a less tangible benefit than savings made from remote sensing. Even so, it was hoped that Mir would eventually become self-financing, and there were great expectations that the Kristall module would transform the complex into a materials factory. But what is the financial value of learning how the human body adapts to the lack of gravity? Pure research is notoriously difficult to subject to reliable cost-benefit analysis.

Despite the early financial arguments in favour of the space station, in the economic collapse that followed the demise of the Soviet Union the programme was re-evaluated in a harsher light, and funding was drastically reduced. Fortunately, just a few years earlier it had been decided to start charging foreign researchers to cover the costs of their visits to Mir. This had two timely beneficial effects. Firstly, since the 'fee' was dependent on contributions made in kind, it resulted in the donation of a great deal of high-technology apparatus. Secondly, as escalating inflation eroded the value of the rouble, it provided a crucial source of hard currency; so much so, in fact, that in 1995, as NASA became involved in the programme, RSA chief Yuri Koptev had to admit that continued operation of Mir would not have been possible without the 350 *billion* roubles from international partners. Even the trivial act of filming an oversized cola can against an Earth backdrop brought in a million dollars. Just as the fragmented Soviet Union started to consider the possibility of abandoning the Mir complex as an unaffordable luxury, its significance as the only orbital facility available capable of sustaining human life was finally recognised by the international community.

There is, however, a curious irony. When NASA was denied the justification for its own space station as a way-station for its return to the Moon, then on to Mars, it argued that the microgravity environment would attract commercial investment, but commercial interest was muted. Investment was secured for Spacehab specifically to offer low-cost testing of equipment intended for later deployment to the space station, but the funding for Spacehab was attracted only after NASA had placed advance orders to guarantee the module's viability. But, of course, this is to be government-funded work, so Spacehab did not really demonstrate American commercial interest in the use of space, only in the infrastructure serving space operations with the government as the customer. It seems, therefore, that in the *absence* of a separate market sector, the integrated economy of the Soviet Union was able to take a broader view; a luxury which it has now been forced to abandon in its effort to become a Western-style free market. To the Soviets, investment in space stations was *clearly* justifiable in terms of the national economy. This is a rich irony indeed.

It is a quarter of a century since Salyut 1 was launched. It was used for less than a month. The Mir complex has been a decade in the making, and it has been inhabited for most of that time. Assembly of the International Space Station is about to start. Will this be a success? If the ISS is still in use in 25 years time, then it *must* be deemed worth the investment. In all likelihood, it will have been greatly expanded, and *nobody* will be able to figure out how the world ever functioned without it.

Table 15.5. Soyuz launches and recoveries

Spacecraft	Launched	MT	Recovered	MT	Days
Soyuz 1	23 Apr 1967	0335	24 Apr 1967	0623	1.12
Soyuz 2	25 Oct 1968	1200	28 Oct 1868	1100	2.96
Soyuz 3	26 Oct 1968	1134	30 Oct 1968	1025	3.95
Soyuz 4	14 Jan 1969	1030	17 Jan 1969	0951	2.97
Soyuz 5	15 Jan 1969	1004	18 Jan 1969	1058	3.04
Soyuz 6	11 Oct 1969	1410	16 Oct 1969	1253	4.95
Soyuz 7	12 Oct 1969	1345	17 Oct 1969	1225	4.95
Soyuz 8	13 Oct 1969	1320	18 Oct 1969	1211	4.95
Soyuz 9	1 Jun 1970	2200	19 Jun 1970	1450	17.71
Soyuz 10	23 Apr 1971	0254	25 Apr 1971	0240	1.99
Soyuz 11	6 Jun 1971	0755	30 Jun 1971	0216	23.76
Soyuz 12	27 Sep 1973	1518	29 Sep 1971	1434	1.97
Soyuz 13	18 Dec 1973	1455	26 Dec 1983	1150	7.87
Soyuz 14	3 Jul 1974	2151	19 Jul 1974	1521	15.73
Soyuz 15	26 Aug 1974	2258	28 Aug 1974	2310	2.00
Soyuz 16	2 Dec 1974	1240	8 Dec 1974	1104	5.93
Soyuz 17	11 Jan 1975	0043	9 Feb 1975	1403	29.55
Soyuz 18	24 May 1975	1758	26 Jul 1975	1718	62.97
Soyuz 19	15 Jul 1975	1520	21 Jul 1975	1351	5.94
Soyuz 20	17 Nov 1975	1737	16 Feb 1976	0542	90.50
Soyuz 21	6 Jul 1976	1509	24 Aug 1976	2133	49.27
Soyuz 22	15 Sep 1976	1248	23 Sep 1976	2242	8.41
Soyuz 23	14 Oct 1976	2040	16 Oct 1976	2046	2.00
Soyuz 24	7 Feb 1977	1912	25 Feb 1977	1236	17.73
Soyuz 25	9 Oct 1977	0540	11 Oct 1977	0626	2.03
Soyuz 26	10 Dec 1977	0419	16 Jan 1978	1422	37.42
Soyuz 27	10 Jan 1978	1526	16 Mar 1978	1419	64.95
Soyuz 28	2 Mar 1978	1828	10 Mar 1978	1645	7.93
Soyuz 29	15 Jun 1978	2317	3 Sep 1978	1440	79.64
Soyuz 30	27 Jun 1978	1827	5 Jul 1978	1631	7.92
Soyuz 31	26 Aug 1978	1751	2 Nov 1978	1405	67.84
Soyuz 32	25 Feb 1979	1454	13 Jun 1979	1918	108.18

Table 15.5. (continued)

Spacecraft	Launched	MT	Recovered	MT	Days
Soyuz 33	10 Apr 1979	2034	12 Apr 1979	1935	1.96
Soyuz 34	6 Jun 1979	2113	19 Aug 1979	1530	73.76
Soyuz-T 1	16 Dec 1979	1530	26 Mar 1980	0050	100.38
Soyuz 35	9 Apr 1980	1638	3 Jun 1980	1807	55.06
Soyuz 36	26 May 1980	2121	31 Jul 1980	1815	65.87
Soyuz-T 2	5 Jun 1980	1719	9 Jun 1980	1541	3.93
Soyuz 37	23 Jul 1980	2133	11 Oct 1980	1250	79.63
Soyuz 38	18 Sep 1980	2211	26 Sep 1980	1854	7.86
Soyuz-T 3	27 Nov 1980	1718	10 Dec 1980	1226	12.79
Soyuz-T 4	12 Mar 1981	2200	26 May 1981	1638	74.77
Soyuz 39	22 Mar 1981	1759	30 Mar 1981	1442	7.86
Soyuz 40	14 May 1981	2017	22 May 1981	1658	7.86
Soyuz-T 5	13 May 1982	1358	27 Aug 1982	1904	106.21
Soyuz-T 6	24 Jun 1982	2029	2 Jul 1982	1821	7.91
Soyuz-T 7	19 Aug 1982	2112	10 Dec 1982	2203	113.03
Soyuz-T 8	20 Apr 1983	1711	22 Apr 1983	1729	2.01
Soyuz-T 9	27 Jun 1983	1312	23 Nov 1983	2258	149.41
Soyuz-T 10	8 Feb 1984	1507	11 Apr 1984	1750	63.11
Soyuz-T 11	3 Apr 1984	1709	2 Oct 1984	1357	181.86
Soyuz-T 12	17 Jul 1984	2141	29 Jul 1984	1655	11.80
Soyuz-T 13	6 Jun 1985	1040	26 Sep 1985	1352	112.13
Soyuz-T 14	17 Sep 1985	1639	21 Nov 1985	1331	64.87
Soyuz-T 15	13 Mar 1986	1533	16 Jul 1986	1634	125.04
Soyuz-TM 1	21 May 1986	1222	30 May 1986	1049	8.93
Soyuz-TM 2	6 Feb 1987	0038	30 Jul 1987	0404	174.15
Soyuz-TM 3	22 Jul 1987	0559	29 Dec 1987	1216	160.26
Soyuz-TM 4	21 Dec 1987	1418	17 Jun 1988	1413	179.00
Soyuz-TM 5	7 Jun 1988	1545	7 Sep 1988	0450	91.54
Soyuz-TM 6	29 Aug 1988	0823	21 Dec 1988	1257	114.16
Soyuz-TM 7	26 Nov 1988	1850	27 Apr 1989	0629	151.48
Soyuz-TM 8	6 Sep 1989	0138	19 Feb 1990	0736	166.25
Soyuz-TM 9	11 Feb 1990	0916	9 Aug 1990	1135	179.10
Soyuz-TM 10	1 Aug 1990	1332	10 Dec 1990	0908	130.83
Soyuz-TM 11	2 Dec 1990	1113	26 May 1991	1404	175.12

Table 15.5. (continued)

Spacecraft	Launched	MT	Recovered	MT	Days
Soyuz-TM 12	18 May 1991	1650	10 Oct 1991	0712	144.52
Soyuz-TM 13	2 Oct 1991	0859	25 Mar 1992	1151	175.08
Soyuz-TM 14	17 Mar 1992	1354	10 Aug 1992	0504	145.63
Soyuz-TM 15	27 Jul 1992	1009	1 Feb 1993	0648	188.86
Soyuz-TM 16	24 Jan 1993	0858	22 Jul 1993	1042	179.08
Soyuz-TM 17	1 Jul 1993	1833	14 Jan 1994	1119	196.70
Soyuz-TM 18	8 Jan 1994	1305	9 Jul 1994	1428	181.94
Soyuz-TM 19	1 Jul 1994	1625	4 Nov 1994	1420	125.92
Soyuz-TM 20	4 Oct 1994	0142	22 Mar 1995	0704	169.22
Soyuz-TM 21	14 Mar 1995	0911	11 Sep 1995	1052	181.07
Soyuz-TM 22	3 Sep 1995	1300	29 Feb 1996	1342	179.03
Soyuz-TM 23	21 Feb 1996	1534	2 Sep 1996	1141	192.84
Soyuz-TM 24	17 Aug 1996	1715	2 Mar 1997	-	196
Soyuz-TM 25	10 Feb 1997	1709	1997	-	-
Soyuz-TM 26	5 Aug 1997	1836	- - -	-	-

Table 15.6. Shuttle–Mir launches and landings

Spacecraft	Launched	GMT	Recovered	GMT	Days	Landing
STS 63 Discovery	3 Feb 1995	0522	11 Feb 1995	1151	8.27	Fla
STS-71 Atlantis	27 Jun 1995	1932	7 Jul 1995	1454	9.81	Fla
STS-74 Atlantis	12 Nov 1995	1231	20 Nov 1995	1701	8.20	Fla
STS-76 Atlantis	22 Mar 1996	0813	31 Mar 1996	1328	9.22	Cal
STS-79 Atlantis	16 Sep 1996	0855	26 Sep 1996	1213	10.14	Fla
STS-81 Atlantis	12 Jan 1997	0927	22 Jan 1997	1423	10.20	Fla
STS-84 Atlantis	15 May 1997	0807	24 May 1997	1328	10.22	Fla

Table 15.7. Progress launches

Spacecraft	Launched	MT
Progress 1	20 Jan 1978	1125
Progress 2	7 Jul 1978	1426
Progress 3	8 Aug 1978	0103
Progress 4	4 Oct 1978	0209
Progress 5	12 Mar 1979	0847

Table 15.7. (continued)

Spacecraft	Launched	MT
Progress 6	13 May 1979	0717
Progress 7	28 Jun 1979	1225
Progress 8	27 Mar 1980	2153
Progress 9	27 Apr 1980	0924
Progress 10	29 Jun 1980	0741
Progress 11	28 Sep 1980	1810
Progress 12	24 Jan 1981	1718
Progress 13	23 May 1982	0957
Progress 14	10 Jul 1982	1358
Progress 15	18 Sep 1982	0859
Progress 16	31 Oct 1982	1420
Progress 17	17 Aug 1983	1608
Progress 18	20 Oct 1983	1259
Progress 19	21 Feb 1984	0946
Progress 20	15 Apr 1984	1213
Progress 21	8 May 1984	0247
Progress 22	28 May 1984	1813
Progress 23	14 Aug 1984	1028
Progress 24	21 Jun 1985	0440
Cosmos 1669	19 Jul 1985	1705
Progress 25	19 Mar 1986	1308
Progress 26	23 Apr 1986	2340
Progress 27	16 Jan 1987	0906
Progress 28	3 Mar 1987	1414
Progress 29	21 Apr 1987	1914
Progress 30	19 May 1987	0802
Progress 31	4 Aug 1987	0044
Progress 32	24 Sep 1987	0344
Progress 33	21 Nov 1987	0247
Progress 34	21 Jan 1988	0152
Progress 35	24 Mar 1988	0005
Progress 36	13 May 1988	0330
Progress 37	19 Jul 1988	0113
Progress 38	10 Sep 1988	0334
Progress 39	25 Dec 1988	0712
Progress 40	10 Feb 1989	1154
Progress 41	16 Mar 1989	2154

Table 15.7. (continued)

Spacecraft	Launched	MT
Progress-M 1	23 Aug 1989	0710
Progress-M 2	20 Dec 1989	0631
Progress-M 3	1 Mar 1990	0211
Progress 42	6 May 1990	0044
Progress-M 4	15 Aug 1990	0801
Progress-M 5	27 Sep 1990	1437
Progress-M 6	14 Jan 1991	1750
Progress-M 7	19 Mar 1991	1606
Progress-M 8	30 May 1991	1204
Progress-M 9	21 Aug 1991	0250
Progress-M 10	17 Oct 1991	0305
Progress-M 11	25 Jan 1992	1050
Progress-M 12	20 Apr 1992	0129
Progress-M 13	30 Jun 1992	2043
Progress-M 14	16 Aug 1992	0219
Progress-M 15	27 Oct 1992	2020
Progress-M 16	21 Feb 1993	2132
Progress-M 17	31 Mar 1993	0834
Progress-M 18	22 May 1993	1042
Progress-M 19	11 Aug 1993	0223
Progress-M 20	12 Oct 1993	0035
Progress-M 21	28 Jan 1994	0512
Progress-M 22	22 Mar 1994	0754
Progress-M 23	22 May 1994	0730
Progress-M 24	25 Aug 1994	1825
Progress-M 25	11 Nov 1994	1022
Progress-M 26	15 Feb 1995	1948
Progress-M 27	9 Apr 1995	2334
Progress-M 28	20 Jul 1995	0704
Progress-M 29	8 Oct 1995	2152
Progress-M 30	18 Dec 1995	-
Progress-M 31	5 May 1996	1104
Progress-M 32	1 Aug 1996	-
Progress-M 33	20 Nov 1996	0221
Progress-M 34	6 Apr 1997	-
Progress-M 35	5 Jul 1997	-

Table 15.8. Miscellaneous Soyuz test flights

Spacecraft	Launched			MT	Recovered			MT	Days
Cosmos 496	26	Jun	1972	1753	2	Jul	1972	1724	5.98
Cosmos 573	15	Jun	1973	(0900)	17	Jun	1973	2012	2.47
Cosmos 613	30	Nov	1973	0820	29	Jan	1974	1012	60.50
Cosmos 656	27	May	1974	1030	29	May	1974	1012	1.98
Cosmos 670	6	Aug	1974	0314	9	Aug	1974	0300	2.99
Cosmos 772	29	Sep	1975	0719	1	Oct	1975	0748	2.02
Cosmos 869	29	Nov	1976	1904	17	Dec	1976	1236	17.73
Cosmos 1001	4	Apr	1978	1805	15	Apr	1978	1500	10.87
Cosmos 1074	31	Jan	1979	1207	1	Apr	1979	1500	60.12

Glossary

Aelita-1   A multipurpose biomedical kit on *Salyut 7* (to supersede the *Polynom-2M*). Specifically designed for use in space, it could measure a wide range of parameters, including cardiovascular, cerebral, circulation and blood flow. Unfortunately, it overheated upon being inadvertently left on overnight, but was eventually repaired. Cosmonauts generally found full medical check ups frustrating because of the time it took to unpack all the sensors, pastes, cables, cuffs and belts, assemble and test the monitoring equipment, then wait to fly within radio range so that the telemetry could be transmitted to the doctors on the ground. Unless the data could be completed within one pass, they would have to break off until the next pass. Finally, the equipment had to be disconnected and packed away again.

Aerosol   An experiment on *Salyut 7*. It used a spectrometer to study the upper layers of the atmosphere at visible and infrared wavelengths. These observations were made by pointing the sensors directly down at the Earth, so that it precisely followed the ground track, rather than by pointing it off to the horizon to monitor the absorption of sunlight at sunrise and sunset. This has the advantage of localising the measurement, and also permitted continuous monitoring over extended periods.

Afamia   An experiment on *Mir* using the *Kristallisator* furnace to grow gallium–antimonide monocrystals.

Agidel   An experimental shaver tested on *Salyut 7*. It incorporated a chamber to collect stubble, but in practice this proved to be inadequate, so a larger vacuum cleaner was also used.

AIAA   American Institute for Aeronautics and Astronautics.

Ainur   An electrophoresis apparatus on the *Kristall* module, used to grow protein crystals. It was said to be capable of processing 100 kg of material per year.

Akustika   An experiment on *Mir* to measure the background noise at various points within the complex. The hum from the environmental system was generally around 80 decibels.

Alice-1   A French experiment on *Mir* that investigated the transport of heat and mass in gas–liquid systems, and phase-change phenomena at their critical points.

Alice-2   A French 64-kg furnace that accurately controlled the temperature of a fluid. It used a *CCD* camera to investigate the fine-scale phenomena of the critical-point phase transition (when the properties of the gaseous and liquid phases were identical). Its data and video output were stored for return to Earth. It was a refinement of the *Alice-1* experiment.

Alissa   A French experiment on the *Priroda* module. It used a lidar at 5270 Å aimed straight down to study the atmosphere. In addition to the vertical structure of clouds, it could detect tropospheric aerosols (and so complement the stratospheric capability of *Ozon-M*). Unfortunately, because its four-laser probe consumed 3 kW its use had to be limited. It had initially been intended for *Salyut 7*, so had been designated l'Atmosphere Lidar Sur Salyut (hence Alissa).

Almaz (Diamond)   The reconnaissance platform designed by the *Chelomei Bureau*, the shell of which was modified by the *Korolev Bureau* to serve as *Salyut*.

Altai-1   An experiment on *Salyut 6*. The *Splav* furnace was used to study the diffusion and mass transfer properties of lead and tin (metals that are soluble in their liquid state) and the effects of convective flows occurring across the thermal gradient.

Altai-2   An experiment on *Salyut 6*. A crystal of vanadium pentoxide (an active semi-conductor widely used in the production of thermistors) was grown in the *Splav* furnace to test whether very homogenous crystals could be created in microgravity conditions.

Altyn   A biotechnology experiment on *Mir* which involved a genetic study of plant (wheat) cells.

Amak-3   A blood analyser on *Salyut 4*.

AMEE (Advanced Materials Exposure Experiment)   A *NASA* package deployed outside *Mir*. Its optical properties experiment can be rotated in different directions so that a video camera and a spectrometer can examine each sample and provide real-time data.

Amplituda   An experiment on *Salyut 6* to determine the structural stability of a three-spacecraft complex. Like the *Resonance* experiment, this involved a crewman jumping on the *KTF* in a rhythm defined by a timing signal from the ground, while a transducer measured the effects.

ANBRE   This analogue biomedical recorder was a skin-tight suit supplied to *Mir* by *ESA*, to study posture dynamics by measuring the motion of the major limbs in weightlessness, using a multiple-camera apparatus.

Anna-3   A telescopic spectrometer on *Salyut 1*. It used a *Cerenkov counter* to measure the flux of gamma-rays with energies in excess of 100 MeV, for astrophysical research. It had a 1-degree pointing accuracy.

Anthropometry   An experiment on *Salyut 6* to determine changes in muscular mass and bone structure (specifically the loss of calcium) during the initial phase of adaptation to weightlessness.

Antibiotik   An experiment on *Salyut 7*. Absorbent pads were used to sample bacteria and microflora from different parts of the body. Once these had developed in the *Cytos-2* incubator, they were subjected to various antibiotics to determine their responses. The biomedical data was transmitted to Earth on the telemetry link.

AO-1   The star tracker in the main compartment of *Salyut 7*. The cosmonauts could orientate the station by reference to the stars using the sextant and star tracker manually. The two instruments had first been coaligned so that when a star was centred in one it was in the centre of the field of view of the other. Sightings were taken as the station passed through the Earth's shadow. The star tracker could be used to determine the orientation of the station, then the sextant used to hold it steady. This was a backup capability in case the *Delta/Kaskad* system failed.

APA (Anticipatory Postural Activity) A *NASA* biomedical experiment on *Mir* to observe posture and muscle coordination during various activities.

APDS An Androgynous Peripheral Docking System. The term 'peripheral' applies because its initial capture mechanism is mounted peripherally (unlike the axially projecting probe of the standard *Soyuz* unit, which, by its nature, requires a male/female combination; it is the absence of this restriction which makes the peripheral unit androgynous). The system devised for *Apollo–Soyuz* had the three guide plates canted outwards. Since this historic link-up took place in 1975, this is now know as the APDS-75 configuration. The port on the *Kristall* module of the *Mir* complex, intended to accommodate the *Buran* shuttle, has inward-canted guide plates, and is known as the APDS-89 configuration. This assembly was bought from *Energiya* by *NASA* and attached to the *ODS* to enable *Atlantis* to dock at the *Kristall* module. As the two sets of guide plates meshed, the capture latches they carried engaged to achieve the soft docking. Springs within the mechanism damped out the residual relative motions as the two spacecraft jostled one another. After about fifteen minutes in this state, *Atlantis* fired its thrusters to force the extended guide ring against that on *Kristall* to ensure proper alignment, then the twelve primary latches around the periphery of two collars were commanded to extend and engage, and the ring was retracted (this released the capture latches) to establish the hard docking which formed a rigid connection and a hermetically sealed tunnel within the mechanism.

Apollo The spacecraft built by *NASA* to fly three astronauts into lunar orbit and return them to Earth. Far larger than *Gemini*, it comprised a blunt-cone command module and a cylindrical service module. It had considerably greater manoeuvring and navigational capability than the contemporary *Soyuz*.

Arabidopsis A wallcress plant, commonly regarded as a weed, widely used in genetic experiments because it has the advantage of rapid growth (barely 40 days per generation), and a comparatively simple genome (only about twenty genes) which is well understood.

Aral-91 A programme to study the movement, concentration, composition, temperature and wind speed of dust and aerosols blowing from the recently exposed bed of the Aral Sea.

Arfa An experiment deployed outside *Kvant 1* to investigate the ionosphere and magnetosphere. Its data was used to test the correlation between charged particles at orbital altitude and seismic activity in the Earth's crust.

Argon 16 The computer introduced on *Salyut 4* to handle navigation, flight dynamics and attitude control. It also comprehensively monitored and displayed the status of its systems to the crew. It was afterwards installed on *Soyuz-T*. The Americans pointed out that this technology was comparable with that used by *Gemini*, and greatly inferior to that used by *Apollo*. Nevertheless, it was a significant advance for the ferry, because it enabled it to manoeuvre independently of ground tracking and computer commands.

Argument A biomedical ultrasound scanner on *Salyut 7* which produced a sectional image of the heart on a screen. It was an adaptation of a device used in emergency clinics in the Soviet Union, and was similar to, but not as easy to use as, the *Echograph*.

ARIS (Active-Rack Isolation System)   A Boeing device with a mechanical suspension system to damp out micro-accelerations to facilitate microgravity experiments on the 'noisy' International Space Station.

ARIZ   An X-ray spectrometer mounted on the *ASPG-M* platform.

Armadeus   A French experiment on *Mir* which involved the erection of a 28-kg model solar panel, to test the articulated deployment mechanism. The structure comprised four motorised winding blades that incorporated Carpentier joints to eliminate friction. It was performed inside *Mir*. The deployment was filmed so that the deployment kinematics could be analysed later.

ASPG-M   A 225-kg Czech-built scan platform (similar to those carried on the Vega planetary probes). It was mounted on *Kvant 2*, and could be pointed by remote control (either by the flight controllers or the crew) to make Earth observations without requiring the complex to be reorientated. It carried a variety of apparatus (including the *ITS-7D* infrared spectrometer, the *ARIZ* X-ray spectrometer, the *MKS-M2* multispectral optical spectrometer and the *Gamma-2* videospectral television cameras), and had a pointing accuracy of better than 0.5 arcmin.

Astra-1   A mass spectrometer to analyse the gaseous environment around *Salyut 7*. It revealed a tenuous stream of matter trailing the station, composed of air vented from the small scientific airlocks and the exhaust products from the attitude control thrusters. After a spacewalk, it observed a stream of gas trailing behind the station resulting from the venting of the airlock.

Astra-2   A spectrometer on the *Spektr* module to measure the constituents of the gaseous environment at orbital altitude.

Astro   An experiment on *Salyut 6*. The *Emission* apparatus was used to study the *cosmic ray* background in the near-Earth environment.

Astro   A *CCD* star tracker built by Jena Optronik of Germany. One 80-kg unit was affixed to each side of the unpressurised compartment of *Kvant 1* during a spacewalk in order to assist in orientating the enlarged complex.

ASU   The toilet on *Mir*; there are two — one in the base block and one in the *Kvant 2* module.

Atlantika-89   A programme carried out jointly with Cuba to investigate a suspected correlation between ozone and the formation of hurricanes. *Mir* measured the amount and distribution of ozone.

Atlantis   A *NASA* space shuttle.

Atlet (Athlete)   A load-inducing suit which incorporated elasticated straps which linked a waist corset to the shoes, with braces running over the shoulders, to make the muscles continuously work to straighten out. It was tested on *Salyut 4*, but abandoned in favour of the *Penguin* suit.

Audimir   An Austrian experiment on *Mir* to study the auditory system.

Audio   An East German experiment on *Salyut 6* to investigate frequency characteristics of sound in the station. It tested the ability to distinguish subtle nuances of sound in weightlessness. The way sound propagates is dependent on air pressure; in partial pressure it does not propagate as far and its tone is altered. Audiograms made with the *Elbe* apparatus to determine auditory threshold over a range of frequencies were compared with tests made before and after the flight to determine whether hearing changed. The results revealed that the hearing threshold is lower in space.

Azolla   A biological experiment on *Salyut 6*, carried out with a fast-growing nitrogen-rich water fern (appropriately, *azolla pinnata* was used) commonly used as fertiliser in rice-growing areas, to see if it held promise for use in a closed-cycle hydroponics system on a future station to properly cycle chemicals in its air.

Balance   An experiment on *Salyut 6* to determine the changes in the body's water and mineral balance by measuring water intake, urine excretion, body mass and concentrations of a range of substances in the blood.

Balaton   A Hungarian apparatus (named after a lake in Hungary) on *Salyut 6*, designed to determine the intellectual and motor performance of a cosmonaut. It was held in the palm of the subject's left hand in such a way that the pulse could be measured from one finger while sensors recorded galvanic skin resistance (that is, the electrical conductivity) of two other fingers as an indication of the level of perspiration. This measured the subject's effort in responding to a series of mental exercises involving patterns of audio tones and lights. Each exercise had a number of solutions of different levels of difficulty, and as the cosmonaut passed from one level to the next the skill factor required was increased. Throughout, the cosmonaut was subjected to a series of disruptive rhythms and tones played over his headset.

Balkan   A Bulgarian experiment on *Salyut 6*. It was a study of the Earth with the *Spektr-15K* spectrometer.

Balkan-1   A *lidar* on the *Spektr* module to determine the altitudes of clouds.

Ballisto   An experiment on *Salyut 7* which monitored chest movements to measure the accelerations of the body caused by the heart's pumping action, and changes in the shape and location of the heart from the body's adaptation to the absence of gravity. The experiment was continued on *Mir*.

Bania   The shower unit on *Salyut 7*. It took about two hours to set up, use, clean, and stow away, as it involved a complicated procedure which began with pumping out the contents of the *ASU* urine collector from the small EDV container, which is the temporary receptacle for fluid waste, into the much larger EDV tank in which it is stored. This would enable the smaller container to be used to take dirty water extracted from the shower unit. The cosmonauts then filled the two Kolos containers in its ceiling, and then pulled the curtain down from its mount in the ceiling of the station and turned on the air filtration unit in the floor of the unit which would extract odours which might overtax the environmental unit if they were allowed to leave the enclosure. Once the water in one of the Kolos tanks had been heated, the first man could don his air tube and goggles which would prevent him from drowning and keep the water out of his eyes, and begin his shower. The Kolos containers, however, held only 5 litres each of hot and cold water, so he did not have very long to get himself clean. It took at least five minutes for the enclosure to fill with a water, air and steam mixture, which was made to flow downwards by pumping air through the enclosure. A sponge soaked in highly abrasive katamine (quarternary amines) was used to literally scrape off the dirt which had accumulated on their bodies during a month in space. Clearly, taking a shower in space was not just a matter of stepping into the compartment and turning on the water, and as a result, the cosmonauts were limited to monthly showers. Unfortunately, there was no room in the station to keep it in place. Once, when pumping out the EDV tank, a plug came free and a mixture of dirty water and urine sprayed out, and when this had been cleaned up the cosmonauts really did need the shower that they had started out to erect.

Batyr　A biomedical experiment on *Mir* to evaluate the effect of breathing exercises as a way of easing the initial phase of adaptation.

BAZK　A star camera on *Salyut 7*.

BCAT (Binary-Colloid Alloy Test)　A *NASA* experiment performed in the *Glovebox* on *Mir* to observe the crystallisation of alloys of colloids. Samples using different relative concentrations were processed for 24 hours to establish their behaviour, and then one was left to grow for three months.

Bealuca　A Hungarian experiment on *Salyut 6* which smelted copper–aluminium in the *Splav* furnace.

Berolina　A series of East German experiments in crystal growth using the *Splav* and *Kristall* furnaces on *Salyut 6*. Of the two experiments using the *Splav* furnace, one boiled beryllium–thorium to produce homogeneous glass of far higher quality than possible under the influence of gravity, and the other used a special quartz matrix to produce crystals of a bismuth–antimony semiconductor. Experiments in the *Kristall* furnace grew lead–telluride crystals for the first time using the sublimation technique which vaporised them by heating one end of the ampoule, and seeded a crystal at the other cooler end. *Kristall* was also used to grow a crystal of bismuth–antimony by stretching the seed. It was processed with the material sandwiched between two plates, within an ampoule. This yielded a tree-structured crystal which was about five times larger than that which can be produced on Earth. It was done this way to compare the result with that produced by the *Splav* furnace.

Beta　A multifunction medical unit on *Salyut 4*. It complemented the electrocardiogram functions of the *Polynom-2M*, but it also measured lung capacity and recorded seismograms to determine the rhythm of the heart and the force of blood pumping.

Big Bird　An umbrella name for several classes of US Department of Defense's photographic reconnaissance satellites, the earliest of which was introduced in the early 1970s, contemporary with the *Almaz*.

Biobloc-3　A French experiment on *Salyut 7* which studied the effects of *cosmic rays* on biological materials by sandwiching samples of artima salina cysts and tobacco seeds between heavy-ion detectors.

Biodose　A French experiment on *Mir* which measured the biological effects of *cosmic rays*, to assess long-term effects of exposure.

Biogravistat　A plant growth apparatus on *Salyut 6*. It comprised two disks, one fixed and the other rotating to simulate gravity, one as the experiment and the other as the control. It was a small centrifuge similar to that which had been flown on the Cosmos 782 Biosat, which generated partial gravity to offer plants that had been unable to grow in weightlessness a more conducive environment. At first it excessively aerated the roots, but once this was rectified it was found that the seeds in the gravity chamber grew at twice the rate of those in the stationary chamber, and the roots aligned themselves in the direction of the imparted force. Once the seeds had germinated, they were transferred to other apparatus for further development. Barley was grown in it. Mushrooms grew well, but developed oddly curled stems. Tests demonstrated that plants were able to react to a force as low as one ten-thousandth of normal gravity.

Biokat   An aquarium first flown on *Soyuz 19*, during the *Apollo–Soyuz* mission.

Biokhim   A biomedical device on *Salyut 7* used to measure blood electrolytes.

Biokryst   An electrophoresis apparatus on *Mir*.

Biosfera (Biosphere)   An experiment on *Salyut 6* to investigate the state of the Earth's environment. It involved making visual observations of the Earth's oceans which would be compared with a specially created chromatic atlas to determine optical properties of the atmosphere under different conditions. This would yield information which would improve photographic techniques by evaluating the ability of different films, filters and exposures to record different types of detail. It would also determine the conditions under which different phenomena were best imaged. Because this was a long-term project, observations would be made by successive crews. Eventually, it would establish global levels of marine and atmospheric pollution.

Biostoykost   An apparatus on *Mir* for making polymers.

Bioterm-1,2,3,4   A series of thermostatic plant growth experiments flown on successive Salyuts.

Biryuza   An apparatus on *Mir* used to study "the dynamics of physio-chemical processes", and "the processes of growing crystals, solutions and molten baths in microgravity".

Black Sea-84   A photographic and spectrographic study of the Black Sea region, organised by *Intercosmos* as the first phase of a multi-year programme. Data from *Salyut 7* was correlated with that from the Cosmos 1500 oceanographic and *Meteor-Priroda* multispectral Earth-resources satellites, aircraft and research ships. It was a multispectral survey of the hydrophysical and biological characteristics of the surface of the Black Sea.

Bodyfluids   An Austrian experiment on *Mir* that studied the composition and distribution of blood and bodily fluids in weightlessness by measuring the speed of sound in blood to determine the dynamics of transient fluid motions after stimulation.

Bosra   An experiment on *Mir* to gather data with which to improve mathematical modelling of the upper layers of the atmosphere and ionosphere.

BPAS (Belt-Pack Amplifier System)   A *NASA* experiment on *Mir*, consisting of a harness carrying sensors to measure muscle stimulation and work output during exercise on the *NASA* ergonometer. The data on bone and muscle density was compared with pre/post-flight scanner imagery.

Braslet (Bracelet)   Elasticated rings introduced on *Salyut 7*. They were worn around the thighs during the initial phase of adaptation to restrict the migration of blood from the legs to the upper torso, trapping blood in the legs. Wearing them for up to an hour, several times a day, during the first few days had much the same effect as spending time in the *Tchibis* unit. For a while it was used by visiting crews.

Brillomir   An Austrian experiment on *Mir* designed to measure critical fluctuations during the decomposition of binary liquid mixtures in the absence of gravity.

BST-1M   The primary instrument on *Salyut 6*. This 650-kg instrument was contained in the massive conical mount used by the *OST-1* on *Salyut 4*. It was a telescope with a 1.5-metre diameter mirror for infrared, ultraviolet and submillimetre observations from 50 μm to 2 mm. Although it operated in the vacuum of space its sensors had to be cooled to −269°C; this was done using an improved version of the closed-cycle cryogenic unit first tested on *Salyut 4*. Helium was produced using a compressor, two gas-refrigerating machines and intermediate heat exchangers. It passed through an expanding throttle valve to lower its temperature. Although it consumed 1.5 kW, once the sensor had been cooled to its operating temperature it could be maintained at that level for extended periods. The telescope could be used only when the station was in the Earth's shadow, and for the rest of the time it had to be protected by a cover. It had a pointing accuracy of 1 arcmin. Infrared observations revealed small-scale variations in the submillimetre flux of the atmosphere in the locations in which cyclonic activity originated.

BTS (BioTechnology System) A *NASA* experiment on *Mir* to grow cell cultures and protein crystal. Mammalian (cow) cartilage cells were suspended in a liquid growth-medium in a bioreactor vessel, delivered within a 'powered transfer' package. It exploited microgravity to grow a 3-D tissue culture (in contrast to 2-D on the surface of a nutrient in a Petri dish) to study cell attachment patterns and interactions. The growth was recorded on video, and samples were extracted at regular intervals and then frozen for later analysis; it was a long-term experiment which ran for three months. The bioreactor kept the sample at the right temperature, fed it nutrient, removed waste and produced a log of parameters. It was found that cells of cartilage grew faster in space than on Earth. The result was frozen for subsequent analysis.

Buket A telescopic gamma-ray spectrometer on the *Kristall* module used to measure the energy spectrum and spatial characteristics of cosmic radiation.

Buran (Snowstorm) The Soviet space shuttle. The new vehicle was called the Vozdushno-Kosmichesky Korabl (VKK) which was best translated as 'atmospheric spacecraft'. Its one and only flight, which was conducted automatically, was in November 1988. It was cancelled in the early 1990s.

C-1,2 The sextants in the forward transfer compartment of *Salyut 7*.

Calibration An experiment on *Salyut 7* to measure the effect of small accelerations on the *Magma-F* furnace to develop a mathematical model of its behaviour in microgravity, where thermo-convective agitation is absent.

Caribe A Cuban experiment on *Salyut 6*. It involved smelting five ampoules in the *Splav* and *Kristall* furnaces. Splav was used to make alloys of gallium arsenide with aluminium and of tin–telluride with germanium–telluride. Kristall was used to produce an alloy of germanium and indium, and of zinc–indium–sulphide (ZIS). In addition to creating semiconductors, these studied nucleation and the mechanisms of crystal growth.

CCD An electronic camera technology which uses a charge-coupled detector integrated circuit to form the image. It has the advantage that its efficiency (its ability to register an impinging photon) is about 80 per cent, but is limited both by the spatial resolution of the cells (the pixels) in a detector array and by the dimensions of the array. It has two other key advantages: it yields its output electronically, and in a form ideal for image processing.

CDS-1 The Crimean Diffraction Spectrometer built into the *OST-1* on *Salyut 4*. Its high-precision optical elements could resolve 600 lines per millimetre and yield an ultraviolet spectrum in the range 800–1,300 Å, producing an average resolution of under 2 Å.

Cerenkov counter   An instrument to detect the secondary radiation (Cerenkov radiation) caused by a very-high-speed charged particle passing through a gaseous medium.

Cervical Shock Absorber   A Mongolian-designed experiment on *Salyut 6*. It applied pressure to the cervical part of the spinal column and restricted head movement, and was able to simulate local loads encountered under normal gravity. It was hoped that this would inhibit the onset of the motion sickness commonly suffered during the initial phase of the process of adapting to weightlessness. It was worn continuously for the first three days, except when sleeping.

CFE   A *NASA* experiment on *Mir*. The Candle Flame Experiment (CFE) was a combustion experiment performed in the *Glovebox* so that its cameras could record the event for later analysis. It was later revised into the Forced Flow Flamespread Test (FFFT), which involved passing an airflow over a flame to study growth of the flame under such circumstances, flammability of samples of cellulose and polyethylene, and the ignition process. In all, these trials consumed 80 candles and eight samples of solid fuel. A flame in microgravity forms a sphere rather than the familiar pear shape, and it soon smothers itself because there is no convection to draw off the carbon dioxide that it creates. A forced air flow will keep it burning.

Chelomei Bureau   The rocket design bureau (OKB-52) established by Vladimir Chelomei.

Chlorella   A Czech-designed experiment on *Salyut 6* which investigated the growth of an algae culture in a nutrient medium. The algae used was, obviously, *chlorella*. This is a rapid-growth seaweed which might play an important part in future station operations because it absorbs carbon dioxide and gives off oxygen.

CHR   A German experiment on *Mir* supplied by the Genetics Institute of Essen University to investigate chromosomal aberration in the lymphocytes. It involved taking samples before and after the flight.

Circe   A French radiation dosimeter on *Mir* which used a low-pressure tissue-equivalent gas proportional counter; it proved very effective. It was used to monitor ionising radiation, and gamma-ray and neutron dosages in the station. Long after the joint mission was over, measurements continued to be taken twice a day.

Claznoye   A biomedical device on *Salyut 7* used to measure blood flow in the eye, and the movement of the blind spot.

Climate   An experiment on *Mir* which used the *EFO-1* to study the optical density of the upper atmosphere as part of a long-term investigation into air pollution.

CNES   French Space Agency.

Cogimir   An Austrian experiment on *Mir* that analysed cognitive functions during the process of adaptation to weightlessness.

Cognilab   A French neurosensory experiment on *Mir*. It involved strapping into an instrumented chair which measured the body's response to muscular stimulation under different conditions.

Columbus   The *ESA* module intended to be attached to International Space Station Alpha.

Comet   A French micrometeoroid detector deployed outside *Salyut 7*. Its two chambers could be operated by remote control from within the station. The first was to be exposed to Comet Giacobini–Zinner, the second to Halley's Comet. In each case it was hoped that it would collect dust ejected by the comet during its passage through the inner Solar System.

Coordination   An experiment on *Salyut 6* to investigate the effects of weightlessness on the body's voluntary motor functions, using a mechanical hand/eye psychomotor test.

Cortex   An experiment on *Salyut 6*. It monitored the brain's electrical activity in the absence of gravity by taking electroencephalograms (EEG) to determine responses to various stimuli. The in-flight EEGs would be compared with pre/post-flight data. The data was collected with an instrumented helmet and was stored on a small tape recorder.

Cosmic radiation   The radiation extant in space. It comprises three components: *solar wind*, *solar flares* and *cosmic rays*, all of which consist primarily of charged atomic nuclei with differing speeds and energy.

Cosmic rays   The most energetic component of *cosmic radiation.* These heavily-ionised atoms (some as heavy as iron) have been accelerated by unknown processes in the Galaxy, far beyond the Solar System, to speeds approaching that of light, so they arrive with energies in excess of 1 billion *electron volts*, sufficient to enable them to pass straight through the radiation shielding used to protect spacecraft. They pose a serious long-term threat to space missions, both to equipment and to their crews. The astronauts on the Apollo lunar missions reported 'seeing' sporadic cosmic rays that passed through their eyes. The transit of a heavy nucleus through a crystalline lattice will leave a microscopic track, so such materials were flown on *Mir* specifically to assess the cosmic ray flux. Other experiments exploited the fact that at narrowly-sublight *invacuo* speeds, a cosmic ray will stimulate *Cerenkov* secondary radiation when passing through a gaseous medium.

CSA   Canadian Space Agency.

CSK-1,2,3,4   The *Kristallisator* furnace.

Cytos   A joint Soviet–French experiment on *Salyut 6*. A variety of microorganisms were delivered in a *Bioterm* at 8°C to inhibit their development. The French experiment used paramecium protozoa and the Soviet experiment used proteidae. These were put in the *Cytos* apparatus, which was maintained at 25°C to facilitate their development, and after 12 hours were reinhibited by being returned to the *Bioterm* for return to Earth for analysis. Over a four-day period, eight consecutive generations of microorganism were grown. This experiment was designed to determine the effects of microgravity and radiation on the process of cell division (the basis of the life process) to study the kinetics of cell division. This was part of an investigation of the long-term ability of organisms to live in space. The French team, which had designed and built the *Cytos* apparatus specifically for this experiment, was led by Professor Hubert Planel of the National Space Research Centre at the University of Toulouse, and the experiment was conducted in collaboration with Professor Yuri Nefedov of the Soviet Institute of Medico-biological Research. The test showed that bacteria behaved as they did on Earth, but that simple microorganisms thrived on the nutrients provided.

Cytos-2   A French incubator on *Salyut 7*. It was used to investigate the structure and functioning of bacteria cells in weightlessness and the effect of a variety of antibiotics against bacteria.

Cytos-3   A French incubator on *Salyut 7.*

Danko   A cassette outside *Mir.* It exposed construction materials, and produced electronic data that could be downloaded periodically.

DARA   Reunified Germany's Aviation and Space Research Institute.

DCAM (Diffusion-controlled Crystallisation Apparatus for Microgravity)   A *NASA* experiment on *Mir* which used a semi-permeable membrane to grow protein monocrystal over a long period. The growth process was filmed, and the results were returned to Earth for study.

Deformatsiya (Deformation)   An experiment on *Salyut 6* which used optical instruments to measure the distortion of the complex's structure due to solar heating. The complex was orientated and maintained in a solar-stabilised attitude which created a 300°C temperature differential. Because this attitude baked one side and froze the other, it could not be held for long without risking damaging its systems, but the results indicated that even several hours of such exposure distorted the complex's primary axis by no more than 0.1 degree.

Delta   A self-contained navigational system tested by *Salyut 4.* It was a significant advance over previous stations, which had required data on the path to be measured on every revolution by a network of tracking stations. The new system used a set of Sun sensors to keep track of the station's position in its orbit by monitoring the rising and setting of the Sun on the Earth's horizon in order to determine the period of each orbit. It incorporated the *Argon 16* computer, which combined this with the readings from the radio altimeter to compute the station's orbital parameters, monitor its progress around the globe and compute the times during which it would be within communications range of the various ground stations. It could pinpoint the station's location over the surface of the Earth to within 3 km, and its altitude to within a few hundred metres. It greatly simplified the ground-based support systems and reduced the loads on the cosmonauts. It was tested on *Salyut 4* and then made operational on *Salyut 6* and *Salyut 7.*

Diagnost   A Hungarian experiment which began during the preparations for launch to *Salyut 6* and was then repeated immediately after landing, to study the effects of space flight on the cardiovascular, respiratory and vestibular systems, and on the hearing threshold.

Diagramma   This experiment on *Mir* used a magnetic discharge transducer on a short boom projecting from the scientific airlock to measure the physical characteristics of the atmosphere at orbital altitude (measuring the aerodynamic flow around the complex would help evaluate aerodynamic drag).

Diffusia (Diffusion)   An experiment on *Salyut 5* to produce a more homogeneous alloy of dibenzyl and toluene than can be achieved under normal gravity.

Diffusion   A French experiment on *Salyut 7.* It employed the *Magma-F* furnace to study the dissolution of a polycrystalline solid alloy in its own liquid when in a state of thermodynamic equilibrium, to develop a mathematical model of the solidification process. It investigated capillary forces during crystallisation of alloys at different temperatures.

Discovery   A *NASA* space shuttle.

Diusa   An apparatus on *Salyut 7.* Capable of measuring minute air pressure variations, it was used to test for suspected cabin leaks.

DM   The Docking Module attached to the axial port on the *Kristall* module to provide extra clearance for *Atlantis* from *Mir*'s projecting solar panels, to provide an extra airlock chamber, and to store cargo. At 4.6 metres long, 2.2 metres in diameter, with a mass of only 4.2 tonnes, it was essentially an extremely stretched *Soyuz* orbital module fitted with an *APDS* port at each end.

DOM   A German experiment on *Mir* which evaluated five different dosimeters as methods of measuring radiation within the complex.

DOS   An acronym for permanent orbital station. This designation was assigned to the scientific *Salyut* adapted by the *Korolev Bureau* from the *Almaz* design:

DOS 1	*Salyut 1*	April 1971
DOS 2	\<failed rocket\>	July 1972
DOS 3	Cosmos 557	May 1973
DOS 4	*Salyut 4*	Dec 1974
DOS 5	*Salyut 6*	Sep 1977
DOS 6	*Salyut 7*	Apr 1982

Although DOS 1 to 4 differed in detail (such as solar panels), they were all basically the same configuration with a single docking port at the front. DOS 5 and 6 represented a significant redesign to bring back features of the *OPS* (notably the rear docking unit, the peripherally-mounted engines, and the gyroscopic attitude control system), and this then served as the basis of the *Mir* base block, which in turn led to the Service Module of the International Space Station.

Dose   A Hungarian experiment on *Salyut 6* to measure radiation doses. A number of sensors were worn by the cosmonauts, and others were distributed around the station. Each day they were analysed using the *Pille* thermoluminescent device to monitor the incremental accumulation of radiation.

Dosimir   An Austrian experiment on *Mir* which tested a TLD dosimeter.

Dosug   An experiment on *Mir* that attempted to formally evaluate the influence of music, video and games on a cosmonaut's morale whilst off duty.

Doza-B   An experiment on *Mir* which involved placing biological samples and radiation sensors at specific places in the complex to investigate exposure levels.

Drosophilae   A fruit fly, widely used in genetic studies because it has the advantage of a short reproductive cycle and a well-understood genome.

DS-1   A French biomedical instrument on *Salyut 7*. It was a doppler device to measure blood flow rate.

Duga   A Bulgarian electrophotometer on *Salyut 6*, attached to one of the portholes, and used while in the Earth's shadow. It was pointed by a cosmonaut using an optical sight. It detected, intensified and measured natural emissions at 6,300 Å, 5,577 Å, 4,278 Å and 6,563 Å, all in the optical band, to study a range of upper atmospheric phenomena including the vertical structure of aurorae, the luminous red ionospheric arcs created by severe magnetic storms and stable luminous arcs in the plasmasphere. Such studies were part of an ongoing programme, so were conducted whenever the opportunity arose. Unfortunately, a fault in the construction of its image converter meant that it displayed an inverted image (a new image converter was later flown up and installed).

It revealed that the atmosphere above the equator glowed. This was most noticeable in the red oxygen line, at 6,300 Å, whose intensity varied by a factor of 10. A similar, but less intensive, glow was detected at 5,577 Å. However, because it was not detected at 4,278 Å or 6,563 Å (generated by electron and proton emissions respectively), it was clear that the observed glow could not have been produced by intense flows of such charged particles but by some other mechanism.

Dynalab   A French experiment on *Mir* to measure the propagation of vibrations through the complex (a variation on the Resonance theme).

Eceq   A French experiment on *Mir* which measured the flux of *cosmic rays*, and its effects on electronic equipment in the complex.

Echantillon (Sample)   A French cassette which exposed materials samples to the space environment. The Comes segment exposed samples of paints, reflectors, adhesives, filament-reinforced composites and optical materials; Mapol exposed samples of polymeric materials thought suitable for creating inflatable structures in space; MCAL exposed materials to determine the evolution of absorptivity and emissivity, and also incorporated a pair of dust detectors, one active and one passive (called DIC and DMC), to collect cosmic dust for subsequent examination. A 16-kg 0.75-metre square box, it was affixed to the outside of *Mir* for a year.

Echograph   A French biomedical instrument on *Salyut 7*; an ultrasound scanner to make a sectional image of the heart. It was later transferred to *Mir*.

Echograph-II   An improved ultrasonic body scanner on *Mir*, to measure blood flow in deep vessels (in particular the main truncus and venous return) as well as the capacity of the heart and other internal organs.

Echography   A French experiment on *Salyut 7* which involved using the *Echograph* ultrasound scanner and the *DS-1* doppler device to measure the cavity dimensions, myocardiac thickness, artery blood content and the speed of blood supply to the vessels. The hypothesis was that, in the absence of gravity, the heart moved up into the chest. The fact that Chrétien initially had some difficulty locating his heart lent support to this supposition. He also used the *DS-1* to determine the distribution of blood in the cranial artieries.

EDLS (Enhanced Dynamic Load Sensors)   A *NASA* technology experiment performed on *Mir*. It involved a push-off pad to measure the force used to move within the complex, and an instrumented foot-restraint to monitor movements whilst nominally working 'in place'. The empirical data was to be used in fitting out the modules of the International Space Station.

EFO-1   A Czech electrophotometer used on board *Salyut 7* to make measurements of the upper atmosphere. It was used to monitor stars as they dropped down to the horizon, to measure the flickering light as it passed through the layers of the Earth's atmosphere. (An intensive programme of such observations was to be made to locate and study layers of aerosols in the upper atmosphere). Usually for these observations, the station was orientated with its axis perpendicular to the velocity vector and parallel to the horizon, and then placed in a slow roll timed to keep the apparatus in the porthole facing the horizon to monitor a succession of stars rising or setting through the layers of the atmosphere. Occasionally, however, the station was put in gravity-gradient mode, and stabilised so that the porthole faced ahead (depending on whether microgravity materials processing was underway at the same time). It was transferred to *Mir*. Its main contribution was to the *Climate* experiment.

EFU-Robot   An electrophoresis unit on *Salyut 7*. It was a pilot-scale form of an automatic 'factory' designed to produce biological preparations which could be useful to the food, health and agricultural agencies. It was transferred to *Mir*.

Elbe   An audiometer. It produced an audiogram which measured the hearing threshold across a range of sound frequencies.

Electron volt, eV, keV, MeV   A unit for measuring the energy of a particle. It corresponds to the energy required to displace a free electron across a potential difference of one volt within an electric field; that is 1.6×10^{-19} Joules.

Electrotopography   An experiment on *Salyut 7* performed in one of the scientific airlocks. Deformations on the surface of selected materials exposed to space for different periods were studied. The aim was to explore by electromagnetic means the surface of samples exposed to space. Each sample was put between two metal plates, with photographic film next to the sample. When a charge of several thousand volts was applied across the plates, the electric field was contoured around microscopic defects in the sample. It could detect variations as small as 10^{-11} metres in the shape of a sample. However, a problem developed on the first trial. As soon as the small scientific airlock was opened, the temperature in it fell rapidly to $-150°C$, as did the apparatus. If the airlock was opened immediately to retrieve the sample, the water vapour in the station's atmosphere condensed on it and ruined the experiment. Yet if the airlock was closed, and the temperature was allowed to rise slowly over a period of hours before it was opened, the sample acquired an electric charge that distorted it, again ruining the test. To overcome this, the cosmonauts orientated the station so that the airlock faced the Sun towards the end of the exposure so that the sunlight would warm it up, and if the timing was right the airlock could be closed at just the time when it was at a comfortable 20°C. The trial was rerun a number of times using more sensitive film and longer exposure periods (in some cases, 40 hours).

Electrotopograph-7K   An apparatus on *Mir* used to measure surface distortions of advanced plastics and high-temperature superconductors exposed to space in the scientific airlock for various times.

Elektron   This produced oxygen by electrolysing a 30 per cent potassium hydroxide (KOH) solution, to maintain the required composition of the gaseous environment within the *Mir* complex without requiring so much liquified air to be delivered by *Progress* ferries. It was connected by flexible tubes which were to be distributed throughout the complex. There were twelve electrolysis cells in the unit, which were cooled by the base block's primary coolant loop. It consumed about 4 kg of water per day; whenever possible water recycled from urine was employed. The hydrogen released was vented.

Eleutheroccus   A drug taken over a prolonged period (a daily dose of 4 ml) towards the end of a long spaceflight to act as a tonic, to stimulate the body to work harder, and thereby increase long-term stamina to help cosmonauts prepare to adapt to gravity.

ELITE   An *ESA* experiment that used four infrared television cameras to monitor the subject's posture in weightlessness.

Elma-1   A French smelting experiment on *Salyut 6* using the *Kristall-2* and *Splav* furnaces. Aluminium–copper, aluminium–tin, aluminium–lead and lead–tin were processed to study the processes of diffusion during melting and subsequent cooling of metal alloys. Magnetic alloys of magnesium, cobalt and other metals which do not mix under a strong gravity field were produced, as were monocrystals of germanium and vanadium oxide. The process of diffusion during melting and subsequent cooling of alloys of tin–lead and aluminium–copper was studied. A sample of gadolinium–cobalt (a magnetic material which has computing applications) was also produced.

Elma-2   A French experiment on *Salyut 7*. It involved creating an alloy of aluminium and indium (which are immiscible in Earth's gravity).

Emission   A Mongolian experiment on *Salyut 6*. It involved using dielectric detectors to monitor the strength and composition of *cosmic rays* in the 10 MeV range. The apparatus was inserted in the scientific airlock, and exposed to space for three days. The tracks bored through the detector substance would be examined after the detector had been returned to Earth. A similar detector (also Mongolian-built) had previously been flown on the unmanned Intercosmos 6 satellite, and had revealed that energetic nuclei with charges up to 28 units constituted over 90 per cent of the incident radiation. Another detector was exposed within the station to measure how well the station's hull was able to block radiation.

Emissiya (Emission)   An experiment on *Salyut 4* involving a photometer and a number of spectrometers at the back of the station which scanned the Earth's horizon to measure the luminescence intensity of the red atomic oxygen spectral line to study processes taking place at a height of between 250 km and 270 km, the most Sun-sensitive part of the ionosphere where the electrons in the Earth's magnetic field interacted with the upper atmosphere. The photometer had been built by the Academy of Sciences. When combined with data gathered by geophysical stations on the ground, the results would assist forecasting of short-term changes in the upper part of the atmosphere and could ultimately facilitate reliable forecasting of global climate change.

Energiya (Energy)   A heavy-lift rocket in the *Saturn V* class. Its first launch carried the *Polyus* platform, and its second the *Buran* space shuttle. Dr Boris Gubanov, the chief designer of the Energiya Bureau, which was responsible for its development, said that it was to be used to launch the building blocks of *Mir 2*.

Energomash   The manufacturer of the engines for the *Semyorka* rocket.

Epsilon   Contained in *Kvant 2*, this was an experiment to assess the thermal protection of the complex.

Eotvos   A Hungarian experiment on *Salyut 6* which used the *Kristall* furnace to produce semiconductors of gallium arsenide, indium antimonide and gallium antimonide, and a crystal of an alloy of gallium arsenide and chromium using the moving solvent method.

Era (Air)   A French technology experiment involving a 240-kg self-deploying articulated framework that was erected outside *Mir*. The experiment comprised a 0.6-metre diameter compressed stack of 1-metre long carbon fibre rods, a support platform, a video unit and a control panel. After the platform had been affixed to the anchor, the stack was affixed to an arm on the platform set at an angle of 45 degrees to keep it away from the surface of the complex. Once a 50-pin umbilical had been connected, the articulated pin-jointed rods were supposed to spring open automatically over a four-second period to create 24 identical prisms forming a thick hexagon some 4 metres wide, but it did not do so. It was nudged a few times to try to shake it loose, but it refused to deploy. Rather than jettison it, the cosmonauts persisted and it eventually deployed. When fully unfolded, a set of accelerometers were attached to it to measure its vibration modes. It was subsequently concluded that the structure must have been locked in its folded position by water vapour which had frozen when exposed to vacuum. After the experiment was complete, the structure was jettisoned. This experiment was to evaluate one of the options under consideration by *CNES* to erect antennas in space, as part of its long-term planning for its *Hermes* mini-shuttle development programme.

Ercos   A French cassette left on *Mir* for six months to determine the effect on computer memory chips of exposure to *cosmic rays*.

Erdem   An experiment on *Salyut 6* to study Mongolia, a vast, inaccessible and largely unexplored country; in this case, it was self-evident that space-based observation was the most cost-effective means of surveying. The *KATE-140* and *MKF-6M* cameras (the latter coordinated with similarly equipped aircraft) were used to survey the Earth resources. The data was used to locate geological fault lines, compile a map of soil conditions, identify subterranean water deposits, and map and assess crops and forests. Appropriate Mongolian sites were photographed using hand-held cameras (for the ongoing *Biosfera* experiment). These observations, together with spectroscopic data, were intended to provide an assessment of pasturelands, map the boundaries between fertile and arid terrain, assess water capacity of rivers, glaciers and snowfields, assess atmospheric pollution, and locate ancient geological ring structures (usually the result of meteor impacts).

Erdenet   A materials experiment on *Salyut 6* which used the food heater to dissolve copper sulphite in water and then recrystallise it; the process was filmed. This was performed to investigate the processes of diffusion and redistribution of impurities in a microgravity environment. This test was undertaken as part of a study of exceptionally high quality cleaning materials.

ERI   An experiment on *Mir* to test methods of depositing galvanic coatings.

ERTS (Earth-Resources Technology Satellite)   A pioneering satellite which demonstrated the multispectral imaging system later used by *NASA*'s Landsat series.

ESA   European Space Agency

ESEF   This *ESA* European Science Exposure Facility comprised four cassettes. Three were passive traps to accumulate ambient particulate debris, and the fourth was an active system that recorded the speed, mass and trajectory of each impact. Umbilicals were connected so that the clam-shell covers of the traps could be opened by remote control from within *Mir*. Each was to be opened for a specific period. One of the dust collectors was opened immediately, and another in October 1995 to trap debris from the Draconid meteor stream associated with Comet Giacobini–Zinner. They were closed while spacecraft were manoeuvring nearby, to preclude contamination. Analysis of the active experiment revealed that the complex passed through a stream of debris every ten hours or so, at which time it suffered 5,000 microscopic impacts in an interval of only one minute.

Etalon   An exposure cassette fitted on *Salyut 7* prior to launch, and retrieved in space. It contained a variety of different optical coatings, some of which had bubbled and peeled off.

Euphrates   A Syrian experiment on *Mir* which involved visual, photographic and spectrographic surveys of Syrian territory to study atmospheric pollution, water deposits and mineral resources, and imaging by the *KATE-140* camera to identify ancient sites.

Extinctia (Extinction)   A joint Soviet–Czech experiment on *Salyut 6* to investigate the micrometeoroid dust layer which exists at between 80 km and 100 km altitude by measuring the changes in the apparent brightness of stars as they dropped down behind the the Earth's dark horizon.

FAI   Federation Aeronautique Internationale. Its rules required that a record be exceeded by 10 per cent in order to be considered officially broken.

Faza   The AFM-2 multispectral radiometer on the *Spektr* module sampled in the 0.4–2.2 µm range to study the transition zones between Earth's surface and the atmosphere and the upper atmosphere and space.

FEK-7   A *cosmic ray* detector which used photographic film to study nuclei of transuranic elements. It was hoped that it would prove the existence of Dirac monopoles, but as yet these hypothetical particles remain elusive. Such detectors were initially flown on the circumlunar *Zond* spacecraft and then on *Salyut 1*.

Fem   An Earth-resources experiment on *Mir* which used the *MKF-6MA* in *Kvant 2* to survey Austrian territory.

Feniks   A spectrometer on the *Spektr* module to study the Earth's surface.

Ferma-1   A 20-kg structure comprising a set of lattice-and-pin frames of aluminium–titanium alloy. When fully erected, outside *Salyut 7,* the 15-metre long girder had a square section, about half a metre on a side. It was developed by the Institute of Electrical Welding, in Kiev.

Ferma-Postroitel (Beam-Builder)   This experiment involved erecting a URS girder developed by the Institute of Electrical Welding in Kiev. This 20-kg structure comprised lattice-and-pin frames of an aluminium–titanium alloy. When erected it formed a 15-metre long rectangular girder with a 40-cm square section. It was a prototype of a structure being evaluated for later use in space. There were three deployment options: manual, semi-automatic and automatic. It was deployed automatically, and then retracted. A flat plate was attached to the top and several experiment packages mounted on this before it was extended again, this time manually. One instrument was the 3 mW LED laser which illuminated a sensor set up in a porthole to measure the stability of the girder. The data would help predict the behaviour of larger framework structures. Then with an improved version of the URI multipurpose toolkit, some of the pins were beam-welded to lock the girder. Later the entire package was disconnected and jettisoned.

Ferrit   A cassette outside *Mir* to investigate the change in structure of ferro-magnetic materials.

Fialka-F   An apparatus on *Mir* to study the ultraviolet radiation in near-Earth space.

Filin-2 (Eagle-Owl)   A telescopic X-ray spectrometer on *Salyut 4*, mounted on the outside of the station. It was sensitive to wavelengths in the range 1–60 Å. The telescope used by the spectrometer was boresighted with a 6-cm diameter sight with a 1-degree field of view. It was essentially autonomous, so when it detected a signal all the operator had to do was use the sight to establish the location of the source with respect to the starfield. Since it could relay its results to Earth, it continued to be used while the station was unoccupied.

FM-107   A Fourier mass spectrometer on *Salyut 7*.

Fon   This studied the gaseous environment in the immediate vicinity of the station.

Foton (Photon)   A class of automated satellite for microgravity materials-processing, typically involving apparatus tested by cosmonauts (the *Splav-2* and *Zona* furnaces, and the *Kashtan* electrophoresis unit have all been used). The first satellite was launched in April 1988. The *Vostok* spacecraft was used in order to exploit the capsule to return its results to Earth, but because this did not incorporate solar panels, materials-processing apparatus had to be powered by storage batteries, which limited both payload and duration.

Freedom   The orbital complex proposed by *NASA* to fulfil President Reagan's order to design and assemble a space station in low orbit.

Functionality   An experiment on *Salyut 6* to assess sensory and motor reactions to a range of acoustic and visual irritants introduced whilst performing a monitored mental task.

Gallar   A furnace on *Kvant 1*, derived from the *Korund-1M*. It produced a variety of semiconductors, and some smelts ran for a week or more.

Gamma-1   A multi-function biomedical test kit on *Mir*. Amongst other things, it monitored cardiovascular activity.

Gamma-2   A multispectral television camera cluster mounted on the *ASPG-M* platform, to enable researchers on the ground to make remote-control observations of the Earth. It was used to assess the pollution levels in industrial zones, with the long-term goal of mapping the spread of pollution from large cities. It was also used on favourable ground passes to assess the ecological state of water basins, heavily forested areas and agricultural land to monitor seasonal variations in crop growth.

Gel   An electrophoresis experiment on *Salyut 7*. It made biological gels (a gel is a twin-colloid) with a purity 100 per cent greater than could be achieved on Earth. It made a polycrylamide gel, which could then be used to make the synthesis of biologically active materials on Earth somewhat more efficient. A similar apparatus was used on *Mir*.

Gemini   The two-seat spacecraft developed by *NASA* to develop rendezvous and docking, spacewalking, and long-duration experience in Earth orbit, during 1965 and 1966, preparatory to *Apollo*. The first US spacewalk was on Gemini 4, the first rendezvous was by Gemini 6 and 7, the first docking with an Agena target was by Gemini 8, and the longest flight (14 days in December 1965) was by Gemini 7, with Frank Borman and James Lovell. The Air Force adapted the design (in a configuration known, appropriately, as Blue Gemini) to operate with the *MOL*.

Genom (Genome)   A biotechnology experiment on *Salyut 7*, involving the use of an electrophoresis unit to separate a DNA solution into different strands. The process was lit by an ultraviolet lamp and filmed.

Geoex-86   Organised by *Intercosmos*; a remote sensing programme taking place within East Germany, for which imagery from *Mir* was correlated with that from aircraft, ground teams and Cosmos 1602.

GFZ-1   GeoForschungsZentrum (GFZ) was a 22-cm diameter, 20-kg sphere incorporating 60 small laser retroreflectors. A geodetic satellite built by Kaiser–Threde in Germany and ejected from *Mir*, it was to provide detailed data on the Earth's gravitational field. Because its orbit was lower than all other geodetic satellites, it was hoped that over its two-year life it would permit the distribution of mass within the planet to be mapped.

Glasar-1 A telescope for ultraviolet spectrography in the wavelength range 1,150–1,350 Å, with a resolution of 2 Å. It had a 40-cm primary mirror, and a 1.3-degree field of view. Although it used an electronic image intensifier, exposures of up to 10 minutes were needed to record faint stars (about magnitude 17) on film. There was a small airlock in the rear transfer compartment of *Kvant 1* to enable the cosmonauts to reload it with film. Exposed film was returned to Earth for processing. It was used to survey bright quasars, active galactic nuclei and stellar associations. It was developed jointly by the Byurakan Astrophysical Observatory in Armenia and the Swiss.

Glasar-2 An ultraviolet telescopic spectrograph on the *Kristall* module.

Globus (Globe) The main control panel of *Salyut 4*; a navigational display that indicated the station's position as it travelled around the world.

Glovebox (GBX) A *NASA* chamber on the *Priroda* module for experiments requiring physical isolation. It was fitted with television cameras to record visible results.

Glucometer A biomedical device on *Salyut 7* used to monitor carbohydrate exchange in cells. It showed that this process operated at a greatly reduced rate in microgravity, suggesting that cellular functioning might be disrupted during very long flights.

Granat A spectrometer on the *Kristall* module used to study the spatial characteristics and energy spectrum of radiation.

Greben An experiment on the *Priroda* module; an altimeter to measure mean sea level directly below the complex's path to an accuracy of 10 cm.

Greenhouse A *NASA* experiment conducted in the *Svet* cultivator to study plant reproduction, metabolism and biochemistry. It was first used to grow a strain of dwarf wheat. Shoots were extracted at regular intervals and frozen for return to Earth. They were 'harvested' when they produced seed, and some of these seeds were planted to make a second generation. It was the first time that a staple completed the growth cycle, and this success had important ramifications for food production and atmospheric processing on future stations. For this experiment, the *Svet* was enhanced by instrumentation from Utah State University to monitor light, temperature, air pressure, the level of carbon dioxide, water vapour, and substrate moisture.

Grif A gamma-ray detector on the *Spektr* module to measure emissions from the complex, generated by its passage through the Earth's magnetic field.

Gyrodynes This attitude control system was tested on *Salyut 3*, used operationally on *Salyut 5* and then built into *Salyut 6*, *Salyut 7* and *Mir*'s *Kvant 1* and *Kvant 2* modules. It used a set of electrically-driven flywheels on magnetic bearings, and converted electricity from the solar panels into torque to create an inertial attitude-control system that could reorientate the station without consuming propellant. It proved to be very responsive to commands. Although the system consumed power (each wheel was rated at 90 W) the propellant saving enhanced the station's sustainability. The two *Mir* modules each had six flywheels (each 165 kg) spinning in pairs, one pair on each cartesian axis. Once the desired orientation had been attained, the flywheels damped out vibrations from other equipment, to hold the station stable.

Gyunesh-84 (Sun)   Organised by *Intercosmos*, this was a photographic and spectrographic study of Azerbaijan. One of the targets was a remote-sensing test site. Data from *Salyut 7* was correlated with that gathered by aircraft and ground teams. It produced maps of the most economical grazing pastures for livestock, assessed the extent of infestation in forest regions, and identified subsurface water deposits.

Halong (a bay on the Vietnamese coastline)   An experiment on *Salyut 6*. It used the *Kristall* furnace to produce an alloy of bismuth, tellurium and selenium and an alloy of bismuth–antimony–telluride (BAT), and to grow a monocrystal of gallium phosphide semiconductor.

Hautey   An experiment on *Salyut 6* to investigate how yeast cells (a monocellular microorganism which grows extremely rapidly) divided as part of an ongoing study of intracellular processes.

Hermes   A French-sponsored project to develop a mini-shuttle for delivering crews to space stations and for servicing satellites in low orbit. In 1993 it was cancelled for financial reasons.

HEXE (High-Energy X-ray Experiment)   A set of four *Phoswich* scintillators on *Kvant 1*, sensitive to X-rays in the 15-200 keV range, and with a 1.6 x 1.6 degree field of view .

Hohmann transfer orbit   The most energy-efficient trajectory to move between any two orbits. Since energy corresponds to propellant consumption, most *Soyuz* orbital manoeuvres follow Hohmann elliptical transfers; the result is a slow rendezvous.

Horizon-Dawn   An experiment on *Salyut 6*. It used the *Spektr-15K* spectrometer to monitor sunlight scattered by transmission through the upper atmosphere at low angles of incidence immediately prior to dawn, and immediately after sunset, to measure the diffusion of various molecules and aerosols at an altitude of 100 km.

HPM   A German experiment on *Mir* to study hormonal changes, involving the collection of blood, saliva and urine samples.

HSA   The temperature and humidity control system on *Salyut 7*. Once, when its NOK-3 condenser pump failed and Lebedev opened the panel to examine it, he found an enormous blob of water floating inside the compartment. After water had precipitated on the cold, porous surface of the dryer, it was supposed to be pumped to a tank in the regeneration system to be recycled, but the pump had failed and the water had accumulated. When the pump was disassembled it was found that the bearing had broken. Ironically, a month earlier, Lebedev, his hearing keenly attuned to the sounds of the various apparatus within the walls of the station, had warned the flight controllers that the pump's bearing was running rough, but they had assured him that there was no problem.

HSD   A German experiment on *Mir* devised by the University of Berlin. It employed the *Tchibis* suit to determine tissue layer thickness and compliance.

Igla (Needle)   A rendezvous radar transponder introduced by *Salyut 6*. As before, it required the station to reorientate itself to face an approaching spacecraft. Once the spacecraft's final transfer orbit brought it within 25 km of a station, *Igla* was activated. It provided data on range, closing rate, line-of-sight angular velocity, and perpendicular deviation from the straight-in approach vector. It automatically controlled the approach until it was 200 metres out, then paused so that the crew could perform a final check. At the 200-metre pause, it rolled the vehicle to align for the final phase. If everything was as it should be, it was standard practice for the cosmonauts to let a ferry perform an automated docking.

Ikar An experiment on the *Priroda* module that was built jointly by Russia and Bulgaria. It comprised three sets of radiometers sampling at six wavelengths in the 0.3–6.0-cm microwave range. The Ikar-N radiometers looked straight down at the ground track, the Ikar-D scanning radiometers had an oblique view and Ikar-P offered a panoramic view. The oblique-looking radiometers viewed a swathe of track up to 750 km wide.

Illuminator This experiment recorded the degradation of the glass of *Salyut 6*'s portholes due to exposure to space. The accumulation of particles on their external surface was recorded by taking photographs with the Pentacon-6M camera. Changes in their transmission characteristics were measured using the *Spektr-15K* spectrometer. During a spacewalk, one of the cosmonauts wiped the glass of one of the portholes with his glove to try to collect some of this dust for analysis (this smear was still clearly noticeable a year later), but noted that it seemed to be embedded in the glass rather than accumulated on its surface. The cosmonauts regularly examined the glass for any signs of impacts by micrometeoroids. The portholes had a double layer of glass, each layer 14 mm thick. A 4-mm diameter crater was found in one porthole, and the micrometeoroid particle appeared to be embedded in the glass.

Illusion A French experiment on *Mir* which used the *Physalie-M* apparatus to investigate how the sensory and motor systems adapt, in a continuation of a previous experiment.

Imitator An experiment on *Salyut 6* which measured the thermal profile of the *Kristall* furnace in its operating environment. This, together with the examination of the malfunctioning *Kristall-2* unit, enabled the designers to develop an optimised smelting procedure that would yield extremely homogeneous crystals.

Immunity An experiment on *Salyut 6* to investigate changes in the proteins and minerals directly related to the body's immune system by determining antibody and immunoglobin levels in blood.

Immunology A French experiment on *Mir* which studied the characteristics of the immune system.

Impulse An apparatus on *Salyut 4* to assess the state of the vestibular apparatus by measuring the threshold of sensitivity of the vestibular system to different stimuli (brain impulses which create the illusion of a yawing or banking motion in the absence of gravity). It was also on *Salyut 5*.

Inquiry A psychological test.

Inkubator-1 An apparatus on *Salyut 6* to study embryo development. Quail eggs developed much more slowly than those on the ground.

Inkubator-2 A Czech-supplied apparatus on *Kvant 2* to further the study of embryonic growth. The quails that hatched after two weeks were the first Earth-based creatures to be born in space. After another two weeks, it was evident that the chicks were not developing properly; they were extremely frail, and appeared to be unable to feed themselves from the food tray in their habitat. Despite this, Dr Ganna Maleshko, of the Institute of Medical Biology in Moscow, reported that they had demonstrated that the embryo had developed normally; this was a significant step forward in the study of adaptation to weightlessness. Such experiments were a vital precursor to the development of closed-environment spaceborne habitats, because fowl could eventually contribute a valuable source of food. Subsequent investigations studied the post-hatching development process.

Intercosmos   A research organisation founded by the Soviet Academy of Sciences. In 1976 a number of Socialist countries signed an accord with the Soviet Union each to fly a cosmonaut to a *Salyut* station under this banner.

Interferon   An experiment on *Salyut 6* to test the possibility of manufacturing interferon (a chemical produced by human cells which interferes with the development of viruses) on a commercial scale in space. One part of this experiment involved injecting a vial of human white corpuscles into an interferon-producing substance to determine whether the rate of production was increased in microgravity. A related blood analysis test determined whether weightlessness affected the generation of interferon within the body. Another test evaluated whether existing interferon pharmaceutical preparations, which had been delivered both in the liquid state and in lyophillised gel, were rendered more or less effective against viruses by being exposed to weightlessness. Subsequently, attempts were made to to determine the capacity of human lymphocytes to synthesise interferon.

Iskra-2 (Spark)   A satellite released from *Salyut 7*. This was the first time that a manned station in Earth orbit had released a satellite (although satellites had been left in lunar orbit by *Apollo* missions). The 28-kg package had been built by students of the Moscow Institute for Aviation. Its hexagonal structure incorporated a simple repeater designed to relay communications between radio hams across the Soviet Union. Its surface was embossed with the emblems of the countries participating in the *Intercosmos* project. It had no propulsion, and was simply ejected from the scientific airlock by a spring and left to drift away from the station. Prior to deployment, the cosmonauts tested its battery and transmitters. It deployed its two side antennas immediately it cleared the airlock. Its orbit rapidly decayed and it re-entered the atmosphere about two months later.

Iskra-3   The second amateur radio satellite deployed from *Salyut 7*. Its orbit decayed within a month.

Isparitel (Evaporator)   A 24-kg electron-beam gun designed by the Ukrainian Academy of Science's Institute of Electrical Welding in Kiev, used in the Vaporiser experiment to deposit thin coatings. It was tested on *Salyut 6* and *Salyut 7*, and was set up in the scientific airlock. Exposures could range between 1 second and 10 minutes, depending on the thickness required. At its finest setting, it could deposit a layer of just a few µm, at its thickest about 1 mm. After some teething problems, this low-voltage electron beam was used to melt tiny granules of either aluminium or silver so that they could be vaporised and sprayed to condense on small disks of carbon and titanium, to determine whether such coatings would be practicable as a way of protecting future stations from the deteriorating effects of the space environment. This test followed directly from the successful respraying of the main mirror of the *OST-1* on *Salyut 4*.

Isparitel-M   Apparatus on *Salyut 7* to extend the *Vaporiser* experiment and further investigate the deposition of thin-layer coatings.

Istok   A long-duration exposure cassette containing connectors and bolts.

Istok-1   An experiment on the *Priroda* module, built jointly by Russia, Poland, Romania and the Czech Republic to study the oceans. Its infrared radiometer sampled in 64 channels between 1.6 µm and 3.6 µm. It incorporated a television camera sampling in the range 0.4–0.75 µm.

ISX A German experiment on *Mir*. It involved wearing knee restraints while performing calf exercises to evaluate the effect of isometric exercises on muscles, blood pressure and heart rate.

ITS-7D An infrared spectrometer mounted on the *ASPG-M* platform. It sampled in the 4–16 μm range, and was used to study the transition zones between the Earth's surface and the lower atmosphere and the upper atmosphere and space.

ITS-K A telescopic infrared spectrometer carried on *Salyut 4* and *Salyut 5*. It could be used to observe the sky, but its primary objective was the Earth's atmosphere. The infrared sensors carried by previous stations had used conventional compressor-based cooling units, which had used large amounts of electricity and been unreliable. In this case, however, a far more advanced cryogenic system was employed which used an 'ice coating' of solid nitrogen. This had been designed by the Kharkov Physical–Technical Institute of Low Temperatures, in the Ukraine. Not only did this draw far less power, it could be run for extended periods at –223°C. The telescope employed a 30-cm diameter mirror and projected a 20-arcmin field of view onto the slit of the spectrometer. It incorporated a fluorite prism which selected radiation in the range 1–7 μm to yield a spectrum with a resolution of 600 lines per millimetre. With the slit parallel to the horizon at sunrise and sunset it was used to measure the temperature of the layers of the upper atmosphere and to determine the density and distribution of water vapour (using the 2.7 μm absorption feature). It was so sensitive that it could detect minute changes in the concentrations of water vapour in the upper atmosphere. To make such observations the spectrometer was aligned exactly on the centre of the Sun's disk, and measurements were taken as the Sun rose or set behind the horizon (the slit of the spectrometer was aligned parallel to the horizon). It was a fleeting opportunity, but the automatic attitude control system was able to add data regularly as the station orbited the Earth every 92 minutes. At such times, it was possible to measure sunlight passing through a 'layer' of air up to 1,000 km deep. The infrared spectrometer measured water vapour absorption. The general level of understanding of the conditions and processes operating in the upper atmosphere were so basic that until the matter could be resolved, two competing models were to be tested by the observations. One model assumed a dry atmosphere, and the other assumed a moist atmosphere. The difference in the levels of possible moisture concentrations according to these two models is of the order of between 200–300 per cent. Although the data from *Salyut 4* argued in favour of the dry model, the observations were only preliminary, applying only to a narrow range of locations and times, and later studies established that the temperature of the upper atmosphere ranged between 400°C and 1,700°C because of heating resulting from the absorption of most of the Sun's ultraviolet radiation. Such data were important for achieving an understanding of global weather systems and long-term climatic changes. It was concluded that nitrogen oxide, formed in the disintegration of the flow of corpuscular particles from the solar wind, was the main source of infrared emission from the upper layers of the atmosphere. It also helped to define the energy spectrum of the Sun across the infrared range (absorption by water vapour in the lower atmosphere prevents this being done from the ground), refined the value of the solar constant (the total amount of energy emitted by the Sun) and enabled the level of carbon monoxide molecules in the solar corona to be determined (it is thought that this plays a major part in the process which heats the corona to one million degrees).

Izvestia (News)   A newspaper in the Soviet Union.

Kaliningrad   The flight control centre (known as TsUP, pronounced "soup") in a north-eastern Moscow suburb, superseding *Yevpatoria*. It was first used in 1973, and controls all *Mir* operations. It was recently named the Sergei *Korolev Control Centre*.

KAP-350   An advanced topographic mapping camera in *Kvant 2*.

Kapillyar (Capillary)   An experiment on *Salyut 6* with the *Kristall* furnace. A crystal of germanium was grown by using capillary action within a molybdenum matrix. One novel test grew a disk-shaped monocrystal of silicon to evaluate the feasibility of producing solar cells in space during the construction of large power systems in orbit.

Kardiokassett   An experiment on *Salyut 6* to evaluate a device to measure cardiovascular parameters for the right-hand part of the heart.

Kardiolider (Cardioleader)   A Polish apparatus on *Salyut 6* designed to record cardiovascular reactions via one chest sensor which monitored cardiac strain and another which monitored the speed of the ergonometer. By matching the level of exertion to the performance of the heart a cosmonaut could be given warning if the exercise was excessive, and this would enable an optimum programme of exercises to be developed for future flights.

Kaskad (Cascade)   An automated attitude control system tested by *Salyut 4*. Orientation was measured with respect to infrared horizon sensors, which established the local vertical, and the *Neytral* velocity-vector sensor which determined the angle of the station with respect to its direction of travel. It greatly assisted observational work (previous crews had spent up to 30 per cent of their time on routine tasks related to orientating earlier stations), and also resulted in "a considerable reduction in fuel consumption". It was made operational on *Salyut 6*. When refined for *Salyut 7*, it could orientate the station to within 1 degree (an order of magnitude better than before), and was considered sufficiently reliable to permit the station to be manoeuvred without crew supervision.

Kasyun   An experiment on *Mir* which made use of the *Kristallisator* furnace to smelt an aluminium–nickel alloy.

KATE-140   A 140-mm focal length large-format topographic mapping camera. From a 350-km orbit, pointed straight down at the ground, its 85-degree field of view could photograph an area of 450×450 km with a resolution of 50 metres. It operated in both the visual and infrared bands. The film cassette held 600 pictures. It was a stereoscopic topographical mapping camera, and it could give individual frames or create an extended strip-image. It could be operated either by the crew or by remote control from the ground. It was carried on *Salyut 6* and *Salyut 7*, and although the one on *Salyut 7* was transferred to *Mir*, a second was soon sent up.

KATE-500   A multispectral camera flown on *Salyut 4*. It had 500-mm focal length.

Kazbek-Y   The couches in the *Soyuz* ferry. Each cosmonaut has a personalised contoured couch liner. When a cosmonaut moves from one ferry to another, the liner and the *Sokol* pressure suit are transferred.

KFA-1000   A film camera with a ground resolution of 5 metres.

KFV   A German experiment on *Mir* that used the applied potential tomography (APT) apparatus and the *Tchibis* suit to investigate changes in the distribution and flow of body fluids. The APT measured the distribution of fluids in different parts of the body.

KGA-1   A holographic camera flown up to *Salyut 6* to further study the deterioration of the portholes. This 5-kg device used a helium–neon laser to create a series of holograms. It was tested by recording a dissolving salt crystal, to reveal how the density of the crystal was distributed through dissolution without convective flaws. The imaging process was extremely sensitive to vibration interference; on Earth, it required an enormous supporting structure to eliminate interference. The laser system had been developed by the Physical–Technical Institute in Leningrad. In the future, holographic cameras would enable engine performance and other processes which are otherwise difficult to monitor to be investigated.

KGA-2   A holographic camera on *Salyut 7*.

Khrunichev   The manufacturer of the *Soyuz*, *Progress*, *Almaz*, *Salyut*, *Mir* and *TKS*-based spacecraft.

Kinesigraph   A French experiment on *Mir* to make stereoscopic images with which to investigate the restitution of movement of the corporal segment.

KL   Two television cameras (KL-103 and KL-140) on the *ASPG-M* scan platform, used to indicate to the remote operator what the primary instruments were looking at.

Klimet–Rubidium   A materials experiment on *Mir* which used the *Kristallisator* furnace using a mixture of rubidium, silver and iodine to test a concept for extremely lightweight batteries and condensers. Because this experiment consumed so much power, it was performed while the cosmonauts slept.

Kolosok   An experiment on *Mir* to investigate aerosol structure in zero gravity.

Koltso (Ring)   A walk-around communications system in *Salyut 6* which cosmonauts could operate without having to return to the control panel to speak to the control centre.

Kometa   An experiment on *Salyut 7*. It involved making observations designed to improve a mathematical model for reproducing the colour of the sea as viewed through the atmosphere so that more accurate colour-based measurements would be able to be made in the future.

Komplast   A cassette outside *Salyut 7*. It contained non-composite materials whose physical properties were to be tested after prolonged exposure to the space environment.

Komza   A Swiss–Russian experiment on the *Spektr* module to sample gas in the immediate vicinity of the station. It employed interchangeable cassettes.

Kondor   A Canadian experiment on *Mir* designed to develop a means of controlling the radiation within the complex.

Kontrol   An experiment to test the structural stability of the docking collar on which *Kvant 2* was mounted.

Konus (Cone)   The conical drogue in the standard probe/drogue docking system. The unit rotated into the station's docking compartment. The multiple docking adaptor on *Mir*, a 2.2-metre diameter sphere, was too cramped to accommodate five drogue assemblies, so, since it was not operationally necessary that they all be so equipped, only two were installed, and the other ports had flat covers. One drogue was kept on the axial port, where spacecraft docked, and the other was relocated to facilitate the movement of modules to the radial ports. (Moving the drogue usually required depressurising the compartment, and this was referred to as an 'internal spacewalk'.)

Korabl–Sputnik   Any spacecraft incorporating a cabin for a crew.

Korolev Bureau   The rocket design bureau (OKB-1) established by Sergei Korolev.

Korolev Control Centre   The control centre at *Kaliningrad*, renamed after Sergei Korolev by President Boris Yeltsin.

Korund   A 200-kg electric furnace on *Salyut 7*. This incorporated a revolving sample holder which could be loaded with up to a dozen ampoules. Although each sample was processed singly, it could be set up to operate either automatically or by remote control from the ground. This meant that a departing crew could prepare it so that it could run after they had gone (when they would no longer be around to vibrate the station). It would also be possible to set up such a semi-automatic system in a free-flying crew-tended module. In contrast to the earlier *Splav* apparatus, in which the sample was fixed and the thermal field was varied to control the growth of a crystal, and *Kristall*, in which the sample was moved while the thermal field was held constant, the Korund furnace was capable of moving the sample (in an ampoule 25 mm in diameter and 30 cm long) at speeds varying from a few millimetres a day to a few centimetres a minute, and could vary the temperature between 20°C and 1,270°C at rates of 0.1°C to 10°C per minute, to produce a monocrystal of up to 1.5 kg. With a fully loaded rotating table of samples, it was able to create 18 kg of monocrystal without human supervision. It was hoped that the second generation technology would demonstrate the feasibility of an orbital manufacturing facility for extremely pure semiconductors which could be used either on an unmanned spacecraft or a later station. Unfortunately, it switched itself off as soon as it attained its operating temperature. The cosmonauts realised that the thermal coefficients of the heaters had been defined by ground tests, where thermal convection had drawn off energy, but in microgravity, where convection does not develop, the elements had heated up much more rapidly, leading the control system to believe that it was overheating. Once reconfigured with more appropriate coefficients it was found to be usable, so they started a crystal-production run to test it. Then the outer compartment overheated (to 100°C) due to a design flaw. Because it was such an important apparatus, however, the cosmonauts figured out a way to repair it, but even so, it could not be used for more than a few hours at a time. Later, the mechanism that moved the ampoules within the thermal chamber jammed and its motor burnt out.

Korund-1M   A 136-kg semi-industrial-scale furnace on *Mir,* for crystal growth. It was an improved form of the apparatus tested on *Salyut 7*. The samples were processed individually for periods ranging between six and 150 hours. Once set up, it was completely automatic, but because it consumed 1 kW, its use had to be limited.

KR-5   A spectrometer on the *Spektr* module.

Krasnaya Zvezda (Red Star)   The newspaper of the armed forces of the Soviet Union.

Krater   A furnace (strictly speaking a melting zone) on the *Kristall* module used to study crystallisation of a high critical-temperature superconductor, to produce epitaxial layers of silicon (that is, growing the crystalline structure on the surface of an existing crystal in such a way that the two lattices are aligned), and to produce monocrystals of exotic alloy (such as barium oxide, yttrium oxide and copper oxide) and semiconductors (notably gallium arsenide). Many smelts ran for ten days.

Kristall (Crystal)   The 'technology' expansion module of the *Mir* complex. It had a press-urised volume of 60 m^3 and two solar panels with a total area of 72 m^2 which provided 9 kW. Its solar panels were retractable, and could be extended out to a maximum 36-metre span. Its primary scientific payload was a bank of furnaces, but it also con-tained Earth-observation apparatus.

Kristall   An experiment on *Salyut 5* to study how crystals grow in microgravity condi-tions. Seed crystals were placed in containers with a water solution of potash and alum. Once the crystals had matured, they were extracted for return to Earth. It was hoped to produce much larger monocrystals than was possible on Earth. The first experiment ran for three weeks. In the second, a dyeing agent was added to the solution of potash alum to study the diffusion of the mixture into the developing crystalline structure.

Kristall-1   An electric furnace used to create monocrystals of semiconductors. Unlike *Splav*, which it superseded, it could be operated within the station (set up in the rear transfer compartment). An improved version of the *Splav*, it used zone-melting to crys-tallise semiconductors. Whereas *Splav* created a temperature gradient to facilitate crys-tallisation, Kristall employed uniform heating and exposed a sample to a steady-state thermal zone at a temperature in the range 400–1,200°C. It could operate in four differ-ent ways: the first method produced monocrystals from the gaseous phase by sublima-tion, which evaporated the sample and then transported the gas to the cooling zone so that it could settle on and enhance the seed; a second produced films of monocrystal using chemical gas-transportation; a third produced monocrystals using a 'moving sol-vent' to obtain a high-temperature solution; and the fourth method employed a seeding technique in which as the temperature was slowly reduced in the crystallisation zone one side of the sample was cooled while the other side was kept hot. This yielded a seed on the cold side of the sample, and as it was slowly stretched across the zone, incremen-tal crystallisation produced a single, highly regular crystal. Also, because the tempera-ture could be controlled more accurately than in *Splav*, the crystals were considerably more homogeneous. The samples were carried in capsules which were 10 mm in diame-ter and 175 mm long, and could be moved through the heating chamber at rates be-tween 0.188 mm and 0.376 mm per minute. Although this was the first time that this apparatus had been flown, when it was tested it produced a monocrystal of gallium arsenide employing the high-temperature solution technique (this semiconductor could be used in the construction of highly efficient solar collectors). It was tested on *Salyut 6*, and worked well, but broke down after successfully performing 40 smelting operations.

Kristall-2   Flown up to *Salyut 6* to replace the earlier model.

Kristall-3   Flown up to *Salyut 6*.

Kristallisator   A Czech-built automated semiconductor furnace on *Salyut 7*. It could maintain heat and pressure constant at temperatures up to 1,000°C, to grow crystals in smelts lasting from several hours to several days. The unit on *Salyut 7* was transferred to *Mir*, and then used for a series of experiments to investigate the process of crystalli-sation in silver–germanium and lead chloride–silver chloride, which are eutectic alloys (that is, they have an extremely low freezing point).

KRT-10   A 10-metre wide radio telescope. The 350-kg unit comprised the antenna and a device to attach it to the rear docking unit of *Salyut 6* in 1979. It was a hexagonal structure of rods deployed by springs. By using shorter rods in the centre and longer rods at the edge, it formed a parabolic dish. A fine mesh over the surface formed the reflector, and a 5-metre tall triangular mount supported the five radiometers (four horns in the 12-cm band and a spiral antenna operating in the 72-cm band). It was attached to the rear docking assembly, projecting, furled, within the *Progress* ferry, so that it was exposed when the ferry withdrew, at which point it was commanded to deploy. This action was televised by the ferry. As it unfurled, it completely hid the view of the station behind. A video tape had been delivered showing the crew how to operate it. They monitored its reaction to a series of pointing commands. On alternate days (to permit the batteries to be recharged) over the ensuing weeks, it was to be used to carry out a series of pioneering observations. It was to be used both to observe the Earth and celestial sources. When aimed down, it could reveal the structure of meteorological phenomena, map sea state, and measure water salinity and soil humidity, but the 12-cm band was limited to 7,000-metre resolution. For the astrophysical part of its programme, it was to work in conjunction with the new 70-metre ground-based radio telescope of the long-range space communications centre in the Crimea (together creating an interferometer with a diameter as wide as the Earth). It gave mixed results, but it was seen as being just an early test of the use of an orbital telescope, and so the very act of trying to use it generated valuable design feedback. For the deep sky observations it had been used in two modes: in one mode the station had been stabilised so that the telescope gave continuous coverage of a given source, and in the other mode it was set in a rotation which had been orientated so that the telescope would scan the Milky Way, so that it could map sources. When aimed at the Earth, it demonstrated that such an instrument would be able to provide useful geological and oceanic data. It had also been used to monitor an eruption by Mount Etna. The dish was jettisoned to clear the rear docking hatch, but, as it slowly drifted away, it began to wobble and became entangled in the apparatus which projected from the back of the station. The mesh could be seen clearly by the television camera on the docking unit. At the suggestion of ground controllers, the station's main engines were fired to try to blast the dish free and leave it behind, but this failed to dislodge it. The next suggestion was that the station should be pitched up and down to shake it free, but this did not work either. Eventually, Vladimir Lyakhov and Valeri Ryumin spacewalked to release it. Although the task involved only four quick snips with the pliers, it was a potentially dangerous operation because Ryumin had to crawl down between the dish and the rear of the station. As a general rule, mission planners regard any item not specifically built to be manipulated by spacewalking cosmonauts as a potential death trap, and Ryumin was reminded of this when he cut a strand of the entangled mesh away and discovered that he had set the heavy structure rotating towards him. Finally, after he had completely cut it free, he pushed it away from the station.

KSS-2   A photographic spectrograph carried on *Soyuz 13* and *Salyut 4*. This 2-channel fast-operating spectrometer was aimed at the centre of the solar disk, its slit parallel to the Earth's horizon, at the day and twilight horizons, to measure the strength and distribution of atmospheric pollutants and the vertical distribution of water vapour in the atmosphere. It could also be used to measure the reflectance spectrum of the Earth's surface under direct insolation and distinguish soils with different moisture content.

KTF   The Kompleksny Trenazher Fisichesky was an integrated physical trainer based around an electrically driven moving belt treadmill almost a metre long by half a metre wide. It was used in conjunction with a customised harness using adjustable elasticated cords which attached to the framework of the track in order to provide a hold-down load of around 50 kg. KTF-1 was installed on *Salyut 1*.

Kurs (Course)   The advanced rendezvous radar transponder introduced by *Mir,* which did not need the entire complex to continually reorientate itself to face an approaching spacecraft; it need only be reorientated so that the spacecraft was within the transponder's field of view. It operated in the S-Band.

Kursk-85   A survey of the Kursk Oblast region. *Salyut 7*'s *MKF-6M* and *KATE-140* data were correlated with the *Meteor-Priroda* satellite and aircraft and ground teams.

Kvant 1 (Quantum)   An expansion module for the *Mir* complex. It was delivered by a *TKS*-based tug and docked at the rear of the base block. It had a pressurised volume of 40 m^3 and contained the Vozdukh and Elektron regenerative life support systems. It had no power-generating capability of its own to run its astrophysical observatory.

Kvant 2   An expansion module for the *Mir* complex. It had a pressurised volume of 60 m^3 and two solar panels with a total area of 50 m^2 which provided 7 kW. It incorporated a large airlock, a *Bania*, a second *ASU*, a second *Elektron* life-support system, and the *Vika* oxygen production system (to replenish the airlock). Its primary scientific payload was for Earth observation (primarily the *ASPG-M* and the *MKF-6MA*), but it also contained astrophysical apparatus.

Kyulong   A Vietnamese experiment on *Salyut 6*. It used the *MKF-6M* and *KATE-140* cameras to carry out Earth-resources observations of the Vietnamese peninsula, to make the first comprehensive survey of its geological, botanical, mineral and coastal resources to yield information on tidal flooding, hydrological features of the Central Plateau and silting in the Mekong and Red River deltas, and assess the effects of defoliants sprayed during the war, and used the *Spektr-15K* spectrometer to assess the health of crops.

Labrint   An experiment on *Mir* which studied the interaction between the visual and vestibular systems in the early phase of adaptation to weightlessness.

Levka (Lion's Cub)   An experiment performed on *Soyuz 13* which measured blood flow to the brain as it adapted to weightlessness. It involved a chest expander which imparted a force of 15 kg when stretched. A cosmonaut was to flex this at a rate of 30 times a minute while electrodes measured the way in which the flow of blood to the brain developed.

Levkoi (Gillyflower)   An apparatus on *Salyut 4* which measured specific features of blood flow within the brain. Also on *Salyut 5*.

Lidar   A radar which uses a laser pulse rather than a radar pulse.

Ljappa   A mechanical arm carried by the modules that docked at the front of the *Mir* complex. It transferred the modules from the axial port, where they docked, to a permanent position on a lateral port. This involved performing a 45-degree rotation, so the axial collar was offset by that amount to ensure that the modules would be properly aligned once finally docked.

Logion　An Austrian experiment on *Mir* to study the characteristics of a liquid metal ion source to evaluate whether it could be used to remove the electrostatic charge that accumulates on a satellite in orbit. It was done to try to find a way of making a magnetospheric research satellite more sensitive, by discharging it so that its own charge would not interfere with extremely fine measurements.

Lotos　An apparatus on *Salyut 6* to mould shapes of structural elements from polyurethane foam.

Luch (Beam)　An as yet incomplete *SDRN* requiring three geostationary satellites to relay signals (voice, telemetry and real-time experiment data) between *Mir* and Kaliningrad to provide continuous communications coverage. More often than not, however, the satellites are leased to relay commercial television.

Lungmon　An Austrian experiment on *Mir* which tested a new electrical heart and lung monitoring unit.

LV-1　The optical viewfinder on *Salyut 7*. It provided a ×6 magnified image, so was often used to observe features on the Earth in somewhat higher resolution than was possible with the naked eye.

Lyulin　An experiment on *Mir* which used a microprocessor (*Zora*) to test psychological and physiological analysis of their reactions, in particular their reaction time.

Lyulin　A dosimeter incorporating a microcomputer processor used to make measurements of the flux of ionising radiation, providing a high-resolution temporal profile around the complex's orbit.

Magma-F　An electric furnace on *Salyut 7*. It was an improved form of the *Kristall* furnace used on *Salyut 6*. An accelerometer and a magnetic sensor had been added to measure vibrations affecting the melting and crystallisation processes, as vibrations tended to disrupt the process. It could accommodate quartz ampoules up to 20 mm in diameter and 200 mm long, could heat samples to 900°C and continuously monitor its temperature at 14 points, and send this data to Earth as part of the telemetry stream. To eliminate perturbations caused by the cosmonauts moving about inside the station, it was found to give better results if it was unbolted and then left to float freely. At one point, cadmium theioselinide (a material with important laser and communications applications) was produced.

Magnetobiostat　A plant apparatus on *Salyut 6*. Popov and Ryumin had made a nationwide appeal for suggestions for better ways of growing plants in the absence of gravity; the use of a magnetic field had been thought an excellent idea, so the necessary apparatus had been hurriedly put together. *Arabidopsis*, crepis and ginseng grew successfully. A similar unit was installed on *Mir*.

MAK-1 (Poppy)　A small (16-kg) satellite equipped to study the upper atmosphere and radio its data direct to Earth. It was ejected from *Mir's* scientific airlock. It was meant to unfurl a wide dish antenna immediately, but did not do so, probably because its battery was flat. It had been intended to carry out a three-day programme. As the first in the series, it had two objectives: firstly, to verify the electrical, temperature regulation and telemetry systems, and secondly, to run the Fokus payload to study phenomena arising from electrons and plasmas at orbital altitude.

MAK-2　A small satellite ejected from *Mir*'s scientific airlock to investigate the physical characteristics of the ionosphere.

Malakhit　A greenhouse apparatus on *Salyut 6*. It used a synthetic soil incorporating capsules which released ion-exchange fertiliser at a controlled rate. It was used to investigate the growth of orchids which, because these were adapted to a dry atmosphere, it was hoped might prove more able to survive in the station's artificial environment. Unfortunately, even though the orchids had already started to flower when they were delivered, they deteriorated immediately and the flowers disintegrated, but once back on Earth they reflowered. However, orchids planted in space developed roots which were long enough to protrude from the apparatus, but they did not flower. In a similar unit on *Salyut 7*, tomatoes, coriander, celantro, radishes, borage and cucumbers were all planted. Unfortunately, the borage and radish did not grow at all; the celantro grew to a height of 7 cm and then died; only the cucumbers thrived.

Marina-2　A gamma-ray telescopic spectrometer on the *Kristall* module.

Mariya　A magnetic spectrometer to study the mechanisms by which flows of high-energy particles, X-rays and gamma rays are generated in near-Earth space. It was first used on *Salyut 7*, and an improved form was installed on *Mir*. Its data was used to evaluate a correlation between concentrations of charged particles streaming in the Earth's magnetic field in space above areas of the Earth's surface subjected to seismic activity. Subsequent terrestrial studies revealed that magnetic fields are indeed associated with tectonically active rock.

Mariya-2　A magnetic spectrometer on the *Kristall* module.

Mars　A visual test performed on *Salyut 7* to assess the ability to distinguish subtle colour variations. (It was planned to employ colour charts to calibrate observations for some experiments).

Maskat　An experiment on *Mir* which tested the ability of given compounds to enrich genetic material.

Medilab　A French experiment on *Mir* which involved taking a series of blood and urine samples to test for hormonal changes.

Medusa　A cassette attached to *Salyut 6* to expose materials to the space environment. It involved a number of biological samples, half of which were attached to the outside of the station, whilst the rest were inside. The external cassettes held quartz ampoules with various mixtures of amino acids and other 'building blocks' of life. Mounting them outside exposed them to the full effect of solar and cosmic radiation in the station's orbital path. These would eventually be retrieved and compared with containers of biopolymers (microcultures which make up any living organism) kept within the station. The experiment, related to 'the problem of the origin of life in the Universe,' was designed to identify any functional changes which an elementary living culture undergoes as a result of being exposed to the ambient space environment over a prolonged period. Analysis demonstrated that exposure to sunlight caused components of nucleic acids to develop into substances similar to nucleotides; a step in the process towards creating nucleic acids. A similar biopolymer cassette was placed outside *Salyut 7* to further study the effect of the radiation environment on the chemical processes fundamental to the creation of life.

MEEP The four exposure cassettes of *NASA*'s **Mir Environmental Effects Package** (MEEP) were affixed to the *DM* by spacewalking astronauts. Two of the cassettes sampled ambient particulate matter (one was passive, but the active one recorded the time, size and trajectory of all particles hitting it). The other two exposed construction materials (including insulation, paints, and optical glass coatings) to be used on International Space Station Alpha.

MEFCE (Mir Electric Field Characterisation Experiment) A *NASA* experiment on *Mir*, set up to sample ambient emissions in the 400 MHz to 18 GHz radio range.

Membrane A biomedical experiment on *Salyut 7* to study calcium loss in bone by measuring the exchange of calcium in cell membranes. It revealed that taking an antioxidant preparation could significantly reduce this calcium metabolism.

Mera A long-range radio-technical system on *Salyut 7,* to assist approaching spacecraft. On its first test, although it was switched on at a distance of 250 km, the scanner did not lock on until the separation was down to 30 km. Nevertheless, it led the ferry in until the *Igla* could take over.

Merkur (Mercury) The three-seat descent module developed for *Almaz* and its associated *TKS* ferry. It had an *Apollo*-style blunt-cone configuration. There was a hatch in the side for entry, and a hatch through the heat shield in the base to gain access to the parent vehicle behind. After separation, thrusters on its nose orientated it for its deorbit burn; releasing the tower exposed the parachute compartment at its base. Several capsules were launched (in 1976, 1978 and 1979) to verify the re-entry control system, and those released by the *TKS* were successfully recovered (one of which returned cargo from *Salyut 7*; it could return 500 kg of material, which was ten times the capacity of the *Soyuz* returning with a crew of two), but it was never flown by cosmonauts. The empty capsule had a mass of 2.5 tonnes.

Meteor A class of meteorological satellite. One of the Meteor-3 series included TOMS, an American-built ozone mapper.

Meteor–Priroda (Nature) A satellite launched in June 1980. Its instrumentation included a ten-channel multispectral camera (Fragment) which transmitted 2,000-km long swath-images 30 km wide, with a resolution varying between 30 metres and 800 metres. During the early 1980s, Earth observations by the *Salyut* crews were often correlated with this data.

MGAS (Metabolic-Gas Analyser System) A *NASA* biomedical monitor on *Mir* to assess metabolic response to exercise. It measured inspired and expired air to yield data on protein metabolism. This was used in conjunction with the *NASA* ergonometer. It had previously been flown on a life sciences Spacelab mission.

Microaccelerator A French experiment on *Mir* which used a video camera to measure microscopic accelerations.

MIDAS (Materials In Devices As Superconductors) A *NASA* experiment on *Mir* to study the electrical properties of high-temperature superconductors.

Migmas An Austrian experiment on *Mir* which tested a mass spectrometer intended for use in microgravity. A similar device was under development for *ESA*'s *Columbus* module as a materials analysis tool.

MIK The building in which *Semyorka* rockets are integrated with *Soyuz* spacecraft and loaded on the rail transporter.

Mikrovib An Austrian experiment on *Mir* which studied skin sensitivity by analysing the spontaneous and stimulated microvibrations of the body's surface in the absence of gravity.

MIM The Microgravity Isolation Mount (MIM) on *Mir* supplied by Canada. It employed a magnetic field to damp out ambient microaccelerations, and isolated apparatus from microvibrations in the range 0.01–100 Hz. It was placed in *NASA*'s *Glovebox* so that residual vibrations transmitted onto the surface of a liquid could be videotaped for its initial trial. *SAMS* simultaneously recorded the ambient vibrations to assess the MIM's ability to filter them out. It was hoped that such an active filter would be able to offer a significantly better vibration environment for microgravity research.

Minidoza An experiment on *Salyut 6* to measure the flux and energy spectrum of *cosmic rays*.

MIPS The *Mir* interface-to-payloads system used a portable computer to enable *NASA*'s apparatus on *Mir* to download data via the station's telemetry link.

Mir (Peace, New World, Community) The 'base block' of the Mir orbital complex; a habitat module with permanent facilities for a crew of two, providing power generation and regulation, navigation and attitude control. It comprised three main compartments (the multiple docking adapter, the main compartment, and the rear transfer compartment), around which was the unpressurised engine compartment. It had a pressurised volume of 90 m^3, two solar panels with a total area of 76 m^2 providing 9 kW, and a third panel of 22 m^2 which was subsequently installed to bring the overall output to 11 kW. Over a 10-year period, it was expanded by the addition of five modules (*Kvant 1, Kvant 2, Kristall, Spektr* and *Priroda*).

Mir 2 A follow-on to *Mir,* envisaged as an elaborate orbital complex comparable with *NASA*'s *Freedom* space station concept.

MIRAS The French–Belgian Mir InfraRed Atmospheric Spectrometer (MIRAS) was aff-ixed to a platform at the far end of the *Spektr* module. It scanned the absorption lines in the atmosphere's infrared reflection spectrum, to identify its trace constituents, to map their distribution, and to monitor the evolution of the atmosphere over a period of at least a year, in order to help understand the interaction between solar illumination and the atmosphere. Its data (which was stored in memory and downloaded daily) was integrated with that from satellites and aircraft. An earlier version of this instrument had been used on the ATLAS-1 space shuttle mission in 1992.

Mirgen An Austrian experiment on *Mir* which used blood analysis to evaluate the effect of space flight on genetic material.

Mirror A mirror-beam furnace on *Mir*. It employed two small lamps and a special optical system to smelt small samples at temperatures up to 1,000°C. It was to be used to experiment with different alloy materials, and tested procedures for furnaces in the *Kristall* module.

MISDE (MIr Structural Dynamics Experiment) A *NASA* experiment on *Mir*. A set of accelerometers and strain gauges were installed throughout the complex to measure the transient stresses resulting from manoeuvres and docking operations, and from thermal effects due to flying into and out of the Earth's shadow, to further characterise the microgravity environment.

MKF-6   A multispectral camera, flown on *Soyuz* 22 in 1976 to test it prior to installation on *Salyut 6*. The 200-kg package was mounted on the front of the orbital module instead of on the docking unit. It had been built by Carl Zeiss, at its factory in Jena in East Germany. From the 250-km orbit, the camera was able to image an area 110 × 150 km with 15-metre ground resolution. By testing the camera in this way it would be possible to identify any improvements which would be necessary before it was deployed operationally. When not in use, the camera's optics were protected by covers. According to Dr Heinz Kautzleben, Director of the Central Institute for Earth Physics in Berlin, the test version of the camera was capable of functioning in space for only a week. It produced 'photographs', each of which comprised six separate frames taken through filters selecting narrow wavelength bands between 0.46 and 0.86 μm (four visual and two in the infrared spectrum). A set of six film cassettes was needed to complete each photographic sequence across the full spectral range. After being developed on Earth, these images would be examined using a four-channel Zeiss MSP-4 multispectral projector. Unfortunately, because each cassette weighed 13 kg only two could be returned in the cramped *Soyuz* descent module, so the use of the camera had to be restricted in this test. Each cassette contained film for up to 1,200 images, however, so it would provide a comprehensive test of its capabilities. It was aimed vertically at the ground. Some observations were to be correlated with data collected simultaneously by Antonov-30 aircraft and ground teams in selected regions of the Soviet Union for the Raduga Earth-resources programme, to help interpret the imagery. Orbital multispectral photography offered a far more economic way of prospecting for scarce resources in the Earth's crust, but for its data to be interpreted its sensors had to be calibrated, so similar observations were made from aircraft flying at around 7,500 metres, and all of this overhead imagery had to be combined with ground observations of soil and atmospheric properties. An orbital platform can survey territory at a far greater rate than can a high-flying aircraft. In ten minutes an orbital camera can photograph 1 million km^2, a task which would take several years for an aircraft to complete.

MKF-6M   An improved form of the *MKF-6*, flown on *Salyut 6*. It was similar to the unit which had been tested on *Soyuz 22*, but upgraded to function for several years. At *Salyut 6*'s higher altitude (350 km) each photograph covered an area of 165 × 220 km with a resolution of 20 metres. There was a 60 per cent overlap between adjacent pictures in a sequence, so stereoscopic views could be generated. Simultaneous photographs were taken in six spectral bands (specifically, 0.46–0.50 μm, 0.52–0.56 μm, 0.58–0.62 μm, 0.64–0.68 μm, 0.70–0.74 μm and 0.78–0.86 μm) on film in cassettes holding 1,200 frames. Because photographic film deteriorated rapidly in the increased radiation environment at orbital altitude, new cassettes of film would have to be flown up, and exposed film would have to be returned before it fogged over, so a series of ferries would have to visit the station on a regular basis. One of its major tasks was to survey the planned route of the Trans-Siberian railway, then being built from Baikal to Amur. During a four-month tour, one crew took 18,000 images (it would have taken two years of surveying from aircraft, or a lifetime of fieldwork, to perform the same survey on Earth). At the request of the Ministry of Agriculture, it regularly photographed the Salsky test site near Rostov, in the Ukraine. This was one of four agricultural areas which had been set up to calibrate multispectral cameras on orbital stations (the other sites were at Voronezh, near Lake Baikal, and in Kirghizia). These had been planted with specific cereals, vegetables and grasses to assess the ability of such a camera to distinguish between various plants under different conditions. A similar apparatus was installed on *Salyut 7*.

MKF-6MA Contained in *Kvant 2*, this multispectral camera was an improved version of *MKF-6M*.

MKS An East German spectrometer flown on the *Intercosmos* 21 satellite.

MKS-M An East German spectrometer–camera carried on *Salyut 7*. It could be pointed with greater freedom than the other fixed cameras (namely the *MKF-6M* and *KATE-140*), and could be fitted with different spectrometers for studying the atmosphere (A) or the biosphere (B). The -AS combination sensed 6 wavelengths in the 0.7–0.8 μm range. The -BS combination sensed 12 wavelengths in the 0.4–0.9 μm range, selected to detect chlorophyll.

MKS-M2 A photoelectric spectrometer mounted on the *ASPG-M* platform on *Kvant 2*. It measured visible and near-infrared wavelengths, and was used to study the transition zone between the Earth's surface and the lower atmosphere. It was remotely operated and its results were transmitted to Earth.

MMK-1 A number of micrometeoroid detection panels with a total area of 4 m^2 were preset on the outside of *Salyut 4*. Each panel incorporated two layers of metal separated by a thin layer of insulation and formed a capacitor. When a micrometeoroid particle hit the outer layer it deformed it and generated a brief electric current which could be counted. The current indicated the size of the particle and its penetrating power. It could detect a particle as small as a billionth of a gramme.

MOL The US Air Force developed a manned orbiting laboratory to serve as a reconnaissance platform. It was to be launched on a Titan-IIIM with a Blue *Gemini* mounted on top. In orbit, the crew would open a hatch through the heat shield and enter the 4-metre long, 3-metre wide pressurised module. It was to have carried crew-operated telescopic cameras. The maximum mission capability was one month. There was no provision for replenishment or crew exchange, so the platform was to be discarded when the crew returned to Earth in their spacecraft. Serious preliminary studies began in 1964, firmed up in 1965 and fixed in 1967. A target date of 1968 was set for an unmanned test of the entire system, and 1969 for the initial operational flight. Development was slow, however, and only the test of the modified spacecraft had been made to verify the integrity of the hatch during re-entry when the project was cancelled in 1969, partly for financial reasons, but mainly because it had become clear that the reconnaissance systems on the automated *Big Bird* spy satellites would offer an equivalent capability and be both cheaper and more flexible.

Molniya (Lightning) A class of communications satellite placed in highly elliptical orbits ranging between 460 km and 39,350 km at 65 degrees so that they would spend most of their time far above the Soviet Union, serving as relays.

MOMS-2P A German experiment on *Mir*; an electro-optical multispectral imager. It had previously been flown on the Spacelab-D2 shuttle mission, and could be operated in various modes, including high resolution surface-relief topography. It was mounted on a scan platform on the *Priroda* module, and carried its own GPS-based package so that the position of the complex (within about 5 metres) and its orientation (to within 10 arcsec) could be assigned to each image.

Monimir   An Austrian experiment on *Mir* which analysed postural reflexes to assist in the development of a computerised neurological analyser. It used the *Motomir* apparatus to investigate the movement of the subject's head and arms.

Monitoring   A test by acoustic sensors to assess the structural stability of the docking port on which *Kvant 2* was mounted; if the hermetic integrity was lost, the entire complex would depressurise.

Morova   A Czech-designed materials processing experiment on *Salyut 6*. It was performed in the *Splav* furnace over a two-day period, and was designed to evaluate the possibility of producing rare metals and alloys in microgravity. Small quartz ampules containing a variety of silver–lead chlorides and copper–lead chlorides had been prepared by the Czechoslovak Academy of Sciences. These were processed to produce extremely pure crystals and semiconductors. Other samples combined glass and metal to create new materials having 'certain electro-optical properties.' Hardened melts of crystalline and glass-forming materials were far more homogeneous than those which could be produced on Earth. In one case, lead chloride and copper chloride was held at 500°C for 20 hours and then cooled at 10°C per hour. Although this produced lead crystals which were both bigger and more homogeneous than possible on Earth, some displayed physical deformities (such as a helical surface) caused by microaccelerations due to the operation of other equipment on the station.

Motomir   An Austrian experiment on *Mir* which involved neurophysiological analysis of human motorics. It used a large four-element ergometer to measure the force and velocity characteristics of the subject's arms and legs during defined movements.

Moz   An experiment on the *Priroda* module developed jointly by Russia and Germany. Its multispectral spectrometer sensed reflected solar insolation using 17 channels at visible and infrared wavelengths between 0.4 cm and 1.0 cm, looking straight down to study the oceans.

MSRE   A Mir Sample Return Experiment (MSRE) developed by *NASA* for micrometeoroid research. It was placed outside *Kvant 2*.

MSS   A photoelectric spectrometer on *Salyut 7*.

MSS-2   A spectrometer on *Salyut 7*. It had a 7-degree field of view and observed the 0.4–0.8 μm range with a resolution of 1,200 lines per millimetre. Although its wide field was suitable for studying terrain features on the ground track, or the ocean, it was inappropriate for making horizon observations. It turned out that using the *KATE-140* to image the site of the spectra was cumbersome.

MSS-2M   An improved form of the *MSS-2* that both showed its output visually to the cosmonauts and also incorporated a boresighted camera to enable them to record the site of an interesting spectrum.

MSU-E   A high-resolution (45-metre) optical scanner on the *Priroda* module. Its CCD sensed three wavelengths between 0.5 and 1.0 μm.

MSU-KS   A medium-resolution optical scanner on the *Priroda* module that sampled five wavelengths between 0.5 and 12.5 μm.

Multiplikator (Multiplier)   A biological experiment on *Salyut 6* to study the growth rate of microorganisms.

Nanovesi An experiment on *Salyut 6*. Samples of structural materials and optical coatings were placed in the scientific airlock to expose them to the space environment to assess their degradation.

NASA The US National Aeronautics and Space Administration.

NASDA National Space Development Agency of Japan.

Nausicca-1 A French experiment on *Mir* which measured radiation in the complex, and correlated this with the complex's movement in its orbit.

Neptune The main control panel of the *Soyuz-TM* ferry.

Neptune A visual test performed on *Salyut 7* to study the degradation in visual acuity during the early phase of adaptation to weightlessness.

Neva-5 A radiometer on the *Spektr* module.

Neytral-Vektor An ion sensor which determined its orientation with respect to its direction of flight through the ambient 'ionic wind'.

Niva A videotape system on *Salyut 7* which enabled the cosmonauts to record visual observations while out of communications range, and then relay them to specialists on the ground once back in range.

NSF The US National Science Foundation

Oasis A hydroponic apparatus on *Salyut 1* for higher plants. The seeds were placed in a nutrient solution under a strong fluorescent lamp. Cameras photographed the chamber at regular intervals to monitor the rates of growth. Khibiny (chinese) cabbage, flax, onion, marrow-stem kale and crepis (actually the hawk's beard strain, an unobtrusive weed producing tiny white flowers) were all tried. Crepis was a common subject for studies of plant genetics. Flax was selected because it has a strong tissue structure which should be unnecessary in space. It was important to find out whether kale would grow and, if so, whether its nutritional value was altered by being grown in the absence of gravity, because it could be eaten and might well serve as the basis for hydroponic gardens in future orbital stations. The final form of the apparatus had lamps, fans to create an airflow, a water pump and layers of woven material emulating soil to aerate the roots. Peas finally grew in the *Salyut 7* Oasis apparatus, which used an ion-exchange resin substrate. They germinated, then grew 20 cm tall (so much that they touched the lamps in the compartment's ceiling, and their pots had to be lowered to give them room to grow). When they were 30 cm tall, the tallest had sprouted six branches, 23 leaves, and a thick tangle of tendrils. Although the leaves were well-extended, they were coated with mould; some white, some brown. The white root systems had expanded out from the artificial soil. When the tendrils started to twist together, they developed buds that slowly unrolled to form leaves. Wheat was also grown successfully. These experiments not only investigated the characteristics of the plants grown in weightlessness; they also evaluated their potential as a source of food on future stations.

Oasis-2 A biological experiment carried on *Soyuz 13* in 1973. It was a closed-cycle system involved in producing proteins. It comprised two interconnected cylinders. In one, water-oxidising bacteria was cultivated. This consumed hydrogen released by water electrolysis, to sustain growth. Oxygen was drawn off into the second cylinder containing urobacteria which absorbed the oxygen and released carbonic acid, which was fed back into the first cylinder to synthesise more biomass. In essence, the waste from one process was the raw material for the other. This symbiotic regenerative process increased the biomass by a factor of 30 in just two days, demonstrating that it would be feasible to produce protein from simple raw materials to drive a food production system on a future orbital station.

ODS   The Orbiter Docking System (ODS) was used by *Atlantis* on visits to *Mir*. Set in a twin-triangular truss spanning the front of the payload bay, it was basically two interconnected tubes, one leading from the mid-deck hatch to a hatch to the transfer tunnel to a Spacelab or a Spacehab module further aft, and the other running upwards through an airlock to the *APDS* at the top to mate with *Mir*.

ODU   The unified propellant system introduced by *Salyut 6* in which both the main orbital manoeuvring engines and the attitude control thrusters used the same propellants (so that it could be more easily replenished than if each engine unit used different propellants). After being installed in the *Progress* cargo ferry, it was fitted to the *Soyuz-T* form of the crew ferry (and thereafter to the *Soyuz-TM*).

OLIPSE   A *NASA* experiment on *Mir*. The *Optizon* liquid phase sintering experiment (OLIPSE) used the *Optizon* furnace in the *Kristall* module to process samples of metals.

Opros (Questionnaire)   A psychological test which involved filling out a comprehensive series of questions concerning the cosmonauts' eating and sleeping habits, level of physical fitness and posture, sense of smell, vision and hearing and other factors which would enable the psychologists to correlate their physical and mental health. Visiting cosmonauts would fill it out every day, the residents once a week.

OPS   Orbiting piloted station, the *Chelomei Bureau's Almaz* military reconnaissance platform configuration.

OPS 1	*Salyut 2*	April 1973
OPS 2	*Salyut 3*	June 1974
OPS 3	*Salyut 5*	June 1976

An improved reconnaissance platform was built for launch in 1979, with its crew flying the *TKS* rather than the *Soyuz*, but this was cancelled. An automated version for commercial purposes was subsequently developed.

OPS 4	Cosmos 1870	July 1987
OPS 5	Almaz 1	March 1991

Its primary instrument was an S-Band (10-cm) synthetic aperture radar that produced images of a 25-km wide swath to the side of the ground track. The radar had 25-metre resolution in the case of Cosmos 1870, and 15-metre resolution in the case of the automated *Almaz*. Because its data were transmitted to Earth there was no need for a film-return capsule. It was the same configuration as the manned platform, but instead of the rear-mounted solar panels it had larger panels on either side of the forward section. These had a total area of 86 m^2, and generated the 10 kW needed to operate the radar. Two radar antennas extended on mounts on the forward compartment, and extended along the length of the body. The attitude was controlled by gyrodynes. It later emerged that the first satellite was to have been tested in the early 1980s, but had been repeatedly postponed; the second was a year late, and the third was apparently cancelled for financial reasons.

Optizon-1   A non-crucible furnace in the *Kristall* module which heated a sample by zonal melting (that is, by focusing radiant energy from an electric light source) to produce semiconductors (notably silicon).

Optokinez   A biomedical apparatus on *Salyut 7* used to investigate the relationship between the vestibular and visual systems during adaptation to weightlessness.

Optovert   An Austrian biomedical device on *Mir* to study the interaction of the visual and vestibular systems to measure eye-movement in response to optokinetic stimulation, and to assess its effect on motoric performance.

Orion-1   This was the primary instrument on *Salyut 1*. It comprised a complex and highly accurate optical and electronic system involving a 280-mm and a 50-mm mirror. One telescope and a spectrograph were mounted on the outside of the station, and another telescope, mounted inside the station, looked through a quartz screen. Being above the Earth's atmosphere, it was able to observe far into the ultraviolet. It recorded stellar spectra on film. It was sensitive in the 2,000–3,000 Å band, with a resolution of about 4 Å, and it had an automatic tracker to follow the desired star with an accuracy of 1 arcsec. It could be used only while the station was in the Earth's shadow. An airlock enabled the film to be replaced, and exposed film was returned to Earth for processing.

Orion-2   An ultraviolet telescope carried on *Soyuz 13*. This was mounted on the front of the orbital module in place of the docking unit. The cosmonauts had received extensive training in how to operate it by the staff of the Byurakan Observatory in Armenia. It was an improved form of the *Orion-1* apparatus tested on *Salyut 1*, but in this case the entire unit was mounted outside the spacecraft. When not in use it was protected by a shroud. It employed a wide-field meniscus telescope which could simultaneously sample an area of about 20 degrees square, and the optical components were made of crystalline quartz. It was sensitive to radiation in the range 2,000–3,000 Å. Designed by Grigor Gurzadyan of Armenia, it was mounted on a fully stabilised platform incorporating a dozen electric motors. Once the spacecraft had been orientated so that it pointed in the general direction of the target and had been provided with reference stars, it could automatically lock onto the desired coordinates and hold itself steady throughout the required exposure, which could last up to about 20 minutes. It used special film supplied by *NASA*. In all, the cosmonauts recorded 10,000 spectrograms of 3,000 stars in the constellations of Auriga, Perseus, Taurus, Gemini and Orion. Most of these stars were of 10th magnitude at visual wavelengths, but some were as faint as 12th magnitude.

Orlan-DM   An improved spacesuit used on *Mir* superseding that carried by *Salyut*. It comprised an aluminium alloy body section with an integral helmet and backpack. It operated at about 6 psi (relatively high pressure compared to *NASA* standard), but its elasticated constant-volume arm and leg joints gave much greater flexibility and dexterity than the previous suit. The operator wore an inner garment containing tubes for water coolant. It was entered by a hatch in the rear (the backpack was hinged and swung aside). All equipment was accessible for maintenance simply by opening the rear access hatch (there were no access ports on the backpack). It required just 30 minutes pre-breathing oxygen at 10 psi before dropping the pressure to 6 psi (this was done with the suit closed, and thus consumed the backpack's resources, so this period had to be kept as brief as possible). This new suit did not use a communications/power umbilical because its backpack supplied power and had transmitters for the voice link and telemetry. The oxygen tank, and the lithium hydroxide scrubber which removed carbon dioxide, could sustain seven hours' operation (the *Salyut* suit had been limited to five hours) in addition to several hours pre/post-activity in the airlock. Another feature, welcomed by the doctors, was the biomedical sensors that relayed data via the telemetry link. Like earlier suits, its systems were rated for about 10 excursions before requiring maintenance.

Orlan-DMA   An improved *Orlan-DM* suit on *Mir*.

Orthostatism   A French experiment on *Mir* that involved a study of orthostatic resistance using an *Echograph* to monitor the changes in cardiovascular capacity and venous circulation, and sampling to investigate hormonal changes due to the absence of gravity. A subsequent refinement of this study used the *Echograph*, *Diuresis* and *Tissue*, and made use of the Haut Schicht Dicke apparatus left by a German visitor.

OST-1   A solar telescope, the main instrument on *Salyut 4*, housed in a tall, broad, conical mount projecting up from the floor of the main compartment. It employed a 25-cm diameter mirror with a 2.5-metre focal length, and incorporated the *CDS-1* diffraction spectrometer. The entire assembly had been designed and built by the Crimean Astrophysical Observatory specifically for the flight, and the crew had received extensive training from the staff of the observatory in how to use it. It had been used by remote control for two weeks before the crew had arrived, and at some point a sensor had been burnt out, disabling the pointing system and preventing further remote control use. Although the cosmonauts could not actually observe its mechanism to align the secondary mirror, they discovered that by listening to its servo mechanism and timing its travel they were able to estimate its position sufficiently accurately to align it so that sunlight reflected directly onto the main mirror. Although this meant that the station would have to be rotated to scan the telescope over its target, at least the programme could now be resumed. Without their intervention the telescope would have remained unusable, and a significant portion of the scientific programme would have been lost. The mirror was resurfaced by the *Zentis* experiment, to extend its useful life. It was used to study dynamic processes on narrow cross-sections of the solar disk, including flocculi and prominences. In addition, at sunrise and sunset, its ultraviolet spectrometer could be used to measure the strength of ozone's absorption lines, to extend investigations by Earth-based instruments. It was a fleeting opportunity, but the automatic attitude control system was able to add data regularly, as the station orbited the Earth every 92 minutes. At such times, it was possible to measure sunlight passing through a sample of air 1,000 km deep. The results helped assess the level of pollutants and, later, contributed to a long-term study of what is now universally referred to as 'ozone depletion.'

OVI   A German experiment on *Mir*. It was devised by the University of Mainz and used a pair of glasses equipped with stimulators and sensors to investigate the influence of vestibular asymmetry and the stimulation of the throat receptors on eye movement and the 'subjective horizontal' in the absence of gravity. It involved a series of psychophysiological tests.

Oxymeter   A Czech experiment on *Salyut 6* to determine the concentration of oxygen in skin tissue in the absence of gravity.

Ozon-M An experiment on the *Priroda* module. Its spectrometer assessed the distribution of ozone and other aerosols in the upper atmosphere by measuring their absorption of sunlight at sunrise and sunset at four wavelengths between 0.12 to 0.26 μm.

Palma (Palm)   An experiment first conducted on *Salyut 4*, designed to evaluate reaction to external irritants. (Psychologists would eventually evaluate a large range of visual and acoustic irritants during tests conducted by the *Intercosmos* crews on later stations). It was also on *Salyut 5*.

Palmyra   An experiment on *Mir*. It mixed two substances to produce a synthetic material with a crystalline structure similar to that of human dental and bone tissue.

Parallax-Zagorka   An image intensifier for the *Rozhen* experiment. It was also used on its own to study the vertical distribution of luminescence in the polar, middle and equatorial latitudes and also the luminescence of the complex itself as it flew through the extremely rarefied atmosphere at its orbital altitude.

PAS (Passive Accelerometer System)   A *NASA* experiment on *Mir* to measure low-level but continuous accelerations on the complex (as opposed to transient shocks transmitted through the structure), to assess the extent to which factors such as air drag decelerated the complex in its orbit, and how differential gravity drew items inside towards the Earth. It involved placing a small metal ball within a reference frame and recording how it drifted over time. These measurements helped to characterise the effects influencing microgravity experiments. These data complemented those from the MISDE and SAMS instrumentation.

Payload Systems   The first American apparatus on *Mir*. Payload Systems Inc's commercial contract was to fly six experiments over several years. The first was a protein crystal growth experiment in which two enzymes (hen egg white lysozyme and D-amino transferase) were submitted to 112 tests by three crystallisation processes (batch, vapour diffusion and boundary-layer diffusion). It was returned after two months.

PCN   A French camera on *Salyut 7*. It was a hand-held low-light camera incorporating a variety of filters. It was intended to be used to make observations of faint upper atmosphere phenomena such as luminescent clouds, and was used to photograph a variety of faint sky phenomena, including the zodiacal light, noctilucent clouds and lightning storms. It was transferred to *Mir*.

Pelican   A remote manipulator outside the *Spektr* module. This 2-metre long twin-segment arm, which had a gripper on its end, was to be used to extract small experimental packages from a scientific airlock and affix them to a number of external anchor points, each of which had a socket for a power, control and telemetry umbilical. It enabled exposure cassettes to be deployed and retrieved without requiring a cosmonaut to go outside. *Energiya* developed a standardised experiment package. With the package in place, four pallets could be unfolded individually, as required, to expose the samples.

Perception   A Cuban experiment on *Salyut 6*. It evaluated visual, tactile and muscular sensitivity using the Cuban-supplied *Contact* apparatus.

Phoswich   A scintillation detector consisting of two crystals, e.g. NaI(Tl) and CsI(Na), optically coupled and forming a PHOSphor sandWICH. The NaI(Tl) acts as an X-ray detector, while the CsI(Na) scintillator acts as an active shield. The scintillation light produced in a phoswich is viewed, through a light guide, by a photomultiplier tube.

Physalie   A French experiment on *Mir* to evaluate the coordination between the body's sensory and motor systems. This involved using a number of devices to record biomedical data whilst being filmed performing a variety of bodily movements.

Physiolob   A French experiment on *Mir* to study the rapid changes to the cardiovascular system which occur in the early phase of adaptation to weightlessness, particularly how the body senses blood pressure and regulates its flow. This involved donning a harness supporting various sensors.

Phyton   A cultivator on *Salyut 6* for growing higher plants, developed by the Ukrainian Institute of Molecular Biology. It incorporated a very powerful lamp, a nutrient medium, and a filter to extract air impurities which, although no danger to the crew, might inhibit the growth of the plant. It was designed so that the roots grew 'down' in the dark, while the shoots grew 'up' towards the lamp. Seeds of onions, cucumbers, tomatoes, garlic and carrots were placed into compartments of a black panel containing a nutrient solution. The objective of the experiment was to grow the plant in the hope that it would flower and, hopefully, produce a seed of its own. This would demonstrate the complete growth cycle from seed, grown plant, and back to seed in weightlessness and thus achieve the goal of almost a decade of effort. It was supplemented by data gained from other experiments investigating the development of individual cells. Unfortunately, the *arabidopsis* plant grew so rapidly that it choked on its own waste products. The onions very quickly grew seed stalks (this was unexpected because on Earth onions do not produce seeds in so short a period). Another type of higher plant was a strain of wheat, characterised by a very short stalk and a very heavy yield of grain; although this grew, it did so more slowly than expected, and, as a result, it had only just reached its most complex stage, and was in the process of producing ripe grain, when it was necessary to cut short the experiment and return to Earth, so it was not possible to proceed to the second generation in space. In the case of fungi, which have a very good geotropic reaction (that is, they react to gravity), the fruit bodies formed were totally disorientated unless they were provided with a light source, in which case they would grow towards it, and so reacquire a measure of orientation.

Phyton-2   A plant-growth apparatus on *Salyut 6*. It incorporated three light sources and interchangeable plant pods containing an ion-exchange nutrient. The plants were given measured quantities of water by an automatic sequencer and ventilated. It subsequently turned out that shoots developed better if the apparatus was put near a porthole so that they could be exposed to sunlight rather than the artificial light. Despite all efforts, however, the peas and wheat still died early in their development cycle.

Phyton-3   A plant-growth apparatus on *Salyut 6*. After many years of experiments, in 1980 a higher plant flowered in space (this was a key step towards the goal of completing the growth cycle, from seed to seed). This was an *arabidopsis*. Grown from its bud, it flowered four days after the control experiment on Earth. The next objective was to have one complete the cycle by producing seeds. *Arabidopsis* in the apparatus on *Salyut 7* podded. The largest plant was 65 mm tall, had 12 green leaves, three tiny pink flowers about 2 mm across and a pod 3 mm long. When the pods ripened, they burst open to reveal the seeds within. It was great news for the biologists, who were keen to receive seeds from a plant grown in space.

PIE (Particle Impact Experiment) This package was supplied by *NASA* for micrometeoroid research. It was placed outside *Kvant 2*.

Pille (Moth) A thermoluminescent ionising radiation dosimeter. Being portable, it was employed to measure the dosage rates at different parts of the *Mir* complex. It was also used to assess the increased dosage resulting from passing through the *South Atlantic Anomaly*. It measured doses in the 10-millirad to 10-rad range.

PILOT A standard *NASA* computer program used to test piloting skills (and thereby retain them) during shuttle missions (in particular to practice landings just before returning to Earth), which Thagard used on *Mir* to evaluate how his performance changed over such a long period.

Pingvin (Penguin) A lightweight work-suit with elasticated straps sewn into its fabric to impose compressional loads on specific muscle groups, to counter the atrophying of the back muscles in weightlessness. It was so called because it caused a test subject on Earth to hunch over and waddle like a penguin. It was tested on *Salyut 1*, and later, without the elasticated straps, was worn as a straightforward work suit.

Pion An apparatus flown up to *Salyut 6* to investigate crystallisation both in and out of solution, and the effect of convection currents in fluids under the influence of microgravity. The mixture to be studied was delivered in a clear disk-shaped container. It incorporated a lamp which illuminated the sample from one side while a camera recorded the transformation from the other side. The temperature and time were recorded on each frame.

Pion-M An improved crystal growth apparatus on *Salyut 7*. It was used to study melting and crystallisation processes of various materials, and heat and matter transport in liquids to assist in the design of an apparatus to process large quantities of biological materials, particularly colloids. It was found that silica aerogel (a type of glass) in suspension formed small saucer-shaped structures, fluoroplastics formed tree-like structures, and glass pellets formed arbitrary but extremely robust lumps. It was transferred to *Mir*, and then used to study the thermocapillary process in microgravity.

Piramig A French camera on *Salyut 7*. A highly sensitive camera which was to be used to study the upper atmosphere, the interplanetary medium and faint galaxies at visible light and near-infrared wavelengths. It was set up in the porthole in the forward transfer compartment normally used by the *MKF-6M*. It could be used only in orbital darkness. The compartment was isolated, with the hatches shut and the lights off (the cosmonauts could use only low-power torches) and with the station's thrusters off (to avoid ejecting glowing debris into the field of view). It used an extremely sensitive camera incorporating fibre optics and photomultipliers, and it had filters operating between 2,000 and 10,000 Å, so could be used for ultraviolet, visible and near-infrared observations of the upper atmosphere, the interplanetary medium and a number of astrophysical sources. The *Kaskad* orientated the station to within 1 degree, then the camera was fine-tuned to achieve an accuracy of about 1 arcsec. It was difficult to keep steady for a five-minute exposure. The station was specifically orientated for such observations for up to six hours on some days, and during 20 sessions about 350 photographs were taken. These turned out to be very good, in particular those of high altitude cloud formations, so the resident cosmonauts were asked to use the hand-held *PCN* camera twice each week to photograph the horizon in the hope of recording more of these clouds (at an altitude of 80 km) and other upper atmospheric phenomena (at altitudes up to their 350 km orbit).

Pirin   A Bulgarian smelting experiment on *Salyut 6*. The *Splav* furnace was used to make 'foam metal' in microgravity. These combine low specific weight with softness and high mechanical strength. In this case, they produced foam aluminium. This was achieved by heating a capsule containing quartz ampoules of silumin, titanium hydride and silicon nitride for 10 minutes at 800°C. Foam steel would have all the strength of normal steel, but be as light as wood. In the future, such materials could be used in the construction of extremely large orbital complexes. Later, another experiment in this series was performed in the *Kristall* furnace to assess the stability and structure of zinc crystals grown using the diffusion process.

Plasma-1   A freezer on *Salyut 7*, used to store body fluid samples for return to Earth.

Pleven-87   An experiment on *Mir* which used fifteen psychological tests, many of which tested the locomotor functions and volition processes. (Its data was processed by the *Zora* computer.)

Plotnust   An ultrasonic device used on *Salyut 4* to measure the density of bone calcium.

Pneumatik   An experiment on *Salyut 6*. It involved using the *Pneumatik-1* apparatus while in the *Tchibis* suit to determine whether a blood distribution appropriate to gravity could be re-established in space, to limit the rush of blood to the cranial regions.

Pneumatik-1   A device to reduce the flow of blood to the upper torso.

Polarizatsiya   An experiment on *Mir* employing Bulgarian apparatus to make photometric observations of stars, galaxies and nebulae.

Polynom-2   An electrocardiograph flown on *Salyut 3*. It was a multifunction blood flow unit to investigate a range of circulatory conditions. It was to be used to measure blood flow before and after exercise to investigate the effect of weightlessness on the heart and brain, and could measure vascular tension. During the first few days in space as the human body adapted to weightlessness, blood gathered in the brain, and previous cosmonauts had reported this as an unpleasant feeling. These tests were intended to measure exactly how the flow of blood changed, in the hope that a way could be found to ameliorate the unpleasant symptoms.

Polynom-2M   A multipurpose electrocardio unit flown on *Salyut 4* to monitor the performance of the heart. In addition to general cardiograms it could isolate the major blood vessels and monitor the various phases of the cardiographic cycle.

Polyus (Polar)   The payload of the first *Energiya* rocket was an engineering mock-up of a spacecraft which Nikolai Gerasimov, head of the Salyut Bureau, said was a multipurpose platform that was to be used to deliver 40 tonnes of cargo to a space station. Dr Vladimir Pallo, the platform's designer, said that it was essentially an enlarged version of the *Mir* core module, both longer and wider, incorporating a *TKS* tug for propulsion and attitude control; it was 38 metres long and had a mass of 80 tonnes. An alternative role would be as an industrial-scale microgravity factory for materials processing and biotechnology which would operate autonomously, but receive occasional visits by cosmonauts to service the equipment, restock the raw materials and take away the product.

Posa   A French apparatus on *Salyut 7,* comprising a set of motion sensors and muscle monitors for the *Posture* experiment.

Posture   A French experiment on *Salyut 7* designed to investigate how the body maintained certain postures in microgravity. The subject donned the *Posa* sensors and used a foot restraint which defined the motion coordinates, and then, with both eyes closed, enacted a series of body movements to exercise sensory and locomotive functions so that the responses could be recorded for later analysis. A variety of sensory organs and muscle groups (different from those used when in motion) are used to maintain posture. The apparatus took considerably longer to set up than had been expected .

Potential   An experiment on *Mir* which investigated the interaction between the body's nervous and muscular systems.

Potok (Flow)   An experiment on *Salyut 5* to evaluate the possibility of building cosmic capillary pumps for liquids which do not need electricity. It comprised two interconnected containers linked by a narrow channel. The capillary action and surface tension caused the fluid in the first container to flow into the second, and the cosmonauts recorded this process. It was an important experiment because the designers hoped that it would lead to a pump without moving or electronic components, in order to simplify the replenishment system of a future tanker spacecraft.

Pravda (Truth)   The newspaper in the Communist Party of the Soviet Union.

Priroda (Nature)   An expansion module of the *Mir* complex. It had a pressurised volume of 66 m^3.

Priroda-5   A multispectral Earth observation system on the *Kristall* module using two *KFA-1000* cameras.

Prognoz   An experiment on *Mir* which assessed the operational performance of the cosmonauts. It involved using a video (on the *Zora* computer) to run a series of sensory-response tests. It used a series of mathematical puzzles and audiovisual cues to assess functional psychological state. If these tests were performed daily, degradations in performance could be tracked. It made use of the *Pleven-87* apparatus.

Prognoz 1 (Forecast)   A class of automated research satellite (the first was launched in 1972) in highly elliptical orbits ranging between 950 km and 200,000 km at 65 degrees so that they spent most of their time high above the Soviet Union. It contained various detectors to study the emissions from the Sun, and their interaction with the Earth's magnetosphere.

Progress   A *Soyuz*-based spacecraft used as a resupply ferry for the *Mir* complex. It was essentially a *Soyuz* with the descent module replaced by a cylindrical compartment with tanks for fluids. Dry cargo was packed in containers bolted into racks within the orbital module. In addition to propellants, water, pressurised oxygen and nitrogen, it delivered food, scientific apparatus, tools, and replacement components for *Mir*'s environmental and thermal regulation systems. Total capacity was 2,500 kg (1,000 kg of wet cargo, and 1,500 kg of dry cargo). It used the same unified *ODU* engine block as the second-generation *Salyut*. Each ferry was kept docked as long as possible to serve as storage, then loaded with rubbish just before its departure. Sometimes the docking assembly would be replaced by an experiment package to be activated while the spacecraft stood alongside the station.

Progress-M   An improved form of the *Progress* cargo/tanker ferry. It reintroduced solar panels. A *Raduga* Earth-return capsule could be inserted into the docking unit to return the results of experiments.

Proton   The *Chelomei Bureau*'s medium rocket with a lifting capacity comparable to *NASA*'s Saturn IB. It burned storable hypergolic propellants (UDMH and nitrogen tetroxide). Its first launch in 1965 was a two-stage configuration which put the 12,200-kg Proton high-energy physics satellite into orbit (the heaviest scientific satellite ever launched). With an upper stage, the rocket was later used to launch advanced planetary probes and send the *Zond* spacecraft on a circumlunar trajectory. In 1968, it put Proton 4 into orbit (at 18 tonnes, this satellite tested the main section of the pressure hull of the *Almaz* reconnaissance platform). It subsequently put the *Salyut* stations, the *Mir* base block, and their *TKS*-based expansion modules in low Earth orbit.

PRV   A precision radio-altimeter on the *Priroda* module operating at 2.25 cm.

PSY   A German experiment on *Mir* which used a portable computer to perform perception, speech and psychomotor tests.

Pulsar X-1   A hard X-ray spectrometer on *Kvant 1*, consisting of four identical *Phoswich* detectors sensitive to X-rays and gamma-rays in the 50-800 keV energy range, and with a 3 x 3 degree field of view. Each of the detectors has 314 cm^2 geometric area.

Pulsar-2   A 1,000-kg X-ray telescope (similar to *Pulsar X-1*) which was to have been delivered by *Buran* and attached to *Kristall*'s lateral *APDS* port.

Pulstrans   An Austrian experiment on *Mir* to analyse pulse transmission and heart frequency while the body was subjected to stress.

Puma   A viewfinder on *Salyut 7*. It had ×15 magnification.

QUELD   A *NASA* experiment on *Mir*. The Queen's University Experiment in Liquid Diffusion (QUELD) was supplied by Canada. It used an isothermal furnace that operated at temperatures up to 1,000°C, and was used to study boundary layer processes by measuring diffusion coefficients for semiconductors, binary-metal materials and glasses.

Raduga (Rainbow)   A small recoverable capsule ejected by a *Progress-M* immediately after the vehicle made its deorbit manoeuvre. It could return 150 kg of compact cargo to Earth. Considering the experiment schedule, and the rate at which this was expected to produce results, it seemed likely that a capsule would be needed on every third or fourth resupply ferry. At 380 kg empty, however, it seriously diminished the cargo capacity of the vehicle which delivered it. It was delivered inside the orbital module, then inserted into the ferry's docking collar instead of the docking probe assembly. The truncated cone, 1.4 metres long with a bottle nose, projected in through the hatch into the cargo compartment (the capsule was 0.6 metres wide at it base and the hatch was 0.8 metres in diameter). After the ferry had withdrawn to make its deorbit burn, tracking stations computed its trajectory in order to compute the optimum time to order it to eject the descent capsule, which had to be jettisoned at an altitude of about 120 km. Obviously, in a change to the previous procedure, instead of deorbiting over the Pacific, the capsule-returning ferry was deorbited over Kazakhstan so that its capsule would descend in the normal recovery zone. A pressure sensor determined when the altitude had decreased to 15 km and then commanded the parachute to release. When at 4 km, it switched on the radio beacon. Following the break-up of the Soviet Union, recoveries were switched to a site near the Urals, on Russian territory. Most capsules were successfully recovered. It has the designation VBK (ballistic descent craft).

Raketa A vacuum cleaner used on *Salyut 4* to suck up the worst of the dust and debris which appeared as soon as a station entered weightlessness.

Rapana A 26-kg, 5-metre long girder akin to *Sofora*, but using a different smart material. It was developed by the Institute of Electrical Welding, in Kiev. This *Ferma-2* truss was erected on top of *Kvant 1*. It was a scaled-down test of a structure which was intended to be used to hold the parabolic dishes of a solar collector away from the body of the proposed *Mir 2* orbital complex.

Reaktsia (Reaction) An experiment on *Salyut 5* which used a chemical energy source to smelt high-grade nickel and manganese solder, then soldered stainless steel pipes, to evaluate techniques being considered for assembly work in space. It was the first time that soldering had been attempted in space. It used two chemical energy sources to smelt high grade nickel and manganese solder to solder stainless steel pipes. The experiment was to reveal how the process of soldering differed in microgravity. On Earth, the solder does not spread evenly over the seam, but this should not happen in the absence of gravity. It was necessary to observe the formation and the crystallisation of the solder seam, and then to establish its mechanical characteristics. If it turned out that joints made in space were stronger than those made on Earth, soldering might well be able to play a part in the construction of structures in space. In the tests, the cosmonauts soldered two stainless steel 15-mm diameter tubes with a wall thickness of 0.001 metres using nickel manganese solder heated by a flame. It would later be demonstrated that these seams were strong and airtight and could withstand a pressure of 500 atmospheres. When the cosmonauts used a soldering iron to effect repairs to electrical equipment, they used a solder which did not include tin.

Rech (Speech) An East German experiment on *Salyut 6*. This simply required reciting a series of specific phrases during each communication with the ground so that any variations in speech pattern could be correlated to state of health, mental attitude and general level of activity. These phrases, such as saying the number '226' in German, required elocutionary dexterity. Even a mundane experiment such as this could shed light on how various activities caused stress.

Reflotron A biomedical kit on *Mir,* supplied by West Germany. It provided instant blood analysis, so that the changes in blood chemistry could be studied as they happened (rather than after analysis following return to Earth). Given a drop of blood, the analyser would automatically measure a wide range of parameters (including a report on haemoglobin, cholesterol, uric acid and glucose) and store the results for later transmission to Earth via the telemetry link.

Refraction A Hungarian experiment conducted on *Salyut 6,* using the *VPA-1M*.

Rekomb This was a biotechnology experiment on *Mir*. It cultivated hybrid microorganisms having properties that would enable them to synthesise biologically active substances on Earth.

Relaks (Relax) A psychological experiment on *Salyut 6* carried out to identify stresses endured by cosmonauts whilst living and working in space on long-duration missions. It tested the influence of various recreational exercises on their psychological condition. It was part of a long-term programme to determine the most favourable conditions for living in space.

REM (Radiation Environment Monitor) An *ESA* experiment set up outside Mir to measure charged-particle flux.

Reograf (Rheograph)   An apparatus on *Salyut 4* which measured the distribution of blood by measuring blood flow in the head, torso and extremities.

Reporter   An East German experiment on *Salyut 6*. It involved conducting a methodical study of different types of film for use both for external photography and within the station. Jahn had two specially modified cameras (a Pentacon-6M and a Praktica EE-2) with a range of lenses for use in these studies.

Reservoir   An experiment on *Mir* which investigated the development of a system for filling and emptying a capillary tension reservoir.

Resistance   An experiment on *Salyut 6* which used a detector to measure the loads imposed by drag from the upper atmosphere.

Resonance   An ongoing engineering experiment which measured the vibration modes of the complex in its various docked configurations, and with different apparatus operating, to identify any vibrations which might adversely affect the structure. In its simplest form, a cosmonaut strapped into the harness of the *KTF* and jumped up and down in time with a control signal. In another case they set up a number of spring-loaded masses and monitored their oscillations.

Resurs   A cassette outside *Salyut 7*, containing structural materials (16 plates of different alloys) whose physical properties were to be tested after prolonged exposure to the space environment.

Rezeda (Mignonette)   A spirometer on *Salyut 4* which determined lung capacity. It was also on *Salyut 5*.

Ritm   A plant growth apparatus that incorporated a radiation dosimeter. It was carried on *Soyuz* 16.

RMS-2   This respiratory measurement system was supplied to *Mir* by *ESA*.

Rodnik (Spring)   An improved water supply system introduced on *Salyut 7*. Two 250-litre spherical tanks in the aft equipment section could be refilled directly by pumping from tanks in a *Progress* ferry (this was an improvement over the previous procedure, in which water was delivered in dozens of tiny spherical bottles, each of which contained 5 kg of water, and had to be transferred by hand). Water was fed to a squirt-nozzle by the table. Although the cosmonauts had hot water on demand for reconstituting dehydrated food packs, the heater could supply only 0.5 litres at a time. Each cosmonaut ingested about two litres of water per day (including that employed in reconstituted food). The Rodnik system was also used by the *Mir* base block, and expansion tanks were mounted externally on the *Kvant 2* and *Kristall* modules.

Roentgen   The group of four X-ray experiments (*TTM, HEXE, Sirene-2 and Pulsar X-1*) in the unpressurised compartment around the rear transfer compartment of *Kvant 1*.

ROK   A German experiment on *Mir*, developed by the Max Planck Institute to study the response to different orientations in weightlessness.

Rost   A plant growth apparatus on *Salyut 7*, used to grow oranges.

Rozhen   A Bulgarian apparatus on *Mir*. This was a sophisticated system incorporating a digitally processed electro-optical telescope (using the *Parallax-Zagorka* image intensifier and the *Therma* photometer). It was being evaluated for an autonomous space observatory to study deep sky sources. Its imagery was processed in orbit, then downlinked.

RS-17   An X-ray telescope on *Salyut 7*, mounted in the rear transfer tunnel. It was sensitive to high-energy photons in the 2,000–800,000 eV range, and was used to determine the temperature profile of the accretion disk in the Cygnus X-1 black hole system, and to study the Crab nebula and X-ray pulsars.

RSA   Russian Space Agency

RSS   A hand-held spectrograph for Earth studies, tested on *Soyuz* 5 and *Soyuz* 7, and used on *Soyuz* 9. It could be used to measure the reflectance spectrum of the Earth's surface under direct insolation and so distinguish soils with different moisture content.

RSS-2   A hand-held spectrograph on *Salyut 3*. It was designed in Leningrad to study the processes affecting the Earth's thermal environment, and to determine trends in climate changes which could be caused by the saturation of the atmosphere with aerosols such as industrial smoke, dust and chemical pollutants. These were important observations because it was believed that the disruption of the optical and radiation properties of the upper atmosphere would create the conditions necessary for climatic changes. It could be used either to observe the atmosphere at times of orbital sunrise or sunset by measuring the selective absorption of sunlight passing through the layers of the atmosphere, or to make observations of the Earth's surface and the oceans by measuring the reflectance spectrum under overhead solar illumination. It was used to measure the distribution of gaseous aerosol pollutants in the upper atmosphere and to measure the distribution of ozone (a thin layer in the upper atmosphere at an altitude of about 90 km) to determine the extent to which it is contaminated by aerosol pollutants. Its output was recorded on film.

RSS-2M   An improved form of the hand-held spectrometer, used on *Salyut 5*. As part of a long-term study of climatic changes, it was used to establish the presence of aerosols (particles such as smoke, dust and chemical pollutants) at different altitudes in the atmosphere to an accuracy of between 500 and 800 metres. It was also used to observe strong currents in the Atlantic. One study concentrated on the Volga region, measuring the distribution of water in the river, and the results correlated with simultaneous studies by ground teams and aircraft. The Soviet Ministry of Land Improvement and Water Conservation had plans to divert some of the outflow of rivers in the northern part of the USSR to areas further south, so these observations were intended to assess the extent to which the flow of southern rivers could be increased. The measurements of water capacity subsequently influenced planning for several hydroengineering projects in the Volga region.

RSS-3   A spectrometer on *Mir*, sensitive in the 0.4–1.1 μm range. Its 600 lines per millimetre diffraction grating gave a resolution of 1 per cent. Although portable, it was usually mounted in a frame in a porthole and left to operate autonomously at sunrise and sunset, to study the atmosphere.

RT-4   An X-ray telescope on *Salyut 4*. It was sensitive to X-rays with wavelengths in the range 44–60 Å (that is, soft X-rays with energies under 1 keV). It used a telescope with a 0.2-metre diameter parabolic mirror to illuminate a sophisticated photon counting system to measure the intensity of a source. Its field of view was not wide enough for it to be used to scan the sky for new sources, but it could be used to make further studies of sources discovered by the *Filin* system.

RT-4M   A mirror-based soft X-ray telescope on *Salyut 7*. It was an improved version of the apparatus tested on *Salyut 4*.

Ruchei (Brook)   An electrophoresis unit installed in *Kvant 1,* used to synthesise interferon and anti-influenza vaccines.

Ryabina   A spectrometer on *Salyut 7,* used to study gamma-rays and charged particles.
Ryabina-2   A telescopic spectrometer on *Kvant 2.*
Ryabina-4P   A telescopic spectrometer on the *Spektr* module.

Salyut 1   This orbital station comprised three compartments, arranged as a series of four
  cylinders with different diameters. The *Soyuz* ferry docked at the front, opposite the
  station's propulsive unit. As the cosmonauts left the tunnel they emerged into the transfer compartment. It was the narrowest section, barely 2 metres across, and contained
  some of the astrophysical apparatus and several control panels. At the other end of the
  transfer compartment was the hatch to the main work volume which extended right
  through the 3-metre diameter cylinder into the 4-metre diameter cylinder at the rear,
  behind which was the propulsive unit, which was the same as that in the *Soyuz* ferry's
  service module. It was 14.6 metres long (about 22 metres with a *Soyuz* docked). The
  main controls were on a number of panels, similar to those installed in the *Soyuz* ferry,
  in a console across the floor, just on the other side of this internal hatch. The rest of the
  station, which formed a long room rather than a cylinder, contained various experimental work stations. The length of the docked combination was about 30 metres and it had
  a mass of about 25 tonnes. The total habitable volume, with the ferry attached, was
  about 100 m^3. Its size was made evident during the first television transmissions, when
  the cosmonauts were shown performing somersaults for the camera. There were eight
  seats, each of which was positioned near a workstation so that the cosmonauts could
  strap themselves in while at work. In all, Salyut 1 incorporated 20 portholes, but only a
  few gave an unobstructed view, and the others were dedicated to various equipment.
  Two pairs of unfolding solar panels, like those used by *Soyuz*, were attached at the front
  on the new transfer section and at the rear on the engine unit. With a *Soyuz* docked, the
  total solar collector area was 36 m^2, which generated just 3.6 kW under full illumination. The fact that these could not be rotated to keep facing the Sun meant that whenever the batteries ran low, the station had to be reoriented to face the Sun so that they
  could be recharged, and this manoeuvre consumed propellant. The station was totally
  dependent on the batteries while it was in the Earth's shadow, so power had to be carefully managed.
Salyut 3, 5   This was the *Almaz* stripped of its *Merkur* capsule. It was 11.6 metres long
  (about 18 metres with a *Soyuz* docked). It had a two-element stepped-cylinder configuration. The docking port was at the rear, set between the two engines. A small film-
  return capsule was accessible by a hatch within the wall of the transfer tunnel (this was
  a 400-kg, 0.85-metre wide drum-shaped spin-stabilised capsule with an ablative shield;
  it could return 120 kg of compact payload, which was far more than could be ferried
  back in a *Soyuz* descent module). The solar panels were stowed alongside the transfer
  tunnel and then unfolded. The primary instrument was a 6-metre focal length folded-
  optics telescopic camera for Earth observation with about 50-cm resolution. Since this
  viewed through the floor of the compartment, the station had to maintain given multi-
  axis rotation to keep the camera facing the Earth as the spacecraft progressed along its
  orbit. Because it could not hold a fixed orientation with respect to the Sun, its solar
  panels had to be able to rotate. To minimise the propellant used in adjusting the attitude
  of the station, an inertial system was installed which used electrically-driven

magnetically-mounted flywheels whose gyroscopic action could finely control the orientation of the station. Salyut 3 introduced a water reclamation system that condensed water from the station's atmosphere. This proved to be very useful, as it could reclaim evaporated water to yield fresh water (1 litre per day per cosmonaut aboard). Recovered water was used for washing and food preparation. It had been an important experiment because recycling would help make a station self-sufficient and would reduce the amount of water which would have to be stored aboard at launch.

Salyut 4   This was the Salyut 1 configuration, but fitted with three large solar panels on the narrower part of the main compartment instead of four small fixed panels. The new panels were capable of tracking the Sun. Their combined area was 60 m^2 and they generated 4 kW. It had a hatch in the side of the transfer compartment. (This compartment could be sealed off so that it could serve an airlock, although it isn't clear why this was deemed necessary, because the orbital module of a *Soyuz* has a hatch and could be used as an airlock in the event that the cosmonauts had to work outside.) It used the water reclamation system which condensed water from the station's atmosphere (introduced by Salyut 3). It was primarily an Earth observation platform.

Salyut 5B   A computer introduced by *Mir* to supersede the *Argon 16*.

Salyut 6, 7   This second-generation configuration combined the best features of the *OPS* and *DOS* designs. To accommodate a docking port at either end, the rear-mounted *Soyuz* propulsion unit was replaced by the *OPS* peripherally-mounted engine system. The clip-on solar panels were 5 metres long and 1.25 metres wide, and each pair added about 1.2 kW to the basic power level (about the same as did a docked *Soyuz-T*). They were primarily Earth-observation platforms.

SamoRoentgen   An experiment on *Salyut 7* using an X-ray detector.

SAMS (Space Acceleration Measurement System)   A *NASA* experiment on *Mir* to measure the microgravity environment. Its three-axis accelerometers continuously measured the vibration modes at the sites on *Mir* where *NASA* intended to install its experiments. It had already been used on shuttle flights that conducted microgravity research.

Saturn V   The *NASA* rocket which sent the *Apollo* spacecraft to the Moon.

SDRN (Satellite Data Relay Network)   The *Luch* relay satellite system.

Seeds   An experiment on *Mir* that simply exposed a bag of tomato seeds to radiation by storing them in the *Kvant 2* airlock, which was a relatively unshielded compartment. They were planted after return to Earth, and examined for genetic irregularities.

Semyorka (Old Number Seven)   Korolev's affectionate name for the R-7 rocket which was developed as an intercontinental ballistic missile, but immediately adapted to launch spacecraft.

Sever   A side-looking camera on *Mir*. It had 'a fixed unit with a portable apparatus' that enabled it to take oblique images to either side of the ground track, to highlight surface relief.

SDI   The Strategic Defence Initiative (SDI), announced by President Ronald Reagan on 23 March 1983, called for the development of a "peace shield" which would protect America from ballistic missiles. Basic research for weapons capable of incapacitating incoming strategic missiles was funded. Many space-based approaches were studied, including kinetic-kill interceptors, X-ray lasers and particle-beams, but no 'Star Wars' systems were deployed. The Department of Defense had (wrongly) said that

Cosmos 1267, which docked with *Salyut 7* in 1981, had converted that station into an "orbital battle station".

Sfera (Sphere)   A furnace on *Salyut 5* to study the process of smelting and resolidifying metals by passing ingots of bismuth, lead, tin and cadmium through a furnace to sealed containers, in an effort to create perfect spheres from the liquid metal. It was designed to study the process of melting and hardening molten metals in microgravity. Small ingots containing bismuth, lead, tin and cadmium were inserted by remote control, and after melting at 60°C the tiny samples, each no bigger than the size of a match-head, were allowed to solidify in free fall in the hope that they would form perfect spheres. In fact, metal spheres cast under microgravity did not yield the smooth surface which had been expected.

ShK-1   The scientific airlock, consisting of two concentric spheres. The outer sphere had two hatches, one of which faced inwards, and the other outwards. The inner sphere, which could rotate inside the outer sphere, had a single hatch. The inner sphere was aligned with its hatch facing inwards so that an item could be inserted, and it was then rotated so that the outer hatch could be opened to expose the item to vacuum.

Silya-4   An experiment on *Salyut 4* which used a light nuclear isotope spectrometer to record the isotopic and chemical composition of *cosmic rays*. The apparatus would be refined for use on further flights, in order to establish a major new field of investigation.

Sirena   A Polish materials-processing experiment on *Salyut 6* in the *Splav* furnace. This used an ampoule of cadmium–mercury–telluride (CMT), prepared by scientists at the Warsaw Institute of Physics, which was to be heated to create a semiconductor which is one of the most sensitive detectors of infrared radiation yet discovered. The process involved fusing cadmium telluride with high density mercury telluride. In Earth's gravity, the process produced an imperfect stratified product. The first experiments, lasting 48 hours, yielded crystals with higher homogeneity than obtainable on Earth, with a purity of 50 per cent, compared with 15 per cent in the most effective terrestrial process. Although initially the process had a very low yield (fewer than 10 per cent of the crystals were suitable for use in industrial applications) subsequent fine tuning of the process substantially increased the yield. Other infrared-sensitive materials produced included tin–lead–tellurium, cadmium–mercury–selenium and lead–selenium–telluride. One run produced a CMT monocrystal with a mass of almost 50 grammes.

Sirene-1 (Lilac)   A French-Soviet X-ray spectrometer on *Salyut 7*. It was installed in the rear transfer compartment, and operated by remote control from a panel in the main compartment. The station was reorientated to aim it. Its data was recorded on magnetic tape.

Sirene-2   A high-pressure gas scinitillation proportional counter (GSPC) on *Kvant 1*, sensitive to X-rays in the 2–100 keV range, and with a 3 x 3 degree field of view. The detector geometric area is 300 cm². It was used to study emissions from extremely high-temperature rarefied cosmic gas sources. It was an improved form of the experiment flown on the EXOSAT observatory, and was supplied by *ESA*.

Skif   A Belorussian experiment on *Salyut 7*. An improved form of the *MSS-2M* spectrometer, it was used to study the vertical structure of the atmosphere.

Skif-M   This spectrometer on *Mir* sampled in five bands in the 0.4–1.1 µm range. Its output was stored on magnetic tape and processed by the *Zora* computer.

SKK-11   An exposure cassette full of construction materials deployed on *Kvant 2*.

SKR-2M   An X-ray spectrometer sensitive in the range 2–25 keV, installed aboard *Salyut 7*.

Smak (Taste)   An experiment (also known as Vkus) on *Salyut 6* devised by the Polish Military Institute of Aviation Medicine to investigate why the sense of taste changed in space. (Some foods which were pleasant on Earth had turned out to be unpleasant in space, and others had been found to be much better when eaten in space.) By measuring the electrical stimulation of the taste buds they were able to correlate subjective reactions to taste with specific data.

SN1987A   A supernova discovered on 23 February 1987 in the Large Magellanic Cloud (LMC). Since it was the nearest such event for several centuries, it was intensely observed by ground-based and satellite-borne telescopes, including the X-ray telescopes in *Kvant 1*.

Sofora   A girder developed by the Institute of Electrical Welding in Kiev. Unlike previous trusses deployed in space, it incorporated thermomechanical joints. Its tubular rods were connected by sleeve joints made from a 'smart' titanium–nickel alloy which contracted to re-establish a predefined shape when heated and, in doing so, extended and locked the truss. Once the first 1.5 × 1.5-metre square-section element had been affixed to a base plate mounted on *Kvant 1*, further elements were added. When the final (twentieth) element was in place, the girder stood 14 metres tall. The assembly process required 16 hours in all, but it demonstrated that orbital 'construction work' was feasible. There was a pivot about a third of the way up the girder, so that the upper part could be swung down to a point just above a ferry docked at the rear port. The *VDU* thruster block was eased out of a *Progress* ferry and mounted on top of the girder, which was then swung 11 degrees beyond vertical so that the thrusters would fire in the same plane as the roll-control thrusters set around the periphery of the rear of the base block.

Sokol (Falcon)   The lightweight (8-kg) pressure suit worn by cosmonauts for launch and re-entry; introduced with the *Soyuz-T*, superseding the pressure suit worn following the accident which killed the *Soyuz 11* crew.

Solar constant   A measure of the energy radiated by the Sun which is received by the Earth: 1.35 kW/m^2.

Solar flare   A flux of ionised plasma ejected from the Sun by a magnetic disturbance on its surface (when seen on the limb of the solar disk, such a storm shows itself as a prominence arcing between sunspots). Although a solar storm typically lasts for only an hour, it spews forth a 'tongue' of plasma which expands out across the Solar System. This plasma is composed primarily of hydrogen and helium nuclei travelling at about half the speed of light, which gives them energies corresponding to up to 100 million *electron volts*. When these charged particles flood into the Earth's magnetosphere, they degrade global communications. It is an important component of *cosmic radiation*, and would pose a threat to *Mir* crews if it were to penetrate the magnetosphere to their low orbit, a risk only near the *South Atlantic Anomaly*.

Solar Power   The nominal power available to orbital station elements at the time of launch, before the output of the solar transducers degraded through exposure to the space environment, before auxiliary panels were added to compensate, and counting only the output from the panels on the station itself (that is, discounting any contribution from docked ferries):

	m²	kW
Salyut 1 24	2.4	
Salyut 3, 5	-	5
Salyut 4	60	4
Salyut 6, 7	60	4
Cosmos 929, 1267, 1443, 1686	40	3
Cosmos 1870	86	10
Mir	76	9
Kvant 2	50	7
Kristall	72	8
Spektr	126	16
FGB	56	4

Solar wind   The flux of ionised plasma emitted continuously by the Sun. It forms the background of *cosmic radiation*. It comprises primarily hydrogen and helium nuclei travelling at speeds of several hundred kilometres per second (slow compared with similar material ejected by a *solar flare*), corresponding to energies of around 1,000 *electron volts*. It poses no threat to a spacecraft in low Earth orbit because the Earth's magnetic field acts as a shield.

Son-K   A Bulgarian experiment on *Mir* which gathered electrophysiological data (the electrical activity in the brain) while a cosmonaut slept, and recorded it on a long-duration cassette tape.

South Atlantic Anomaly   A region above the South Atlantic where the innermost of the Earth's radiation belts, which forms a torus around equatorial latitudes at an altitude of 1,000–5,000 km, dips several hundred kilometres towards the surface, posing a threat to spacecraft.

Soyuz (Union)   The *Korolev Bureau's* flagship spacecraft, designed to be a modular system which could be configured to suit a variety of roles. It comprised three modules: the service module contained the propulsion system, the descent module took the crew to orbit and back again, and the orbital module provided accommodation and equipment. The descent and orbital modules had an overall pressurised volume of 10 m³. Soyuz 1 to Soyuz 11 provided a shirt-sleeve environment for a crew of three and the solar panels gave an endurance of several weeks. Following the depressurisation accident which killed the crew of Soyuz 11, a seat was deleted to accommodate an independent life-support system for the remaining crew of two, who wore Sokol pressure suits during launch, docking manoeuvres and re-entry. In this form (excepting Soyuz 13, Soyuz 16, Soyuz 19 and Soyuz 22; it took the series to Soyuz 40) the solar panels were replaced by batteries. This limited its endurance to two days, and it was used only as a transport to *Salyut* stations. The Soyuz-T variant reintroduced the solar panels and the third seat, and it had the *Argon 16* computer of *Salyut 4* and the unified *ODU* engine of the second-generation *Salyut*. The solar panels had an area of 12 m² and gave a peak of 1.2 kW. Soyuz-T could carry a crew of three and 150 kg of cargo to *Mir*, but return only 50 kg of cargo. The Soyuz-TM increased the return capacity to 120 kg. Although the to-orbit cargo could be stowed in racks in the orbital module, the return cargo had to be squeezed into the tiny descent module. Despite being 'old technology', the latest configuration is admirably suited to its solitary role of transporting crews to and from the *Mir* complex, and it is highly reliable.

Spektr (Spectrum)　An expansion module for the *Mir* complex. It had a pressurised volume of 62 m³, and had four solar panels with a total area of 126 m² (in effect a set of base block panels and a set of *Kvant 2* panels) providing a total of 16 kW.

Spektr　An experiment on *Salyut 4* which involved a variety of apparatus installed around the station to investigate the properties of the atmosphere at orbital altitudes. This expanded on studies made on previous flights. The various instruments measured the density, composition and temperature of flows of neutral gas and plasma encountered by the station as it followed its orbit. The hull of the station actually became charged as it passed through such plasma and this charge could interfere with some of the station's systems and experiments. These impacts also contributed to the drag which eroded the station's orbit. The results of these studies would influence the design of future spacecraft.

Spektr-15K　A portable (11-kg) multispectral spectrometer on *Salyut 6,* developed by the Bulgarian Academy of Sciences to study oceanic plankton, crop yield, and the propagation of atmospheric pollution from industrial sites. It measured the reflection of solar insolation at 15 wavelengths in the visible and near-infrared range, between 4,500 and 8,500 Å, and stored its data on magnetic tape.

Spektr-15M　A spectrometer on *Salyut 7.*

Spektr-256　A spectrometer on *Mir* developed by Bulgaria to study the atmosphere. It sampled 256 channels in the 0.4–0.8 micron range with a holographic grating and a CCD detector, and its data was processed by the *Zora* computer.

Spectrum　An experiment on *Salyut 6.* It used the *Spektr-15K* spectrometer to study the physical, chemical and biological characteristics of Cuba and the surrounding sea.

Spin-6000　A 256-channel X-ray and gamma ray spectrometer designed by the Radium Institute of Leningrad to study materials. It was used during a *Mir* spacewalk to assess the extent to which the structure radiated gamma rays as a result of its passage through the Earth's magnetic field.

Spiral　A cassette outside *Salyut 7.* It contained structural materials (springs, seals, threaded connectors, pipe fittings conditioned to expand and contract as their temperature varied, cables, and various metals under constant stress) which were to be tested following prolonged exposure to the extremes of temperature, levels of radiation, X-rays, ultraviolet light, *cosmic rays* and micrometeoroids of the space environment, to determine whether they were suitable for 'assembly work in space.'

Splav-1 (Alloy)　A 23-kg electric furnace on *Salyut 6.* It had a separate control panel. Physically, the furnace took the form of a disk attached to a narrow cylindrical stem. It was placed in the airlock so that it could radiate its excess heat to vacuum. It had hot, cool and thermal-gradient compartments. The 0.3 kW heating chamber operated at up to 1,100°C, and its electronic control system could maintain this temperature to within 5°C. It incorporated a set of molybdenum reflectors to focus heat onto a capsule 170 mm long and 21 mm in diameter, which contained three crystal ampoules. Upon being heated in the hot chamber, the ampoules would fuse. They were automatically transferred into a 'cooling chamber' which maintained a linear thermal gradient to ensure optimum conditions for the formation of a monocrystal. When the sample had cooled to 650°C, it would be transferred into a third chamber which was held at this temperature to facilitate three dimensional crystallisation. The crystallisation process, which would typically take several days, would be recorded by time-lapse photography. Samples were returned to Earth for further analysis. It was also used to study the process of diffusion in molten metal in the absence of gravity; capsules of copper–indium, aluminium–magnesium and indium–antimonide were heated, then the process of crystallisation observed. Such an experiment lasted 14 hours, during which time the station remained in gravity-gradient stabilised orientation, with its axis pointed at the ground, and the attitude control system was switched off so that it drifted freely (otherwise its tiny manoeuvres might interfere with the test). Despite this precaution, the cosmonauts reported that they could see small imperfections caused by vibrations as they had moved about within the complex. This prompted concern that the 'microgravity factory' of the future might have to be a free-flying module which, although it would be visited by cosmonauts from time to time to carry out maintenance, would function automatically. Other experiments saw the production of alloys to investigate the interaction between solid and liquid metals as a step towards understanding the processes of welding and soldering in the absence of gravity. In one novel experiment the station was set rotating to create a centrifugal force that would cause the sample to undergo directional solidification. This contrasted with the earlier practice of trying to hold the station stable so as to minimise disturbances.

Splav-2   Used on *Salyut 6*. A similar unit was subsequently flown on Cosmos 1841 (a *Vostok*-based spacecraft devoted to microgravity research) in 1987.

Sprut-5   A spectrometer deployed outside *Kvant 2* to measure the flux of charged particles. It relayed its data to Earth with the telemetry stream.

Sputnik (Fellow Traveller)   Any spacecraft.

SRVK   The water reclamation system which condensed water vapour from the station's air. This water was used for food preparation. It could condense 1 litre of water vapour from the air per day per person aboard.

SSAS (Solid-Sorbent Air Sampler)   A *NASA* experiment on *Mir* to sample the air in the cabin for analysis of microbial lifeforms that might pose a long-term threat to its habitability.

STR   *Salyut 4* tested this sophisticated thermal regulation system. The outer surface was covered with 'screen vacuum heat insulation' composed of layers of synthetic film sprayed with aluminium that minimised heat loss. The station incorporated an intricate set of radiators which collected solar heat on the sunward side and radiated excess heat on the shaded side. These enabled the thermal regulation system to control heat transfer within a wide range of temperatures. Individual elements of the multiple-loop system, which comprised heating and cooling units, had three or more backups to provide a high degree of redundancy, and so safety. The computer commanded the entire system, correlating the heaters and coolers with the station's orientation with respect to the Sun and the Earth's shadow. Water could be discharged into space, so that the evaporation would rapidly reduce the thermal energy of the station, in an emergency. While a ferry was powered down docked to the station its thermal regulation systems were put under the station's control. An improved version of *STR* was used by *Salyut 6*, and again on *Salyut 7*.

Stratokinetika   An experiment on *Mir* which studied the body's movement in the absence of gravity.

Strela (Arrow)   The general name for the *Mir* base block's onboard computer complex. It included an information system which monitored and reported on the status of the station's various systems, and provided on-line documentation (previously, limited information had been available in written form) that could be automatically updated by the ground controllers (this last feature superseded the *Stroka* teletype uplink).

Strela   A pair of telescopic cranes installed on either side of the *Mir* base block. Although the 45-kg crane was only 2 metres long when packed, it could be extended to a length of 14 metres. Operated by a pair of hand-cranks, it could be raised above or lowered below the base block, and be rotated around the outside arc to reach as far back as *Kvant 1*. It could transfer a load of 750 kg between any two points on its side of the complex. It was 'parked' canted against *Kvant 2*, so that the cosmonauts could ascend it on their way back to the airlock, and then slip down it at the beginning of the next spacewalk to access the crank. The cranes were mounted on fixtures originally used to support the launch shroud.

Stroka   An uplink teleprinter to send lengthy communications to the crew. Introduced by *Salyut 4*, and used on all subsequent stations.

Strombus   This *Ferma-3* girder was erected on top of *Kvant 1*. It was a four-segment, 6-metre long truss.

Struktura   A materials processing experiment on *Salyut 6* which used the *Pion* apparatus to investigate heat exchange and mass transfer during crystal formation in an aqueous solution. A later experiment on *Mir* produced alloys of aluminium–copper–iron, aluminium–tungsten and aluminium–copper.

Sugar   An experiment on *Salyut 6* to grow four monocrystals of sucrose in different solutions. They grew surprisingly quickly. It was the first time that organic crystals had been grown in space.

Superconductor   A French experiment on *Mir*. It employed the *Krater* furnace to investigate the crystallisation of a high-critical-temperature superconductor in microgravity.

Supercooling   An experiment on *Salyut 7*. A 3-mm sphere of an extremely pure alloy of silver and germanium was melted by first cooling it with liquid helium and then heating it with a laser. It was hoped that this would produce a much stronger material.

Superpocket   A French experiment on *Mir* to study the neurosensory system and the reconditioning of postural reflexes.

Support   A Cuban experiment on *Salyut 6*. It involved wearing a specially-designed adjustable shoe (dubbed the Cuban Boot) for six hours each day. This was designed to impose a load on the arch of the foot to simulate the forces the foot feels when standing, to determine whether this affected the ability of the vestibular system to adapt to the absence of gravity. This was based on the theory that on Earth the state of the muscles in the foot contribute to the sense of balance, so when this sense is denied in space it produces a vestibular reaction. This test would also enable doctors to investigate the recovery and readaptation of locomotive stability on returning to gravity, because cosmonauts had been observed to develop gait and postural peculiarities whilst recovering from their flight.

SUR   A German experiment on *Mir*. It studied whether the body's circadian rhythm changed during the process of adaptation to the absence of gravity.

Svet (Light)   A Bulgarian cultivator on the *Kristall* module for growing higher plants. It used a substrate infused with nutrient, and a high-intensity lamp. Radish and lettuce grew rapidly. It was adapted by *NASA* for its *Greenhouse* experiment.

Svet   A spectrometer on the *Spektr* module.

Svet-1   An apparatus on *Salyut 7*. Its mechanism provided an infinitely variable colouriser, for describing visual Earth observations. It replaced the booklet specifying 1,000 colours that had previously been used. To calibrate it, various dyes were released into the Black Sea.

Svetlana   An 800-kg industrial-scale electrophoresis processing system in the unpressurised compartment of *Kvant 1*. It was an advanced, semi-automated form of *Tavriya*, named after Svetlana Savitskaya, who had tested the prototype. It was said to be able to process 100 kg of material per year, but this capability was never exploited. Its first trial was in separating microorganisms to produce agricultural antibiotics to assist in stock rearing.

Svetobloc (Light Box)   A cultivator on *Salyut 7* which worked on the principle of a hothouse. Orchids and tomatoes were successfully grown in it.

Svetobloc-M   A greenhouse in the *Kristall* module used to cultivate high-order crops such as lettuce and radishes.

Svezhest (Freshness)   An experiment on *Salyut 6* which was designed to ionise the air in the station and, it was hoped, make the station slightly more comfortable. The cosmonauts reported that they liked the fact that it made the station smell of a pine forest.

Synergies   A French experiment on *Mir* which studied the function of the vestibular system in controlling the dynamic equilibrium and the stabilisation of references in body synergy during a series of complex movements.

Tamponazh   An experiment on *Salyut 7* designed to test how different sealants solidified. It involved setting up small pipes filled with prepared concrete solutions which were then left to set. It was sponsored by the oil and gas industry to see how pores formed in materials such as concrete (used to seal wells), because pores permit oil and gas to pass.

Tass   The Soviet News Agency.

Taurus   An X-ray detector on the *Spektr* module to measure emissions from the complex, generated by its passage through the Earth's magnetic field.

Tavriya   An experimental processing system for biological materials, tested by Svetlana Savitskaya on *Salyut 7* in 1984. It separated biologically active substances by passing an electric current through a fluid medium (that is, by electrophoresis). It was basically a column of biological compounds which separated into homogeneous fractions, each of which had the same characteristics. The main chamber of the apparatus was almost 1 metre long, and incorporated 230 needles positioned to draw material off from the different layers along its length. On its evaluation trial, human blood proteins (albumin and haemoglobin) were separated. The cosmonauts were able to see the process, which was filmed for later analysis. Once verified, it was used to process several samples. In one run, it was used to purify urokinase (an enzyme present in human urine). Later, for the first time, interferon was processed using the electrophoresis technique (the electric current acted regardless of molecular weight). The trial was extremely encouraging, as the apparatus proved to be hundreds of times more productive than a comparable unit on Earth, and its product was of 10 to 15 times greater purity. It was later used to purify albumen (protein solution) and produced enough pure protein from membranes of an influenza virus to satisfy the Pasteur Institute in Leningrad for many months. Even the small quantities of materials produced by these experiments represented significant results. Test production of vaccine refined from the early results was begun immediately. Later runs synthesised an anti-infection preparation produced by genetic engineering and an antibiotic for agricultural use which, when added to animal and poultry fodder, would increase the weight of the animals by up to 20 per cent. The electrophoresis experiments proved so successful that the design of an industrial-scale unit was begun. It was hoped that on later missions this would be able to yield extremely pure vaccines and other pharmaceutical products on a semi-commercial basis. Such apparatus was also carried on *Vostok*-based spacecraft. (In 1987 for example, the *Kashtan* apparatus on Cosmos 1841, a precursor for the *Foton* materials-processing satellites, synthesised alpha-1 thymosin by purifying thymus hormone, and produced interferon with which to treat viral and tumour diseases, for the Institute of Biomedical Technology in Moscow.)

Tchibis (Lapwing)   A Lower-Body Negative Pressure (LBNP) suit comprising reinforced rubberised leggings with a seal at the waist, from which air could be extracted to create a pressure below the ambient (selectable down to 70 per cent ambient) to draw blood from the upper body. By measuring pulse and blood pressure at different suit pressures it is possible to determine cardiovascular capacity. It is worn for about an hour a day during the initial phase of adaptation and for half an hour once a week, then during the final ten days or so is once again worn on a daily basis. Although this has proven to be effective in ameliorating the unpleasant effects of adaptating to weightlessness, it is insufficient preparation for the return to gravity, so is supplemented in the final phase by exercises to increase cardiovascular capacity, drugs to increase stamina and saline solution for rehydration.

TDRS (Tracking and Data Relay System)   *NASA*'s geostationary relay satellite network.

Teleassistance   A French experiment on *Mir*. It repeated the Orthostatisme experiment and evaluated how a link-up to an expert on the ground could provide a cosmonaut with technical backup during a complex task in orbit. One obvious difficulty was that without the *Luch* network, the limited communications with *Mir* seriously limited such interactive science.

TeleGeo-87   An Earth-observation programme organised by *Intercosmos* (such programmes were run each year during the summer), this time concentrating on Poland. The *Mir* cosmonauts participated.

TEPC (Tissue-Equivalent Proportional Counter)   A *NASA* experiment on *Mir*. A sophisticated radiation monitor that stored its data and downloaded it at regular intervals.

Terra-K   A commercial programme in which agricultural or industrial sites could request orbital investigations. *Mir* provided photography and spectrometry. Various high-resolution cameras were used to provide mapping imagery, and the *MKF-6MA* camera and the *Spektr-256* spectrometer provided multispectral analysis of the growth of vegetation, assessed water purity, and tracked the spread of pollution.

TES   A German experiment on *Mir* that employed the *Kristallisator* furnace to measure the heat capacity of a supercooled metallic melt (in this case antimony, an alloy of silver and germanium, and two sapphire samples) and its variation with temperature. The study of the specific heat of supercooling fusions exploited the fact that in microgravity, contact between the material and the walls of the furnace can be eliminated. Determining the heat capacity enabled other thermophysical properties to be derived.

Therma   An impulse photometer used for the *Rozhen* experiment.

Tien-Shan-88   An *Intercosmos* study of Tadjikistan and Kirghizistan. The *Mir* cosmonauts contributed imagery.

Titus   A German-built six-zone tubular furnace capable of temperatures of up to 1,250°C, used by *ESA* to process semiconductors, alloys and glass.

TKS   The ferry for the *Almaz* platform. It comprised the FAB (the universal auxiliary block) and the VA (the *Merkur* descent module). The entire vehicle was 17.5 metres long, but much of this was taken up by the escape tower of the capsule. To fit the *Proton* rocket, it had a 4.15-metre diameter base. TKS was an acronym for Transportnaya Korabl Snabscheniya—transporter and logistics spacecraft. It was tested as Cosmos 929 in 1977, and Cosmos 1267 docked with *Salyut 6* in 1981, after that station had concluded its main programme, to test its ability to control the orientation of the joint complex. When operational, Cosmos 1443 delivered cargo to *Salyut 7* in 1983, and Cosmos 1686 followed in 1985 with a cluster of scientific instruments. It was referred to by the Western press as a 'Heavy Cosmos' initially and later as a 'Star' module. It was basically a 3-metre diameter compartment around which the engines and propellant tankage were grouped. The *Merkur* capsule was accessed via a hatch in the end of the module, via a short tunnel that led to a hatch in the capsule's heatshield. At the other end the structure flared out to an adapter ring that mated to the *Proton* rocket. On the far side of the ring was a short conical cap which incorporated the docking unit. Overall, it had a mass of 20,000 kg (the main vehicle was 14 tonnes, including the 4,000 kg of miscellaneous cargo; the *Merkur* and associated propulsion unit was 6,000 kg). A stripped-down version (without a *Merkur* capsule) was the 'tug' that delivered *Kvant 1* to *Mir* in 1987. The FAB was expanded (by adding integral pressurised compartments, as TKM) to serve as the basis for the modules used to expand the *Mir* complex between 1989 and 1996. A similar vehicle is to serve as the FGB, on the International Space Station in 1997. All forms used a pair of 400-kg thrust engines burning UDMH and N_2O_4, but their position and the capacity of the externally-mounted propellant tanks varied. The basic vehicle deployed a pair of solar panels (like those on the second-generation *Salyuts*) spanning 16 metres, with a total area of 40 m^2, and 3 kW capacity, and had a single cylindrical compartment with a typical internal volume of about 50 m^3.

TMS   A cassette outside *Salyut 7*. It contained thermomechanical compounds whose physical properties were to be tested after prolonged exposure to the space environment.

TON   A German experiment on *Mir*. Supplied by the University of Hamburg, it used a specially devised sensor to determine the interior pressure of the eye.

Tonus-2 A sophisticated electromechanical muscle stimulator used on *Salyut 4*.

Torsion   An experiment on *Salyut 7* to investigate the strength of various materials when exposed to space.

TORU   A system which enabled a cosmonaut on *Mir* to fly an automated ferry by remote control. A pair of hand controllers for rotational and translational motions were installed on the main control panel, to perform the same function as did those on a *Soyuz*, and a screen presented the view from the camera in the ferry's docking camera with the same data-overlay. The cosmonaut flew the ferry just as if physically onboard. It was tested with Progress-M 15, then used to dock the errant Progress-M 24.

Travers   An experiment on the *Priroda* module. This two-channel synthetic aperture radar radiated at 23 cm and 92 cm, and had medium (30-metre) resolution ground-imaging capability.

Trek   A *cosmic ray* detector supplied by scientists at the University of California. It was installed outside *Kvant 2*. The 1 m^2 plate was a passive detector which contained layers of phosphate glass to record the passage of the super heavy nuclei component of the *cosmic ray* flux. It was retrieved after two years. It was only the second American experiment to be delivered to the *Mir* complex.

Tropex-74   An oceanic study of the Atlantic in which observations by *Salyut 3* were correlated with data from the *Meteor* weather satellites and ship-based studies.

Tropics   An experiment on *Salyut 6*. It involved using the *MKF-6M* to assess Cuba's natural resources. It was actually Tropics-3, and continued earlier observations of other regions. As always, the orbital photography was coordinated with data gathered by airborne and ground teams. Unfortunately, although the weather in the Caribbean was excellent throughout the flight, the timing of the flight meant that there would be few suitable daylight passes, so the residents performed most of this work after the visitors had returned to Earth. The results showed that at both its eastern and western extremities the island of Cuba is criss-crossed by intersecting networks of faults.

Trud (Labour)   A newspaper in the Soviet Union.

TTM   The wide-angle COded-Mask Imaging Spectrometer (COMIS) on *Kvant 1*, sensitive to X-rays in the 2-30 keV range, and with a 7.8 x 7.8 degree field of view. It used a coded aperture mask to determine the location of X-ray sources. Supplied by the Netherlands Space Research Organisation, Utrecht and Birmingham University, it was an improved form of apparatus flown on the Spacelab 2 shuttle mission. It developed an intermittent failure in late 1987, suffering from severe electron precipitation in the vicinity of the *South Atlantic Anomaly*, so the detector was replaced during a spacewalk.

UBD   This ultrasonic bone densitometer was supplied to *Mir* by *ESA* to measure microgravity-induced calcium loss from bones in the lower leg, as a means of assessing countermeasures.

Utrof-Pannoniya   A Hungarian Earth observation programme on *Salyut 6*, involved in using the *MKF-6M* camera and the *Spektr-15K* spectrometer. Their data were to be used to compile a geomorphological map of the Carpathian Basin and the Tisza River Basin, to assess the effect of the Kishkere reservoir on soil salination in the inland waterways linked to the Danube, and to assess the ecological state of Lake Balaton. The data were to be correlated with that simultaneously gathered by an Antonov-30 flying at 6,500 metres, an Antonov-2 flying at 2,000 metres, a low-level helicopter, and several ground teams. Of three faults which they identified in Hungary, one was found, within months, to have oil and gas reserves.

URI   The 30-kg Universal Manual Toolkit (URI) tested on *Salyut 7* was a development of the *Isparitel* electron-beam apparatus. Developed by the Institute of Electrical Welding in Kiev, it incorporated a variety of tools designed to process metal in space, including a small hand-held non-contacting infrared thermometer to measure the temperature of the materials. Savitskaya first used an electron beam to cut 0.5-mm titanium and stainless steel plates. She reported this to be very easy and noted that she could actually see the beam. Then she joined two stainless steel and four titanium samples using tack-welding, and reported that she had produced three good seams. Next she soldered two metal plates together using solder composed of tin and lead, but reported that she did not think that the seam looked very good. Finally, she heated a small silver granule in a crucible (it took only 45 seconds) and then sprayed a silver coating onto an anodised black aluminium plate. She complained that it was more awkward than she had expected because it was hard to see the spray. Academician Paton, the toolkit's designer, reported that these simple tests demonstrated that in the future large space structures could be assembled by robotic systems equipped with such construction tools.

Vaporiser   An experiment on *Salyut 6* using the *Isparitel* apparatus. Hundreds of disks (made of glass, carbon, titanium and various other metals) were sprayed with a variety of thin coatings (including gold, silver, copper, aluminium and various polymers). In one case a multiple-layer coating was formed in which one layer was sprayed whilst the airlock was exposed to direct sunlight, and the other in darkness.

Vazon   A plant cultivation experiment on *Salyut 6*. A similar experiment (ginseng, onion, and chlorella) was undertaken on *Mir*.

VDU   A 700-kg thruster block affixed to the *Sofora* girder above *Kvant 1* to facilitate propellant-efficient roll-control. Rolling the complex consumed a lot of propellant using the thrusters on the periphery of the base block. An 85 per cent saving in propellant resulted from controlling the roll of the complex using these thrusters, because they were so much further from the axis of the complex. It was hoped that this would reduce the frequency with which cargo ferries had to be dispatched to replenish the base block's tanks. It was delivered in a specially-configured *Progress* ferry which carried the oblong box in an unpressurised central section instead of the usual wet-cargo compartment. It was a completely self-contained unit, with its own propellant tanks. It required only a power and command umbilical to be connected, and this was run down the length of the girder and plugged into the Kvant module. Since no provision had been made for replenishment, it had been loaded with sufficient propellant for several years of use.

Vektor   An experiment on *Salyut 7*. It used an Indian-built electrocardiograph to study the cardiovascular system.

Vibrogal   An experiment on *Mir* to study microscopic accelerations shaking the complex. It characterised the small dynamic loads resulting from the operation of the equipment. The data were collected near the *Gallar* furnace.

Vibroseismograph   An experiment on *Mir* to measure the microscopic accelerations acting on the complex.

Vika   An oxygen production unit included to satisfy a sudden demand, most frequently to repressurise the airlock on *Mir* following a spacewalk. When thermally decomposed at high temperature (1,000°C), sodium chlorate ($NaClO_3$) releases oxygen. An exothermic charge initiated the short-lived reaction, and generated a surge of oxygen production. A stock of cartridges was maintained, as the once-only cartridges had to be replaced after use.

Vinimal-92   A French experiment on *Mir* which studied sensor-motor performance in perception and orientation tests. It used the *Echograph-II* monitor screen to present cues, so that a cosmonaut could use a hand-operated key in response, to test visual acuity. When subsequently rerun, it used a miniature flight simulator to assess the effects of adaptation to absence of gravity.

Vita   A biotechnology experiment on *Mir* that used animal cells to cultivate protein compounds which were to be used later on Earth to produce pharmaceutical agents. It studied the growth dynamics of cells which produced luciferase, a biologically active albumen.

Voal   A materials experiment on *Mir* to produce an alloy of wolfram (a raw form of tungsten) and aluminium.

VOG   A German experiment on *Mir* that used video-oculography cameras to capture correlations between eye movement and the vestibular system.

Volkov   A spectrometer on the *Spektr* module.

Volna-2 (Wave)   Contained in *Kvant 2*, this 250-kg apparatus was designed to investigate fluid flows in capillaries, to assist in the design of propellant tanks. It involved setting up a special hydrotank, a control unit and photographic apparatus to record the results of the tests.

Vorotnik   An experiment on *Salyut 6* to test an aid to adapting to microgravity. It involved wearing a collar to create an artificial load on the cervical vertebrae and limit the rotation of a cosmonaut's head, in the hope of eliminating vestibular disorientation.

Voskhod (Sunrise)   A version of the *Vostok* capsule adapted to accommodate several cosmonauts.

Vostok (East)   The 2-metre diameter spherical spacecraft that carried Yuri Gagarin on his pioneering Earth orbit. It could control its attitude but it had no orbital manoeuvring capability. It remained attached to the upper stage of the rocket, which performed the deorbit burn and was then jettisoned. The capsule was covered with ablative. It made a ballistic re-entry, and the occupant ejected once the parachute had opened, to make a separate landing.

Vozdukh   A 150-kg carbon dioxide scrubber in *Kvant 1* which superseded the lithium hydroxide unit in the *Mir* base block. It took the complex another step towards a closed-cycle environment, eliminating the need to ferry up lithium hydroxide canisters in *Progress* ferries (once saturated a LiOH cartridge was discarded). This regenerative system used two desiccant and two regenerative molecular sieve beds of an absorbent similar to zeolite. The amount of gas that can be absorbed by the zeolite is directly proportional to the gas/absorbent contact area (so it operated by gas-capillary action). It is also a function of the pressure until the absorbent is saturated. However, it is inversely dependent on temperature, so the air has to be chilled first. Saturated zeolite is regenerated by exposing it to vacuum, which vents the CO_2. Fans drew air from the cabin and fed it through a silica-gel dryer that removed the water vapour. The heat released was removed by the base block's coolant loop. The dry but now freezing

$(-50^{\circ}C)$ air was fed through the zeolite to remove the CO_2. The clean air was then heated to about 90°C (the exact temperature depending on the environmental control system settings) and fed into another silica-gel container where the hot dry air vaporised water in the silica-gel, simultaneously regenerating the silica-gel and rehumidifying and cooling the air for return to the cabin. Once the in-flow silica-gel unit was saturated with water, and the out-flow unit was dry, the two flows were switched. Similarly, there were two zeolite chambers. While one was in use, the other was leaking its gas to vacuum. Although it was a closed loop, about 0.25 kg of air was lost per day when the zeolite chambers were swapped over, but this was a negligible loss.

VPA-1 The visual polarisation analyser on *Salyut 4*, used to measure optical polarisation of the horizon as part of a study of the upper atmosphere.

VPA-1M An improved form of *VPA-1* flown on *Salyut 6*.

Vremya (Time) An East German experiment on *Salyut 6* which used the *Rula* timing device to measure the ability to react to stimuli, estimate specific periods of time and compare two intervals.

VTL The veloergometer, designed by the Lykachov Motor Works in Moscow, was a stationary bicycle which could be set to impose a load of up to 20 kg. In the series, VTL-1 was on *Salyut 4*, VTL-2 was on *Salyut 5*, and VTL-3 was on *Salyut 6*.

Vulkan The welding apparatus tested on *Soyuz 6*.

Weightlessness The common (and misleading) term for 'zero gravity', the effective absence of gravity associated with orbital motion; this too is incorrect, however, because microscopic vibrations caused by the routine operation of a spacecraft's systems induce accelerations indistinguishable from gravitation, so the seemingly 'weightless' state of everything aboard is more properly defined as microgravity.

VSK-3, Vzor (Visor) The 15-degree field of view optical periscope viewfinder of the *Soyuz* ferry.

Yakor (Anchor) A series of work stations attached to the outside of a station to act as a foot restraint, and on which to set up apparatus for experiments during spacewalks.

Yantar (Amber) An electron beam on *Mir* which extended earlier work with the *Isparitel* apparatus. It was set up in the scientific airlock and used to apply thin metallic coatings of alloys (such as silver–palladium and tungsten–aluminium) onto polymer film.

Yelena-F A 22-kg gamma ray detector on *Salyut 6*. This measured the gamma-rays and charged particles in the near-Earth environment. It was designed by the Moscow Engineering and Physics Institute. It had a 30-degree field of view, and its detector was sensitive to energy in the 30–500 MeV energy range and drew only 10 W, so it could be left running for prolonged periods. It had a *Cerenkov counter*, eight scintillation counters and sixteen photomultipliers, and produced its results photographically. As expected, it recorded higher readings when the station passed the location of the *South Atlantic Anomaly*. It would be used to measure the strength, distribution and processes contributing to the gamma-ray background at orbital altitude. Observations were sometimes coordinated with similar detectors flown on radio-equipped high-altitude balloons. The results mapped high energy electrons flowing in the *South Atlantic Anomaly*, and demonstrated that the background flux was highly dependent on latitude; its level varied by a factor of 10 from the equator, where it was least, to the latitudes corresponding to the furthest extent of its 51-degree inclination orbital track. A similar unit was carried on *Salyut 7*.

Yevpatoria   The original flight control centre in the Crimea.

YMK   The 'Icarus' autonomous manoeuvring unit developed by the *Zvezda* Bureau for free-flying during spacewalks (YMK is an acronym for Yustroistvo Manevrirovania Kosmonautov, which translates as the cosmonaut manoeuvring unit). It was similar to the MMU built by *NASA*, but was somewhat larger and was covered with a thermal blanket. It had four T-shaped thruster packs, each of which had eight pressurised nitrogen thrusters (the use of a harmless propellant meant that there were was no corrosive exhaust to contaminate externally-mounted apparatus). A control panel was mounted on each armrest (left for translational, right for rotational control) along with a group of toggle switches and a joystick for specifying motions. It could be manoeuvred either manually, or in one of two semi-automatic modes, one for rapid response, the other to maximise propellant efficiency. The nitrogen bottles in the back-pack had to be replaced after use, and only a few had been supplied. It was not intended to be used on a regular basis, however. It was tested on *Mir* for subsequent use by spacewalking *Buran* cosmonauts. It had been designed to be compatible with the *Orlan-DM* spacesuit (without a suit capable of independent operation, it would be difficult to test the YMK properly); its frame fit snugly around the suit's backpack, and it was fastened by a waist belt. The docking port apertures were too narrow for a cosmonaut wearing it to pass through, so the test had to wait until *Kvant 2*, with its 1-metre wide airlock hatch, was available. It proved to be extremely stable, and responsive to commands. After the test, it was mounted on a frame outside so as not to clutter up the airlock.

Yusa   A radiometer on the *Spektr* module.

Zarya (Dawn)   A Hungarian experiment on *Salyut 6* using the *Spektr-15K* spectrometer to study absorption lines at sunrise and sunset to determine the density and temperature of the air in the stratosphere and the troposphere.

Zentis   Because microscopic debris is knocked out of the skin of the station by particles of matter and molecules in the station's orbital path, an orbital station creates a cloud of debris around itself. After several weeks of exposure to this, sufficient dust had accumulated on the mirror of the *OST-1* and tarnished its highly reflective coating. Taking advantage of the natural vacuum, Grechko used an automatic system to respray them. This was achieved by passing an electric current through a tungsten wire to melt a blob of aluminium, so that a vaporised spray of metal would settle on the mirror. Since it proved to be so easy to restore the mirror to its original reflective condition, this opened the way for the fitting of evermore sophisticated mirror-based systems on future stations.

Ziemia (Earth)   A Polish experiment on *Salyut 6* to make Earth-resources observations of Poland by using the *MKF-6M*.

Znamya (New Light, or Banner)   A space-mirror experiment. The 40-kg package was attached to the docking assembly of a *Progress* ferry which initiated a fast roll manoeuvre, so that the centrifugal force dragged out eight triangular petals which formed a 20-metre diameter reflector made of 5 μm aluminium-coated Kevlar. The orientation of the ferry had been chosen so that the mirror beamed sunlight towards the Earth, with the result that a spot of light 4,000 metres wide travelled along the ground track. This was to assess the feasibility of using orbital mirrors to illuminate extreme northerly regions which suffered extended darkness.

Zona-2   A crucible-less furnace in the *Kristall* module to produce monocrystals of semi-conductors.

Zona-3   A crucible-less furnace located in the Kristall module.

Zond   A circumlunar variant of the *Soyuz* spacecraft using only the descent module and a modified service module, launched by *Proton*. Several automated flights were made, but it never carried cosmonauts.

Zone   A Cuban experiment on *Salyut 6*. It studied the dissolution of monocrystals of sucrose. Cuba had developed apparatus specifically to enable photographs to be taken of processes taking place in the *Kristall* furnace.

Zora   A 3-kg portable computer delivered to *Mir* to analyse data in space.

Zvezda (Star)   The manufacturer of the *Orlan* spacesuits and the *YMK* autonomous manoeuvring system.

Bibliography

Although any study of space station development will ultimately rely upon the contemporary record of events published in *Spaceflight*, the *Journal of the British Interplanetary Society, Flight International, Aviation Week & Space Technology*, the excellent but short-lived *Spaceflight News*, and more recently the Internet, the following books offer useful snapshots of interpretation. They have been listed in order of publication to provide a sense of advancement.

Clarke, Arthur C., *The Exploration of Space*, Temple Press, 1951

von Braun, Werner, *Across the Space Frontier*, a series of articles, with artwork by Chesley Bonestell, published in *Collier's Magazine*, New York, 1952; subsequently reprinted, and edited by Cornelius Ryan, in *Across The Space Frontier*, Viking Press, 1952, and *Conquest of the Moon*, Viking Press, 1953.

Poole, Lynn, *Your Trip into Space*, Lutterworth Press, 1954.

Bates, D. R. (*ed.*), *Space Research and Exploration*, Eyre & Spottiswoode, 1957.

Burgess, Eric, *Satellites and Spaceflight*, Chapman & Hall, 1957.

Beard, R.B. and Rotherman, A.C., *Space Flight and Satellite Vehicles,* Newnes, 1957.

Burchett, Wilfred, and Purdy, Anthony, *Cosmonaut Yuri Gagarin: First Man in Space*, Panther, 1961.

Emme, Eugene, *A History of Space Flight*, Holt, Rinehart & Winston, 1965.

Shelton, William, *Soviet Space Exploration: the First Decade,* Arthur Barker, 1969.

Vladimirov, Leonid, *The Russian Space Bluff*, Tom Stacey, 1971.

Stoiko, Michael, *Soviet Rocketry: the First Decade of Achievement*, David & Charles, 1971.

Smolder, Peter, *Soviets in Space: the Story of the Salyut and the Soviet Approach to Present and Future Space Travel*, Lutterworth Press, 1973.

O'Neill, Gerard, *The Colonisation of Space,* in *Physics Today*, September 1974.

US Senate, *Soviet Space Programs: 1971–75*, Senate Committee on Aeronautical and Space Sciences, US Government Printing Office, 1976.

O'Neill, Gerard, *The High Frontier: Human Colonies in Space*, Jonathan Cape, 1977.

Heppenheimer, T.A., *Colonies in Space*, Warner Books, 1978.

Cooper, Henry, *A House in Space: the First True Account of the Skylab Experience*, Panther, 1978.

Baker, David, *The History of Manned Space Flight*, Cavendish, 1981.

Furniss, Tim, *The Story of the Space Shuttle*, Hodder & Stoughton, 1982.

Shapland, David and Rycroft, Michael, *Spacelab: Research in Earth Orbit*, Cambridge University Press, 1984.

Furniss, Tim, *Spaceflight: the Records*, Guinness, 1985.

Smith, Melvyn, *Space Shuttle*, Foulis–Haynes, 1985.

US Senate, *Soviet Space Programs: 1976–80*, Senate Committee on Commerce, Science and Transportation, US Government Printing Office, 1985.

Furniss, Tim, *Manned Spaceflight Log*, Jane's, 1986.

Peebles, Curtis, *Guardians: Strategic Reconnaissance Satellites*, Ian Allan, 1987.

Bond, Peter, *Heroes in Space: from Gagarin to Challenger*, Blackwell, 1987.

Shkolenko, Yuri, *The Space Age*, Progress Publishing, Moscow, 1987.

Oberg, James and Oberg, Alcestis, *Pioneering Space: Living on the Next Frontier*, McGraw–Hill, 1987.

Lebedev, Valentin, *Diary of a Cosmonaut: 211 Days in Space*, Phytoresource Research, 1988.

Glushko, Valentin, *Soviet Cosmonautics: Questions and Answers*, Novosti, 1988.

Clark, Phillip, *The Soviet Manned Space Programme*, Salamander, 1988.

Harvey, Brian, *Race into Space: the Soviet Space Programme,* Ellis Horwood, 1988.

Gatland, Kenneth, *The Illustrated Encyclopedia of Space Technology*, Salamander, 1989.

Aldrin, Buzz, and McConnell, Malcolm, *Men from Earth*, Bantam, 1989.

Spangerburg, Ray, and Moser, Diane, *Space Exploration: Opening the Space Frontier*, Facts-on-File, 1989.

Spangerburg, Ray, and Moser, Diane, *Space Exploration: Living and Working in Space*, Facts-on-File, 1989.

Asimov, Isaac, *Beginnings: the Story of Origins – of Mankind, the Earth, the Universe*, Berkley, 1989.

Hooper, Gordon, *The Soviet Cosmonaut Team*, GRH Publications, 1990.

Baker, James, *Planet Earth: the View from Space*, Harvard University Press, 1990.

Newkirk, Dennis, *Almanac of Soviet Manned Space Flight*, Gulf Publishing Company, 1990.

Rycroft, Michael (*ed.*), *The Cambridge Encyclopedia of Space*, Cambridge University Press, 1990.

Calder, Nigel, *Spaceship Earth*, Channel 4 Books, 1991.

Semenov, Yuri, Rymin, Valeri, Popov, Viktor, Pivnyuk, Vladimir, Gilberg, Lev, Novokshchenov, Nikolai, and Rebov, Mikhail, *Cosmonautics 1991*, Cosmos Books, 1992.

NASA, *Space Station Freedom Media Handbook*, NASA, 1992.

Sharman, Helen, and Priest, Christopher, *Seize the Moment: an Autobiography of Britain's First Astronaut*, Victor Gollancz, 1993.

Bond, Peter, *Reaching for the Stars: the Illustrated History of Manned Spaceflight,* Cassell, 1993.

Matson, Wayne, *Cosmonautics: a Colorful History*, Cosmos Books, 1994.

Pivnyuk, Vladimir, and Bockman, Mark, *Space Station Handbook: Mir User's Manual,* Cosmos Books, 1994.

Johnson, Nicholas, *The Soviet Reach for the Moon*, Cosmos Books, 1994.

US Senate, *US–Russian Coooperation in Space,* US Congressional Office of Technology Assessment, US Government Printing Office, 1995.

Jenkins, Dennis, *Space Shuttle: the History of Developing the National Space Transportation System*, Jenkins, 1996.

Schmidt, Stanley, and Zubrin, Robert (*eds.*), *Islands in the Sky: Bold New Ideas for Colonizing Space*, Wiley, 1996; articles from *Analog* magazine.

Bizony, Piers, *Islands in the Sky: Building the International Space Station*, Aurum Press, 1996.

Gribbin, John, and Gribbin, Mary, *Fire on Earth: in Search of the Doomsday Asteroid*, Simon & Schuster, 1996.

Kondratyev, K.Ya., Buznikov, A.A. and Pokrovsky, O.M., *Global Change and Remote Sensing*, Wiley–Praxis, 1996.

Harvey, Brian, *The New Russian Space Programme: from Competition to Collaboration*, Wiley–Praxis, 1996.

Index[†]

[†] A **bold** reference is to a page with a substantial reference.

WILEY-PRAXIS SERIES IN SPACE SCIENCE AND TECHNOLOGY

Forthcoming Titles

SOLAR POWER SATELLITES: A Space Energy System for Earth

Peter E. Glaser, Vice President (retired), Arthur D. Little, Inc., USA, Frank P. Davidson, Coordinator, Macro-Engineering Research Group, Massachusetts Institute of Technology, USA, Katinka I. Csigi, Principal Consultant, ERIC International, USA

THE SPACE SHUTTLE: Roles, Missions and Accomplishments

David M. Harland, formerly Visiting Professor, University of Strathclyde, UK

THE CHINESE SPACE PROGRAMME: From Conception to Future Capabilities

Brian Harvey, M.A., H.D.E., F.B.I.S.

THE SPACE DEBRIS ENVIRONMENT: Hazard and Risk Assessment

Nicholas L. Johnson, et al., NASA Johnson Space Center, Houston, Texas, USA   *£60 – July 1999*

SOLAR SAILING: Technology, Dynamics and Mission Applications

Colin R. McInnes, Department of Aerospace Engineering, University of Glasgow, UK

SATELLITE-BORNE INSTRUMENTS FOR EARTH OBSERVATION   *– £80   no date yet (Aug. 98)*

C.B. Pease, formerly Royal Aerospace Establishment, Farnborough, UK and David Carter, Matra Marconi Space, Portsmouth, UK

SPACE GOVERNANCE: A Blueprint for Future Activities

George S. Robinson, Attorney-at-Law, President, Ocean-Space Services, Adjunct Professor, George Mason University, Institute of International Transactions, Virginia, USA and Declan O'Donnell, Attorney-at-Law, President of the World Bar Association, President and Founder of the United Societies in Space, USA

THE MOON: RESOURCES, FUTURE DEVELOPMENTS AND COLONIZATION

David G. Schrunk, formerly Radiation Safety Officer and Head of Nuclear Medicine, Polomar-Pomerado Hospital District, Escondido, California, Founder and Chairman, Science of Laws Institute, San Diego, California, USA; Burton L. Sharpe, System Sales Engineer, Communications Corporation, St Louis, MO, formerly Resident Site Engineer, NASA/Jet Propulsion Laboratory, supporiting US Transportation Command, Scott AFB, IL, USA; Bonnie L. Cooper, Scientist, Oceaneering Space Systems, Houston, Texas, USA

ROCKET AND SPACECRAFT PROPULSION: Principles, Practice and New Developments   *Aug 1999*
no price yet

M.J.L. Turner, Principal Research Fellow, Department of Physics and Astronomy, University of Leicester, UK   *Aug '98*